FIGURATE NUMBERS

FIGURATE NUMBERS

Elena Deza
Moscow State Pedagogical University, Russia

Michel Marie Deza
Ecole Normale Superieure, Paris, France

World Scientific

NEW JERSEY • LONDON • SINGAPORE • BEIJING • SHANGHAI • HONG KONG • TAIPEI • CHENNAI

Published by

World Scientific Publishing Co. Pte. Ltd.
5 Toh Tuck Link, Singapore 596224
USA office: 27 Warren Street, Suite 401-402, Hackensack, NJ 07601
UK office: 57 Shelton Street, Covent Garden, London WC2H 9HE

British Library Cataloguing-in-Publication Data
A catalogue record for this book is available from the British Library.

FIGURATE NUMBERS

ISBN-13 978-981-4355-48-3
ISBN-10 981-4355-48-8

Typeset by Stallion Press
Email: enquiries@stallionpress.com

Printed in Singapore by B & Jo Enterprise Pte Ltd

Contents

Notations

- $\mathbb{N} = \{1, 2, 3, \ldots\}$ - the set of positive integers.
- $\mathbb{Z} = \{\ldots, -3, -2, -1, 0, 1, 2, 3, \ldots\}$ - the set of integers.
- $b|a$ - a non-zero integer b divides an integer a: $a = bc$, where $c \in \mathbb{Z}$.
- $gcd(a_1, \ldots, a_n)$ - *greatest common divisor* of integers $a_1, \ldots, a_n$, at least one of which is non-equal to 0, i.e., the greatest integer, dividing $a_1, \ldots, a_n$. If $gcd(a_1, \ldots, a_n) = 1$, the numbers $a_1, \ldots, a_n$ are called *relatively primes*; if $gcd(a_i, a_j) = 1$ for any distinct $i, j \in \{1, \ldots, n\}$, the numbers $a_1, \ldots, a_n$ are called *pairwise relatively primes*.
- $lcm(a_1, \ldots, a_n)$ - *least common multiple* of non-zero integers $a_1, \ldots, a_n$, i.e., the least positive integer divided by $a_1, \ldots, a_n$.
- $rest(a, b)$ - the *reminder* of an integer a after its *integer division* by an positive integer b: $a = bq + rest(a, b)$, where $q, rest(a, b) \in \mathbb{Z}$, and $0 \leq rest(a, b) < b$.
- $P = \{2, 3, 5, 7, 11, 13, 17, 19, \ldots\}$ - the set of *prime numbers*, i.e., the positive integers, having exactly two positive integer divisors; $p, q, p_1, p_2, \ldots, p_k, \ldots, q_1, q_2, \ldots, q_s, \ldots$ - prime numbers; $n = p_1^{\alpha_1} \cdot \ldots \cdot p_k^{\alpha_k}$ - *prime factorization* of a positive integer $n > 1$, i.e., its representation as a product of positive integer powers of different prime numbers $p_1, \ldots, p_k$.
- $S = \{4, 6, 8, 9, 10, 12, 14, 15, 16, 18, \ldots\}$ - the set of *composite numbers*, i.e., the positive integers, having more than two positive integer divisors.

- $n = (c_s c_{s-1} \ldots c_1 c_0)_g = c_s g^s + c_{s-1} g^{s-1} + \cdots + c_1 g + c_0$, $g \in \mathbb{N} \backslash \{1\}$, $0 \leq c_i \leq g-1$, $c_s \neq 0$ - representation of a positive integer n in *base* g; for instance, $279 = 1B3_{12} = 1 \cdot 12^2 + 11 \cdot 12 + 3$.
- $\lfloor x \rfloor$, $x \in \mathbb{R}$ - *floor function*: the largest integer less than or equal to x.
- $\{x\}$, $x \in \mathbb{R}$ - *fractional value* of x: $\{x\} = x - \lfloor x \rfloor$.
- $\lceil x \rceil$, $x \in \mathbb{R}$ - *ceiling function*: the smallest integer greater than or equal to x.
- $\phi(n)$, $n \in \mathbb{N}$ - *Euler's totient function*: the number of positive integers that are relatively prime to n: $\phi(n) = |\{x \in \mathbb{N} : x \leq n, gcd(x,n) = 1\}|$.
- $\mu(n)$, $n \in \mathbb{N}$ - *Möbius function*: $\mu(1) = 1$, $\mu(n) = (-1)^k$, if $n = p_1 \cdot \ldots \cdot p_k$ is a product of k distinct primes, and $\mu(n) = 0$, if n has repeated prime factors.
- $\tau(n) = \sum_{d|n} 1$, $n \in \mathbb{N}$ - *tau function* (or *number of divisors' function*): the number of positive integer divisors of a positive integer n.
- $\sigma(n) = \sum_{d|n} d$, $n \in \mathbb{N}$ - *sum of divisors' function.*
- $\sigma_k(n) = \sum_{d|n} d^k$, $n \in \mathbb{N}, k \in \mathbb{C}$ - *divisor function*: the sum of the k-th powers of positive integer divisors of a positive integer n. In particular, $\sigma_0(n) = \tau(n)$, and $\sigma_1(n) = \sigma(n)$.
- $a \equiv b(mod\ n)$ - an integer a is *congruent* to an integer b modulo n, $n \in \mathbb{N}$, i.e., $n|(a-b)$.
- $\mathbf{a}_n = \{x \in \mathbb{Z} : x \equiv a(mod\ n)\} = \{\ldots, a-2n, a-n, a, a+n, a+2n, a+3n, \ldots\}$ - *residue class (of a) modulo n*: the set of all integers, congruent to a modulo n. Any representative r_a of $\mathbf{a}_n$ is called *residue* of a modulo n. The smallest non-negative representative of $\mathbf{a}_n$ is called *smallest non-negative residue* of a modulo n; it is the *reminder* $rest(a,n)$ of a after its *integer division* by n. The smallest in absolute value representative of $\mathbf{a}_n$ is called *minimal residue* of a modulo n.
- $(\frac{a}{p})$ - *Legendre symbol*: $(\frac{a}{p}) = 1$, if an integer a, relatively prime to an odd prime number p, is a *quadratic residue modulo* p (i.e., the congruence $x^2 \equiv a(mod\ p)$ has a solution $x_0 \in \mathbb{Z}$), $(\frac{a}{p}) = -1$, if an integer a, relatively prime to an odd prime number p, is a *quadratic*

non-residue modulo p (i.e., the congruence $x^2 \equiv a(mod\ p)$ does not have a solution), and $(\frac{a}{p}) = 0$, if $p|a$.

- $(\frac{a}{n}) = (\frac{a}{p_1})^{\alpha_1} \cdot \ldots \cdot (\frac{a}{p_k})^{\alpha_k}$ for odd positive integer $n = p_1^{\alpha_1} \cdot \ldots \cdot p_k^{\alpha_k}$ - *Jacobi symbol.* When n is a prime, the Jacobi symbol $(\frac{a}{n})$ reduces to the Legendre symbol.
- $P_n(a)$ - *multiplicative order* (or *modulo order*) of an integer a (relatively prime to n) modulo n: the smallest positive integer exponent γ for which $a^\gamma \equiv 1(mod\ n)$.
- $[a_0, a_1, \ldots, a_n, \ldots] = a_0 + \frac{1}{a_1 + \frac{1}{\ldots \frac{1}{a_n + \ldots}}}$ - *continued fraction*: here $a_0 \in \mathbb{Z}$, $a_1, \ldots, a_n \in \mathbb{N}$, and the last term a_n, if it exists, is greater than 1.
- $\delta_k = [a_0, a_1, \ldots, a_k] = \frac{P_k}{Q_k}$, $k = 0, 1, \ldots, n, \ldots$ - the *convergants* of the continued fraction $[a_0, a_1, \ldots, a_n, \ldots]$.
- $x^2 - Dy^2 = \pm 1$, where D is a square-free positive integer - a *Pell's equation.*
- $x^2 - Dy^2 = \pm c$, where $c \in \mathbb{N}\backslash\{1\}$, and D is a square-free positive integer - a *Pell-like equation.*
- a_1, $a_2 = a_1 + d$, $a_3 = a_2 + d = a_1 + 2d$, $\ldots$, $a_n = a_{n-1} + d = a_1 + (n-1)d$, $\ldots$, where $a_1, d \in \mathbb{R}$ - *arithmetic progression* with *difference* d; $a_1 + \cdots + a_n = \frac{n(a_1+a_n)}{2}$.
- b_1, $b_2 = b_1 \cdot q$, $b_3 = b_2 \cdot q = b_1 \cdot q^2$, $\ldots$, $b_n = b_{n-1} \cdot q = b_1 \cdot q^{n-1}$, $\ldots$, where $b_1, q \in \mathbb{R}\backslash\{0\}$ - *geometric progression* with *common ratio* q; $b_1 + \cdots + b_n = \frac{b_1(q^n-1)}{q-1}$ for $q \neq 1$.
- $f(x) = c_0 + c_1x + c_2x^2 + \cdots + c_nx^n + \cdots, |x| < r$ - *generating function* of the sequence $c_0, c_1, c_2, \ldots, c_n, \ldots$.
- $b_0c_{n+k} + b_1c_{n+k-1} + \cdots + b_nc_k = 0$, $b_0, \ldots, b_n \in \mathbb{R}$, $b_0 \neq 0$ - *linear recurrent equation of n-th order* for a sequence $c_0, c_1, c_2, \ldots, c_{n-1}$, $c_n = -\frac{b_1}{b_0}c_{n-1} - \ldots - \frac{b_n}{b_0}c_0$, $c_{n+1} = -\frac{b_1}{b_0}c_n - \ldots - \frac{b_n}{b_0}c_1$, $\ldots$, $c_{n+k} = -\frac{b_1}{b_0}c_{n+k-1} - \ldots - \frac{b_n}{b_0}c_k$, $\ldots$ with the *initial values* $c_0, c_1, \ldots, c_{n-1}$.
- $A = ((a_{ij}))$, $1 \leq i, j \leq n$ - square $n \times n$ matrix with real entries a_{ij}. The matrix A is called *symmetric*, if $a_{ij} = a_{ji}$ for all $i, j \in \{1, 2, \ldots, n\}$. The matrix A is called *identity matrix* and denoted by I_n, if $a_{ii} = 1$ and $a_{ij} = 0$ for $i \neq j$. The matrix $A^T = ((a_{ji}))$ is

called *transpose* of A. The matrix A^{-1}, such that $A \cdot A^{-1} = I_n$, is called *inverse* of A.

- $det\,A$ or $|A|$ - *determinant* of a given $n \times n$ matrix $A = ((a_{ij}))$. For $n = 2$ it holds $det\,A = a_{11}a_{22} - a_{21}a_{12}$; for any $n \geq 3$ it holds $det\,A = \sum_{j=1}^{n}(-1)^{1+j}a_{1j} \cdot det\,A_{1j}$, where A_{1j} is the $(n-1)\times(n-1)$ matrix, obtained by deletion the first row and the j-th column of A.
- $n!$ - *factorial* of a non-negative integer n: $n! = 1 \cdot 2 \cdot \ldots \cdot n$, $n \in \mathbb{N}$; $0! = 1$.
- $x^{\underline{n}}$, $x \in \mathbb{R}$, $n \in \mathbb{N}$ - *falling factorial* of x: $x^{\underline{n}} = x \cdot (x-1) \cdot \ldots \cdot (x - n + 1)$.
- $x^{\overline{n}}$, $x \in \mathbb{R}$, $n \in \mathbb{N}$ - *raising factorial* of x: $x^{\overline{n}} = x \cdot (x+1) \cdot \ldots \cdot (x + n - 1)$.
- $\binom{n}{m}$ - *binomial coefficients*: $\binom{n}{m} = \frac{n!}{m!(n-m)!}$, $0 \leq m \leq n$.
- $F_n = 2^{2^n} + 1$, $n = 0, 1, 2, \ldots$ - *Fermat numbers*.
- $M_n = 2^n - 1$ $n = 1, 2, 3, \ldots$ - *Mersenne numbers*.
- $u_1, u_2, \ldots, u_n, \ldots$ - *Fibonacci numbers*: $u_{n+2} = u_{n+1} + u_n$, $u_1 = u_2 = 1$.
- $L_1, L_2, \ldots, L_n, \ldots$ - *Lucas numbers*: $L_{n+2} = L_{n+1} + L_n$, $L_1 = 1$, $L_2 = 3$.
- $C_n = \frac{1}{n}\binom{2n-2}{n-1}$, $n = 1, 2, 3, \ldots$ - *Catalan numbers*.
- $S(n, m) = \frac{1}{m!}\sum_{i=0}^{m}(-1)^i\binom{m}{i}(m-i)^n$, $n \geq 1$, $1 \leq m \leq n$ - *Stirling numbers of the second kind*.
- $B(n) = \sum_{k=0}^{n-1}\binom{n-1}{k}B(k)$ with $B(0) = 1$ - *Bell numbers*.
- $B_n = -\frac{1}{n+1}\sum_{k=1}^{n}\binom{n+1}{k+1}B_{n-k}$ with $B_0 = 0$ - *Bernoulli numbers*.
- $S_m(n) = \frac{n((m-2)n-m+4)}{2}$ - n-th *m-gonal number*, i.e., m-gonal number with *index* n.
- $CS_m(n) = \frac{mn^2-mn+2}{2}$ - n-th *centered m-gonal number*.
- $P_m(n) = mn^2 - mn + 1$ - n-th *m-gram number*; in particular, $P_6(n) = S(n) = 6n^2 - 6n + 1$ - n-th *star number*.
- $P(n) = n(n+1)$ - n-th *pronic number*.
- $S_m^3(n) = \frac{n(n+1)((m-2)n-m+5)}{6}$ - n-th *m-gonal pyramidal number*.
- $C(n) = n^3$ - n-th *cubic number*.
- $O(n) = \frac{n(2n^2+1)}{3}$ - n-th *octahedral number*.
- $I(n) = \frac{n(5n^2-5n+2)}{2}$ - n-th *icosahedral number*.

- $D(n) = \frac{n(9n^2-9n+2)}{2}$ - n-th *dodecahedral number.*
- $SO(n) = n(2n^2 - 1)$ - n-th *stella octangula number.*
- $RD(n) = 4n^3 - 6n^2 + 4n - 1$ - n-th *rhombic dodecahedral number.*
- $CS_m^3(n) = \frac{mn^3+n(6-m)}{6}$ - n-th *centered m-gonal pyramidal number.*
- $PCS_m^3(n) = \frac{n(mn^2-mn+2)}{2}$ - n-th *centered m-gonal prism number.*
- $S_m^k(n) = \frac{n(n+1)\dots(n+k-2)((m-2)n-m+k+2)}{k!}$ - n-th *k-dimensional m-gonal pyramidal number*; in particular,

$$S_3^k(n) = \frac{n(n+1)\dots(n+k-1)}{k!}$$

  - n-th *k-dimensional hypertetrahedron number.*
- $C^k(n) = n^k$ - n-th *k-dimensional hypercube number.*
- $O^k(n) = \sum_{j=0}^{k-1}(-1)^j\binom{k-1}{j}2^{k-j-1}S_3^{k-j}(n)$ - n-th *k-dimensional hyperoctahedron number.*
- $N^k(n) = (n+1)^{k+1} - n^{k+1}$ - n-th *k-dimensional nexus number.*
- $TS_4(n)$, $TCS_m(n)$, $TS_3^3(n)$, $TC(n)$, $TO(n)$, $TI(n)$ - n-th *truncated square, truncated centered m-gonal, truncated tetrahedral, truncated cubic, truncated octahedral, truncated icosahedral number*, respectively.
- $\overline{C}(n), \overline{S}_3^3(n)$, $\overline{O}(n)$, $\overline{S}_3^k(n)$, $\overline{C}^k(n)$, $\overline{O}^k(n)$ - n-th *centered cube, centered m-gonal pyramid, centered octahedron, k-dimensional centered hypertetrahedron, k-dimensional centered hypercube, k-dimensional centered hyperoctahedron number*, respectively.
- ${}^-S_m(n) = S_m(-n)$, ${}^-CS_m(n) = CS_m(-n)$, ${}^-S_m^3(n) = S_m^3(-n)$, ${}^-S_3^k(n) = S_3^k(-n)$, ${}^-C^k(n) = C^k(-n)$, ${}^-O^k(n) = O^k(-n)$, $n \in \mathbb{N}$ - n-th *generalized m-gonal, generalized centered m-gonal, generalized m-gonal pyramidal, generalized k-dimensional hypertetrahedron, generalized k-dimensional hypercube, generalized k-dimensional hyperoctahedron number with negative index*, respectively.

Preface

Figurate numbers, as well as a majority of classes of special numbers, have long and rich history. They were introduced in Pythagorean school (VI-th century BC) as an attempt to connect Geometry and Arithmetic. Pythagoreans, following their credo "all is number", considered any positive integer as a set of points on the plane.

In general, a *figurate number* is a number that can be represented by regular and discrete geometric pattern of equally spaced points. It may be, say, a *polygonal*, *polyhedral* or *polytopic number* if the arrangement form a regular polygon, a regular polyhedron or a reqular polytope, respectively. Figurate numbers can also form other shapes such as centered polygons, L-shapes, three-dimensional (and multidimensional) solids, etc.

In particular, polygonal numbers generalize numbers which can be arranged as a triangle (*triangular numbers*), or a square (*square numbers*), to an m-gon for any integer $m \geq 3$.

Beyond classical polygonal numbers, *centered polygonal numbers* can be constructed in the plane from points (or balls). Each centered polygonal number is formed by a central dot, surrounded by polygonal layers with a constant number of sides. Each side of a polygonal layer contains one dot more than any side of the previous layer.

In Chapter 1, we consider above and other plane figurate numbers with multitude of their properties, interconnections and interdependence.

Putting points in some special order in the space, instead of the plane, one obtains *space figurate numbers*. The most known are *pyramidal numbers*, corresponding to triangular, square, pentagonal, and, in general, m-gonal pyramids. They are given as sums of corresponding polygonal numbers. Among these objects, are *reliable in practice* (i.e., the corresponding geometrical arrangement of balls is stable) only triangular and square pyramids, constructed from balls, and Ancient Greeks considered only those two classes of space figurate numbers. *Cubic numbers* correspond to cubes, which are constructed from balls. The *octahedral*, *dodecahedral* and *icosahedral numbers* correspond to the three remaining Platonic solids.

Often one considers *centered space figurate numbers*. Their construction is similar to the one for the centered polygonal numbers. Are considered also the numbers, which can be obtained by adding or subtracting pyramidal numbers of smaller size. It corresponds to truncation of corresponding polyhedra or to putting pyramids on their faces, as in constructing of star polyhedra.

These and other classes of space figurate numbers are collected in Chapter 2.

Similarly, one can construct *multidimensional figurate numbers*, i.e., figurate numbers of higher dimensions k, but for $k \geq 4$ it loses the practical sense. In the dimension four, the most known figurate numbers are *pentatope numbers*, which are four-dimensional analogue of triangular and tetrahedral numbers and correspond to an four-dimensional simplex, and *biquadratic numbers*, which are four-dimensional analogue of square numbers and cubic numbers.

The elements of the theory of multidimensional figurate numbers are given in Chapter 3.

The theory of figurate numbers does not belong to the central domains of Mathematics, but the beauty of these numbers attracted the attention of many scientists during thousands years. The list (not full) of famous mathematicians, who worked in this domain, consists of Pythagoras of Samos (ca. 582 BC–ca. 507 BC), Hypsicles of Alexandria (190 BC–120 BC), Plutarch of Chaeronea (ca. 46–ca. 122), Nicomachus of Gerasa (ca. 60–ca. 120), Theon of Smyrna (70–135), Diophantus of Alexandria (ca. 210–ca. 290), Leonardo of Pisa, also known as Leonardo Fibonacci (ca. 1170 –ca. 1250), Michel

Stifel (1487–1567), Gerolamo Cardano (1501–1576), Claude Gaspard Bachet de Méziriac (1581–1638), René Descartes (1596–1650), Pierre de Fermat (1601–1665), John Pell (1611–1685), Blaise Pascal (1623–1662), Leonhard Euler (1707–1783), Joseph-Louis Lagrange (1736–1813), Adrien-Marie Legendre (1752–1833), Carl Friedrich Gauss (1777–1855), Augustin-Louis Cauchy (1789–1857), Carl Gustav Jacob Jacobi (1804–1851), Wacław Franciszek Sierpiński (1882–1969), Barnes Wallis (1887–1979).

Moreover, many mathematical facts have deep connections with figurate numbers, and a lot of well-known theorems can be formulated in terms of these numbers. In particular, figurate numbers are related to many other classes of special positive integers, such as *binomial coefficients*, *Pythagorean triples*, *perfect numbers*, *Mersenne and Fermat numbers*, *Fibonacci and Lucas numbers*, etc.

A large list of such special numbers, as well as other areas of Number Theory, related to figurate numbers, is given in Chapter 4.

Figurate numbers were studied by the ancients, as far back as the Pythagoreans, but nowadays the interest in them is mostly in connection with the following *Fermat's polygonal number theorem*. In 1636, Fermat proposed that every number can be expressed as the sum of at most m m-gonal numbers. In a letter to Mersenne, he claimed to have a proof of this result but this proof has never been found. Lagrange (1770) proved the square case, and Gauss proved the triangular case in 1796. In 1813, Cauchy proved the proposition in its entirety.

In Chapter 5, we give full and detailed proof of the Fermat's polygonal number theorem and related results in full generality, since it was scattered in many, often difficult to find publications.

In a small Chapter 6, we collected some remarkable individual figurate-related numbers.

The Chapter 7 consists of exercises and hints for their solutions.

Finally, the huge Index lists all classes of special numbers mentioned in the text.

The main purpose of this book is to give a systematic and complete presentation of the theory of figurate numbers, presenting much of their properties, facts and theorems with full proofs. We intended to find and organize much of scattered material, as well as present updated material with all details, in clear and unified way.

The target audience consists of teachers and students (especially, of University) interested in Number Theory, General Algebra, Cryptography and related fields, as well as general audience of amateurs of Mathematics.

The book is so organized that it can be used as a source material for individual scientific work by the students. It permits to select the degree of difficulty, need for collecting information and the rigor of the proofs, In fact, the material of the book was already used for many student's bachelor and magister (i.e., master) degrees. Specifically, Chapters 1, 2, and 3 are accessible to undergraduate students and general readers of Mathematics, but Chapters 2 and 3 are slightly more difficult. The Chapter 4 and, especially, Chapter 5, are more involved and require some (still basic) mathematical culture.

Chapter 1

Plane figurate numbers

In this introductory chapter, the basic figurate numbers, **polygonal numbers**, are presented. They generalize **triangular** and **square numbers** to any regular m-gon.

Besides classical polygonal numbers, we consider in the plane **centered polygonal numbers**, in which layers of m-gons are drown centered about a point.

Finally, we list many other two-dimensional figurate numbers: **pronic numbers**, **trapezoidal numbers**, **polygram numbers**, **truncated plane figurate numbers**, etc.

1.1 Definitions and formulas

1.1.1. Following to ancient mathematicians, we are going now to consider the sets of points forming some geometrical figures on the plane.

Starting from a point, add to it two points, so that to obtain an equilateral triangle. Six-points equilateral triangle can be obtained from three-points triangle by adding to it three points; adding to it four points gives ten-points triangle, etc. So, by adding to a point two, three, four etc. points, then organizing the points in the form of an equilateral triangle and counting the number of points in each such triangle, one can obtain the numbers 1, 3, 6, 10, 15, 21, 28, 36, 45, 55, ... (Sloane's A000217, [Sloa11]), which are called **triangular numbers**.

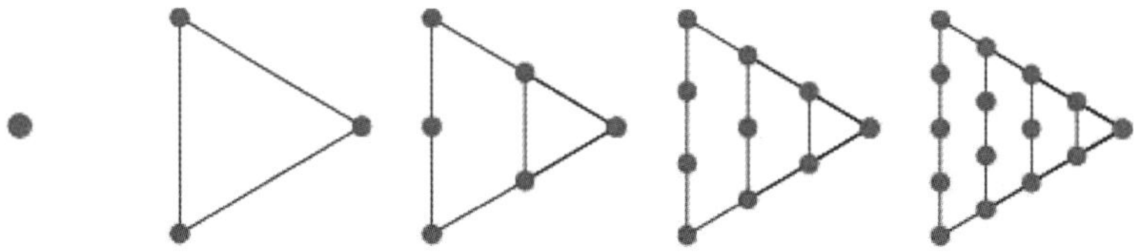

Similarly, by adding to a point three, five, seven etc. points and organizing them in the form of a square, one can obtain the numbers 1, 4, 9, 16, 25, 36, 49, 64, 81, 100, ... (Sloane's A000290), which are called **square numbers**.

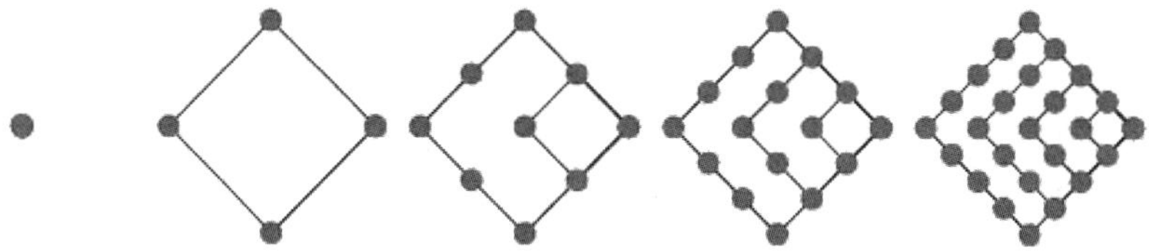

By adding to a point four, seven, ten etc. points and forming from them a regular pentagon, one can construct **pentagonal numbers** 1, 5, 12, 22, 35, 51, 70, 92, 117, 145, ... (Sloane's A000326).

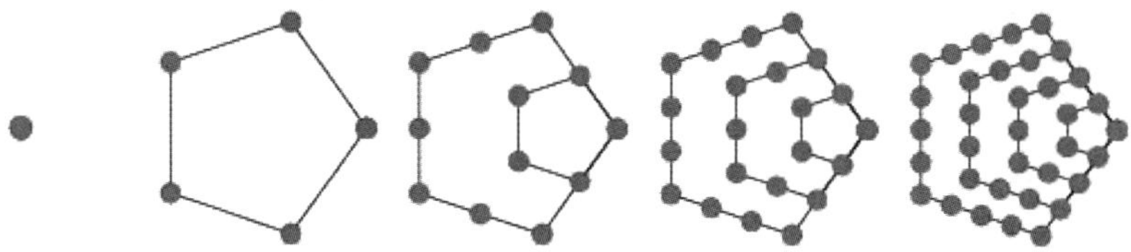

Following this procedure, we can construct **hexagonal numbers** 1, 6, 15, 28, 45, 66, 91, 120, 153, 190, ... (Sloane's A000384),

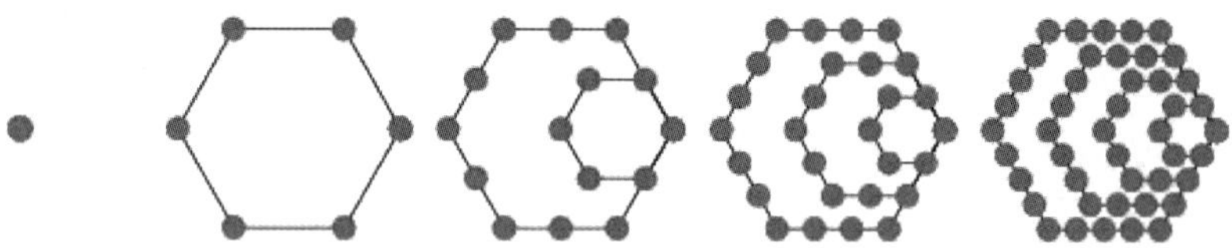

heptagonal numbers 1, 7, 18, 34, 55, 81, 112, 148, 189, 235, ... (Sloane's A000566), **octagonal numbers** 1, 8, 21, 40, 65, 96, 133, 176, 225, 280, ... (Sloane's A000567), **nonagonal numbers** 1, 9, 24, 46, 75, 111, 154, 204, 261, 325, ... (Sloane's A001106), **decagonal**

numbers 1, 10, 27, 52, 85, 126, 175, 232, 297, 370, ... (Sloane's A001107), **hendecagonal numbers** 1, 11, 30, 58, 95, 141, 196, 260, 333, 415, ... (Sloane's A051682), **dodecagonal numbers** 1, 12, 33, 64, 105, 156, 217, 288, 369, 460, ... (Sloane's A051624), etc.

So, we have constructed several simplest classes of **polygonal numbers** — positive integers, corresponding to an arrangement of points on the plane, which forms a regular polygon. One speaks about **m-gonal numbers** if the arrangement forms a regular m-gon.

1.1.2. Polygonal numbers were a concern of Pythagorean Geometry, since Pythagoras is credited with initiating them, and originating the notion that these numbers are generated from a *gnomon* or *basic unit*. A *gnomon* is a shape which, when added to a figure, yields another figure similar to the original. So, in our case, a gnomon is the piece which needs to be added to a polygonal number to transform it to the next bigger one.

The gnomon of a triangular number is the positive integer of the general form $n+1$, $n = 1, 2, 3, \ldots$: starting with n-th triangular number, one obtains $(n+1)$-th triangular number adjoining the line with $n+1$ elements. For instance, the 21-point triangle, composed of gnomons, looks like this:

$$\begin{array}{ccccccccccc} & & & & & 1 & & & & & \\ & & & & 2 & & 2 & & & & \\ & & & 3 & & 3 & & 3 & & & \\ & & 4 & & 4 & & 4 & & 4 & & \\ & 5 & & 5 & & 5 & & 5 & & 5 & \\ 6 & & 6 & & 6 & & 6 & & 6 & & 6 \end{array}$$

The gnomon of the square number is the odd number of the general form $2n+1$, $n = 1, 2, 3, \ldots$: in order to get the $(n+1)$-th square from n-th square, we should adjoin $2n+1$ elements, one to the end of each column, one to the end of each row, and a single one to the corner. The square of size 6×6, composed of gnomons, looks like this:

$$\begin{array}{cccccc} 6 & 6 & 6 & 6 & 6 & 6 \\ 5 & 5 & 5 & 5 & 5 & 6 \\ 4 & 4 & 4 & 4 & 5 & 6 \\ 3 & 3 & 3 & 4 & 5 & 6 \\ 2 & 2 & 3 & 4 & 5 & 6 \\ 1 & 2 & 3 & 4 & 5 & 6 \end{array}$$

The general rule for enlarging the regular polygon to the next size is to extend two adjacent sides by one point and then to add the required extra sides between those points. So, to transform n-th m-gonal number into the $(n+1)$-th m-gonal number, one adjoins $(m-2)n+1$ elements.

Therefore, the triangular numbers are obtained as consecutive sums of elements of the arithmetic progression $1, 2, 3, 4, \dots, n, \dots$; namely, they are $1 = 1$, $3 = 1+2$, $6 = 3+3$, $10 = 6+4$, ... The square numbers are obtained as consecutive sums of elements of the arithmetic progression $1, 3, 5, 7, \dots, 2n+1, \dots$: $1 = 1$, $4 = 1+3$, $9 = 4+5$, $16 = 9+7$, ... The pentagonal numbers are obtained by summation of the elements of the arithmetic progression 1, 4, 7, $10, \dots, 3n+1, \dots$: $1 = 1$, $5 = 1+4$, $12 = 5+7$, $22 = 12+10$, ... The hexagonal numbers are obtained by summation of the elements of the arithmetic progression $1, 5, 9, 13, \dots, 4n+1, \dots$: $1 = 1$, $6 = 1+5$, $15 = 6+9$, $28 = 15+13$, ..., and so on.

Putting the points in one line, one can speak about **linear numbers**. In fact, any positive integer is a linear number. Similarly to the above construction, the linear numbers have, as a gnomon, the number 1, and are obtained as consecutive sums of elements of the sequence $1, 1, 1, 1, \dots, 1, \dots$: $1 = 1$, $2 = 1+1$, $3 = 2+1$, $4 = 3+1$, ... In fact, the linear numbers are one-dimensional analogues of two-dimensional polygonal numbers.

1.1.3. The first general definition of m-gonal numbers was given by Hypsicles of Alexandria in II-th century BC and was quoted by Diophantys in his tract *On polygonal numbers* (see [Diop], [Heat10]): *if there are as many numbers as we please beginning with one and increasing by the same common difference, then when the common difference is 1, the sum of all the terms is a triangular number; when 2, a square; when 3, a pentagonal number; and the number of the angles is called after the number exceeding the common difference by 2, and the side after the number of terms including 1.*

In contemporary mathematical language, it has the following form: *n-th m-gonal number $S_m(n)$ is the sum of the first n elements of the arithmetic progression*

$$1, 1+(m-2), 1+2(m-2), 1+3(m-2), \dots, \quad m \geq 3.$$

So, by definition, it holds

$S_m(n) = 1+(1+(m-2))+(1+(m-2))+\cdots+(1+(m-2)(n-1)).$

In particular, we get

$$S_3(n) = 1+2+\cdots+n,\ S_4(n) = 1+3+\cdots+(2n-1),$$
$$S_5(n) = 1+4+\cdots+(3n-2),$$
$$S_6(n) = 1+5+\cdots+(4n-3),\ S_7(n) = 1+6+\cdots+(5n-4),$$
$$S_8(n) = 1+7+\cdots+(6n-5).$$

Above expression implies the following recurrent formula for m-gonal numbers:

$$S_m(n+1) = S_m(n) + (1+(m-2)n), \quad S_m(1) = 1.$$

In particular, we get

$$S_3(n+1) = S_3(n) + (n+1),\ S_4(n+1) = S_4(n) + (2n+1),$$
$$S_5(n+1) = S_5(n) + (3n+1),$$
$$S_6(n+1) = S_5(n) + (4n+1),\ S_7(n+1) = S_7(n) + (5n+1),$$
$$S_8(n+1) = S_8(n) + (6n+1).$$

For many applications it is convenient to add the value $S_m(0) = 0$ to the list.

Since the sum of the first n elements of an arithmetic progression $a_1, \ldots, a_n, \ldots$ is equal to $\frac{a_1+a_n}{2} \cdot n$, one obtains the following general formula for n-th m-gonal number:

$$S_m(n) = \frac{n((m-2)n-m+4)}{2} = \frac{(m-2)}{2}(n^2-n)+n$$
$$= \frac{(m-2)n^2-(m-4)n}{2} = \frac{1}{2}m(n^2-n)-n^2+2n.$$

In particular, one has

$$S_3(n) = \frac{n(n+1)}{2},\ S_4(n) = \frac{n\cdot(2n)}{2} = n^2,\ S_5(n) = \frac{n(3n-1)}{2},$$
$$S_6(n) = \frac{n(4n-2)}{2} = n(2n-1),\ S_7(n) = \frac{n(5n-3)}{2},$$
$$S_8(n) = \frac{n(8n-4)}{2} = n(4n-2).$$

These formulas for m-gonal numbers with $3 \le m \le 30$, as well as the first few elements of the corresponding sequences and the numbers of these sequences in the Sloane's *On-Line Encyclopedia of Integer Sequences* (OEIS, [Sloa11]) classification, are given in the table below.

Name	Formula												Sloane
Triangular	$\frac{1}{2}(n^2+n)$	1	3	6	10	15	21	28	36	45	55	66	A000217
Square	$\frac{1}{2}(n^2-0\cdot n)$	1	4	9	16	25	36	49	64	81	100	121	A000290
Pentagonal	$\frac{1}{2}(3n^2-1\cdot n)$	1	5	12	22	35	51	70	92	117	145	176	A000326
Hexagonal	$\frac{1}{2}(4n^2-2n)$	1	6	15	28	45	66	91	120	153	190	231	A000384
Heptagonal	$\frac{1}{2}(5n^2-3n)$	1	7	18	34	55	81	112	148	189	235	286	A000566
Octagonal	$\frac{1}{2}(6n^2-4n)$	1	8	21	40	65	96	133	176	225	280	341	A000567
Nonagonal	$\frac{1}{2}(7n^2-5n)$	1	9	24	46	75	111	154	204	261	325	396	A001106
Decagonal	$\frac{1}{2}(8n^2-6n)$	1	10	27	52	85	126	175	232	297	370	451	A001107
Hendecagonal	$\frac{1}{2}(9n^2-7n)$	1	11	30	58	95	141	196	260	333	415	506	A051682
Dodecagonal	$\frac{1}{2}(10n^2-8n)$	1	12	33	64	105	156	217	288	369	460	561	A051624
Tridecagonal	$\frac{1}{2}(11n^2-9n)$	1	13	36	70	115	171	238	316	405	505	616	A051865
Tetradecagonal	$\frac{1}{2}(12n^2-10n)$	1	14	39	76	125	186	259	344	441	550	671	A051866
Pentadecagonal	$\frac{1}{2}(13n^2-11n)$	1	15	42	82	135	201	280	372	477	595	726	A051867
Hexadecagonal	$\frac{1}{2}(14n^2-12n)$	1	16	45	88	145	216	301	400	513	640	781	A051868
Heptadecagonal	$\frac{1}{2}(15n^2-13n)$	1	17	48	94	155	231	322	428	549	685	836	A051869
Octadecagonal	$\frac{1}{2}(16n^2-14n)$	1	18	51	100	165	246	343	456	585	730	891	A051870
Nonadecagonal	$\frac{1}{2}(17n^2-15n)$	1	19	54	106	175	261	364	484	621	775	946	A051871
Icosagonal	$\frac{1}{2}(18n^2-16n)$	1	20	57	112	185	276	385	512	657	820	1001	A015872
Icosihenagonal	$\frac{1}{2}(19n^2-17n)$	1	21	60	118	195	291	406	540	693	865	1056	A015873
Icosidigonal	$\frac{1}{2}(20n^2-18n)$	1	22	63	124	205	306	427	568	729	910	1111	A015874
Icositrigonal	$\frac{1}{2}(21n^2-19n)$	1	23	66	130	215	321	448	596	765	955	1166	A015875
Icositetragonal	$\frac{1}{2}(22n^2-20n)$	1	24	69	136	225	336	469	624	801	1000	1221	A015876
Icosipentagonal	$\frac{1}{2}(23n^2-21n)$	1	25	72	142	235	351	490	652	837	1045	1276	
Icosihexagonal	$\frac{1}{2}(24n^2-22n)$	1	26	75	148	245	366	511	680	873	1090	1331	
Icosiheptagonal	$\frac{1}{2}(25n^2-23n)$	1	27	78	154	255	381	532	708	909	1135	1386	
Icosioctagonal	$\frac{1}{2}(26n^2-24n)$	1	28	81	160	265	396	553	736	945	1180	1441	
Icosinonagonal	$\frac{1}{2}(27n^2-25n)$	1	29	84	166	275	411	574	764	981	1225	1496	
Triacontagonal	$\frac{1}{2}(28n^2-26n)$	1	30	87	172	285	426	595	792	1017	1270	1551	

1.1.4. There are many different methods to obtain above formulas.

For example, the geometrical illustration for $n = 4$ on the picture below shows that n-th triangular number is one half of rectangle with the edges n and $n+1$. Hence, $S_3(n) = \frac{n(n+1)}{2}$.

```
* * * * ·
* * * · ·
* * · · ·
* · · · ·
```

On the other hand, this formula can be obtained by induction using the fact, that the triangular number with the index $n+1$ is obtained from the triangular number with the index n by addition of the number $n+1$. For $n = 1$ one has $S_3(1) = 1 = \frac{1\cdot(1+1)}{2}$. Going from n to $n+1$, one obtains $S_3(n+1) = S_3(n) + (n+1) = \frac{n(n+1)}{2} + (n+1) = \frac{(n+1)(n+2)}{2}$.

The special summation of the form

$$2S_3(n) = \begin{array}{c} 1 + \cdots + n + \\ n + \cdots + 1 \end{array}$$

gives $2S_3(n) = (n+1) + \cdots + (n+1) = n(n+1)$, i.e., one gets $S_3(n) = \frac{n(n+1)}{2}$.

Finally, another way to derive the formula $\frac{n(n+1)}{2}$ for $S_3(n)$ (which, obviously, is a polynomial of degree two) is to use the first three triangular numbers in order to find the coefficients A, B and C of the general second degree polynomial $An^2 + Bn + C$. It is $A + B + C = 1$, $4A + 2B + C = 3$, $9A + 3B + C = 6$ for $n = 1, 2, 3$, respectively. This leads to $A = \frac{1}{2}$, $B = \frac{1}{2}$, $C = 0$ and implies $S_3(n) = \frac{1}{2}n^2 + \frac{1}{2}n = \frac{n(n+1)}{2}$.

For square numbers, the geometrical interpretation in the case $n = 3$ shows that two n-th square numbers form a rectangle with edges n and $2n$; therefore, $S_4(n) = \frac{2n \cdot n}{2} = n^2$.

$$\begin{array}{cccccc} \cdot & * & * & * & * & * \\ \cdot & \cdot & \cdot & * & * & * \\ \cdot & \cdot & \cdot & \cdot & \cdot & * \end{array}$$

On the other hand, $(n+1)$-th square number can be obtained from n-th square number by addition of the number $2n+1$, and one can prove the formula $S_4(n) = n^2$ by induction. For $n = 1$ one has $S_4(1) = 1 = 1^2$. Going from n to $n+1$, one obtains

$$S_4(n+1) = S_4(n) + (2n+1) = n^2 + (2n+1) = (n+1)^2.$$

The special summation of the form

$$2S_4(n) = \begin{array}{ccccc} 1 & + \quad 3 & + \cdots + & (2n-1) & + \\ (2n-1) & + (2n-3) & + \cdots + & 1 & \end{array}$$

gives $2S_4(n) = 2n + \cdots + 2n = 2n^2$. It yields $S_4(n) = n^2$.

Finally, another way to derive the formula n^2 for $S_4(n)$ is to use the first three square numbers in order to find the coefficients A, B, and C of the general second degree polynomial $An^2 + Bn + C$.

In fact, for $n = 1$ one has $A + B + C = 1$, for $n = 2$ we obtain $4A + 2B + C = 4$, and for $n = 3$ it is $9A + 3B + C = 9$. This leads to $A = 1$, $B = 0$, $C = 0$ and implies $S_4(n) = n^2$.

The similar considerations can be applied to any m-gonal number. In a geometrical setting, two copies of the sum $a_1+a_2+\cdots+a_{n-1}+a_n$ of the first n elements of an arithmetic progression $a_1,\ldots,a_n,\ldots$ corresponds to the rectangle with the edges a_1+a_n and n. It reflects the fact, that the sums a_1+a_n, a_2+a_{n-1}, a_3+a_{n-2}, $\ldots$ are equal. In our case, $a_1=1$ and $a_n=1+(m-2)(n-1)$. It yields $2S_m(n)=(1+(1+(m-2)(n-1)))n=((m-2)n-m+4)n$, and $S_m(n)=\frac{((m-2)n-m+4)n}{2}$. Let as prove the formula $S_m(n)=\frac{n((m-2)n-m+4)}{2}$ by induction. It holds for $n=1$ since $S_m(1)=1=\frac{1\cdot((m-2)\cdot 1-m+4)}{2}$. Going from n to $n+1$, one obtains

$$\begin{aligned}S_m(n+1)&=S_m(n)+(1+(m-2)n)\\&=\frac{n((m-2)n-m+4)}{2}+(1+(m-2)n)\\&=\frac{n^2(m-2)+n(4-m)+2n(m-2)+2}{2}\\&=\frac{(n^2+n+1)(m-2)+(n+1)(4-m)}{2}\\&=\frac{(n+1)((m-2)(n+1)-m+4)}{2}.\end{aligned}$$

The special summation[1] has the form

$$2S_m(n)=$$

$$\begin{array}{ccccccccc}1&+&1+(m-2)&+&\cdots&+&1+(m-2)(n-1)&+\\1+(m-2)(n-1)&+&1+(m-2)(n-2)&+&\cdots&+&1.&\end{array}$$

It gives $2S_m(n)=(2+(m-2)(n-1))+\cdots+(2+(m-2)(n-1))=n(2+(m-2)(n-1))$, and $S_m(n)=\frac{n((m-2)n-m+4)}{2}$.

Finally, another way to derive the general formula for $S_m(n)$ is to use the first three m-gonal numbers to find the coefficients A, B, and C of the general 2-nd degree polynomial An^2+Bn+C. For $n=1$

[1]According to legend, at age 10 Gauss was told by his teacher to sum up all the numbers from 1 to 100. He reasoned that each number i could be paired up with $101-i$, to form a sum of 101, and if this was done 100 times, it would result in twice the actual sum, since each number would get used twice due to the pairing. Hence, the sum would be $1+\cdots+100=\frac{100\cdot 101}{2}$.

one has $A + B + C = 1$, for $n = 2$ we obtain $4A + 2B + C = m$, and for $n = 3$ it is $9A + 3B + C = 3m - 3$. This leads to $A = \frac{m-2}{2}$, $B = \frac{4-m}{2}$, $C = 0$, and implies above formula for $S_m(n)$.

1.1.5. The *generating function* for the sequence $S_m(1), S_m(2), \ldots, S_m(n), \ldots$ of the m-gonal numbers has the form $f(x) = \frac{x((m-3)x+1)}{(1-x)^3}$, i.e., it holds

$$\frac{x((m-3)x+1)}{(1-x)^3} = S_m(1)x + S_m(2)x^2 + S_m(3)x^3 + \cdots + S_m(n)x^n + \cdots, \ |x| < 1.$$

In particular, one gets

$$\frac{x}{(1-x)^3} = x + 3x^2 + 6x^3 + \cdots + S_3(n)x^n + \cdots, \quad |x| < 1;$$
$$\frac{x(x+1)}{(1-x)^3} = x + 4x^2 + 9x^3 + \cdots + S_4(n)x^n + \cdots, \quad |x| < 1;$$
$$\frac{x(2x+1)}{(1-x)^3} = x + 5x^2 + 12x^3 + \cdots + S_5(n)x^n + \cdots, \quad |x| < 1;$$
$$\frac{x(3x+1)}{(1-x)^3} = x + 6x^2 + 15x^3 + \cdots + S_6(n)x^n + \cdots, \quad |x| < 1.$$

In order to obtain the above formula, let us consider two polynomials

$$f(x) = a_0 + a_1x + \cdots + a_mx^m \quad \text{and} \quad g(x) = b_0 + b_1x + \cdots + b_nx^n$$

with real coefficients and $m < n$. It follows (see, for example, [DeMo10]), that *the rational function* $\frac{f(x)}{g(x)}$ *is the generating function of the sequence* $c_0, c_1, c_2, \ldots, c_n, \ldots$, *which is a solution of the linear recurrent equation* $b_0c_{n+k} + b_1c_{n+k-1} + \cdots + b_nc_k = 0$ *of n-th order with coefficients* $b_0, b_1, \ldots, b_n$.

In fact, one has the decomposition $\frac{f(x)}{g(x)} = c_0 + c_1x + c_2x^2 + \cdots + c_nx^n + \cdots$ if $|x| < r$, and $r = \min_{1 \le i \le n} |x_i|$, where $x_1, \ldots, x_n$ are the roots of the polynomial $g(x)$. It yields that the rational function $\frac{f(x)}{g(x)}$ is the generating function of the obtained sequence $c_0, c_1, c_2, \ldots, c_n, \ldots$. Moreover, one gets $f(x) = g(x)(c_0 + c_1x +$

$c_2x^2 + \cdots + c_nx^n + \cdots)$. In other words, it holds

$$a_0 + a_1x + a_2x^2 + \cdots + a_mx^m = (b_0 + b_1x + b_2x^2 + \cdots + b_nx^n) \times (c_0 + c_1x + c_2x^2 + \cdots + c_nx^n + \cdots).$$

It is easy to check now the following equalities:

$$\begin{aligned} a_0 &= b_0c_0, \quad a_1 = b_0c_1 + b_1c_0, \\ a_2 &= b_0c_2 + b_1c_1 + b_2c_0, \ldots, a_m = b_0c_m + \cdots + b_mc_0, \\ 0 &= b_0c_{m+1} + \cdots + b_{m+1}c_0, \ldots, \quad 0 = b_0c_n + \cdots + b_nc_0, \\ 0 &= b_0c_{n+1} + \cdots + b_nc_1, \ldots, \quad 0 = b_0c_{n+k} + \cdots + b_nc_k, \ldots . \end{aligned}$$

So, the sequence $c_0, c_1, \ldots, c_n, \ldots$ is a solution of the *linear recurrent equation* $b_0c_{n+k} + \cdots + b_nc_k = 0$ *of n-th order with coefficients* $b_0, \ldots, b_n$. Moreover, one can find the first n elements of this sequence using the first n above equalities: $a_0 = b_0c_0$, $a_1 = b_0c_1 + b_1c_0, \ldots, a_{n-1} = \sum_{k=0}^{n-1} b_kc_{n-1-k}$.

On the other hand, let the sequence $c_0, c_1, c_2, \ldots, c_n, \ldots$ be a solution of a linear recurrent equation $b_0c_{n+k} + \cdots + b_nc_k = 0$ of n-th order with coefficients $b_0, \ldots, b_n$. Let us define numbers $a_0, \ldots, a_{n-1}$ by the formulas $a_i = \sum_{k=0}^{i} b_kc_{i-k}$, $i = 0, 1, 2, \ldots, n-1$, using the *initial values* $c_0, c_1, \ldots, c_n$ of the given sequence. It yields the equality

$$a_0 + a_1x + a_2x^2 + \cdots + a_{n-1}x^{n-1} = (b_0 + b_1x + b_2x^2 + \cdots + b_nx^n) \times (c_0 + c_1x + c_2x^2 + \cdots + c_nx^n + \cdots).$$

In other words, one gets

$$\frac{a_0 + a_1x + a_2x^2 + \cdots + a_{n-1}x^{n-1}}{b_0 + b_1x + b_2x^2 + \cdots + b_nx^n} = c_0 + c_1x + c_2x^2 + \cdots + c_nx^n + \cdots .$$

So, the generating function of the sequence $c_0, c_1, \ldots, c_n, \ldots$ has the form $\frac{f(x)}{g(x)}$, where

$$g(x) = b_0 + b_1x + \cdots + b_nx^n, \quad \text{and} \quad f(x) = a_0 + \cdots + a_{n-1}x^{n-1},$$

with $a_0 = b_0c_0$, $a_1 = b_0c_1 + b_1c_0, \ldots, a_{n-1} = b_0c_{n-1} + b_1c_{n-2} + \cdots + b_{n-1}c_0$.

Now we can find the generating function for the sequence of m-gonal numbers. Let us consider the recurrent equation $S_m(n+1) = S_m(n) + (1 + (m-2)n)$. Going from n to $n+1$, one obtains $S_m(n+2) = S_m(n+1) + (1 + (m-2)(n+1))$. Subtracting first equality from second one, we get $S_m(n+2) - S_m(n+1) = S_m(n+1) - S_m(n) + (m-2)$, i.e., $S_m(n+2) = 2S_m(n+1) - S_m(n) + (m-2)$.

Similarly, one obtains $S_m(n+3) = 2S_m(n+2) - S_m(n+1) + (m-2)$, and

$$S_m(n+3) - S_m(n+2) = 2S_m(n+2) - 2S_m(n+1) - S_m(n+1) + S_m(n),$$

i.e., $S_m(n+3) = 3S_m(n+2) - 3S_m(n+1) + S_m(n)$. Hence, we get for the sequence of the m-gonal numbers the following linear recurrent equation:

$$S_m(n+3) - 3S_m(n+2) + 3S_m(n+1) - S_m(n) = 0.$$

It is a linear recurrent equation of 3-rd order with coefficients $b_0 = 1, b_1 = -3, b_2 = 3, b_3 = -1$. Its initial values are $S_m(1) = 1$, $S_m(2) = m$, $S_m(3) = 3m - 3$. Denoting $S_m(n+1)$ by c_n, one can rewrite the above equation as

$$c_{n+3} - 3c_{n+2} + 3c_{n+1} - c_n = 0, \quad c_0 = 1,\ c_1 = m,\ c_2 = 3m - 3.$$

Therefore, the generating function for the sequence of m-gonal numbers has the form

$$\frac{f(x)}{g(x)} = \frac{a_0 + a_1 x + a_2 x^2}{b_0 + b_1 x + b_2 x^2 + b_3 x^3},$$

where $b_0 = 1$, $b_1 = -3$, $b_2 = 3$, $b_3 = -1$, and $a_0 = b_0 c_0 = 1$, $a_1 = b_0 c_1 + b_1 c_0 = 1 \cdot m + (-3) \cdot 1 = m - 3$, $a_2 = b_0 c_2 + b_1 c_1 + b_2 c_0 = 1 \cdot (3m-3) + (-3)m + 3 \cdot 1 = 0$. Since $g(x) = 1 - 3x + 3x^2 - x^3 = (1-x)^3$ has three coinciding roots $x_1 = x_2 = x_3 = 1$, the generating function for the sequence of the m-gonal numbers obtains the form

$$\frac{1 + (m-3)x}{(1-x)^3} = S_m(1) + S_m(2)x + S_m(3)x^2 + \cdots + S_m(n)x^{n-1} + \cdots, \quad |x| < 1.$$

In other terms, it holds

$$\frac{x(1 + (m-3)x)}{(1-x)^3} = S_m(1)x + S_m(2)x^2 + S_m(3)x^3 + \cdots + S_m(n)x^n + \cdots, \quad |x| < 1.$$

1.2 Main properties of polygonal numbers

The considerations, similar to above ones, give many interesting properties of polygonal numbers.

1.2.1. For example, Theon of Smyrna proved in the II-th century AC that *the sum of two consecutive triangular numbers is a square number*, obtaining the following formula, called now the *Theon formula* (see [Theo]):

$$S_3(n) + S_3(n-1) = S_4(n).$$

In fact, one has $S_3(n) + S_3(n-1) = \frac{n(n+1)}{2} + \frac{(n-1)n}{2} = n^2 = S_4(n)$.

Alternatively, it can be demonstrated diagrammatically:

$$\begin{array}{cccc} * & * & * & * \\ * & * & * & \cdot \\ * & * & \cdot & \cdot \\ * & \cdot & \cdot & \cdot \end{array}$$

In the above example, constructed for $n = 4$, the square is formed by two interlocking triangles. Also, one can prove this formula by induction. For $n = 2$, it holds $S_3(2) + S_3(1) = 3 + 1 = 4 = S_4(2)$. Going from n to $n + 1$, one obtains

$$\begin{aligned} S_3(n+1) + S_3(n) &= S_3(n) + (n+1) + S_3(n-1) + n \\ &= S_4(n) + 2n + 1 = n^2 + 2n + 1 \\ &= (n+1)^2 = S_4(n+1). \end{aligned}$$

A special summation of the form

$$S_3(n) + S_3(n-1) = \begin{array}{r} 1 + 2 + 3 + \cdots + \quad n \quad + \\ 1 + 2 + \cdots + (n-1) \end{array}$$

gives $S_3(n) + S_3(n-1) = 1 + 3 + 5 + \cdots + (2n-1) = S_4(n)$.

1.2.2. Similarly, we can construct triangular numbers, using as inner blocks some triangular numbers of smaller size. For example, a triangular number with even index can be constructed using the following formula:

$$S_3(2n) = 3S_3(n) + S_3(n-1).$$

In fact, one has $3S_3(n)+S_3(n-1) = 3\cdot\frac{n(n+1)}{2} + \frac{(n-1)n}{2} = \frac{n}{2}(4n+2) = \frac{2n(2n+1)}{2} = S_3(2n)$.

The geometrical illustration of this property for $n = 3$ is given in the picture below.

By induction, one has, for $n = 2$, that $3S_3(2) + S_3(1) = 3 \cdot 3 + 1 = 10 = S_3(4)$, and, going from n to $n+1$, we obtain

$$\begin{aligned} 3S_3(n+1) + S_3(n) &= S_3(n) + S_3(n-1) + 3(n+1) + n \\ &= S_3(2n) + (2n+1) + (2n+2) \\ &= S_3(2n+1) + (2n+2) = S_3(2n+2) \\ &= S_3(2(n+1)). \end{aligned}$$

Finally, a special summation of the form

$$3S_3(n) + S_3(n-1) = \begin{array}{cccccccc} 1 + \cdots + n + 1 + & 2 & + \cdots + & n & + \\ & n + (n-1) & + \cdots + & 1 & + \\ & 1 & + \cdots + & (n-1) & \end{array}$$

implies $3S_3(n) + S_3(n-1) = 1 + \cdots + n + (n+1) + \cdots + 2n = S_3(2n)$.

1.2.3. A triangular number with odd index can be constructed using the following similar formula:

$$S_3(2n+1) = 3S_3(n) + S_3(n+1).$$

In fact, one has $3S_3(n) + S_3(n+1) = \frac{3n(n+1)}{2} + \frac{(n+1)(n+2)}{2} = \frac{n}{2}(4n+2) = \frac{2n(2n+1)}{2} = S_3(2n+1)$.

The geometrical illustration of this property for $n = 2$ is given below.

By induction, we have, for $n = 1$, that $3S_3(1) + S_3(2) = 3 \cdot 1 + 3 = 6 = S_3(3)$, and, going from n to $n+1$, we obtain

$$\begin{aligned} 3S_3(n+1) + S_3(n+2) &= 3S_3(n) + S_3(n+1) + 3(n+1) + (n+2) \\ &= S_3(2n+1) + (2n+2) + (2n+3) \\ &= S_3(2n+3) = S_3(2(n+1)+1). \end{aligned}$$

Finally, a summation of the special form

$$3S_3(n) + S_3(n+1) =$$

$$\begin{array}{rcccl} 1 + \cdots + n + 1 + & 2 & + \cdots + & n & + \\ n + & (n-1) & + \cdots + & 1 & + \\ & 1 & + \cdots + & (n-1) & + n + (n+1) \end{array}$$

gives $3S_3(n) + S_3(n+1) = 1 + \cdots + n + (n+1) + \cdots + 2n + (2n+1) = S_3(2n+1)$.

1.2.4. The next formula needs more triangles for the construction a mosaic:

$$S_3(3n-1) = 3S_3(n) + 6S_3(n-1).$$

In fact, we have $3S_3(n) + 6S_3(n-1) = \frac{3n(n+1)}{2} + \frac{6(n-1)n}{2} = \frac{n}{2}(9n-3) = \frac{(3n-1)3n}{2} = S_3(3n-1)$.

The geometrical illustration of this property for $n = 3$ is given below.

$$\begin{array}{cccccccc} & & & & & & & \ast \\ & & & & & & \ast & \ast \\ & & & & & \ast & \ast & \ast \\ & & & & \cdot & \star & \star & \bullet \\ & & & \cdot & \cdot & \star & \bullet & \bullet \\ & & \ast & \circ & \circ & \diamond & \diamond & \ast \\ & \ast & \ast & \circ & \star & \diamond & \ast & \ast \\ \ast & \ast & \ast & \star & \star & \ast & \ast & \ast \end{array}$$

By induction, we have, for $n = 2$, that $3S_3(2) + 6S_3(1) = 15 = S_3(5)$, and, going from n to $n+1$, we obtain

$$\begin{aligned} 3S_3(n+1) + 6S_3(n) &= 3S_3(n) + 6S_3(n-1) + 3(n+1) + 6n \\ &= S_3(3n-1) + 3n + (3n+1) + (3n+2) \\ &= S_3(3n+2) = S_3(3(n+1)-1). \end{aligned}$$

Finally, a special summation of the form

$$3S_3(n) + 6S_3(n+1) =$$

$$\begin{array}{ccccccccccccc}
1+\cdots+n+ & 1 & + & 2 & +\cdots+ & n & + & & & & & & \\
 & n & + & (n-1) & +\cdots+ & 1 & + & & & & & & \\
 & & & 1 & +\cdots+ & (n-1) & + & 1 & + & 2 & +\cdots+ & (n-1) & + \\
 & & & & & & & 1 & + & 2 & +\cdots+ & (n-1) & + \\
 & & & & & & & (n-1) & + & (n-2) & +\cdots+ & 1 & + \\
 & & & & & & & (n-1) & + & (n-2) & +\cdots+ & 1 & + \\
 & & & & & & & 1 & + & 2 & +\cdots+ & (n-1) &
\end{array}$$

shows, that

$$3S_3(n) + 6S_3(n-1) = 1 + \cdots + n + (n+1) + \cdots + 2n$$

$$+(2n+1) + \cdots + (3n-1) = S_3(3n-1).$$

1.2.5. The following property is called the *Diophantus' formula* (or, sometimes, the *Plutarch formula*):[2]

$$S_4(2n+1) = 8S_3(n) + 1.$$

In fact, we have $8S_3(n)+1 = \frac{8n(n+1)}{2}+1 = 4n^2+4n+1 = (2n+1)^2 = S_4(2n+1)$.

The geometrical illustration for $n = 2$ is given below.

$$\begin{array}{ccccc}
\ast & \ast & \cdot & \cdot & \star \\
\ast & \bullet & \cdot & \star & \star \\
\bullet & \bullet & \diamond & \ast & \ast \\
\cdot & \cdot & \star & \ast & \bullet \\
\cdot & \star & \star & \bullet & \bullet
\end{array}$$

By induction, we have, for $n = 1$, that $8S_3(1) + 1 = 9 = S_4(3)$, and, going from n to $n + 1$, we obtain

$$\begin{aligned}
8S_3(n+1) + 1 &= (8S_3(n) + 1) + 8(n+1) = S_4(2n+1) + 8(n+1) \\
&= (2n+1)^2 + 8(n+1) = 4n^2 + 12n + 9 = (2n+3)^2 \\
&= S_4(2n+3) = S_4(2(n+1)+1).
\end{aligned}$$

[2]This formula was known by Plutarch [Plut], a contemporary of Nicomachus, and Diophantus [Diop] (about 250 AC) generalized this theorem, proving by a cumbersome geometric method that $8(m-2)S_m(n)+(m-4)^2 = ((m-2)(2n-1)+2)^2$, and spoke of this result as a new definition of polygonal numbers equivalent to that of Hypsicles.

Finally, a special summation of the form

$$8S_3(n)+1=$$

$$\begin{array}{l}
1+2+\cdots+\quad n \quad +\\
\quad 1+\cdots+(n-1)+n+\\
\qquad\qquad\qquad\quad \begin{array}{cccccc}
1+ & 2 & +\cdots+ & n & +\\
1+ & 2 & +\cdots+ & n & \\
n+ & (n-1) & +\cdots+ & 1 & +\\
n+ & (n-1) & +\cdots+ & 1 & +\\
1+ & 2 & +\cdots+ & n & +\\
 & 1 & +\cdots+ & (n-1) & +
\end{array}\\
\qquad\qquad\quad n+\\
\qquad\qquad\quad 1
\end{array}$$

gives $8S_3(n)+1 = 1+3+\cdots+(2n-1)+(2n+1)+(2n+3)+\cdots+(4n+1)=S_4(2n+1)$.

1.2.6. What happens, if we consider now two consecutive triangle numbers with even (or odd) indices? The answer is given by the following formula:

$$S_3(n-1)+S_3(n+1)=2S_3(n)+1.$$

In fact, we have $S_3(n-1)+S_3(n+1) = \frac{(n-1)n}{2}+\frac{(n+1)(n+2)}{2} = \frac{2n^2+2n+2}{2} = 2\cdot\frac{n(n+1)}{2}+1 = 2S_3(n)+1$.

The geometrical illustration of this property for $n=3$ is given below.

By induction, we have, for $n=2$, that $S_3(1)+S_3(3)=1+6=7=2\cdot 3+1=2S_3(2)+1$, and, going from n to $n+1$, we obtain

$$\begin{aligned}
S_3(n)+S_3(n+2) &= S_3(n-1)+S_3(n+1)+n+(n+2)\\
&= 2S_3(n)+1+2(n+1) = 2(S_3(n)+(n+1))+1\\
&= 2S_3(n+1)+1.
\end{aligned}$$

The special summation of the form

$$S_3(n-1)+S_3(n)=\begin{array}{r}1+\cdots+(n-1)+\\ 1+\cdots+(n-1)+n+\\ n+1\end{array}$$

gives $S_3(n-1)+S_3(n+1)=2(1+\cdots+n)+1=2S_3(n)+1$.

1.2.7. The similar relations exist for other polygonal numbers. For example, one has the following property:

$$S_5(n)=S_4(n)+S_3(n-1).$$

In fact, it holds $S_4(n)+S_3(n-1)=n^2+\frac{(n-1)n}{2}=\frac{3n^2-n}{2}=\frac{n(3n-1)}{2}=S_5(n)$.

One can easy obtain a geometrical interpretation of this property, noting that $S_4(n)$ is the sum of the first n odd numbers. The corresponding picture for $n=4$ is given below.

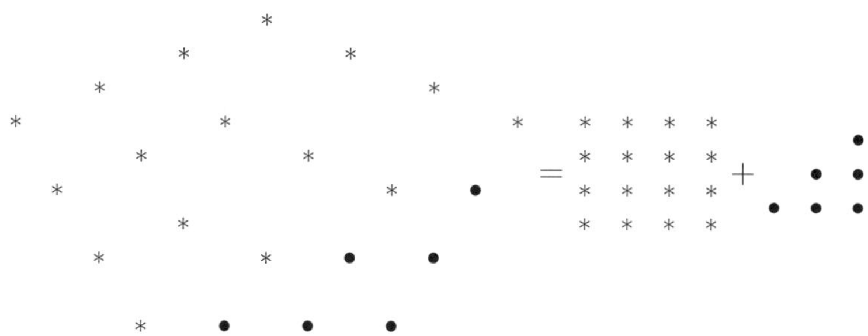

By induction, we have, for $n=2$, that $S_4(2)+S_3(1)=4+1=5=S_5(1)$, and, going from n to $n+1$, we obtain

$$\begin{aligned}S_4(n+1)+S_3(n)&=S_4(n)+S_3(n-1)+(2n+1)+n\\&=S_5(n)+(3n+1)=S_5(n+1).\end{aligned}$$

A special summation of the form

$$S_4(n)+S_3(n-1)=\begin{array}{r}1+3+\cdots+(2n-1)+\\ 1+\cdots+\ (n-1)\end{array}$$

gives $S_4(n)+S_3(n-1)=1+4+\cdots+(3n-2)=S_5(n)$.

1.2.8. The following property connects triangular and hexagonal numbers:

$$S_6(n)=S_3(n)+3S_3(n-1).$$

In fact, it is easy to see that $S_3(n)+3S_3(n-1)=\frac{n(n+1)}{2}+3\frac{(n-1)n}{2}=\frac{n}{2}(4n-2)=S_6(n)$.

The geometrical illustration of this property also is easy to obtain; below it is given for $n = 3$.

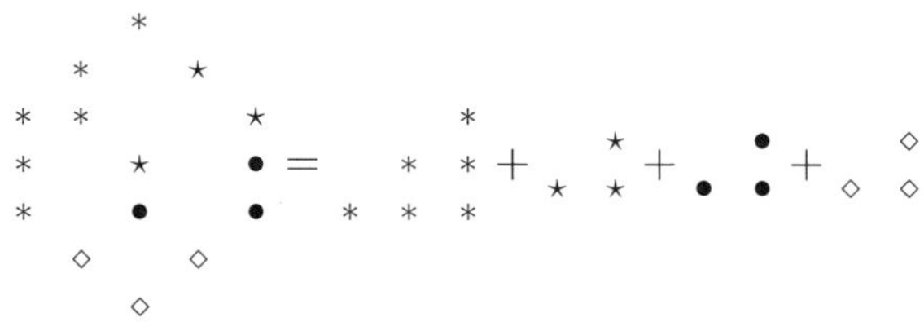

By induction, we have, for $n = 2$, that $S_3(2) + 3S_3(1) = 3 + 3 \cdot 1 = 6 = S_6(1)$, and, going from n to $n + 1$, we obtain

$$\begin{aligned} S_3(n+1) + 3S_3(n) &= S_3(n) + 3S_3(n-1) + (n+1) + 3n \\ &= S_6(n) + (4n+1) = S_6(n+1). \end{aligned}$$

A special summation of the form

$$S_3(n) + 3S_3(n-1) = \begin{array}{r} 1 + 2 + \cdots + (n-1) + 1 \\ 1 + \cdots + (n-1) + \\ 1 + \cdots + (n-1) + \\ 1 + \cdots + (n-1) \end{array}$$

gives $S_3(n) + S_3(n-1) = 1 + 5 + \cdots + (4(n-1)+1) = S_6(n)$.

1.2.9. The next property is widely known as the *hexagonal number theorem*:

$$S_6(n) = S_3(2n-1),$$

i.e., *every hexagonal number is a triangular number.*

The simplest way to get a proof of this fact is to compare two formulas:

$$S_3(2n-1) = \frac{(2n-1)2n}{2}, \quad \text{and} \quad \frac{n(4n-2)}{2} = S_6(n).$$

The geometrical illustration of this property for $n = 3$ is given below.

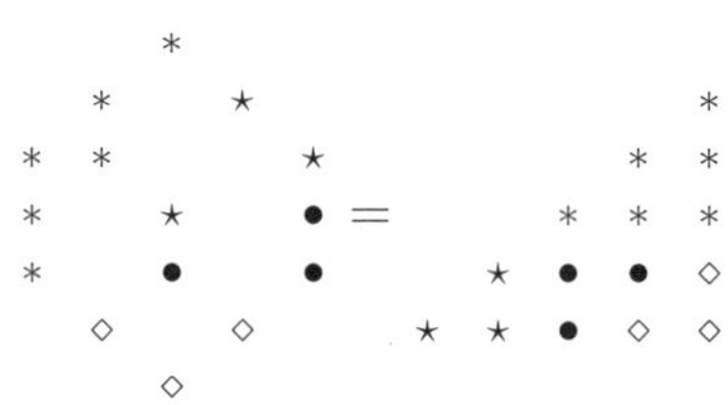

It is easy to see that we just rearranged the four small triangles on the previous picture into one big triangle.

Of course, one can prove it by induction: for $n = 1$, it holds $S_6(1) = 1 = S_3(2 \cdot 1 - 1)$, and, going from n to $n+1$, we obtain

$$\begin{aligned} S_6(n+1) &= S_6(n) + (4n+1) = S_3(2n-1) + (4n+1) \\ &= \frac{(2n-1)2n}{2} + (4n+1) = \frac{4n^2+6n+2}{2} \\ &= \frac{(2n+1)(2n+2)}{2} = S_3(2n+1) = S_3(2(n+1)-1). \end{aligned}$$

A special summation of the form

$$S_3(2n-1) = \begin{array}{c} 1+2+4+\cdots+(2n-2)+ \\ 3+5+\cdots+(2n-1) \end{array}$$

also gives $S_3(2n-1) = 1+5+9+\cdots+(4n-3) = S_6(n)$.

1.2.10. The *octagonal number theorem* shows a connection between octagonal and triangular numbers:

$$S_8(n) = 6S_3(n-1) + n.$$

In fact, one has $6S_3(n-1) + n = 6\frac{(n-1)n}{2} + n = (3n^2 - 3n) + n = 3n^2 - 2n = \frac{6n^2-4n}{2} = S_8(n)$.

The geometrical interpretation of this property for $n = 4$ is given below:

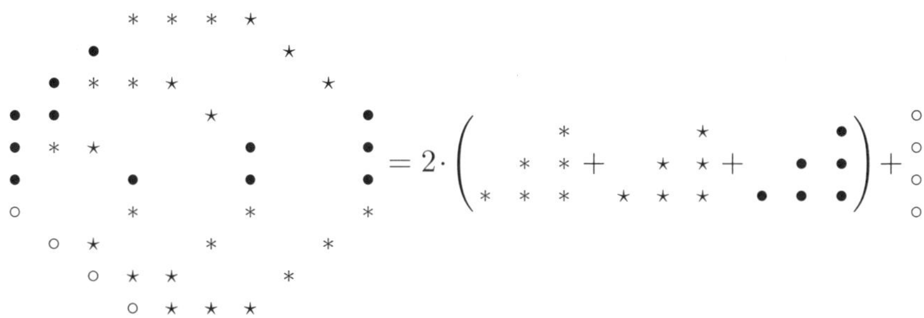

By induction, we have, for $n = 2$, that $S_8(2) = 8 = 6 \cdot S_3(1) + 2$, and, going from n to $n+1$, we obtain

$$\begin{aligned} S_8(n+1) &= S_8(n) + (6n+1) = (6S_3(n-1) + n) + (6n+1) \\ &= 6\frac{(n-1)n}{2} + 6n + (n+1) = (3n^2+3n) + (n+1) \\ &= 6\frac{n(n+1)}{2} + (n+1) = 6S_3(n) + (n+1). \end{aligned}$$

A special summation of the form

$$6S_3(n-1)+n=\begin{array}{l}1+2+\cdots+(n-1)+\\ \quad 1+\cdots+(n-2)+(n-1)+\\ \quad 1+\cdots+(n-2)+(n-1)+\\ \quad 1+\cdots+(n-2)+(n-1)+\\ \quad 1+\cdots+(n-2)+(n-1)+\\ \quad 1+\cdots+(n-2)+(n-1)+\\ \qquad\qquad\qquad\qquad n\end{array}$$

gives $6S_3(n-1)+n=1+7+\cdots+(6n-11)+(6n-5)=S_8(n)$.

1.2.11. The following formula of Nicomachus of Alexandria was obtained in the I-st century BC in his *Introduction to Arithmetic* (see [Nico]): *any figurate number is equal to the sum of the figurate number of previous name, and taking in the row the same place, and the triangle number, taking in the row the previous place.*

In other words, the difference between n-th m-gonal number and n-th $(m-1)$-gonal number is the $(n-1)$-th triangular number. So, one has the following *Nicomachus formula*:

$$S_m(n)=S_{m-1}(n)+S_3(n-1).$$

In fact, since $S_m(n)=\frac{n((m-2)n-m+4)}{2}$, one has

$$S_3(n-1)+S_{m-1}(n)=\frac{(n-1)n}{2}+\frac{n((m-3)n-(m-1)+4)}{2}$$
$$=\frac{n}{2}((m-2)n-m+4)=S_m(n).$$

By induction, we have, for $n=2$, that $S_{m-1}(2)+S_3(1)=(m-1)+1=m=S_m(1)$, and, going from n to $n+1$, we obtain

$$S_{m-1}(n+1)+S_3(n)=(S_{m-1}(n-1)+S_3(n-1))+((m-3)n+1)$$
$$+n=S_m(n)+((m-2)n+1)=S_m(n+1).$$

A special summation of the form

$$S_{m-1}(n)+S_3(n-1)=\begin{array}{l}1+(m-1)+\cdots+((m-3)(n-1)+1)+\\ \qquad\quad 1\quad+\cdots+\qquad\qquad (n-1)\end{array}$$

gives $S_{m-1}(n)+S_3(n-1)=1+m+\cdots+((m-2)(n-1)+1)=S_m(n)$.

The geometrical illustrations of this property for $m=4$ and $m=5$ were given in the proofs of the Theon formula $S_4(n)=S_3(n)+S_3$

$(n-1)$ and the formula $S_5(n) = S_4(n) + S_3(n-1)$, which are particular cases of the Nicomachus formula.

1.2.12. Bachet de Méziriac in his supplement of two books on the polygonal numbers of Diophantus (see [Dick05]) obtained the property, which allows to get any polygonal number using only triangular numbers: *any m-gonal number is equal to the sum of the triangular number, taking in the row the same place, and $m-3$ triangular numbers, taking in the row the previous place.* So, one has the following *Bachet de Méziriac formula:*[3]

$$S_m(n) = S_3(n) + (m-3)S_3(n-1).$$

One can obtain this equation, using the Diophantus' formula consecutively for $(m-1)$-gonal, $(m-2)$-gonal, etc., numbers:

$$S_m(n) = S_{m-1}(n) + S_3(n-1),\ S_m(n) = S_{m-2}(n) + 2S_3(n-1), \dots,$$

$$S_m(n) = S_4(n) + (m-4)S_3(n),\ S_m(n) = S_3(n) + (m-3)S_3(n-1).$$

Of course, one can check this formula by direct computation: $S_3(n) + (m-3)S_3(n-1) = \frac{n(n+1)}{2} + (m-3)\frac{(n-1)n}{2} = \frac{n}{2}((m-3)(n-1) + n + 1) = \frac{n((m-2)n-m+4)}{2} = S_m(n)$.

By induction, we have, for $n = 2$, that $S_3(2) + (m-3)S_3(1) = 3 + (m-3)\cdot 1 = m = S_m(1)$, and, going from n to $n+1$, we obtain

$$\begin{aligned} &S_3(n+1) + (m-3)S_3(n) \\ &\quad = (S_3(n) + (m-3)S_3(n-1)) + (n+1) + (m-3)n \\ &\quad = S_m(n) + ((m-2)n+1) = S_m(n+1). \end{aligned}$$

A special summation of the form

$$S_3(n) + (m-3)S_3(n-1) = \begin{array}{r} 1 + 2 + \cdots + \quad n \quad + \\ 1 + \cdots + (n-1) + \\ 1 + \cdots + (n-1) + \\ \cdots \quad\quad \\ 1 + \cdots + (n-1) \quad \end{array}$$

gives $S_3(n) + (m-3)S_3(n-1) = 1 + (m-1) + \cdots + ((m-2)(n-1)+1) = S_m(n)$.

[3]It means that *any polygonal number is a linear combination of triangular numbers with non-zero coefficients.*

A geometrical illustration of this property for $m = 6$ was given in the proof of the formula $S_6(n) = S_3(n) + 3S_3(n-1)$, which is an particular case of the Bachet de Méziriac formula.

1.2.13. An additional useful equation has the following form:

$$S_m(n) = n + (m-2)S_3(n-1).$$

It can be obtained from the Bachet de Méziriac formula, using that $S_3(n) = S_3(n-1) + n$: $S_m(n) = S_3(n) + (m-3)S_3(n-1) = (S_3(n-1) + n) + (m-3)S_3(n-1) = n + (m-2)S_3(n-1)$.

Of course, one can check this formula by direct computation: $n + (m-2)S_3(n-1) = n + (m-2)\frac{(n-1)n}{2} = \frac{n}{2}((m-2)(n-1)+2) = \frac{n((m-2)n-m+4)}{2} = S_m(n)$.

By induction, we have, for $n = 2$, that $2 + (m-2)S_3(1) = 2 + (m-2) \cdot 1 = m = S_m(1)$, and, going from n to $n+1$, we obtain

$$\begin{aligned}(n+1) + (m-2)S_3(n) &= (n + (m-2)S_3(n-1)) + 1 + (m-2)n \\ &= S_m(n) + ((m-2)n+1) = S_m(n+1).\end{aligned}$$

A special summation of the form

$$n + (m-2)S_3(n-1) = \begin{array}{l} 1 + 2 + \cdots + (n-1) + \\ \quad 1 + \cdots + (n-2) + (n-1) + \\ \quad 1 + \cdots + (n-2) + (n-1) + \\ \quad \cdots \\ \quad 1 + \cdots + (n-2) + (n-1) + \\ \qquad\qquad\qquad + \quad n \end{array}$$

gives $n + (m-2)S_3(n-1) = 1 + (m-1) + \cdots + ((m-2)(n-2) + 1) + ((m-2)(n-1)+1) = S_m(n)$.

The geometrical illustration of this property for $m = 8$ was given in the proof of the octagonal number theorem $S_8(n) = n + 6S_3(n-1)$ which is the most known case of it.

1.3 Square triangular numbers

1.3.1. A **square triangular number** (or **triangular square number**) is a number which is both, a triangular number and a perfect square. The first few square triangular numbers are 1, 36, 1225,

41616, 1413721, ... (Sloane's A001110). The indices of the corresponding square numbers are 1, 6, 35, 204, 1189, ... (Sloane's A001109), and the indices of the corresponding triangular numbers are 1, 8, 49, 288, 1681, ... (Sloane's A001108).

Since every triangular number is of the form $\frac{1}{2}u(u+1)$, and every square number is of the form v^2, in order to find find all square triangular numbers, one seeks positive integers u and v, such that $\frac{1}{2}u(u+1) = v^2$.

The problem of finding square triangular numbers reduces to solving the Pell's equation $x^2 - 2y^2 = 1$ in the following way:

$$\begin{aligned}\frac{1}{2}u(u+1) = v^2 \Leftrightarrow u^2 + u = 2v^2 \Leftrightarrow 4u^2 + 4u + 1 \\ = 8v^2 + 1 \Leftrightarrow (2u+1)^2 - 2(2v)^2 = 1.\end{aligned}$$

So, having a solution (u, v) of the equation $\frac{1}{2}u(u+1) = v^2$ and taking $x = 2u + 1$ and $y = 2v$, one obtains a solution (x, y) of the Pell's equation $x^2 - 2y^2 = 1$. On the other hand, from a solution (x, y) of the Pell's equation $x^2 - 2y^2 = 1$ one can obtain a solution (u, v) of the equation $\frac{1}{2}u(u+1) = v^2$ with $u = \frac{x-1}{2}$, and $v = \frac{y}{2}$. Since the method of finding all solutions of a Pell's equation is well-known, one obtains the method of finding all triangular numbers, which are also square numbers.

1.3.2. There are infinitely many triangular numbers that are also square numbers. It was shown by Euler in 1730 (see [Dick05]). More exactly, Euler proved the following theorem: *all positive integer solutions of the equation $\frac{1}{2}u(u+1) = v^2$ can be obtained by the formulas*

$$u_n = \frac{(3+2\sqrt{2})^n + (3-2\sqrt{2})^n - 2}{4}, \quad \textit{and}$$
$$v_n = \frac{(3+2\sqrt{2})^n - (3-2\sqrt{2})^n}{4\sqrt{2}},$$

where n is any positive integer.

In fact (see, for example, [Buch09]), all non-negative integer solutions of the Pell's equation $x^2 - Dy^2 = 1$ can be obtained by the formulas $x = P_{kt-1}$, $y = Q_{kt-1}$, where P_i and Q_i are numerators and denominators, respectively, of convergents in the decomposition

$$\sqrt{D} = a_0 + \cfrac{1}{a_1 + \frac{1}{a_2 + \cdots}}$$
$$= [a_0, a_1, \ldots, a_k, a_1, \ldots, a_k, a_1, \ldots, a_k, \ldots] = [a_0, (a_1, \ldots, a_k)]$$

of $\sqrt{D}$ into a continued fraction, k is the length of the period of this decomposition, and t is any positive integer for which kt is even. For $D = 2$, it holds

$$\sqrt{2} = 1 + \cfrac{1}{2 + \frac{1}{2 + \cdots}} = [1, 2, 2, 2, 2, \ldots] = [1, (2)],$$

i.e., $k = 1$, and all positive integer solutions of the equation $x^2 - 2y^2 = 1$ have the form $x = P_{2n-1}$, $y = Q_{2n-1}$, $n \in \mathbb{N}$.

The numerators and denominators of the convergents of a continued fraction $a_0 + \frac{1}{a_1 + \frac{1}{a_2 + \cdots}} = [a_0, a_1, \ldots, a_n, \ldots]$ have the following property (see [Buch09]):

$$P_n = a_n P_{n-1} + P_{n-2}, \quad \text{and} \quad Q_n = a_n Q_{n-1} + Q_{n-2}$$

for any positive integer n, $n \geq 2$. In particular, for the decomposition $\sqrt{2} = [1, (2)]$ of $\sqrt{2}$ into the continued fraction, we have:

- $P_0 = 1$, $Q_0 = 1$ (since $\frac{P_0}{Q_0} = 1 = \frac{1}{1}$);
- $P_1 = 3$, $Q_2 = 2$ (since $\frac{P_1}{Q_1} = 1 + \frac{1}{2} = \frac{3}{2}$);
- $P_n = 2P_{n-1} + P_{n-2}$, $Q_n = 2Q_{n-1} + Q_{n-2}$ for $n \geq 2$ (since $a_n = 2$ for $n \geq 1$).

Now it is easy to obtain the sequence $P_1 = 3, Q_1 = 2$, $P_3 = 17, Q_3 = 12$, $P_5 = 99$, $Q_5 = 70$, $P_7 = 577$, $Q_7 = 408$, ..., giving the sequence $(3, 2)$, $(17, 12)$, $(99, 70)$, $(577, 408)$, $(3363, 2378)$, ... of all positive integer solutions $(x_n, y_n), n \in \mathbb{N}$, of the Pell's equation $x^2 - 2y^2 = 1$. Using above relations $u_n = \frac{x_n - 1}{2} = \frac{P_{2n-1} - 1}{2}$ and $v_n = \frac{y_n}{2} = \frac{Q_{2n-1}}{2}$, one obtains the sequence

$$(1, 1),\ (8, 6),\ (49, 35),\ (288, 204),\ (1681, 1189),\ \ldots$$

of all positive integer solutions $(u_n, v_n), n \in \mathbb{N}$, of the equation $\frac{1}{2}u\,(u + 1) = v^2$.

It is easy to show (for example, by induction), that the numerators P_n and denominators Q_n of the convergents of the continued fraction $1 + \frac{1}{2+\frac{1}{2+\cdots}} = [1, (2)] = \sqrt{2}$ have the following form:

$$P_n = \frac{(1+\sqrt{2})^{n+1} + (1-\sqrt{2})^{n+1}}{2}, \quad \text{and}$$

$$Q_n = \frac{(1+\sqrt{2})^{n+1} - (1-\sqrt{2})^{n+1}}{2\sqrt{2}}.$$

In fact, for $n = 1$, one has

$$\frac{(1+\sqrt{2})^2 + (1-\sqrt{2})^2}{2} = 3 = P_1, \quad \text{and}$$

$$\frac{(1+\sqrt{2})^2 - (1-\sqrt{2})^2}{2\sqrt{2}} = 2 = Q_1.$$

For $n = 2$, one gets

$$\frac{(1+\sqrt{2})^3 + (1-\sqrt{2})^3}{2} = 17 = P_3, \quad \text{and}$$

$$\frac{(1+\sqrt{2})^3 - (1-\sqrt{2})^3}{2\sqrt{2}} = 12 = Q_3.$$

Going from $n-2$ and $n-1$ to n, one has

$$\begin{aligned} P_n &= a_n P_{n-1} + P_{n-2} \\ &= 2 \cdot \frac{(1+\sqrt{2})^n + (1-\sqrt{2})^n}{2} + \frac{(1+\sqrt{2})^{n-1} + (1-\sqrt{2})^{n-1}}{2} \\ &= \frac{(1+\sqrt{2})^{n-1}(2(1+\sqrt{2})+1) + 2(1-\sqrt{2})^{n-1}(2(1-\sqrt{2})+1)}{2} \\ &= \frac{(1+\sqrt{2})^{n-1}(1+\sqrt{2})^2 + (1-\sqrt{2})^{n-1}(1-\sqrt{2})^2}{2}, \end{aligned}$$

i.e., $P_n = \frac{(1+\sqrt{2})^{n+1}+(1-\sqrt{2})^{n+1}}{2}$ holds. Similarly, we get

$$\begin{aligned} Q_n &= a_n Q_{n-1} + Q_{n-2} \\ &= 2 \cdot \frac{(1+\sqrt{2})^n - (1-\sqrt{2})^n}{2\sqrt{2}} + \frac{(1+\sqrt{2})^{n-1} - (1-\sqrt{2})^{n-1}}{2\sqrt{2}} \\ &= \frac{(1+\sqrt{2})^{n-1}(2(1+\sqrt{2})+1) - (1-\sqrt{2})^{n-1}(2(1-\sqrt{2})+1)}{2\sqrt{2}} \\ &= \frac{(1+\sqrt{2})^{n-1}(1+\sqrt{2})^2 - (1-\sqrt{2})^{n-1}(1-\sqrt{2})^2}{2\sqrt{2}}, \end{aligned}$$

i.e., $Q_n = \frac{(1+\sqrt{2})^{n+1}-(1-\sqrt{2})^{n+1}}{2\sqrt{2}}$ holds.

Using the above equalities for odd indices, we obtain

$$P_{2n-1} = \frac{(1+\sqrt{2})^{2n} + (1-\sqrt{2})^{2n}}{2}, \quad \text{and}$$

$$Q_{2n-1} = \frac{(1+\sqrt{2})^{2n} - (1-\sqrt{2})^{2n}}{2\sqrt{2}}.$$

These formulas with the identities $(1 \pm \sqrt{2})^2 = 3 \pm 2\sqrt{2}$ yield now[4] the formulas for u_n and v_n:

$$\begin{aligned} u_n &= \frac{P_{2n-1} - 1}{2} \\ &= \frac{(1+\sqrt{2})^{2n} + (1-\sqrt{2})^{2n} - 2}{4} = \frac{(3+2\sqrt{2})^n + (3-2\sqrt{2})^n - 2}{4}, \\ v_n &= \frac{Q_{2n-1}}{2} = \frac{(1+\sqrt{2})^{2n} - (1-\sqrt{2})^{2n}}{4\sqrt{2}} \\ &= \frac{(3+2\sqrt{2})^n - (3-2\sqrt{2})^n}{4\sqrt{2}}. \end{aligned}$$

For small values of n we obtain the following results:

* for $n = 1$, one has $u_1 = 1$, $v_1 = 1$, and $S_{4,3}(1) = S_3(1) = S_4(1) = 1$;
* for $n = 2$, one has $u_2 = 8$, $v_2 = 6$, and $S_{4,3}(2) = S_3(8) = S_4(6) = 36$;
* for $n = 3$, one has $u_3 = 49$, $v_3 = 35$, and $S_{4,3}(3) = S_3(49) = S_4(35) = 1225$;
* for $n = 4$, one has $u_4 = 288$, $v_4 = 204$, and $S_{4,3}(4) = S_3(288) = S_4(204) = 41616$;
* for $n = 5$, one has $u_5 = 1681$, $v_5 = 1189$, and $S_{4,3}(5) = S_3(1681) = S_4(1189) = 1413721$;
* for $n = 6$, one has $u_6 = 9800$, $v_6 = 6930$, and $S_{4,3}(6) = S_3(9800) = S_4(6930) = 48024900$;

---

[4] Of course, the same result can be obtained using the following well-known rule: *if* (x_0, y_0) *is the smallest positive integer solution of the Pell's equation* $x^2 - Dy^2 = 1$, *then all positive integer solutions* (x, y) *of the equation are given by* $x + y\sqrt{D} = \pm(x_0 + y_0\sqrt{D})^n$, $n \in \mathbb{N}$; *therefore,* $x = \frac{(x_0+y_0\sqrt{D})^n + (x_0-y_0\sqrt{D})^n}{2}$, *and* $y = \frac{(x_0+y_0\sqrt{D})^n - (x_0-y_0\sqrt{D})^n}{2\sqrt{D}}$, $n \in \mathbb{N}$. In our case $D = 2$, $(x_0, y_0) = (3, 2)$, and we have $x = \frac{(2+3\sqrt{2})^n + (2-3\sqrt{2})^n}{2}$, $y = \frac{(2+3\sqrt{2})^n - (2-3\sqrt{2})^n}{2\sqrt{2}}$.

* for $n = 7$, one has $u_7 = 57121, v_7 = 40391$, and $S_{4,3}(7) = S_3(57121) = S_4(40391) = 1631432881$.

1.3.3. Using the above formulas, one obtains now, that *n-th square triangular number $S_{4,3}(n)$ has the following form:*

$$S_{4,3}(n) = \frac{1}{32}((17 + 12\sqrt{2})^n + (17 - 12\sqrt{2})^n - 2).$$

In fact, $S_{4,3}(n) = v_n^2$. Noting that $(3 \pm 2\sqrt{2})^2 = 17 \pm 12\sqrt{2}$, one gets

$$S_{4,3}(n) = \left(\frac{(3 + 2\sqrt{2})^n - (3 - 2\sqrt{2})^n}{4\sqrt{2}}\right)^2$$
$$= \frac{1}{32}((17 + 12\sqrt{2})^n + (17 - 12\sqrt{2})^n - 2).$$

1.3.4. As n becomes larger, the ratio $\frac{u_n}{v_n}$ approaches $\sqrt{2} = 1.4142\ldots$:

$$\frac{u_1}{v_1} = \frac{1}{1} = 1; \quad \frac{u_2}{v_2} = \frac{8}{6} = 1.3333\ldots; \quad \frac{u_3}{v_3} = \frac{49}{35} = 1.4;$$
$$\frac{u_4}{v_4} = \frac{288}{204} = 1.4117\ldots; \quad \frac{u_5}{v_5} = \frac{1681}{1189} = 1.4137\ldots;$$
$$\frac{u_6}{v_6} = \frac{9800}{6930} = 1.4141\ldots; \quad \frac{u_7}{v_7} = \frac{57121}{40391} = 1.4142\ldots.$$

In general, it holds $\frac{u_n}{v_n} = \frac{P_{2n-1}-2}{Q_{2n-1}} = \frac{P_{2n-1}}{Q_{2n-1}} - \frac{2}{Q_{2n-1}}$. Since $Q_t \to \infty$ and $\frac{P_t}{Q_t} \to \sqrt{2}$ for $t \to \infty$, we get $\lim_{n\to\infty} \frac{u_n}{v_n} = \sqrt{2}$.

Similarly, the ratio $\frac{S_{4,3}(n+1)}{S_{4,3}(n)}$ of successive square triangular numbers approaches $17 + 12\sqrt{2} = 33.9705\ldots$ (see [Wiki11]):

$$\frac{36}{1} = 36; \quad \frac{1225}{36} = 34.0277\ldots; \quad \frac{41616}{1225} = 33.9722\ldots;$$
$$\frac{1413721}{41616} = 33.9706\ldots; \quad \frac{48024900}{1413721} = 33.9705\ldots.$$

1.3.5. There is an another way to obtain the consecutive members of the sequence of square triangular numbers (see [CoGu96]): *n-th square triangular number $S_{4,3}(n)$ is equal to $(P_nQ_n)^2$, where $\frac{P_n}{Q_n}$ is n-th convergent in the decomposition of $\sqrt{2}$ into the continued fraction.*

In order to prove this fact, let us show that the numerators P_n and denominators Q_n of the convergents in the decomposition $\sqrt{2} = [1, (2)]$ have the following property:

$$Q_{2n-1} = 2P_{n-1}Q_{n-1} \quad \text{for } n \in \mathbb{N}.$$

In fact, for $n = 1$, one has $Q_1 = 2 = 2 \cdot 1 \cdot 1 = 2P_0Q_0$. For $n > 1$, it holds

$$\begin{aligned} &2\frac{(1+\sqrt{2})^n + (1-\sqrt{2})^n}{2} \cdot \frac{(1+\sqrt{2})^n - (1-\sqrt{2})^n}{2\sqrt{2}} \\ &\quad = \frac{(1+\sqrt{2})^{2n} - (1-\sqrt{2})^{2n}}{2\sqrt{2}}. \end{aligned}$$

Hence, we get $S_{4,3}(n) = v_n^2 = \left(\frac{Q_{2n-1}}{2}\right)^2 = (P_nQ_n)^2$. Remind that $P_0 = 1$, $Q_0 = 1$, $P_1 = 3$, $Q_1 = 2$, and $P_n = 2P_{n-1} + P_{n-2}$, $Q_n = 2Q_{n-1} + Q_{n-2}$ for any $n \geq 2$. So, we start with the fractions $\frac{P_0}{Q_0} = \frac{1}{1}$ and $\frac{P_1}{Q_1} = \frac{3}{2}$. Moreover, for any $n \geq 2$ we construct n-th convergent $\frac{P_n}{Q_n}$ by doubling the $(n-1)$-th convergent and "adding" it to the $(n-2)$-th convergent. The first few elements of the described sequence are:

$$\frac{1}{1}, \frac{3}{2}, \frac{2\cdot 3+1}{2\cdot 2+1} = \frac{7}{5}, \frac{2\cdot 7+3}{2\cdot 5+2} = \frac{17}{12}, \frac{2\cdot 17+7}{2\cdot 12+5} = \frac{41}{29},$$

$$\frac{2\cdot 41+17}{2\cdot 29+12} = \frac{99}{70}, \frac{2\cdot 99+41}{2\cdot 70+29} = \frac{239}{169}.$$

The numerators and denominators of the obtained fractions permit to construct the first few square triangular numbers:

$$\begin{aligned} &(1\cdot 1)^2 = 1,\ (3\cdot 2)^2 = 36,\ (7\cdot 5)^2 = 1225,\ (17\cdot 12)^2 = 41616, \\ &(41\cdot 29)^2 = 1413721,\ (99\cdot 70)^2 = 48024900, \\ &(239\cdot 169)^2 = 1631432881. \end{aligned}$$

1.3.6. Conversely, if one knows the value of n-th square triangular number $S_{4,3}(n)$, which is the v_n-th perfect square $S_4(v_n) = v_n^2$ and the u_n-th triangular number $S_3(u_n) = \frac{u_n(u_n+1)}{2}$, one can easily obtain the indices v_n and u_n of corresponding square and triangular numbers:

$$v_n = v_n(S_{4,3}(n)) = \sqrt{S_{4,3}(n)}, \quad \text{and}$$
$$u_n = u_n(S_{4,3}(n)) = \left\lfloor \sqrt{2S_{4,3}(n)} \right\rfloor.$$

The first equality is obvious, while the second one follows from the equality $u_n(u_n+1) = 2S_{4,3}(n)$ and the chain of simple inequalities:

$$\begin{aligned} u_n &= \sqrt{u_n^2} < \sqrt{u_n(u_n+1)} \\ &= \sqrt{u_n^2+u_n} < \sqrt{u_n^2+2u_n+1} = \sqrt{(u_n+1)^2} = u_n+1. \end{aligned}$$

Hence, one has $u_n < \sqrt{2S_{4,3}(n)} < u_n + 1$, i.e., u_n is the greatest integer less then or equal to $\sqrt{2S_{4,3}(n)}$. So, we get $u_n = \lfloor\sqrt{2S_{4,3}(n)}\rfloor$.

1.3.7. The above result of Euler, giving all indices v_n and u_n of the square and triangular numbers, which correspond to n-th square triangular number $S_{4,3}(n) = S_4(v_n) = S_3(u_n)$, is very beautiful, but almost useless to apply for big values of n. However, there are other ways for calculation indices v_n and u_n. In particular, *all positive integer solutions u_n and v_n of the equation $\frac{1}{2}u(u+1) = v^2$ can be obtained by the following recurrent formulas*:

$$u_{n+1} = 6u_n - u_{n-1} + 2, \; u_1 = 1, \; u_2 = 8, \quad \text{and}$$
$$v_{n+1} = 6v_n - v_{n-1}, \; v_1 = 1, \; v_2 = 6.$$

In fact, it holds $v_n = \frac{(3+2\sqrt{2})^n-(3-2\sqrt{2})^n}{4\sqrt{2}}$. Let $v_n = a - b$, where $a = \frac{(3+2\sqrt{2})^n}{4\sqrt{2}}$, and $b = \frac{(3-2\sqrt{2})^n}{4\sqrt{2}}$. It implies $v_{n+1} = a(3+2\sqrt{2}) - b(3-2\sqrt{2}) = 3(a-b) + 2\sqrt{2}(a+b)$, and

$$\begin{aligned} v_{n-1} &= \frac{a}{3+2\sqrt{2}} - \frac{b}{3-2\sqrt{2}} = a(3-2\sqrt{2}) - b(3+2\sqrt{2}) \\ &= 3(a-b) - 2\sqrt{2}(a+b). \end{aligned}$$

Hence, one gets $v_{n-1} + v_{n+1} = 6(a-b) = 6v_n$. In other words, $v_{n+1} = 6v_n - v_{n-1}$ holds.

Similarly, one has $u_n = \frac{(3+\sqrt{2})^n+(3-2\sqrt{2})^n-2}{4}$. Let $u_n = c + d - \frac{1}{2}$, where $c = \frac{(3+\sqrt{2})^n}{4}$, and $d = \frac{(3-2\sqrt{2})^n}{4}$. It implies $u_{n+1} = c(3+2\sqrt{2}) +$

$d(3-2\sqrt{2})-\frac{1}{2}=3(c+d)+2\sqrt{2}(c-d)-\frac{1}{2}$, and

$$u_{n-1}=\frac{c}{3+2\sqrt{2}}+\frac{d}{3-2\sqrt{2}}-\frac{1}{2}$$
$$=c(3-2\sqrt{2})+d(3+2\sqrt{2})-\frac{1}{2}=3(c+d)-2\sqrt{2}(c-d)-\frac{1}{2}.$$

Hence, one gets $u_{n+1}+u_{n-1}=6(c+d)-1=6(c+d-\frac{1}{2})+2=6u_n+2$. In other words, $u_{n+1}=6u_n-u_{n-1}+2$ holds (see [Weis11]).

So, starting from $v_1=1$ and $v_2=6$, one obtains $v_3=35$, $v_4=204$, $v_5=1189$, $v_6=6930$, $v_7=40391$, etc. Starting from $u_1=1$ and $u_2=8$, one obtains $u_3=49$, $u_4=288$, $u_5=1681$, $u_6=9800$, $u_7=57121$, etc.

1.3.8. Now we can prove that *the sequence of square triangular numbers can be obtained by the following recurrent equation:*

$$S_{4,3}(n)=34S_{4,3}(n-1)-S_{4,3}(n-2)+2,\ S_{4,3}(1)=1,\ S_{4,3}(2)=36.$$

In fact, we have $S_{4,3}(n)=v_n^2=(a-b)^2$, $S_{4,3}(n-1)=v_{n-1}^2=(3(a-b)-2\sqrt{2}(a+b))^2$, $S_{4,3}(n+1)=v_{n+1}^2=(3(a-b)+2\sqrt{2}(a+b))^2$. Since $ab=\frac{(3+2\sqrt{2})^n}{4\sqrt{2}}\cdot\frac{(3-2\sqrt{2})^n}{4\sqrt{2}}=\frac{1}{32}$, we obtain

$$S_{4,3}(n+1)+S_{4,3}(n-1)=18(a-b)^2+16(a+b)^2$$
$$=34(a-b)^2+64ab=34S_{4,3}(n)+2.$$

In other words, the equation $S_{4,3}(n)=34S_{4,3}(n-1)-S_{4,3}(n-2)+2$ holds (see [Weis11]).

1.3.9. An infinite subsequence $S_{4,3}(k)$ of square triangular numbers can be generated (see [PSJW62]) by a more simple recurrent formula

$$S_{4,3}(k+1)=4S_{4,3}(k)(8S_{4,3}(k)+1),\ S_{4,3}(1)=1.$$

In fact, let $S_3(u)=\frac{1}{2}u(u+1)=v^2=S_4(v)$. Then one gets

$$4S_3(u)(8S_3(u)+1)=2u(u+1)(4u(u+1)+1)$$
$$=\frac{4u(u+1)(4u(u+1)+1)}{2}=S_3(4u(u+1)).$$

On the other hand,

$$4S_4(v)(8S_4(v)+1)=4v^2(4u(u+1)+1)=4v^2(2u+1)^2=S_4(v(2u+1)).$$

So, starting from the unity, we obtain by above recurrent formula an infinite subsequence of square triangular numbers 1, 36, 41616, 55420693056, 9828650300920574145845 76, ...

1.3.10. The above recurrent formula gives a more simple proof of the fact, that there are infinitely many square triangular numbers. Another proof of this fact can be obtained using the theory of *Pythagorean triples* (see [Sier03]).

At first, let us obtain a bijection between the set (u, v) of positive integer solutions of the equation $\frac{1}{2}u(u+1) = v^2$ and the set (x, z) of positive integer solutions of the equation $x^2+(x+1)^2 = z^2$. In fact, let (x, z) be a positive integer solution of the equation $x^2+(x+1)^2 = z^2$. Then the pair (u, v) with $u = z - x - 1$ and $v = \frac{1}{2}(2x + 1 - z)$ gives a solution of the equation $\frac{1}{2}u(u + 1) = v^2$:

$$\frac{1}{2}(z - x - 1)(z - x) = \left(\frac{1}{2}(2x + 1 - z)\right)^2 \Leftrightarrow$$

$$2(z - x - 1)(z - x) = (2x + 1 - z)^2 \Leftrightarrow$$

$$2z^2 + 2x^2 - 4xz - 2z + 2x = 4x^2 + z^2 + 1 - 4xz + 4x - 2z$$

$$\Leftrightarrow z^2 = 2x^2 + 2x + 1 \Leftrightarrow z^2 = x^2 + (x + 1)^2.$$

It is easy to see that it is a positive integer solution. As $z^2 = x^2 + (x + 1)^2$, then $z^2 > (x + 1)^2$, and $z > x + 1$, i.e., $u = z - x - 1 \in \mathbb{N}$. Similarly, $z^2 = 2x^2 + 2x + 1 < 4x^2 + 2x + 1 = (2x + 1)^2$, and $z < 2x + 1$, i.e., $2x + 1 - z > 0$; as z^2 is a sum of two integers x^2 and $(x + 1)^2$ of different parity, it is an odd number, and, hence, z itself is an odd number. Therefore, $2x + 1 - z$ is an even number, and $v = \frac{1}{2}(2x + 1 - z) \in \mathbb{N}$.

Conversely, let the pair (u, v) be a positive integer solution of the equation $\frac{1}{2}u(u + 1) = v^2$. Then the pair (x, y) of positive integers $x = u + 2v$ and $z = 2u + 2v + 1$ gives a solution of the equation $x^2 + (x + 1)^2 = z^2$:

$$(u + 2v)^2 + (u + 2v + 1)^2 = (2u + 2v + 1)^2 \Leftrightarrow$$

$$2u^2 + 8v^2 + 8uv + 1 + 2u + 4v = 4u^2 + 4v^2 + 8uv + 1 + 4u + 4v$$

$$\Leftrightarrow 4v^2 = 2u^2 + 2u \Leftrightarrow v^2 = \frac{1}{2}u(u + 1).$$

Now we are going to show that the equation $x^2 + (x+1)^2 = z^2$ has infinitely many positive integer solutions (x_n, z_n), $n \in \mathbb{N}$. Namely, starting from the pair $(3, 5)$, we note, that for a given solution (x, z) of the equation $x^2 + (x+1)^2 = z^2$ the pair $(3x + 2z + 1, 4x + 3z + 2)$ also is a solution of this equation:

$$\begin{aligned}
&(3x + 2z + 1)^2 + (3x + 2z + 2)^2 \\
&\quad = (4x + 3z + 2)^2 \Leftrightarrow 18x^2 + 8z^2 + 5 + 24xz + 18x + 12z \\
&\quad = 16x^2 + 9z^2 + 4 + 24xz + 16x + 12z \Leftrightarrow 2x^2 + 2x + 1 \\
&\quad = z^2 \Leftrightarrow x^2 + (x+1)^2 = z^2.
\end{aligned}$$

Obviously, $3x + 2z + 1 > x$ and $4x + 3z + 2 > z$, so, we get an infinite recurrent sequence (x_n, z_n), $n \in \mathbb{N}$, of solutions of the equation $x^2 + (x+1)^2 = z^2$:

$$x_{n+1} = 3x_n + 2z_n + 1, \quad \text{and} \quad z_{n+1} = 4x_n + 3z_n + 2, \quad x_1 = 3, \ z_1 = 5.$$

In fact, these pairs give all positive integer solutions of the equation $x^2 + (x+1)^2 = z^2$. In order to prove it, let us consider a solution (x, z) of our equation with $x > 3$ and check that the pair $(3x - 2z + 1, 3z - 4x - 2)$ also gives a positive integer solution of this equation, but in this case it holds $3z - 4x - 2 < z$. So, we should check the inequalities $3x - 2z + 1 > 0$ and $0 < 3z - 4x - 2 < z$, or, which is the same, the inequalities $2z < 3x + 1$, $3z > 4x + 2$, and $2z < 4x + 2$. As $3x + 1 < 4x + 2$ for $x > 3$, one should check just first two above inequalities. Since $z^2 = 2x^2 + 2x + 1$, so $4z^2 = 8x^2 + 8x + 1 < 9x^2 + 6x + 1 = (3x+1)^2$, and $2z < 3x + 1$. Similarly, $9z^2 = 18x^2 + 18x + 9 > 16x^2 + 16x + 4 = (4x+2)^2$, and $3z > 4x + 2$. So, using the operation $g(x, z) = (3x - 2z + 1, 3z - 4x - 2)$, which delimitates the value z, we necessary come to the pair $(3, 5)$; it means that for some positive integer n, one gets $g^n(x, z) = (3, 5)$. Furthermore, the operation $f(x, z) = (3x + 2z + 1, 4x + 3z + 2)$ is connected with the operation g by the formula $f \cdot g(x, z) = f(3x - 2z + 1, 3z - 4x - 2) = (x, z)$, and, hence, we obtain, that, for any positive integer k, it holds $f^k \cdot g^k(x, z) = (x, z)$. So, it is proven, that for any positive integer solution (x, z) of the equation

$x^2+(x+1)^2 = z^2$, there exists some positive integer n (for which $g^n(x,z) = (3,5)$), such that $f^n(3,5) = (x,z)$. In other words, all positive integer solutions of our equation belong to the following tree with the root $(3,5)$:

$$(3,5) \to (20,29) \to (119,169) \to (696,985) \to \cdots$$

As it was proven above, any solution (x_n, z_n) of the equation $x^2 + (x+1)^2 = z^2$ corresponds to a solution (u_n, v_n) of the equation $\frac{1}{2}u(u+1) = v^2$ with $u_n = z_n - x_n - 1$ and $v_n = \frac{1}{2}(2x_n + 1 - z_n)$. So, $u_1 = z_1 - x_1 - 1 = 5-3-1 = 1$, $v_1 = \frac{1}{2}(2x_1+1-z_1) = \frac{1}{2}(2\cdot 3+1-5) = 1$, and $u_{n+1} = z_{n+1} - x_{n+1} - 1 = (4x_n+3z_n+2)-(3x_n+2z_n+1) = x_n+z_n+1$, $v_{n+1} = \frac{1}{2}(2x_{n+1}+1-z_{n+1}) = \frac{1}{2}(2(3x_n+2z_n+1)+1-(4x_n+2z_n+2)) = x_n + \frac{z_n+1}{2}$. Since the sequences x_n and z_n are increasing, so are the sequences u_n and v_n, and, therefore, we can get all positive integer solutions of the equation $\frac{1}{2}u(u+1) = v^2$ in the form of an infinite sequence (u_n, v_n), $n \in \mathbb{N}$, depending on sequence (x_n, y_n), $n \in \mathbb{N}$:

$$u_1 = 1, \quad v_1 = 1, \quad \text{and} \quad u_{n+1} = x_n + z_n, \; v_{n+1} = x_n + \frac{z_n+1}{2},$$

where

$$x_1 = 3, \; z_1 = 5, \; \text{ and } \; x_n = 3x_{n-1} + 2z_{n-1} + 1,$$
$$z_n = 4x_{n-1} + 3z_{n-1} + 2 \; \text{ for } n \geq 2.$$

So, we have proved, that there exist infinitely many triangular numbers $S_3(u_n)$, which are simultaneously the square numbers $S_4(v_n)$. Moreover, we get a method to find indices u_n and v_n of such numbers, using the tree $(3,5) \to (20,29) \to (119,169) \to (696,985) \to \cdots$ of all positive integer solutions of the Diophantine equation $x^2+(x+1)^2 = z^2$. So, $u_1 = 1$ and $v_1 = 1$. Using the pair $(x_1, z_1) = (3,5)$, we obtain $u_2 = 3+5 = 8$, and $v_2 = 3 + \frac{5+1}{2} = 6$. Using the pair $(x_2, z_2) = (20,29)$, we obtain $u_3 = 20+29 = 49$, and $v_3 = 20 + \frac{29+1}{2} = 35$. Using the pair $(x_3, z_3) = (119,169)$ we get $u_4 = 119+169 = 288$, and $v_4 = 119 + \frac{169+1}{2} = 3204$. Using the pair $(x_4, z_4) = (696,985)$, we get $u_5 = 696+985 = 1681$, and $v_5 = 696 + \frac{985+1}{2} = 1189$, and so on (see also **4.3.1.**).

1.3.11. The generating function for the sequence of the square triangular numbers has the form $f(x) = \frac{x(1+x)}{(1-x)(1-34x+x^2)}$ (see [SlP195]).

More exactly, it holds

$$\frac{x(1+x)}{(1-x)(1-34x+x^2)} = S_{4,3}(1)x + S_{4,3}(2)x^2 + S_{4,3}(3)x^3 + \cdots + S_{4,3}(n)x^n + \cdots, |x| < 4\sqrt{18} - 17.$$

In fact, let us consider the recurrent equation $S_{3,4}(n+2) = 34S_{3,4}(n+1) - S_{3,4}(n) + 2$. Going from n to $n+1$, one obtains $S_{3,4}(n+3) = 34S_{3,4}(n+2) - S_{3,4}(n+1) + 2$. Subtracting first equality from the second one, we get

$$S_{3,4}(n+3) - S_{3,4}(n+2) = 34S_{3,4}(n+2) - 34S_{3,4}(n+1) - S_{3,4}(n+1) + S_{3,4}(n).$$

Hence, we get the following linear recurrent equation:

$$S_{3,4}(n+3) - 35S_{3,4}(n+2) + 35S_{3,4}(n+1) - S_{3,4}(n) = 0.$$

It is a linear recurrent equation of 3-rd order with coefficients $b_0 = 1, b_1 = -35, b_2 = 35$, and $b_3 = -1$. Its initial values are $S_{3,4}(1) = 1$, $S_{3,4}(2) = 36$, and $S_{3,4}(3) = 1225$. Denoting $S_{3,4}(n+1)$ by c_n, one obtains a linear recurrent equation

$$c_{n+3} - 35c_{n+2} + 35c_{n+1} - c_n = 0, \quad c_0 = 1, \quad c_1 = 36, \quad c_2 = 1225.$$

So, the generating function for the sequence of the square triangular numbers has the form

$$\frac{f(x)}{g(x)} = \frac{a_0 + a_1x + a_2x^2}{b_0 + b_1x + b_2x^2 + b_3x^3},$$

where $b_0 = 1$, $b_1 = -35$, $b_2 = 35$, $b_3 = -1$, and $a_0 = b_0c_0 = 1$, $a_1 = b_0c_1 + b_1c_0 = 1 \cdot 36 + (-35) \cdot 1 = 1$, $a_2 = b_0c_2 + b_1c_1 + b_2c_0 = 1 \cdot 1225 + (-35) \cdot 36 + 35 \cdot 1 = 0$.

The polynomial $g(x) = 1 - 35x + 35x^2 - x^3 = (1-x)(1-34x+x^2)$ has three real roots $x_1 = 1$, $x_2 = 17 + 4\sqrt{18}$, and $x_3 = 17 - 4\sqrt{18}$. So, one has $\min\{|x_1|, |x_2|, |x_3|\} = 4\sqrt{18} - 17 = 0.029\ldots$. Hence, the generating function for the sequence of the square triangular numbers has the form

$$\frac{1+x}{(1-x)(1-34x+x^2)} = S_{3,4}(1) + S_{3,4}(2)x + S_{3,4}(3)x^2 + \cdots + S_{4,3}(n+1)x^n + \cdots, \quad |x| < 4\sqrt{18} - 17.$$

1.4 Other highly polygonal numbers

Square triangular numbers are the most known **highly polygonal numbers**, i.e., positive integers, which are polygonal in two or more ways. However, there are many similar classes of such polygonal numbers (see [Weis11]).

1.4.1. A **pentagonal triangular number** is a number which is simultaneously pentagonal and triangular. Such numbers correspond to the positive integer solutions of the Diophantine equation

$$\frac{1}{2}u(3u-1)=\frac{1}{2}v(v+1).$$

Completing the square gives $(6u-1)^2-3(2v+1)^2=-2$. Substituting $x=6u-1$ and $y=2v+1$ gives the Pell-like Diophantine equation $x^2-3y^2=-2$, which has positive integer solutions $(x,y)=(5,3)$, $(19,11)$, $(71,41)$, $(265,153)$, $\ldots$ (see, for example, [Nage51]). In terms of (u,v), these solutions give $(u,v)=(1,1)$, $(\frac{10}{3},5)$, $(12,20)$, $(\frac{133}{3},76)$, $(165,285)$, $\ldots$, of which the integer solutions are

$$(u,v)=(1,1),\ (12,20),\ (165,285),\ (2296,3976),\ (31977,55385),\ \ldots$$

(Sloane's A046174 and A046175). They correspond to the pentagonal triangular numbers (Sloane's A014979)

$$1, 210, 40755, 7906276, 1533776805, \ldots.$$

1.4.2. A **pentagonal square number** is a number which is simultaneously pentagonal and square. Such numbers correspond to the positive integer solutions of the Diophantine equation

$$\frac{1}{2}u(3u-1)=v^2.$$

Completing the square gives $(6u-1)^2-24v^2=1$. Substituting $x=6u-1$ and $y=2v$ gives the Pell's equation $x^2-6y^2=1$. It has positive integer solutions $(x,y)=(5,2)$, $(49,20)$, $(495,198)$, $\ldots$. In terms of (u,v), these give $(u,v)=(1,1)$, $(\frac{25}{3},10)$, $(81,99)$, $(\frac{2401}{3},980)$, $(7921,9701)$, $\ldots$, of which the integer solutions are

$$(u,v)=(1,1),\ (81,99),\ (7921,9701),\ (776161,950599),\\ (76055841,93149001),\ \ldots$$

(Sloane's A046172 and A046173). They correspond to the pentagonal square numbers (Sloane's A036353)

$$1, 9801, 94109401, 903638458801, 8676736387298001, \ldots.$$

1.4.3. A **pentagonal square triangular number** is a number that is simultaneously pentagonal, square and triangular. This requires a solution to the system of Diophantine equations

$$\begin{cases} \frac{1}{2}l(3l-1) = v^2 \\ \frac{1}{2}u(u+1) = v^2. \end{cases}$$

Solutions of this system can be found by checking pentagonal triangular numbers up to some limit to see if any are also square, other than the trivial case 1. Using this approach shows that none of the first 9690 pentagonal triangular numbers are square. So, there is no other pentagonal square triangular number less than 10^{22166}. It is almost certain, therefore, that no other solution exists, although no proof of this fact appeared in print yet (see [Weis11]).

1.4.4. Formally, an **hexagonal triangular number** is a number which is both hexagonal and triangular. However, it was proven, that $S_6(n) = S_3(2n-1)$, i.e., any hexagonal number is triangular, and the situation is trivial.

1.4.5. An **hexagonal square number** is a number which is both hexagonal and square. Such numbers correspond to the positive integer solutions of the Diophantine equation

$$u(2u-1) = v^2.$$

Completing the square and rearranging gives $(4u-1)^2 - 8v^2 = 1$. Substituting $x = 4u - 1$ and $y = 2v$ gives the Pell's equation $x^2 - 2y^2 = 1$. It has positive integer solutions $(x, y) = (3, 2)$, $(17, 12)$, $(99, 70)$, $(577, 408)$, ... In terms of (u, v), these give $(u, v) = (1, 1)$, $(\frac{9}{2}, 6)$, $(25, 35)$, $(\frac{289}{2}, 204)$, ..., of which the integer solutions are

$$(u, v) = (1, 1), (25, 35), (841, 1189), (28561, 40391),$$
$$(970225, 1372105), \ldots$$

(Sloane's A008844 and A046176). They correspond to the hexagonal square numbers (Sloane's A046177)

$$1, 1225, 1413721, 1631432881, 1882672131025, \ldots.$$

1.4.6. An **hexagonal pentagonal number** is a number which is simultaneously pentagonal and hexagonal. Such numbers correspond to the positive integer solutions of the Diophantine equation

$$\frac{1}{2}u(3u-1) = v(2v-1).$$

Completing the square and rearranging gives $(6u-1)^2 - 3(4v-1)^2 = -2$. Substituting $x = 6u-1$ and $y = 4v-1$ gives the Pell-like equation $x^2 - 3y^2 = -2$. Its first few positive integer solutions are $(x, y) = (1, 1), (5, 3), (19, 11), (71, 74), (265, 153), (989, 571), \ldots$. In terms of (u, v), these give the solutions $(u, v) = (\frac{1}{3}, \frac{1}{2}), (1, 1), (\frac{10}{3}, 3), (12, \frac{21}{2}), (\frac{133}{3}, \frac{77}{2}), (165, 143), \ldots$, of which the integer solutions are

$$(u, v) = (1, 1), (165, 143), (31977, 27693), (6203341, 5372251), (1203416145, 1042188953), \ldots$$

(Sloane's A046178 and A046179). They correspond to the hexagonal pentagonal numbers (Sloane's A046180)

$$1, 40755, 1533776805, 57722156241751, 2172315626468283465, \ldots.$$

1.4.7. An **heptagonal triangular number** is a number which is simultaneously heptagonal and triangular. Such numbers correspond to the positive integer solutions of the Diophantine equation

$$\frac{1}{2}u(5u-3) = \frac{1}{2}v(v+1).$$

Completing the square and rearranging gives $(10u-3)^2 - 5(2u+1)^2 = 4$. Substituting $x = 10u-3$ and $y = 2v+1$ gives the Pell-like equation $x^2 - 5y^2 = 4$. It has the positive integer solutions $(x, y) = (3, 1), (7, 3), (18, 8), (47, 21), (322, 144), \ldots$. The integer solutions in u and v are given by

$$(u, v) = (1, 1), (5, 10), (221, 493), (1513, 3382), (71065, 158905), \ldots$$

(Sloane's A046193 and A039835). They correspond to the heptagonal triangular numbers (Sloane's A046194)

$$1, 55, 121771, 5720653, 12625478965, \ldots.$$

1.4.8. An **heptagonal square number** is a number which is simultaneously heptagonal and square. Such numbers correspond to the positive integer solutions of the Diophantine equation

$$\frac{1}{2}u(5u-3)=v^2.$$

Completing the square and rearranging gives $(10u-3)^2-40v^2=9$. Substituting $x=10u-3$ and $y=2v$ gives the Pell-like equation $x^2-10y^2=9$. It has positive integer solutions $(x,y)=(7,2),(13,4),(57,18),(253,80),(487,154),\ldots$. The integer solutions in u and v are given then by

$$(u,v)=(1,1),\ (6,9),\ (49,77),\ (961,1519),\ (8214,12987),\ \ldots$$

(Sloane's A046195 and A046196). They correspond to the heptagonal square numbers (Sloane's A036354)

$$1,81,5929,2307361,168662169,\ldots.$$

1.4.9. An **heptagonal pentagonal number** is a number which is simultaneously heptagonal and pentagonal. Such numbers correspond to the positive integer solutions of the Diophantine equation

$$\frac{1}{2}u(5u-3)=\frac{1}{2}v(3v-1).$$

Completing the square and rearranging gives $3(10u-3)^2-5(6v-1)^2=22$. Substituting $x=10u-3$ and $y=6v-1$ gives the Pell-like equation $3x^2-5y^2=22$. It has the positive integer solutions $(x,y)=(3,1),\ (7,5),\ (17,13),\ (53,41),\ (133,103),\ \ldots$. The integer solutions in u and v are given then by

$$(u,v)=(1,1),\ (42,54),\ (2585,3337),\ (160210,206830),\ (9930417,12820113),\ \ldots$$

(Sloane's A046198 and A046199). They correspond to the heptagonal pentagonal numbers (Sloane's A048900)

$$1,4347,16701685,64167869935,246532939589097,\ldots.$$

1.4.10. An **heptagonal hexagonal number** is a number which is simultaneously heptagonal and hexagonal. Such numbers correspond to the positive integer solutions of the Diophantine equation

$$\frac{1}{2}u(5u-3) = v(2v-1).$$

Completing the square and rearranging gives $(10u - 3)^2 - 5(4v-1)^2 = 4$. Substituting $x = 10u - 3$ and $y = 4v - 1$ gives the Pell-like equation $x^2 - 5y^2 = 4$. It has positive integer solutions $(x, y) = (3, 1), (7, 3), (18, 8), (47, 21), (123, 55), \ldots$. The integer solutions in u and v are given by

$$(u, v) = (1, 1),\ (221, 247),\ (71065, 79453),\ (22882613, 25583539),\ (7368130225, 8237820025),\ \ldots$$

(Sloane's A048902 and A048901). They correspond to the heptagonal hexagonal numbers (Sloane's A048903)

$$1, 121771, 12625478965, 1309034909945503, 135723357520344181225, \ldots.$$

1.4.11. An **octagonal triangular number** is a number which is simultaneously octagonal and triangular. Such numbers correspond to the positive integer solutions of the Diophantine equation

$$u(3u-2) = \frac{v(v+1)}{2}.$$

Completing the square and rearranging gives $8(3u - 1)^2 - 3(2v+1)^2 = 5$. Substituting $x = 2(2u - 1)$ and $y = 2v + 1$ gives the Pell-like equation $2x^2 - 3y^2 = 5$. It has the positive integer solutions $(x, y) = (2, 1)$, $(4, 3)$, $(16, 13)$, $(38, 31)$, $(158, 129)$, $(376, 307)$, $\ldots$. These give the solutions $(u, v) = (\frac{2}{3}, 0)$, $(1, 1)$, $(3, 6)$, $(\frac{20}{3}, 15)$, $(\frac{80}{3}, 64)$, $(63, 153)$, $\ldots$, of which the integer solutions are

$$(u, v) = (1, 1),\ (3, 6),\ (63, 153),\ (261, 638),\ (6141, 15041),\ \ldots$$

(Sloane's A046181 and A046182). They correspond to the octagonal triangular numbers (Sloane's A046183)

$$1, 21, 11781, 203841, 113123361, \ldots.$$

1.4.12. An **octagonal square number** is a number which is simultaneously octagonal and square. Such numbers correspond to the quadratic Diophantine equation

$$u(3u-2)=v^2.$$

Completing the square and rearranging gives $(3u-1)^2-3v^2=1$. Substituting $x=3u-1$ and $y=v$ gives the Pell's equation $x^2-3y^2=1$. Its first few positive integer solutions are $(x,y)=(2,1)$, $(7,4)$, $(26,15)$, $(97,56)$, $(362,209)$, ... These give the solutions $(u,v)=(1,1)$, $(\frac{8}{3},4)$, $(9,15)$, $(\frac{98}{3},56)$, $(121,209)$, ..., of which the integer solutions are

$$(u,v)=(1,1),\ (9,15),\ (121,209),\ (1681,2911),\ (23409,40545),\ \ldots$$

(Sloane's A046184 and A028230). They correspond to the octagonal square numbers (Sloane's A036428)

$$1,225,43681,8473921,1643897025,\ldots.$$

1.4.13. An **octagonal pentagonal number** is a number which is simultaneously octagonal and pentagonal. Such numbers correspond to the Diophantine equation

$$u(3u-2)=\frac{1}{2}v(3v-1).$$

Completing the square and rearranging gives $(6v-1)^2-8(3u-1)^2=-7$. Substituting $x=6v-1$ and $y=2(3u-1)$ gives the Pell-like equation $x^2-2y^2=-7$. Its first few positive integer solutions are $(x,y)=(1,2)$, $(5,4)$, $(11,8)$, $(31,22)$, $(65,46)$, ... These give the solutions $(u,v)=(\frac{1}{3},\frac{2}{3})$, $(1,1)$, $(2,\frac{5}{3})$, $(\frac{16}{3},4)$, $(11,8)$, ..., of which the integer solutions are

$$(u,v)=(1,1),\ (11,8),\ (1025,725),\ (12507,8844),\ (1182657,836265),\ \ldots$$

(Sloane's A046187 and A046188). They correspond to the octagonal pentagonal numbers (Sloane's A046189)

$$1,176,1575425,234631320,2098015778145,\ldots.$$

1.4.14. An **octagonal hexagonal number** is a number which is simultaneously octagonal and hexagonal. Such numbers correspond to the positive integer solutions of the Diophantine equation

$$u(3u-2)=v(2v-1).$$

Completing the square and rearranging gives $8(3u-1)^2-3(4u-1)^2=5$. Substituting $x=2(3u-1)$ and $y=4v-1$ gives the Pell-like equation $2x^2-3y^2=5$. The first few positive integer solutions are $(x,y)=(2,1)$, $(4,3)$, $(16,13)$, $(38,31)$, $(158,129)$, $(376,307)$, ... These give the solutions $(u,v)=(\frac{2}{3},\frac{1}{2})$, $(1,1)$, $(3,\frac{7}{2})$, $(\frac{20}{3},8)$, $(\frac{80}{3},\frac{65}{2})$, $(63,77)$, ...,of which the integer solutions are

$$(u,v)=(1,1),\ (63,77),\ (6141,7521),\ (601723,736957),\\ (58962681,72214241),\ \ldots$$

(Sloane's A046190 and A046191). They correspond to the octagonal hexagonal numbers (Sloane's A046192)

$$1, 11781, 113123361, 1086210502741, 10429793134197921, \ldots.$$

1.4.15. An **octagonal heptagonal number** is a number which is simultaneously octagonal and heptagonal. Such numbers correspond to the Diophantine equation

$$\frac{1}{2}u(5u-3)=v(3v-2).$$

Completing the square and rearranging gives $3(10u-3)^2-40(3v-1)^2=-13$. Substituting $x=10u-3$ and $y=2(3v-1)$ gives the Pell-like equation $3x^2-10y^2=-13$. Its first few positive integer solutions are $(x,y)=(3,2)$, $(7,4)$, $(73,40)$, $(157,86)$, ... The integer solutions in u and v are given then by

$$(u,v)=(1,1),\ (345,315),\ (166145,151669),\ (80081401,73103983),\\ (38599068993,35235967977),\ \ldots$$

(Sloane's A048904 and A048905). They correspond to the octagonal heptagonal numbers (Sloane's A048906)

$$1, 297045, 69010153345, 16032576845184901,\\ 3724720317758036481633, \ldots.$$

1.4.16. A **nonagonal triangular number** is a number which is simultaneously nonagonal and triangular. Such numbers correspond to the positive integer solutions of the Diophantine equation

$$\frac{1}{2}u(7u-5)=\frac{1}{2}v(v+1).$$

Completing the square and rearranging gives $(14u - 5)^2 - 7(2v+1)^2 = 18$. Substituting $x = 14u-5$ and $y = 2v+1$ gives the Pell-like equation $x^2-7y^2=18$. It has positive integer solutions $(x,y) = (5,1)$, $(9,3)$, $(19,7)$, $(61,23)$, $(135,51)$, $(299,113)$, $(971,367)$, $\ldots$. The integer solutions in u and v are given then by

$$(u,v)=(1,1),\ (10,25),\ (154,406),\ (2449,6478),\ (39025,103249),\ \ldots$$

(Sloane's A048907 and A048908). They correspond to the nonagonal triangular numbers (Sloane's A048909)

$$1, 325, 82621, 20985481, 5330229625, \ldots.$$

1.4.17. A **nonagonal square number** is a number which is simultaneously nonagonal and square. Such numbers correspond to the Diophantine equation

$$\frac{1}{2}v(7v-5)=u^2.$$

Completing the square and rearranging gives $(14u-5)^2-56v^2=25$. Substituting $x = 14u-5$ and $y = 2v$ gives the Pell-like equation $x^2-14y^2=25$. It has positive integer solutions $(x,y)=(9,2)$, $(23,6)$, $(75,20)$, $(247,66)$, $(681,182)$, $(2245,600)$, $\ldots$. The corresponding integer solutions in u and v are

$$(u,v)=(1,1),\ (2,3),\ (18,33),\ (49,91),\ (529,989),\ \ldots$$

(Sloane's A048910 and A048911). They give the nonagonal square numbers (Sloane's A036411)

$$1, 9, 1089, 8281, 978121, \ldots.$$

1.4.18. A **nonagonal pentagonal number** is a number which is simultaneously nonagonal and pentagonal. Such numbers correspond to the positive integer solutions of the Diophantine equation

$$\frac{1}{2}u(7u-5)=\frac{1}{2}v(3v-1).$$

Completing the square and rearranging gives $3(14v - 5)^2 - 7(6u - 1)^2 = 68$. Substituting $x = 14v - 5$ and $y = 6u - 1$ gives the Pell-like equation $3x^2 - 7y^2 = 68$. Its positive integer solutions (x, y) correspond to positive integer solutions in u and v when

$$(u, v) = (1, 1), (14, 21), (7189, 10981), (165026, 252081), (86968201, 132846121), \ldots$$

(Sloane's A048913 and A048914). It gives the nonagonal pentagonal numbers (Sloane's A048915)

$$1, 651, 180868051, 95317119801, 26472137730696901, \ldots.$$

1.4.19. A **nonagonal hexagonal number** is a number which is simultaneously nonagonal and hexagonal. Such numbers correspond to the positive integer solutions of the Diophantine equation

$$\frac{1}{2}u(7u - 5) = v(2v - 1).$$

Completing the square and rearranging gives $(14v-5)^2-7(4u-1)^2 = 18$. Substituting $x = 14v - 5$ and $y = 4u - 1$ gives the Pell-like equation $x^2 - 7y^2 = 18$. It has positive integer solutions $(x, y) = (5, 1)$, $(9, 3)$, $(19, 17)$, $(61, 23)$, $(135, 51)$, $(509, 193)$, ... The corresponding integer solutions in u and v are

$$(u, v) = (1, 1), (10, 13), (39025, 51625), (621946, 822757), (2517635809, 3330519121), \ldots$$

(Sloane's A048916 and A048917). They give the nonagonal hexagonal numbers (Sloane's A048918)

$$1, 325, 5330229625, 1353857339341, 22184715227362706161, \ldots.$$

1.4.20. A **nonagonal heptagonal number** is a number which is simultaneously nonagonal and heptagonal. Such numbers correspond to the positive integer solutions of the Diophantine equation

$$\frac{1}{2}u(7u - 5) = \frac{1}{2}v(5v - 4).$$

Completing the square and rearranging gives $(14v - 5)^2 - 7(10u - 3)^2 = 62$. Substituting $x = 14v - 5$ and $y = 10u - 3$ gives the Pell-like equation $x^2 - 7y^2 = 62$. Its positive integer solutions (x, y)

correspond to positive integer solutions in u and v when

$$(u, v) = (1, 1),\ (88, 104),\ (12445, 14725),\ (1767052, 2090804),\ (250908889, 296879401),\ \dots$$

(Sloane's A048919 and A048920). It gives the nonagonal heptagonal numbers (Sloane's A048921)

$$1, 26884, 542041975, 10928650279834, 220343446399977901, \dots.$$

1.4.21. A **nonagonal octagonal number** is a number which is simultaneously nonagonal and octagonal. Such numbers correspond to the positive integer solutions of the Diophantine equation

$$\frac{1}{2}u(7u - 5) = v(3v - 2).$$

Completing the square and rearranging gives $(14v - 5)^2 - 56(3u - 1)^2 = 19$. Substituting $x = 14v - 5$ and $y = 3u - 1$ gives the Pell-like equation $3x^2 - 56y^2 = 19$. Its positive integer solutions (x, y) correspond to the following positive integer solutions in u and v:

$$(u, v) = (1, 1),\ (425, 459),\ (286209, 309141),\ (192904201, 208360351),\ (130017145025, 140434567209),\ \dots$$

(Sloane's A048922 and A048923). It gives the nonagonal octagonal numbers (Sloane's A048924)

$$1, 631125, 286703855361, 130242107189808901, 59165603001256545014625, \dots.$$

1.4.22. The table below summarizes the set of m-gonal k-gonal numbers for small values of m and k.

m	k	Sequence	Sloane
4	3	1, 36, 1225, 41616, 1413721, ...	A001110
5	3	1, 210, 40755, 7906276, 1533776805, ...	A014979
5	4	1, 9801, 94109401, 903638458801, 8676736387298001, ...	A036353
6	3	1, 6, 15, 28, 45, ... (i.e., all hexagonal numbers)	A000384
6	4	1, 1225, 1413721, 1631432881, 1882672131025, ...	A046177
6	5	1, 40755, 1533776805, 57722156241751, 2172315626468283465, ...	A046180
7	3	1, 55, 121771, 5720653, 12625478965 ...	A046194
7	4	1, 81, 5929, 2307361, 168662169, ...	A036354
7	5	1, 4347, 16701685, 64167869935, 246532939589097, ...	A048900
7	6	1, 121771, 12625478965, 1309034909945503, 135723357520344181225, ...	A048903
8	3	1, 21, 11781, 203841, 113123361, ...	A046183
8	4	1, 225, 43681, 8473921, 1643897025, ...	A036428
8	5	1, 176, 1575425, 234631320, 2098015778145, ...	A046189
8	6	1, 11781, 113123361, 1086210502741, 10429793134197921, ...	A046192
8	7	1, 297045, 69010153345, 16032576845184901, 3724720317758036481633, ...	A048906
9	3	1, 325, 82621, 20985481, 5330229625, ...	A048909
9	4	1, 9, 1089, 8281, 978121, ...	A036411
9	5	1, 651, 180868051, 95317119801, 26472137730696901, ...	A048915
9	6	1, 325, 5330229625, 1353857339341, 22184715227362706161, ...	A048918
9	7	1, 26884, 542041975, 10928650279834, 220343446399977901, ...	A048921
9	8	1, 631125, 286703855361, 130242107189808901, 59165603001256545014625, ...	A048924

1.5 Amount of a given number in all polygonal numbers

1.5.1. The problem about amount of a given number in all polygonal numbers was formulated by Diophantus in his treatise *On Polygonal Numbers* ([Diop], [Dick05]), of which only a part survives: *find how many times a given number N is contained among all polygonal numbers.*

In other words, for a given positive integer N, one must find all positive integers m and n for which $N = S_m(n)$. Obviously, $N = S_N(2)$, and without loss of generality we can restrict the consideration to $m \geq 3$ and $n \geq 2$. Since $S_m(n) = \frac{n((m-2)n-m+4)}{2}$, one obtains a chain of the following equalities:

$$N = \frac{(m-2)n^2 - (m-2)n + 2n}{2},$$

$$2N = (m-2)n^2 - (m-2)n + 2n,$$

$$2N - 2n = (m-2)(n^2 - n), \quad m - 2 = \frac{2N - 2n}{n(n-1)}.$$

Decomposition of the fraction $\frac{2N-2n}{n(n-1)}$ into a sum of elementary fractions $\frac{2N-2n}{n(n-1)} = \frac{2N-2}{n-1} - \frac{2N}{n}$ gives

$$m - 2 = \frac{2N-2}{n-1} - \frac{2N}{n}.$$

The positive integer numbers n and $n-1$ are relatively prime. So, in above formula for the positive integer $m-2$, the number $2N-2$ is divided by $n-1$, and the number $2N$ is divided by n.

Hence, for finding all polygonal numbers which coincide with N, one can use the following algorithm:

- *find all positive integer divisors of the number* $2N$;
- *find all positive integer divisors of the number* $2N-2$;
- *from first sequence choose numbers, which are on one unity greater than some number from second sequence: these numbers correspond to* n;
- *find* $m = \frac{2N-2}{n-1} - \frac{2N}{n} + 2$.

Let, for example, $N = 7$. All positive integer divisors of $2N = 14$ are $1, 2, 7, 14$. All positive integer divisors of $2N - 2 = 12$ are $1, 2, 3, 4, 6, 12$. Let us choose from the first set numbers $2 = 1 + 1$ and $7 = 6 + 1$. So, one has $n \in \{2, 7\}$. For $n = 2$, one obtains $m = \frac{12}{1} - \frac{14}{2} + 2 = 12 - 7 + 2 = 7$, and $S_m(n) = S_7(2) = 7$; for $n = 7$, it holds $m = \frac{12}{6} - \frac{14}{7} + 2 = 2$, and one can put $S_m(n) = S_2(7) = 7$ (remaind, that $S_2(n) = n$ are linear numbers).

If $N = 105$, then $2N = 210$, and $2N - 2 = 208$. All positive integer divisors of 210 are 1, 2, 3, 5, 6, 7, 10, 14, 15, 21, 30, 35, 42, 70, 105, 210. All positive integer divisors of 208 are $1, 2, 4, 8, 13, 16, 26, 52, 104, 208$. Then one gets $n \in \{2, 3, 5, 14, 105\}$. For $n = 2$, one has $m = \frac{208}{1} - \frac{210}{2} + 2 = 105$; so N is $S_{105}(2)$. For $n = 3$, it holds $m = \frac{208}{2} - \frac{210}{3} + 2 = 36$; so N is $S_{36}(3)$. For $n = 5$, one gets $m = \frac{208}{4} - \frac{210}{5} + 2 = 12$; so N is $S_{12}(5)$. For $n = 14$, $m = \frac{208}{13} - \frac{210}{14} + 2 = 3$; so N is $S_3(14)$. For $n = 105$, $m = \frac{208}{104} - \frac{210}{105} + 2 = 2$; so N is $S_2(105)$.

1.5.2. Call a positive integer ***k*-highly polygonal number**, if it is m-polygonal in k or more ways out of $m = 3, 4, \ldots$ up to some limit.

Then the first few **2-highly polygonal numbers** up to $m = 16$ are 1, 6, 9, 10, 12, 15, 16, 21, 28, 36, ... (Sloane's A090428). In fact, 1 is m-gonal for any m, 6 is 3- and 6-gonal, 9 is 4- and 9-gonal, 10 is 3- and 10-gonal, 12 is 5- and 12-gonal, 15 is 3-, 6- and 15-gonal, 16 is 4- and 16-gonal, 21 is 3- and 8-gonal, 28 is 3- and 6-gonal, 36 is 3-, 4- and 13-gonal, etc.

Similarly, the first few **3-highly polygonal numbers** up to $m = 16$ are 1, 15, 36, 45, 325, 561, 1225, 1540, 3025, 4186, ... (Sloane's A062712). In fact, 1 is m-gonal for any m, 15 is 3-, 6- and 15-gonal, 36 is 3-, 4- and 13-gonal, 45 is 3-, 6- and 16-gonal, 325 is 3-, 6- and 9-gonal, 561 is 3-, 6- and 12-gonal, 1225 is 3-, 4- and 6-gonal, 1540 is 3-, 6- and 10-gonal, 3025 is 4-, 12- and 15-gonal, 4186 is 3-, 6- and 13-gonal, etc.

There are no **4-highly polygonal numbers** of this type (i.e., up to $m = 16$) less than 10^{12}, except 1 (see [Weis11]).

1.5.3. Consider now one more simple question: how to check whether a number N is an m-gonal number $S_m(n)$ for some n? An arbitrary number N can be checked for m-gonality as follows. It is easy to prove the following *Diophantus' identity*:

$$8(m-2)S_m(n) + (m-4)^2 = (2n(m-2) - (m-4))^2.$$

In fact, since $S_m(n) = \frac{n((m-2)n-m+4)}{2}$, one obtains

$$\begin{aligned}
&8(m-2)S_m(n) + (m-4)^2 \\
&= 4(m-2)n((m-2)n - m + 4) + (m-4)^2 \\
&= 4(m-2)^2n^2 - 4(m-2)n(m-4) + (m-4)^2 \\
&= (2n(m-2) - (m-4))^2.
\end{aligned}$$

This identity implies that the value $8(m-2)S_m(n) + (m-4)^2$ is a perfect square for any m-gonal number $S_m(n)$. So, in order to check if an arbitrary positive integer N is an m-gonal number, one can check if the value $8(m-2)N + (m-4)^2$ is a perfect square. If it is not, the number N can not be m-gonal. If it is a perfect square S^2, then solving equation $S = 2n(m-2) - m + 4$ for n gives $n = \frac{S+m-4}{2(m-2)}$ (see [Dick05], [Weis11]).

For example, in order to check if the number 1540 is an 10-gonal number, let as consider the number $8(m-2)N+(m-4)^2$ for $m = 10$. In fact, this number $8 \cdot 8 \cdot 1540 + 6^2 = 98596$ is a perfect square: $98596 = 314^2$, i.e., $S = 314$. The equation $S = 2n(m-2) - m + 4$ with $S = 314$ and $m = 10$ obtains the form $314 = 16n - 6$, giving $n = 20$. Therefore, the number $N = 1540$ is the 20-th 10-gonal number: $1540 = S_{10}(20)$.

On the other hand, it is easy to show that the number 1540 is not an 8-gonal number. In fact, the number $8(m-2)N + (m-4)^2$ for $m = 8$ has the form $8 \cdot 6 \cdot 1540 + 4^2 = 73936$ and is not a perfect square: $73936 \neq a^2$, $a \in \mathbb{Z}$.

1.6 Centered polygonal numbers

Behind classical polygonal numbers, there are many other numbers, which can be constructed in the plane from points (or balls). The centered polygonal numbers form the next important class of such numbers.

1.6.1. The **centered polygonal numbers** (or, sometimes, **polygonal numbers of the second order**) form a class of figurate numbers, in which layers of polygons are drawn centered about a point. Each centered polygonal number is formed by a central dot, surrounded by polygonal layers with a constant number of sides. Each side of a polygonal layer contains one dot more than any side of the previous layer, so starting from the second polygonal layer each layer of a **centered m-gonal number** contains m more points than the previous layer.

So, a **centered triangular number** represents a triangle with a dot in the center and all other dots surrounding the center in successive triangular layers. The following image shows the building of the centered triangular numbers using the associated figures: at each step the previous figure is surrounded by a triangle of new points.

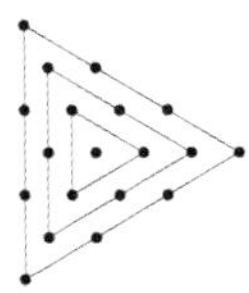

The first few centered triangular numbers are 1, 4, 10, 19, 31, 46, 64, 85, 109, 136, ... (Sloane's A005448).

A **centered square number** is consisting of a central dot with four dots around it, and then additional dots in the gaps between adjacent dots.

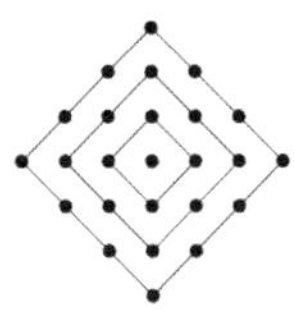

The first few centered square numbers are 1, 5, 13, 25, 41, 61, 85, 113, 145, 181, ... (Sloane's A001844).

A **centered pentagonal number** represents a pentagon with a dot in the center and all other dots surrounding the center in successive pentagonal layers.

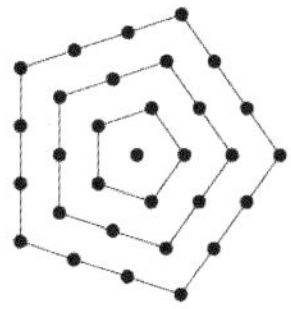

The first few centered pentagonal numbers are 1, 6, 16, 31, 51, 76, 106, 141, 181, 226, ... (Sloane's A005891).

A **centered hexagonal number** represents a hexagon with a dot in the center and all other dots surrounding the center dot in a hexagonal lattice.

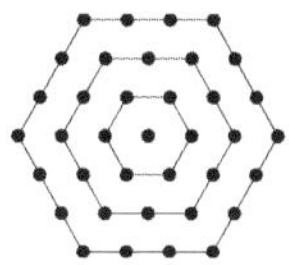

The first few centered hexagonal numbers are 1, 7, 19, 37, 61, 91, 127, 169, 217, 271, ... (Sloane's A003215). Centered hexagonal numbers are most known among centered polygonal numbers. Usually, they are called **hex numbers**.

Following this procedure, we can construct **centered heptagonal numbers** 1, 8, 22, 43, 71, 106, 148, 197, 253, 316, ... (Sloan's A069099), **centered octagonal numbers** 1, 9, 25, 49, 81, 121, 169, 225, 289, 361, ... (Sloane's A016754), **centered nonagonal numbers** 1, 10, 28, 55, 91, 136, 190, 253, 325, 406, ... (Sloane's A060544), **centered decagonal numbers** 1, 11, 31, 61, 101, 151, 211, 281, 361, 451, ... (Sloane's A062786), **centered hendecagonal numbers** 1, 12, 34, 67, 111, 166, 232, 309, 397, 496, ... (Sloane's A069125), **centered dodecagonal numbers** 1, 13, 37, 73, 121, 181, 253, 337, 433, 541, ... (Sloane's A003154), etc.

1.6.2. Algebraically, n-th centered m-gonal number $CS_m(n)$ is obtained as the sum of the first n elements of the sequence $1, m, 2m, 3m, \ldots$. So, by definition, it holds

$$CS_m(n) = 1 + m + 2m + \cdots + (n-1)m.$$

In particular, we get

$$\begin{aligned}
CS_3(n) &= 1+3+6+9+\cdots+3(n-1),\\
CS_4(n) &= 1+4+8+12+\cdots+4(n-1),\\
CS_5(n) &= 1+5+10+15+\cdots 5(n-1),\\
CS_6(n) &= 1+6+12+18+\cdots+6(n-1),\\
CS_7(n) &= 1+7+14+21+\cdots+7(n-1),\\
CS_8(n) &= 1+8+16+24+\cdots+8(n-1).
\end{aligned}$$

The above formula implies the following recurrent formula for the centered m-gonal numbers:

$$CS_m(n+1) = CS_m(n) + nm, \quad CS_m(1) = 1.$$

In particular, we get

$$\begin{aligned}
CS_3(n+1) &= CS_3(n) + 3n, \quad CS_4(n+1)\\
&= CS_4(n) + 4n, \quad CS_5(n+1) = CS_5(n) + 5n,\\
CS_6(n+1) &= CS_6(n) + 6n, \quad CS_7(n+1)\\
&= CS_7(n) + 7n, \quad CS_8(n+1) = CS_8(n) + 8n.
\end{aligned}$$

Since $m+2m+\cdots+(n-1)m = m(1+2+\cdots+(n-1)) = m\frac{(n-1)n}{2}$, one obtains the following general formula for n-th centered m-gonal number:

$$CS_m(n) = 1 + m\frac{(n-1)n}{2} = \frac{mn^2 - mn + 2}{2}.$$

In particular, we have

$$\begin{aligned}
CS_3(n) &= \frac{3n^2-3n+2}{2}, \quad CS_4(n) = 2n^2-2n+1,\\
CS_5(n) &= \frac{(5n^2-5n+2)}{2}, \quad CS_6(n) = 3n^2-3n+1,\\
CS_7(n) &= \frac{7n^2-7n+2}{2}, \quad CS_8(n) = 4n^2-4n+1.
\end{aligned}$$

These formulas for centered m-gonal numbers with $3 \le m \le 30$, as well as the first few elements of the corresponding sequences and the numbers of these sequences in the Sloane's *On-Line Encyclopedia of Integer Sequences* (OEIS, [Sloa11]) classification, are given in the table below.

Name	Formula											Sloane
Cent. triangular	$\frac{1}{2}(3n^2-3n+2)$	1	4	10	19	31	46	64	85	109	136	A005448
Cent. square	$2n^2-2n+1$	1	5	13	25	41	61	85	113	145	181	A001844
Cent. pentagonal	$\frac{1}{2}(5n^2-5n+2)$	1	6	16	31	51	76	106	141	181	226	A005891
Cent. hexagonal	$3n^2-3n+1$	1	7	19	37	61	91	127	169	217	271	A003215
Cent. heptagonal	$\frac{1}{2}(7n^2-7n+2)$	1	8	22	43	71	106	148	197	253	316	A069099
Cent. octagonal	$4n^2-4n+1$	1	9	25	49	81	121	169	225	289	361	A016754
Cent. nonagonal	$\frac{1}{2}(9n^2-9n+2)$	1	10	28	55	91	136	190	253	325	406	A060544
Cent. decagonal	$5n^2-5n+1$	1	11	31	61	101	151	211	281	361	451	A062786
Cent. hendecagonal	$\frac{1}{2}(11n^2-11n+2)$	1	12	34	67	111	166	232	309	397	496	A069125
Cent. dodecagonal	$6n^2-6n+1$	1	13	37	73	121	181	253	337	433	541	A003154
Cent. tridecagonal	$\frac{1}{2}(13n^2-13n+2)$	1	14	40	79	131	196	274	365	469	586	A069126
Cent. tetradecagonal	$7n^2-7n+1$	1	15	43	85	141	211	295	393	505	631	A069127
Cent. pentadecagonal	$\frac{1}{2}(15n^2-15n+2)$	1	16	46	91	151	226	316	421	541	676	A069128
Cent. hexadecagonal	$8n^2-8n+1$	1	17	49	97	1611	241	337	449	577	721	A069129
Cent. heptadecagonal	$\frac{1}{2}(17n^2-17n+2)$	1	18	52	103	171	256	358	477	613	766	A069130
Cent. octadecagonal	$9n^2-9n+1$	1	19	55	109	181	271	379	505	649	811	A069131
Cent. nonadecagonal	$\frac{1}{2}(19n^2-19n+2)$	1	20	58	115	191	286	400	533	685	856	A069132
Cent. icosagonal	$10n^2-10n+1$	1	21	61	121	201	301	421	561	721	901	A069133
Cent. icosihenagonal	$\frac{1}{2}(21n^2-21n+2)$	1	22	64	127	211	316	442	589	757	946	A069178
Cent. icosidigonal	$11n^2-11n+1$	1	23	67	133	221	331	463	617	793	991	A069173
Cent. icositrigonal	$\frac{1}{2}(23n^2-23n+2)$	1	24	70	139	231	346	484	645	829	1036	A069174
Cent. icositetragonal	$12n^2-12n+1$	1	25	73	145	241	361	505	673	865	1081	A069190
Cent. icosipentagonal	$\frac{1}{2}(25n^2-25n+2)$	1	26	76	151	251	376	526	701	901	1126	
Cent. icosihexagonal	$\frac{1}{3}n^2-13n+1$	1	27	79	157	261	391	546	728	936	1170	
Cent. icosiheptagonal	$\frac{1}{2}(27n^2-27n+2)$	1	28	82	163	271	406	568	757	972	1216	
Cent. icosioctagonal	$14n^2-14n+1$	1	29	84	169	281	421	589	785	1009	1261	
Cent. icosinonagonal	$\frac{1}{2}(29n^2-29n+2)$	1	30	88	175	291	436	610	813	1045	1306	
Cent. triacontagonal	$15n^2-15n+1$	1	31	91	181	301	451	631	841	1081	1351	

1.6.3. The generating function for the sequence $CS_m(1)$, $CS_m(2)$, ..., $CS_m(n)$, ... of the centered m-gonal numbers has the form $f(x) = \frac{x(1+(m-2)x+x^2)}{(1-x)^3}$ (see [Sloa11]), i.e., it holds

$$\frac{x(1+(m-2)x+x^2)}{(1-x)^3} = CS_m(1)x + CS_m(2)x^2 + CS_m(3)x^3 + \cdots + CS_m(n)x^n + \cdots, \quad |x| < 1.$$

In particular, one gets

$$\frac{x(x^2+x+1)}{(1-x)^3} = x + 4x^2 + 10x^3 + \cdots + CS_3(n)x^n + \cdots, \quad |x| < 1;$$

$$\frac{x(x+1)^2}{(1-x)^3} = x + 5x^2 + 13x^3 + \cdots + CS_4(n)x^n + \cdots, \quad |x| < 1;$$

$$\frac{x(x^2+3x+1)}{(x-1)^3} = x + 6x^2 + 16x^3 + \cdots + CS_5(n)x^n + \cdots, \quad |x| < 1;$$

$$\frac{x(x^2+4x+1)}{(1-x)^3} = x + 7x^2 + 19x^3 + \cdots + CS_6(n)x^n + \cdots, \quad |x| < 1.$$

In fact, let us consider the recurrent equation $CS_m(n+1) = CS_m(n) + nm$. Going from n to $n+1$, one obtains $CS_m(n+2) = CS_m(n+1) + (n+1)m$. Subtracting first equality from second one, we get $CS_m(n+2) - CS_m(n+1) = CS_m(n+1) - CS_m(n) + m$, i.e., $CS_m(n+2) = 2CS_m(n+1) - CS_m(n) + m$. Similarly, one obtains $CS_m(n+3) = 2CS_m(n+2) - CS_m(n+1) + m$, and

$$CS_m(n+3) - CS_m(n+2) = 2CS_m(n+2) - 2CS_m(n+1) - CS_m(n+1) + CS_m(n),$$

i.e., $CS_m(n+3) = 3CS_m(n+2) - 3CS_m(n+1) + CS_m(n)$. Hence, we get for the sequence of the centered m-gonal numbers the following linear recurrent equation:

$$CS_m(n+3) - 3CS_m(n+2) + 3CS_m(n+1) - CS_m(n) = 0.$$

It is a linear recurrent equation of 3-rd order with coefficients $b_0 = 1, b_1 = -3, b_2 = 3, b_3 = -1$. Its initial values are $CS_m(1) = 1$, $CS_m(2) = m + 1$, $CS_m(3) = 3m + 1$. Denoting $CS_m(n+1)$ by c_n, one can rewrite above equation as

$$c_{n+3} - 3c_{n+2} + 3c_{n+1} - c_n = 0, \ c_0 = 1, \ c_1 = m + 1, \ c_2 = 3m + 1.$$

Therefore, the generating function for the sequence of the centered m-gonal numbers has the form

$$\frac{f(x)}{g(x)} = \frac{a_0 + a_1x + a_2x^2}{b_0 + b_1x + b_2x^2 + b_3x^3},$$

where $b_0 = 1$, $b_1 = -3$, $b_2 = 3$, $b_3 = -1$, and $a_0 = b_0c_0 = 1$, $a_1 = b_0c_1 + b_1c_0 = 1 \cdot (m+1) + (-3) \cdot 1 = m - 2$, $a_2 = b_0c_2 + b_1c_1 + b_2c_0 = 1 \cdot (3m+1) + (-3)(m+1) + 3 \cdot 1 = 1$. Since $g(x) = 1 - 3x + 3x^2 - x^3 = (1-x)^3$ has three coinciding roots $x_1 = x_2 = x_3 = 1$, the generating function for the sequence of the centered m-gonal numbers has the form

$$\frac{1 + (m-2)x + x^2}{(1-x)^3} = CS_m(1) + CS_m(2)x + CS_m(3)x^2 + \cdots + CS_m(n)x^{n-1} + \cdots, \quad |x| < 1.$$

1.6.4. Now we are going to consider some interesting properties of centered polygonal numbers. (see [Weis11], [Wiki11], [CoGu96], [SlPl95], [Gard88]).

I. Obviously, *n-th centered m-gonal number $CS_m(n)$ can be made up of a central point and m copies of the $(n-1)$-th triangular number $S_3(n-1)$ surrounding the central dot:*

$$CS_m(n) = 1 + mS_3(n-1).$$

In fact, this property follows from the formula $CS_m(n) = 1 + m\frac{n(n-1)}{2}$ for the centered polygonal numbers, using that $\frac{n(n-1)}{2} = S_3(n-1)$.

The geometrical interpretation of this property is very natural. On the pictures below it is given, with $n = 4$, for centered triangular and centered square numbers.

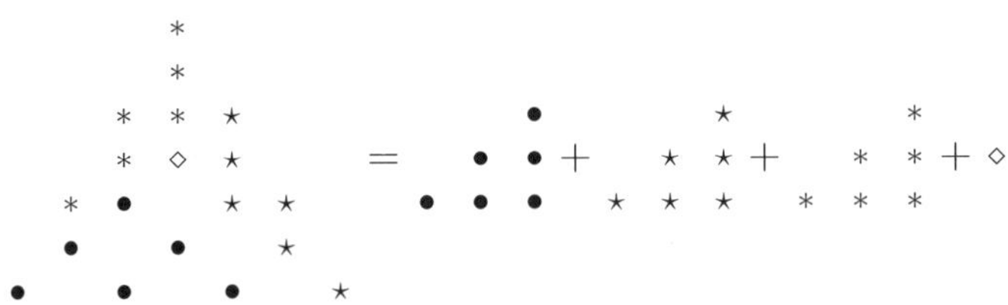

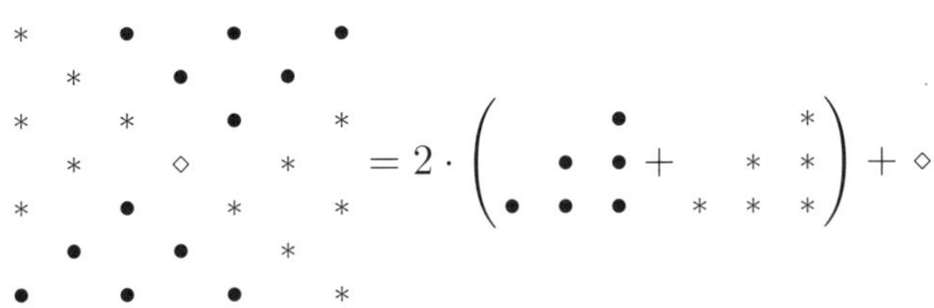

II. It is easy to see that *each centered triangular number from 10 onwards is the sum of three consecutive ordinary triangular numbers:*

$$CS_3(n) = S_3(n) + S_3(n-1) + S_3(n-2), \quad n \geq 3.$$

In fact, $\frac{1}{2}((n-2)(n-1) + (n-1)n + n(n+1)) = \frac{1}{2}(3n^2 - 3n + 2)$.

The geometrical illustration of this property is given below for $n = 4$.

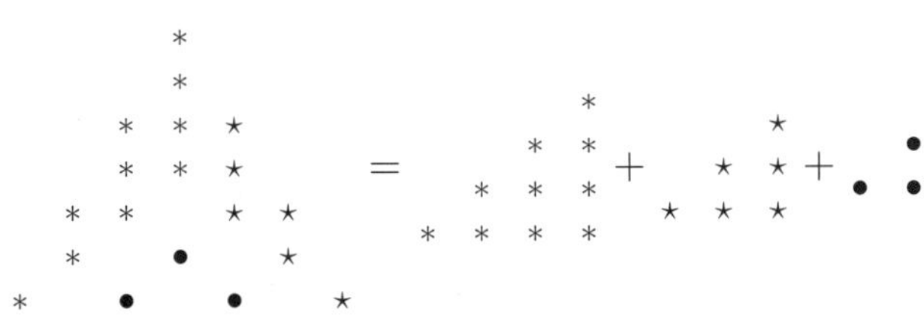

III. Similarly, *any centered square number is the sum of two consecutive square numbers*:

$$CS_4(n) = S_4(n) + S_4(n-1).$$

In fact, the formula $CS_4(n) = 2n^2 - 2n + 1$ for n-th centered square number can be rewritten as $CS_4(n) = n^2 + (n^2 - 2n + 1) = n^2 + (n-1)^2$.

The geometrical illustration of this property can be easy obtained, if we remind, that n-th square number is the sum of the first n odd positive integers. In the picture below it is given for $n = 4$.

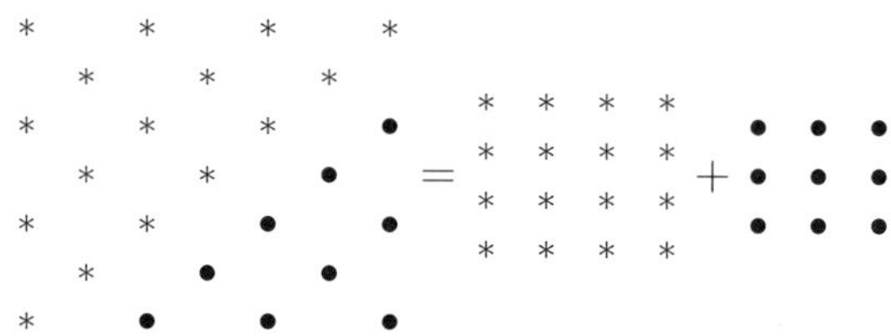

However, one can find both squares $S_4(n)$ and $S_4(n-1)$ properly, as shown on the following picture:

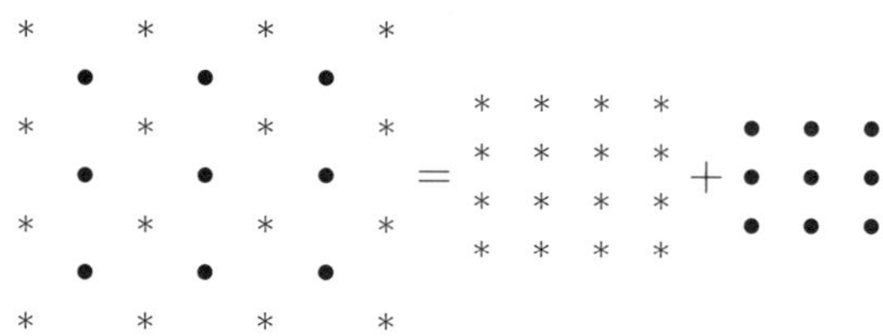

IV. Centered square numbers can be obtained using square and triangular numbers together:

$$CS_4(n) = S_4(2n-1) - 4S_3(n-1).$$

In fact, $S_4(2n-1) = (2n-1)^2 = 4n^2 - 4n + 1 = (2n^2 - 2n + 1) + (2n^2 - 2n) = (2n^2 - 2n + 1) + 4\frac{n(n-1)}{2} = CS_4(n) + 4S_3(n-1)$.

The geometrical illustration of this fact, rewritten in the form $S_4(2n-1) = CS_4(n) + 4S_3(n-1)$, is implied by the fact that any centered square number is made up by four copies of a given triangular number and one central point. On the picture below, constructed for $n = 3$, it is shown, as 5-th square number can be constructed from eight copies of $S_3(2)$ plus one point, giving the 3-rd centered square number (four copies of $S_3(2)$ plus one point) and four "free" copies of $S_3(2)$.

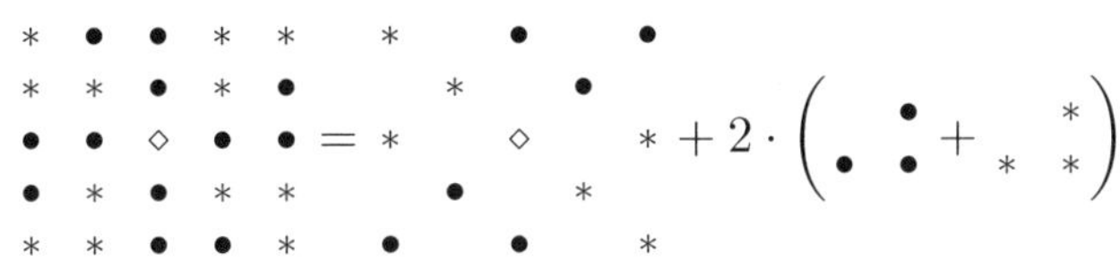

V. The similar connection exists between hex and triangular numbers:

$$CS_6(n) = S_3(3n-2) - 3S_3(n-1).$$

In fact, $S_3(3n-2) = \frac{(3n-2)(3n-1)}{2} = \frac{1}{2}(9n^2 - 9n + 2) = \frac{6n^2-6n+2}{2} + 3\frac{n(n-1)}{2} = CS_6(n) + 3S_3(n-1)$.

The geometrical illustration of this fact, rewritten in the form $S_3(3n-2) = CS_6(n) + 3S_3(n-1)$, is implied by the fact that any centered hexagonal number is made up by six copies of a given triangular number and one central point. On the picture below, constructed for $n = 3$, it is shown, as 7-th triangular number can be constructed from the nine copies of $S_3(2)$ plus one point, giving 3-rd centered hexagonal number (six copies of $S_3(2)$ plus one point) and three "free" copies of $S_3(2)$.

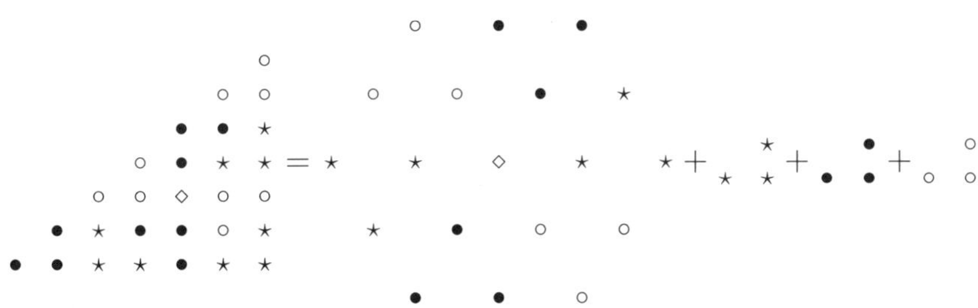

VI. Moreover, *any centered hexagonal number is the difference of two consecutive perfect cubes*:

$$CS_6(n) = n^3 - (n-1)^3.$$

In fact, the formula $CS_6(n) = 3n^2 - 3n + 1$ for n-th centered hexagonal number can be seen as $CS_6(n) = n^3 - (n^3 - 3n^2 + 3n - 1) = n^3 - (n-1)^3$.

VII. Viewed from the opposite perspective, above property yields that *the sum of the first n centered hexagonal numbers is a perfect cube*:

$$CS_6(1) + CS_6(2) + \cdots + CS_6(n) = n^3.$$

It follows immediately from the equation $CS_6(n) = n^3 - (n-1)^3$, using the *telescoping summation*:

$$\begin{aligned} &CS_6(1) + CS_6(2) + CS_6(3) + \cdots + CS_6(n) \\ &\quad = (1^3 - 0^3) + (2^3 - 1^3) + (3^3 - 2^3) + \cdots + (n^3 - (n-1)^3) \\ &\quad = n^3 - 0^3 = n^3. \end{aligned}$$

VIII. Moreover, the centered hexagonal numbers satisfy the recurrence equation

$$CS_6(n) = 2CS_6(n-1) - CS_6(n-2) + 6.$$

In fact, $2CS_6(n-1) - CS_6(n-2) + 6 = 2(3(n-1)^2 - 3(n-1) + 2) - (3(n-2)^2 + 3(n-2) - 1) = 3n^2 - 3n + 1 = CS_6(n)$.

IX. It is easy to check that *any centered octagonal number is equal to a square number with odd index*:

$$CS_8(n) = S_4(2n-1).$$

In fact, the formula $4n^2 - 4n + 1$ for n-th centered octagonal number $CS_8(n)$ can be seen as $(2n-1)^2$.

The geometrical illustration below, given for $n = 3$, shows, that centered octagonal number $CS_8(n)$ and the squared number $S_4(2n-1)$ are numerically equal, consisting from one central point, and several equal sets of the points, belonging to the concentring circles, surrounding this central point, but these "circles" are differently arranged on the plane.

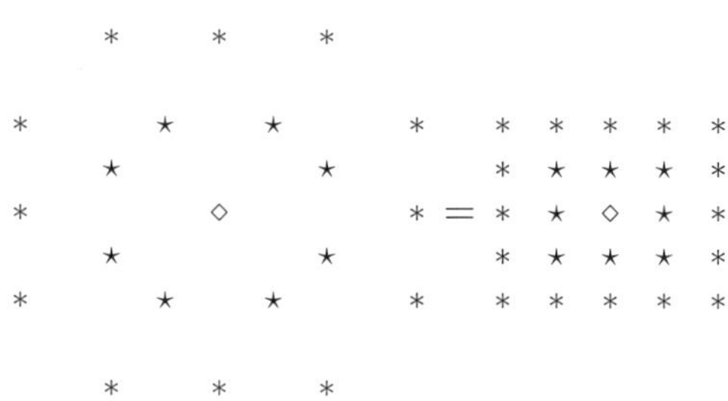

X. Similarly, *every centered nonagonal number is a triangular number*:

$$CS_9(n) = S_3(3n - 2).$$

In fact, $CS_9(n) = \frac{9n^2-9n+2}{2} = \frac{(3n-2)(3n-1)}{2} = S_3(3n-2)$. So, the sequence $CS_9(1), CS_9(2), \ldots, CS_9(n), \ldots$ produces every third triangular number, starting with 1.

XI. The n-th centered dodecagonal number $CS_{12}(n) = 1 + 12S_3(n)$ can be made up of a central point and 12 copies of the $(n-1)$-th triangular number. It corresponds to the number of cells of generated *Chinese checker's board* (or *centered hexagram*), which is called **star number** and denoted by $S(n)$. So, *any centered dodecagonal number coincides with the corresponding star number*:

$$CS_{12}(n) = S(n).$$

In fact, by definition, n-th *star number* $S(n)$ is constructed as n-th centered hexagonal number $CS_6(n) = 1 + 6S_3(n-1)$ with six copies of the $(n-1)$-th triangular number $S_3(n-1)$ appended to each side. The classical Chinese checker's board, shown on the picture below, has a total of $121 = S(5)$ holes.

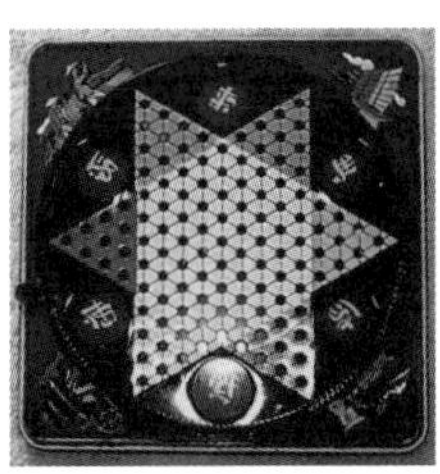

So, we have $S(n) = CS_6(n) + 6S_3(n-1) = (1 + 6S_3(n-1)) + 6S_3(n-1) = 1 + 12S_3(n-1)$. Hence, n-th star number is numerically equal to n-th centered dodecagonal number, but is differently arranged.

The geometrical illustration of this fact for $n = 3$ is given below.

The first few star numbers are 1, 13, 37, 73, 121, 181, 253, 337, 433, 541, ... (Sloane's A003154).

XII. The centered dodecagonal numbers (i.e., star numbers) satisfy the linear recurrent equation

$$CS_{12}(n) = CS_{12}(n-1) + 12(n-1).$$

In fact, it holds $CS_{12}(n-1) + 12(n-1) = (1 + 12\frac{(n-1)(n-2)}{2}) + 12(n-1) = 1 + 6(n-1)(n-2) + 12(n-1)1 + 6(n-1)(n-2+2) = 1 + 6(n-1)n = CS_{12}(n)$.

XIII. Since $CS_6(n) = 3n^2 - 3n + 1$ and $CS_{12}(n) = 6n^2 - 6n + 1$, we obtain that $CS_6(n) = \frac{CS_{12}(n)+1}{2}$. So, one gets $CS_6(n)CS_{12}(n) = \frac{CS_{12}(n)(CS_{12}(n)+1)}{2}$. In other words, *the product of n-th centered hexagonal and n-th centered dodecagonal number is always a triangular number*:

$$CS_6(n)CS_{12}(n) = S_3(CS_{12}(n)).$$

1.6.5. Let us consider now some centered polygonal numbers, which in the same time belong to other classes of figurate numbers.

I. At first, we list the centered m-gonal numbers, which are the classical m-gonal numbers. Just as in the case with classical polygonal numbers, the first centered m-gonal number is 1. Thus, for any m, the number 1 is both, m-gonal and centered m-gonal. The next number, which is both m-gonal and centered m-gonal, can be found using the formula $\frac{m^3-m^2+2}{2}$, which tells us that 10 is both triangular and centered triangular, 25 is both square and centered square, 51 is both pentagonal and centered pentagonal, and so on (see [Wiki11]).

In general, this formula shows, that *the m-gonal number with index $m+1$ coincides with the centered m-gonal number with index m*:

$$S_m(m+1) = CS_m(m).$$

In fact, n-th m-gonal number has the form $S_m(n) = \frac{n(n(m-2)-m+4)}{2}$, and the k-th centered m-gonal number has the form $CS_m(k) = 1 + m\frac{(k-1)k}{2}$. For $n = m+1$ and $k = m$ one obtains:

$$\begin{aligned}\frac{(m+1)((m+1)(m-2)-m+4)}{2} &= \frac{(m+1)(m^2-2m+2)}{2}\\ &= \frac{m^3-m^2+2}{2}, \quad \text{and}\\ 1+m\frac{m(m-1)}{2} = \frac{2+m^2(m-1)}{2} &= \frac{m^3-m^2+2}{2}.\end{aligned}$$

II. A **triangular centered triangular number** is a number, which is simultaneously triangular and centered triangular. The first few such numbers are (Sloane's A128862)

$$1, 10, 136, 1891, 26335, \ldots.$$

In fact, $S_3(u) = CS_3(v)$ for $u = 1, 4, 16, 61, 229, \ldots$ (Sloane's A133161), and $v = 1, 3, 10, 36, 133, \ldots$ (Sloanes A102871). These indices are found by solving the Diophantine equation

$$\frac{u^2+u}{2} = \frac{3v^2-3v+2}{2}.$$

III. A **square centered triangular number** is a number, which is simultaneously square and centered triangular. The first few such numbers are

$$1, 4, 64, 361, 6241, \ldots.$$

In fact, $S_4(u) = CS_3(v)$ for $u = 1, 2, 8, 19, 79, \ldots$ (Sloane's A129445), and $v = 1, 2, 7, 16, 65, \ldots$ (see Sloane's A129444). These indices are found by solving the Diophantine equation

$$u^2 = \frac{3v^2-3v+2}{2}.$$

IV. A **triangular centered hexagonal number** (or **triangular hex number**) is a number, which is simultaneously triangular and centered hexagonal. The first few such numbers are (Sloane's A006244)

$$1, 91, 8911, 873181, 85562821, \ldots.$$

In fact, $S_3(u) = CS_6(v+1)$ for $u = 1, 13, 133, 1321, 13081, \ldots$ (Sloane's A031138), and $v = 0, 5, 54, 539, 5340$ (Sloane's A087125). These indices are found by solving the Diophantine equation

$$\frac{1}{2}u(u+1) = 3v^2 + 3v + 1.$$

V. A **square centered hexagonal number** (or **square hex number**) is a number, which is simultaneously square and centered hexagonal. The first few such numbers are (Sloane's A006051)

$$1, 169, 32761, 6355441, 1232922769, \ldots.$$

In fact, $S_4(u) = CS_6(v+1)$ for $u = 1, 13, 181, 2521, 35113, \ldots$ (Sloane's A001570), and $v = 0, 7, 104, 1455, 20272, \ldots$ (Sloane's A001921). These indices are found by solving the Diophantine equation

$$u^2 = 3v^2 + 3v + 1.$$

The only centered hexagonal number that is both square and triangular is 1 ([Weis11]).

VI. A **triangular star number** is a number, which is simultaneously triangular and centered dodecagonal. The first few such numbers are (Sloane's A006060)

$$1, 253, 49141, 9533161, 1849384153, \ldots.$$

In fact, $S_3(u) = CS_6(v)$ for $u = 1, 22, 313, 4366, 60817, \ldots$ (Sloane's A068774) and $v = 1, 7, 91, 1261, 17557, \ldots$ (Sloane's A068775). These indices are found by solving the Diophantine equation

$$\frac{u(u+1)}{2} = 6v^2 - 6v + 1.$$

VII. A **square star number** is a number, which is simultaneously square and centered dodecagonal. The first few such numbers are (Sloane's A006061)

$$1, 121, 11881, 1164241, 114083761, \ldots.$$

In fact, $S_4(u) = CS_6(v)$ for $u = 1, 11, 109, 1079, 10681, \ldots$ (Sloane's A054320), and $v = 1, 5, 45, 441, 4361, \ldots$ (Sloane's A068778). These indices are found by solving the Diophantine equation

$$u^2 = 6v^2 - 6v + 1.$$

VIII. A **hex star number** (or **centered hexagonal star number**) is number, which is simultaneously centered hexagonal and centered dodecagonal. The first few such numbers are

(Sloane's A006062)

$$1, 37, 1261, 42841, 1455337, \ldots.$$

In fact, $CS_6(u) = CS_{12}(v)$ for $u = 1, 4, 21, 120, 697, \ldots$ (see Sloane's A046090), and $v = 1, 3, 15, 85, 493, \ldots$ (see Sloane's A011900). These indices are found by solving the Diophantine equation

$$3u^2 - 3u + 1 = 6v^2 - 6v + 1.$$

1.7 Other plane figurate numbers

1.7.1. The **pronic number** (or **heteromecic number**) is a positive integer, which can be represented as a product of two consecutive integers. So, by definition, n-th pronic number $P(n)$ has the following form:

$$P(n) = n(n+1).$$

These numbers are sometimes called **oblong numbers** because they are represented by a rectangle with the sides n and $n+1$, i.e., are figurate in this way:

$$\begin{array}{cccccc}
\ast & \ast & \ast & \ast & \ast & \ast \\
\ast & \ast & \ast & \ast & \ast & \ast \\
\ast & \ast & \ast & \ast & \ast & \ast \\
\ast & \ast & \ast & \ast & \ast & \ast \\
\ast & \ast & \ast & \ast & \ast & \ast
\end{array}$$

The first few pronic numbers are 2, 6, 12, 20, 30, 42, 56, 72, 90, 110, ... (Sloane's A002378).

Obviously, *n-th pronic number $P(n)$ is twice n-th triangular number*:

$$P(n) = 2S_3(n).$$

The geometrical illustration of this fact for $n = 5$ is given below.

$$\begin{array}{cccccccccccccccccc}
\ast & \star & \star & \star & \star & \star & & & & & & \star & & & & & & \ast \\
\ast & \ast & \star & \star & \star & \star & & & & & \star & \star & & & & & \ast & \ast \\
\ast & \ast & \ast & \star & \star & \star & = & & & \star & \star & \star & + & & & \ast & \ast & \ast \\
\ast & \ast & \ast & \ast & \star & \star & & & \star & \star & \star & \star & & & \ast & \ast & \ast & \ast \\
\ast & \ast & \ast & \ast & \ast & \star & & \star & \star & \star & \star & \star & & \ast & \ast & \ast & \ast & \ast
\end{array}$$

Therefore, *n-th pronic number is the sum of the first n even integers*:

$$P(n) = 2 + 4 + \cdots + 2n.$$

This fact implies the following recurrent formula for the pronic numbers:

$$P(n+1) = P(n) + 2(n+1), \quad P(1) = 2.$$

It is easy to see that *n-th pronic number is the difference between $(2n+1)$-th square number and $(n+1)$-th centered hexagonal number*:

$$P(n) = S_4(2n+1) - CS_6(n+1).$$

In fact, $S_4(2n+1) - CS_6(n+1) = (2n+1)^2 - (3(n+1)^2 - 3(n+1) + 1) = n^2 + n = n(n+1) = P(n)$. The geometrical illustration of this fact, rewritten in the form $S_4(2n-1) = CS_6(n) + P(n-1)$, is implied by the fact that any centered hexagonal number is made up by six copies of a given triangular number and one central point. On the picture below, constructed for $n = 4$, it is shown, as 7-th square number can be constructed from 3-rd pronic number and six copies of 3-rd triangular number plus one point, which, in turn, form the 4-th centered hexagonal number.

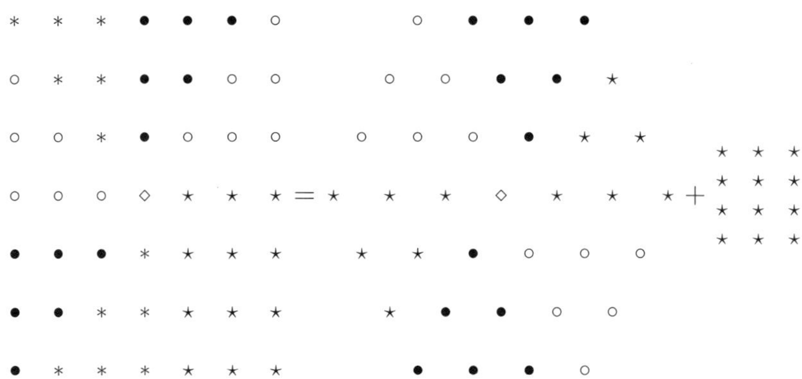

The generating function of the sequence of the pronic numbers is $f(x) = \frac{2x}{(1-x)^3}$ (see [Weis11]), i.e., it holds

$$f(x) = \frac{2x}{(1-x)^3} = P(1)x + P(2)x^2 + P(3)x^3 + \cdots + P(n)x^n + \cdots, \quad |x| < 1.$$

The simplest way to derive this formula is just to multiply by two the generating function $f(x) = \frac{x}{(1-x)^3} = S_3(1) \cdot x + S_3(2)x^2 + S_3(3)x^3 + \cdots + S_3(n)x^n + \cdots$ for triangular numbers. However, it is easy to obtain this result, following the standard procedure.

1.7.2. Using similar arguments, we can consider all composite numbers as **rectangular numbers**, since any composite number $n = a \cdot b$, $1 < a \leq b < n$, can be represented as a non-trivial rectangle with the sides a and b. So, the sequence of the rectangular numbers starts with the entries $4, 6, 8, 9, 10, 12, 14, 15, 16, 18, \ldots$ (Sloane's A002808).

In general, a composite number can be represented as a rectangle in several ways. The number of such non-trivial different rectangles is equal to the number of the above representations of n and can be obtained by the formula $\frac{\tau(n)-2}{2}$ for non-square numbers n. For the squares the formula has the form $\frac{\tau(n)-3}{2} + 1$.

It is interesting to show that *for two different rectangular representations $a \cdot b$ and $c \cdot d$ of a given composite number, four squares a^2, b^2, c^2 and d^2, constructed on the sides of the above rectangles, can be used to construct a new non-trivial rectangle.* For example, the number 12 has two non-trivial rectangular representations $2 \cdot 6$ and $3 \cdot 4$. Then the sum of four squares $2^2, 6^2, 3^2$ and 4^2 is equal to the composite number 65 and, hence, has a rectangular representation, for example, $5 \cdot 12$.

Speaking formally, one can check that *if $N = ab = cd$, then the number $a^2 + b^2 + c^2 + d^2$ is composite.*

In fact, if $ab = cd$, then c divides ab. Let $c = mn$, where m divides a, and n divides b. Then there are p and q such that $a = mp$, and $b = nq$. Solving $(mp)(nq) = (mn)d$ for d gives $d = pq$. It then follows that $a^2 + b^2 + c^2 + d^2 = (mp)^2 + (nq)^2 + (mn)^2 + (pq)^2 = (m^2 + q^2)(n^2 + p^2)$, and, therefore the number $a^2 + b^2 + c^2 + d^2$ is always composite. The more general result that $a^k + b^k + c^k + d^k$ is never prime for $k \geq 0$ an integer also holds ([Hons91]).

1.7.3. The **trapezoidal numbers** are positive integers, which can be represented as an isosceles trapezoid. For example, a trapezoidal representation of 18 is given on the picture below.

```
      *   *   *
    *   *   *   *
  *   *   *   *   *
*   *   *   *   *   *
```

The first trapezoidal numbers are 5, 7, 9, 11, 12, 13, 14, 15, 17, 18, ... (Sloane's A165513).

By definition, any trapezoidal number is a sum of two ore more consecutive positive integers, greater than 1:

$$Tr(n,k) = n + (n+1) + (n+2) + \cdots + (n + (k-1)), \ n \neq 1, \ \ k \neq 1.$$

So, one obtains, that *any trapezoidal number is a difference of two non-consecutive triangular numbers*:

$$\begin{aligned} Tr(n,k) &= n + (n+1) + (n+2) + \cdots + (n + (k-1)) \\ &= (1 + 2 + \cdots + (n + (k-1))) - (1 + 2 + \cdots + (n-1)) \\ &= S_3(n+k-1) - S_3(n-1). \end{aligned}$$

Moreover, the above definition implies that *any trapezoidal number is a particular case of polite number* ([Smit97], [JoLo99]).

In fact, a **polite number** is a positive integer that can be represented as the sum of two or more consecutive positive integers. So, if such *polite representation* starts with 1, we obtain a triangular number, otherwise one gets a trapezoidal number. The first polite numbers are 1, 3, 5, 6, 7, 9, 10, 11, 12, 13, ... (Sloane's A138591).

Obviously, a given positive integer can have several polite representations. In the case of 18, one more such representation is given below.

```
    *   *   *   *   *
  *   *   *   *   *   *
*   *   *   *   *   *   *
```

The *politeness* of a positive integer number is the number of ways it can be expressed as a sum of consecutive positive integers. It is easy to show that *for any positive integer x, the politeness of x is equal to the number of odd divisors of x that are greater than one.*

In fact, suppose a number x has an odd divisor $y > 1$. Then y consecutive integers, centered on $\frac{x}{y}$ (so that their average value is $\frac{x}{y}$), have x as their sum:

$$x = \left(\frac{x}{y} - \frac{y-1}{2}\right) + \cdots + \frac{x}{y} + \cdots + \left(\frac{x}{y} + \frac{y-1}{2}\right).$$

Some of the terms in this sum may be zero or negative. However, if a term is zero, it can be omitted, and any negative terms may be used to cancel positive ones, leading to a polite representation for x. The requirement that $y > 1$ corresponds to the requirement that a polite representation have more than one term. For instance, the polite number $x = 18$ has two non-trivial odd divisors, 3 and 9. It is therefore the sum of 3 consecutive numbers centered at $\frac{18}{3} = 6$: $18 = 5 + 6 + 7$. On the other hand, it is the sum of 9 consecutive integers centered at $\frac{18}{9} = 2$: $18 = (-3)+(-1)+0+1+2+3+4+5+6$, or $18 = 3 + 4 + 5 + 6$.

Conversely, every polite representation of x can be formed by this construction. If a representation has an odd number y of terms, then the middle term can be written as $\frac{x}{y}$, and $y > 1$ is a non-trivial odd divisor of x. If a representation has an even number $2l$ of terms and its minimum value is m, it may be extended, in an unique way, to a longer sequence with the same sum and an odd number $y = 2(m + l) - 1$ of terms, by including $2m - 1$ numbers $-(m - 1), -(m - 2), \dots, -1, 0, 1, \dots, m - 2, m - 1$. After this extension the middle term of the new sequence can be written as $\frac{x}{y}$, and $y > 1$ is a non-trivial odd divisor of x. By this construction, the polite representations of a number and its odd divisors greater than one may be placed into an one-to-one correspondence, that completes the proof (see [SyFr82]).

From this consideration it follows, that the **inpolite numbers**, i.e., positive integers which are not polite, are exactly the powers of two. So, the first few inpolite numbers are 1, 2, 4, 8, 16, 32, 64, 128, 256, 512, ... (Sloane's A000079), and the politeness of the numbers 1, 2, 3, 4, 5, 6, 7, 8, 9, 10, ... is 0, 0, 1, 0, 1, 1, 1, 0, 2, 1, ... (Sloane's A069283).

Therefore, the only polite numbers that may be non-trapezoidal are the triangular numbers with only one non-trivial odd divisor, because for those numbers, according to the bijection described above, the odd divisor corresponds to the triangular representation and there can be no other polite representation. Thus, polite non-trapezoidal numbers must have the form of a power of two multiplied by a prime number. It is easy to show (see Chapter 5 for details), that there are exactly two types of triangular numbers with this form:

* the even perfect numbers $2^{k-1}(2^k - 1)$ formed by the product of a Mersenne prime $2^k - 1$ with half the nearest power of two;
* the products $2^{k-1}(2^k + 1)$ of a Fermat prime $2^k + 1 = 2^{2^n} + 1$ with half the nearest power of two.

1.7.4. The **gnomonic numbers** are closely associated with the square numbers. A gnomonic number is a figurate number of the L-shape, representing the area of the square gnomon obtained by removing a square of side $n-1$ from the square of side n. Since $n^2-(n-1)^2 = 2n-1$, the gnomonic numbers are equivalent to the odd numbers $2n-1$, $n \in \mathbb{N}$, and the first few are 1, 3, 5, 7, 9, 11, 13, 15, 17, 19, ... (Sloane's A005408). On the picture below an geometrical illustration for $n = 1, 2, 3, 4, 5$ is given.

So, by definition, n-th gnomonic number $Gn(n)$ has the form

$$Gn(n) = 2n - 1$$

and can be written as the difference of two consecutive square numbers:

$$Gn(n) = S_4(n) - S_4(n-1).$$

Clearly, the gnomonic numbers can be obtained by the following recurrent formula:

$$Gn(n+1) = Gn(n) + 2, \quad Gn(1) = 1.$$

The generating function for the gnomonic numbers is $f(x) = \frac{x(1+x)}{(1-x)^2}$, i.e., it holds

$$f(x) = \frac{x(1+x)}{(1-x)^2} = Gn(1)x + Gn(2)x^2 + Gn(3)x^3 + \cdots + Gn(n)x^n + \cdots, \quad |x| < 1.$$

The simplest way to get this result is to consider the decomposition

$$\frac{1}{1-x} = 1 + x + x^2 + \cdots + x^n + \cdots, \quad |x| < 1,$$

which can be seen as the sum of the infinite geometric progression $1, x, x^2, \ldots, x^n, \ldots$ with common ratio x. It is convergent for $|x| < 1$

and represents the generating function for the sequence 1, 1, 1, ..., 1, ... The direct multiplication gives

$$\begin{aligned}\frac{1}{(1-x)^2} &= (1+x+x^2+\cdots+x^n+\cdots)\\ &\qquad\cdot(1+x+x^2+\cdots+x^n+\cdots)\\ &= 1+(1+1)x+(1+1+1)x^2+\cdots\\ &= 1+2x+3x^2+\cdots+(n+1)\cdot x^n+\cdots,\end{aligned}$$

i.e., we obtain the following generating function for the sequence 1, 2, 3, ..., n, ...:

$$\frac{1}{(1-x)^2} = 1+2x+3x^2+\cdots+(n+1)\cdot x^n+\cdots, \qquad |x|<1.$$

Multiplication by two gives the generating function for the sequence 2, 4, 6, ..., $2n$, ...:

$$\frac{2}{(1-x)^2} = 2+4x+6x^2+\cdots+2(n+1)\cdot x^n+\cdots, \qquad |x|<1.$$

Finally, the substraction

$$\begin{aligned}\frac{2}{(1-x)^2}-\frac{1}{1-x} &= (2+4x+6x^2+\cdots+2(n+1)\cdot x^n+\cdots)\\ &\quad-(1+x+x^2+\cdots+x^n+\cdots)\\ &= 1+3x+5x^2+\cdots+(2n+1)x^n+\cdots, \qquad |x|<1\end{aligned}$$

gives the equality

$$\frac{1+x}{(1-x)^2} = x+3x^2+5x^3+\cdots+(2n+1)x^n+\cdots, \qquad |x|<1.$$

Of course, one can obtain the proof of this fact by the standard procedure, leading to the the linear recurrent equation $Gn(n+2)-2Gn(n+1)+Gn(n)=0$ with initial values $Gn(1)=1$ and $Gn(2)=3$.

1.7.5. Consider now the notion of **truncated plane figurate numbers**, which can be obtained by cutting figurate numbers of smaller size at each vertex of a given plane figurate number.

I. In the case of **truncated polygonal numbers** $TS_m(n)$ we are going to consider only triangular and square numbers due to their symmetry.

The **truncated triangular numbers** are obtained by cutting triangular numbers of smaller size at each vertex of a given triangular number. More exactly, n-th truncated triangular number $TS_3(n)$ is obtained after cutting from the triangular number $S_3(3n-2)$ three triangular numbers $S_3(n-1)$, one of each vertex of the triangle of the size $3n-2$, and therefore, it is given by the formula

$$TS_3(n) = S_3(3n-2) - 3S_3(n-1).$$

The first values of the sequence of truncated triangular numbers are $1, 7, 19, 37, 61, 91, 127, 169, 217, 271, \ldots$ (Sloane's A003215).

One can see that we get the sequence of hex numbers. This fact can be easy explaned if we remind the formula $CS_6(n) = S_3(3n-2) - 3S_3(n-1)$, which was proven before. So, n-th truncated triangular number is numerically equal to n-th hex number: $TS_3(n) = CS_6(n)$.

Therefore, the general formula for the truncated triangular numbers is

$$TS_3(n) = 3n^2 - 3n + 1,$$

the recurrent equation is

$$TS_3(n+1) = TS_3(n) + 6n, \quad TS_3(1) = 1,$$

while the generating function for this sequence is $f(x) = \frac{x(x^2+4x+1)}{(1-x)^3}$:

$$\frac{x(x^2+4x+1)}{(1-x)^3} = TS_3(1)x + TS_3(2)x^2 + TS_3(3)x^3 + \cdots + TS_3(n)x^n + \cdots, \quad |x| < 1.$$

But we got already the formula $S_3(2n-1) = S_3(n) + 3S_3(n-1)$. So, we have the identity

$$S_3(n) = S_3(2n-1) - 3S_3(n-1),$$

and one can consider n-th triangular number as n-th truncated triangular number, which is obtained after cutting of the triangular number $S_3(2n-1)$ three triangular numbers $S_3(n-1)$, one of each vertex of the triangle of the size $2n-1$.

The **truncated square numbers** are obtained by cutting triangular numbers of smaller size at each vertex of a given square number. More exactly, n-th truncated square number $TS_4(n)$ is obtained after cutting of the square number $S_4(3n-2)$ four triangular numbers $S_3(n-1)$, one of each its vertex, and therefore is given by the formula

$$TS_4(n) = S_4(3n-2) - 4S_3(n-1).$$

The first values of the sequence of truncated square numbers are 1, 12, 37, 76, 129, 196, 277, 372, 481, 604, ... (Sloane's A005892). It is easy to check that the general formula for n-th truncated square number has the form

$$TS_4(n) = 7n^2 - 10n + 4,$$

the recurrent equation is

$$TS_4(n+1) = TS_4(n) + (14n-3), \quad TS_4(1) = 1,$$

while the generating function is $f(x) = \frac{x(1+9x+4x^2)}{(1-x)^3}$, i.e.,

$$\frac{x(1+9x+4x^2)}{(1-x)^3} = TS_4(1)x + TS_4(2)x^2 + TS_4(3)x^3 + \cdots + TS_4(n)x^n + \cdots, \quad |x| < 1.$$

The last result can be obtained by the standard procedure, leading to the linear recurrent equation $TS_4(n+2) - 2TS_4(n+1) + TS_4(n) = 0$ with initial conditions $TS_4(1) = 1$, and $TS_4(2) = 12$.

But we got already the formula $S_4(2n-1) = CS_4(n) + 4S_3(n-1)$, and, therefore, we have the following identity:

$$CS_4(n) = S_4(2n-1) - 4S_3(n-1).$$

It allows to consider n-th centered square number as n-th truncated square number, which is obtained after cutting of the square number $S_4(2n-1)$ four triangular numbers $S_3(n-1)$, one of each its vertex.

II. The **truncated pronic numbers** are obtained by cutting triangular numbers of smaller size at each vertex of a given pronic number. More exactly, n-th truncated pronic number $TP(n)$ is obtained after cutting of the pronic number $P(3n-2)$ four triangular numbers $S_3(n-1)$, one of each vertex of $(3n-2) \times (3n-1)$ rectangle, and therefore is given by the formula

$$TP(n) = P(3n-2) - 4S_3(n-1).$$

The first values of the sequence of truncated pronic numbers are 2, 16, 44, 86, 142, 208, 292, 390, 502, 628, ... It is easy to check that the general formula for n-th truncated pronic number is

$$TP(n) = 7n^2 - 7n + 2,$$

the recurrent equation is

$$TP(n+1) = TP(n) + 14n, \quad TP(1) = 2,$$

while the generating function is $f(x) = \frac{x(2+10x+2x^2)}{(1-x)^3}$:

$$\frac{x(2+10x+2x^2)}{(1-x)^3} = TP(1)x + TP(2)x^2 + TP(3)x^3 + \cdots + TP(n)x^n + \cdots, \quad |x| < 1.$$

The last result can be obtained by the standard procedure, leading to the linear recurrent equation $TP(n+2) - 2TP(n+1) + TP(n) = 0$ with initial conditions $TP(1) = 1$, and $TP(2) = 16$.

In general, in order to obtain truncated pronic numbers we should delete four triangular numbers, relating to the four angles of the corresponding rectangles.

So, for $P(2k) = 2k(2k+1)$, we can construct truncated pronic numbers $P(2k) - 4S_3(t)$ for any $t = 1, 2, \ldots, k$. For the biggest, $t = k$, we get $2k(2k+1) - 2k(k+1) = 2k^2 = 2S_4(k)$, i.e., such truncated numbers give the doubled squares. For $P(2k+1) = (2k+1)(2k+2)$ we can construct truncated pronic numbers $P(2k+1) - 4S_3(t)$ for any $t = 1, 2, \ldots, k$. For the biggest, $t = k$, we get $(2k+1)(2k+2) - 2k(k+1) = 2k^2 + 4k + 2 = 2(k^2 + 2k + 1) = 2(k+1)^2 = 2S_4(k+1)$, i.e., such truncated numbers also give the doubled squares.

III. To obtain **truncated centered polygonal numbers**, one should cut squares of smaller size, corresponding to each angle of a given m-gon, since, in the geometrical settings, the deleted small triangles have $1 = 1^2$, $1 + 3 = 2^2$, $1 + 3 + 5 = 9^2$, ... points each.

So, the **truncated centered m-gonal numbers** are obtained by cutting square numbers of smaller size at each vertex of a given centered m-gonal number. More exactly, n-th truncated centered m-gonal number $TCS_m(n)$ is obtained by cutting of the centered m-gonal number $CS_m(3n-2)$ m square numbers $S_4(n-1)$, one of each its vertex, and so, it is given by the formula

$$TCS_m(n) = CS_m(3n-2) - mS_4(n-1).$$

It is easy to check that the general formula for n-th truncated centered m-gonal number is

$$TCS_m(n) = 1 + \frac{m}{2}(7n^2 - 11n + 4),$$

the recurrent equation is

$$TCS_m(n+1) = TCS_m(n) + m(7n-2), \quad TCS_m(1) = 1,$$

while the generating function is $f(x) = \frac{x(1+(5m-2)x+(1+2m)x^2)}{(1-x)^3}$:

$$\begin{aligned}\frac{x(1+(5m-2)x+(1+2m)x^2)}{(1-x)^3}\\ &= x + (1+5m)x + (1+17m)x^2\\ &\quad + (1+36m)x^3 + (1+62m)x^4 + \cdots\\ &= TCS_m(1)x + TCS_m(2)x^2 + TCS_m(3)x^3\\ &\quad + \cdots + TCS_m(n)x^n + \cdots, \quad |x| < 1.\end{aligned}$$

The last result can be obtained by the standard procedure, leading to the linear recurrent equation $TCS_m(n+2) - 2TCS_m(n+1) + TCS_m(n) = 0$ with initial conditions $TCS_m(1) = 1$, and $TCS_m(2) = 1 + 5m$.

In particular, the general formula for n-th **truncated centered triangular number** is

$$TCS_3(n) = \frac{21n^2 - 33n}{2} + 7,$$

giving the first values 1, 16, 52, 109, 187, 286, 406, 547, 709, 892, ... The recurrent equation is

$$TCS_3(n+1) = TCS_3(n) + (21n - 6), \quad TCS_3(1) = 1,$$

while the generating function is $f(x) = \frac{x(1+13x+7x^2)}{(1-x)^3}$:

$$\begin{aligned}\frac{x(1+13x+7x^2)}{(1-x)^3} &= TCS_3(1)x + TCS_3(2)x^2 + TCS_3(3)x^3\\ &\quad + \cdots + TCS_3(n)x^n + \cdots, \quad |x| < 1.\end{aligned}$$

The general formula for n-th **truncated centered square number** is

$$TCS_4(n) = 14n^2 - 22n + 9,$$

giving the first values 1, 21, 69, 145, 249, 381, 541, 729, 945, 1189, ... The recurrent equation is

$$TCS_4(n+1) = TCS_4(n) + (28n - 8), \quad TCS_4(1) = 1,$$

while the generating function is $f(x) = \frac{x(1+18x+9x^2)}{(1-x)^3}$:

$$\frac{x(1+18x+9x^2)}{(1-x)^3} = TCS_4(1)x + TCS_4(2)x^2 + TCS_4(3)x^3 + \cdots + TCS_4(n)x^n + \cdots, \quad |x| < 1.$$

The general formula for n-th **truncated centered pentagonal number** is

$$TCS_5(n) = \frac{35n^2 - 55n}{2} + 3,$$

giving the first values 1, 26, 86, 181, 303, 468, 668, 903, 1173, 1478, ... The recurrent equation is

$$TCS_5(n+1) = TCS_5(n) + (35n - 10), \quad TCS_5(1) = 1,$$

while the generating function is $f(x) = \frac{x(1+8x+11x^2)}{(1-x)^3}$:

$$\frac{x(1+8x+11x^2)}{(1-x)^3} = TCS_5(1)x + TCS_5(2)x^2 + TCS_5(3)x^3 + \cdots + TCS_5(n)x^n + \cdots, \quad |x| < 1.$$

The general formula for n-th **truncated centered hexagonal number** (or **truncated hex number**) is

$$TCS_6(n) = 21n^2 - 33n + 13,$$

giving the first values 1, 31, 103, 217, 373, 571, 811, 1093, 1417, 1783, ... The recurrent equation is

$$TCS_6(n+1) = TCS_6(n) + (42n - 12), \quad TCS_6(1) = 1,$$

while the generating function is $f(x) = \frac{x(1+28x+13x^2)}{(1-x)^3}$:

$$\frac{x(1+28x+13x^2)}{(1-x)^3} = TCS_6(1)x + TCS_6(2)x^2 + TCS_6(3)x^3 + \cdots + TCS_6(n)x^n + \cdots, \quad |x| < 1.$$

Using the standard procedure to construct truncated objects, more general concept of truncated centered polygonal numbers can be obtained. In fact, truncated centered m-gonal numbers come by the formulas $CS_m(2n)-mk^2$, $k=1,2,\ldots,n$, and $CS_m(2n+1)-mk^2$, $k=1,2,\ldots,n$. So, the truncated centered triangular numbers can be obtained by the formulas $CS_3(2n)-3k^2$, $k=1,2,\ldots,n$, and $CS_3(2n+1)-3k^2$, $k=1,2,\ldots,n$. The truncated centered square numbers come by the formulas $CS_4(2n)-4k^2$, $k=1,2,\ldots,n$, and $CS_4(2n+1)-4k^2$, $k=1,2,\ldots,n$. The truncated hex numbers come by the formulas $CS_6(2n)-6k^2$, $k=1,2,\ldots,n$, and $CS_6(2n+1)-6k^2$, $k=1,2,\ldots,n$.

1.7.6. The **polygram numbers** (or **centered star polygonal numbers**) correspond to the *polygrams* (or *star polygons*) and can be obtained as corresponding centered polygonal numbers with attached triangular numbers of corresponding size to each side of a given polygon. In order to obtain the m**-gram numbers**, we use such procedure for each term of the sequence of the centered m-gonal numbers. The most known figurate numbers of such kind, the star numbers, corresponding to the centered hexagram, and, hence, to the hex numbers, were considered before.

Formally, n-th m**-gram number** $P_m(n)$, $m\geq 3$, is defined as n-th centered m-gonal number $CS_m(n)$ with m copies of the $(n-1)$-th triangular number $S_3(n-1)$ appended to each side:

$$P_m(n)=CS_m(n)+mS_3(n-1)=mn^2-mn+1.$$

As $CS_m(n)=1+m\cdot S_3(n-1)$, we get

$$\begin{aligned}P_m(n)&=CS_m(n)+m\cdot S_3(n-1)\\&=(1+m\cdot S_3(n-1))+m\cdot S_3(n-1)\\&=1+(2m)\cdot S_3(n-1)=CS_{2m}(n).\end{aligned}$$

Hence, n-th m-gram number $P_m(n)=mn^2-mn+1$ is numerically equal to n-th centered $(2m)$-gonal number, but is differently arranged.

For $m=3$, and 4 this procedure is trivial: attaching to the centered triangular number $CS_3(n)$ three triangular numbers $S_3(n-1)$ just gives the shape of n-th centered hexagonal number; attaching to the centered square number $CS_4(n)$ four triangular numbers $S_3(n-1)$ gives the shape of the $(2n+1)$-th square number, which can be seen as n-th octagonal number.

For $m \geq 5$, the polygram numbers correspond to the classical *regular polygrams* (or *star polygons*), where a *regular* $\{p/q\}$ *polygram*, with positive integers p, q, is a figure formed by connecting every q-th point out of p regularly spaced points lying on a circumference. In fact, m-gram numbers correspond to the *regular* $\{m/2\}$ *polygram.*

In particular, the **pentagram numbers** are connected with classical pentagram. By definition, n-th pentagram number $P_5(n)$ is constructed as n-th centered pentagonal number $CS_5(n)$ with five copies of the $(n-1)$-th triangular number $S_3(n-1)$ appended to each side. So, it holds

$$P_5(n) = CS_5(n) + 5S_3(n-1).$$

Hence, n-th pentagram number $P_5(n) = 5n^2 - 5n + 1$ is numerically equal to n-th centered dodecagonal number $CS_{10}(n)$, but is differently arranged.

The star numbers, corresponding to centered hexagrams, can be considered as **hexagram numbers**.

1.7.7. The **Aztec diamond numbers** arise by *Aztec diamond construction.* The *Aztec diamond of n-th order* is the region obtained from four staircase shapes of height n by gluing them together along the straight edges. On the picture below the geometrical illustration of this notion is given for $n = 1, 2, 3$ and 4.

The Aztec diamond of n-th order can be considered also as the union of unit squares in the plane whose edges lie on the lines of a square grid and whose centers (x, y) satisfy to $|x-0.5|+|y-0.5| \leq n$.

The n-th **Aztec diamond number** $Az(n)$, i.e., the number of squares in the Aztec diamond of n-th order, has the form

$$Az(n) = 2n(n+1),$$

giving the values 4, 12, 24, 40, 60, 84, 112, 144, 180, 220, ... (Sloane's A046092).

So, for n-th Aztec diamond number $Az(n)$ it holds

$$Az(n) = 4S_3(n), \quad \text{and} \quad Az(n) = 2P(n),$$

i.e., *n-th Aztec diamond number is four times n-th triangular number and doubled n-th pronic number.*

It is easy to see that the recurrent formula for the Aztec diamond numbers has the form

$$Az(n+1) = Az(n) + 4(n+1), \quad Az(1) = 4,$$

while the generating function is $f(x) = \frac{4x}{(1-x)^3}$, i.e.,

$$\frac{4x}{(1-x)^3} = Az(1)x + Az(2)x^2 + Az(3)x^3 + \cdots + Az(n)x^n + \cdots, \quad |x| < 1.$$

On the other hand, the classical *diamond of n-th order* can be considered as the set of squares whose centers satisfy the inequality $|x| + |y| \leq n$. On the picture below the geometrical illustration for $n = 1, 2, 3, 4$ is given.

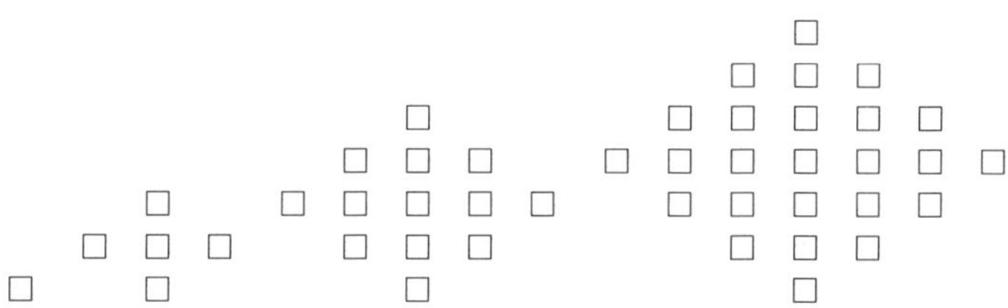

The n-th **diamond number**, giving the number of union squares in n-th order damond, is $2n(n-1)+1$, that is precisely n-th centered square number $CS_4(n)$, giving values 1, 5, 13, 25, 41, 61, 85, 113, 145, 181, ... (Sloane's A001844). So, n-th diamond number is numerically equal to n-th centered square number, and can be seen geometrically as the corresponding square, rotated on 45^0.

1.7.8. The **cross numbers** are defined as numbers of the form $4n-3$, $n = 1, 2, 3, \ldots$, which can be represented geometrically as a cross on the plane. So, n-th cross number $Cr(n)$ has the form

$$Cr(n) = 4n - 3,$$

and the sequence starts from 1, 5, 9, 13, 17, 21, 25, 29, 33, 37, ... (Sloane's A016813). On the figure below we construct $Cr(n)$

for $n = 1, 2, 3, 4$.

The recurrent equation for these numbers is

$$Cr(n+1) = Cr(n) + 4, \; Cr(1) = 1,$$

and the generating function has the form $f(x) = \frac{x(1+3x)}{(1-x)^2}$, i.e.,

$$\frac{x(1+3x)}{(1-x)^2} = Cr(1)x + Cr(2)x^2 + Cr(3)x^3 + \cdots + Cr(n)x^n + \cdots, \; |x| < 1.$$

The last result can be obtained via the standard procedure, giving the linear recurrent equation $Cr(n+2) - 2Cr(n+1) + Cr(n) = 0$ with initial values $Cr(1) = 1$ and $Cr(2) = 5$. On the other hand, the function $f(x)$ can be calculated using the sum of the generating function $\frac{4x}{(1-x)^2} = 4x + 8x^2 + 12x^3 + \cdots$ for the numbers of the form $4n$, and the generating function $\frac{1}{1-x} = 1 + x + x^2 + \cdots$ for the sequence $1, 1, 1, \ldots$.

The positive integers of such form have many interesting properties. For example, *any prime cross number can be represented as a sum of two squares, and a cross number can be represented as a sum of two squares if and only if it has a prime divisors of the form* $4n+3$ *only in even powers* (see [Sier64]).

1.8 Generalized plane figurate numbers

The **generalized plane figurate numbers** are defined as all values of the standard formulas for a given class of the plane figurate numbers, taken for any integer value of the index.

1.8.1. The **generalized polygonal numbers** or, specifically, the **generalized m-gonal numbers**, are defined as all values of the formula

$$S_m(n) = \frac{n((m-2)n-m+4)}{2} = \frac{(m-2)}{2}(n^2-n)+n$$

for n-th m-gonal number, taken for any integer value of n.

For positive integers n, we get ordinary polygonal numbers. For $n=0$, one has $S_m(0)=0$. Finally, $S_m(-n) = \frac{n((m-2)n+m-4)}{2} = \frac{m-2}{2}(n^2+n)-n$ for negative values of the argument, i.e., for any positive integer n.

Let us consider the properties of the above numbers $S_m(-n)$, $n \in \mathbb{N}$. For convenience, denote $S_m(-n)$ by ${}^-S_m(n)$, i.e., by definition,

$$\begin{aligned} {}^-S_m(n) = S_m(-n) &= \frac{n((m-2)n+m-4)}{2} \\ &= \frac{m-2}{2}(n^2+n)-n, \quad n \in \mathbb{N}. \end{aligned}$$

In particular, one has

$$\begin{gathered} {}^-S_3(n) = \frac{n(n-1)}{2},\ {}^-S_4(n) = n^2,\ {}^-S_5(n) = \frac{n(3n+1)}{2}, \\ {}^-S_6(n) = n(2n+1),\ {}^-S_7(n) = \frac{n(5n+3)}{2},\ {}^-S_8(n) = n(4n+2). \end{gathered}$$

So, the **generalized triangular numbers** with negative indices coincide with ordinary triangular numbers of smaller index:

$${}^-S_3(n) = S_3(-n) = S_3(n-1) = \frac{(n-1)n}{2}.$$

They have values 0, 1, 3, 6, 10, ... So, the generalized triangular numbers are

$$\ldots, 10, 6, 3, 1, 0, 0, 1, 3, 6, 10, 15, \ldots,$$

forming sequence 0, 1, 0, 3, 1, 6, 3, 10, 6, ... for $n = 0, \pm1, \pm2, \pm3, \pm4, \ldots$ (see Sloane's A008795).

Obviously, the **generalized square numbers** with negative indices coincide with ordinary square numbers:

$${}^-S_4(n) = S_4(-n) = S_4(n) = n^2.$$

They have values 1, 4, 9, 16, 25, ... So, the generalized square numbers are

$$\ldots, 25, 16, 9, 4, 1, 0, 1, 4, 9, 16, 25, \ldots$$

forming sequence $0, 1, 1, 4, 4, 9, 9, 16, 16, \ldots$ for $n = 0, \pm1, \pm2, \pm3, \pm4, \ldots$ (see Sloane's A008794).

The **generated pentagonal numbers** with negative indices have the form

$$^{-}S_5(n) = S_5(-n) = \frac{3n^2 + n}{2}.$$

They have values $2, 7, 15, 26, 40, \ldots$. So, the generalized pentagonal numbers are

$$\ldots, 40, 26, 15, 7, 2, 0, 1, 5, 12, 22, 35, \ldots$$

forming sequence $0, 1, 2, 5, 7, 12, 15, 22, 26, \ldots$ for $n = 0, \pm 1, \pm 2, \pm 3, \pm 4, \ldots$ (Sloane's A001318).

The generalized pentagonal numbers are the most known class of the generalized figurate numbers; for example, they play an important role in the theory of the unrestricted partitions (see Chapter 4).

The **generalized hexagonal numbers** with negative indices have the form

$$^{-}S_6(n) = S_6(-n) = 2n^2 + n.$$

They have values $3, 10, 21, 36, 55, \ldots$. In other words, the numbers $^{-}S_6(n)$ coincide with the triangular numbers of even indices: $^{-}S_6(n) = 2n^2 + n = \frac{2n(2n+1)}{2} = S_3(2n)$. On the other hand, as it was shown before, the ordinary hexagonal numbers are triangular numbers with odd indices: $S_6(n) = S_3(2n - 1)$. So, the set of the generalized hexagonal numbers

$$\ldots, 55, 36, 21, 10, 3, 0, 1, 6, 15, 28, 45, \ldots$$

contains only triangular numbers, forming the standard sequence $0, 1, 3, 6, 10, 15, 21, 28, 36, \ldots$ (Sloane's A000217) of all ordinary triangular numbers for $n = 0, \pm 1, \pm 2, \pm 3, \pm 4, \ldots$.

The recurrent formula for the sequence $^{-}S_m(1), ^{-}S_m(2), ^{-}S_m(3), \ldots$ of generalized m-gonal numbers with negative indices has the form

$$^{-}S_m(n+1) = ^{-}S_m(n) + ((m-2)(n+1) - 1), ^{-}S_m(1) = m - 3.$$

In fact, one obtains from the recurrent formula $S_m(n+1) = S_m(n) + (1 + (m-2)n)$ for the ordinary m-gonal numbers, that $S_m(n) = S_m(n+1) - (m-2)n - 1$, or $S_m(n-1) = S_m(n) + (m-2)(1-n) - 1$. So, $S_m(0) = 0$, and, for negative integers, one has $S_m(-(n+1)) = S_m(-n) + (m-2)(1+n) - 1$, $n \in \mathbb{N}$.

The generating function for the sequence ${}^{-}S_m(1), {}^{-}S_m(2), {}^{-}S_m(3), \dots$ of the generalized m-gonal numbers with negative indices, i.e., for the sequence $S_m(-1), S_m(-2), S_m(-3), \dots$, has the form $f(x) = \frac{x(x+(m-3))}{(1-x)^3}$; so, it holds

$$\frac{x(x+(m-3))}{(1-x)^3} = S_m(-1)x + S_m(-2)x^2 + S_m(-3)x^3 + \cdots + S_m(-n)x^n + \cdots, \quad |x| < 1.$$

In particular, we have

$$\begin{aligned}
\frac{x^2}{(1-x)^3} &= S_3(-1)x + S_3(-2)x^2 + S_3(-3)x^3 + \cdots + S_3(-n)x^n + \cdots \\
&= S_3(0)x + S_3(1)x^2 + S_3(2)x^3 + \cdots \\
&\quad + S_3(n-1)x^n + \cdots, \quad |x| < 1; \\
\frac{x(x+1)}{(1-x)^3} &= S_4(-1)x + S_4(-2)x^2 + S_4(-3)x^3 + \cdots + S_4(-n)x^n + \cdots \\
&= S_4(1)x + S_4(2)x^2 + S_4(3)x^3 + \cdots \\
&\quad + S_4(n)x^n + \cdots, \quad |x| < 1; \\
\frac{x(x+2)}{(1-x)^3} &= S_5(-1)x + S_5(-2)x^2 \\
&\quad + S_5(-3)x^3 + \cdots + S_5(-n)x^n + \cdots, \quad |x| < 1; \\
\frac{x(x+3)}{(1-x)^3} &= S_6(-1)x + S_6(-2)x^2 \\
&\quad + S_6(-3)x^3 + \cdots + S_6(-n)x^n + \cdots, \quad |x| < 1.
\end{aligned}$$

This fact can be obtained via the standard procedure, leading to the recurrent linear equation ${}^{-}S_m(n+3) - 3\,{}^{-}S_m(n+2) + 3\,{}^{-}S_m(n+1) - {}^{-}S_m(n) = 0$ with initial conditions ${}^{-}S_m(1) = m-3$, ${}^{-}S_m(2) = 3m-8$, and ${}^{-}S_m(3) = 6m-15$.

However, for $m = 3$ and $m = 4$ the situation is much more simple. In fact, the generating function for the sequence of the ordinary triangular numbers is $f(x) = \frac{x}{(1-x)^3}$, i.e.,

$$\frac{x}{(1-x)^3} = S_3(1)x + S_3(2)x^2 + S_3(3)x^3 + \cdots + S_3(n)x^n + \cdots, \quad |x| < 1.$$

Since $S_3(n) = {}^{-}S_3(n+1)$, $n \geq 1$, and ${}^{-}S_3(1) = 0$, we get

$$\begin{aligned}\frac{x}{(1-x)^3} &= {}^{-}S_3(2)x + {}^{-}S_3(3)x^2 + \cdots + {}^{-}S_3(n+1)x^n + \cdots \\ &= {}^{-}S_3(1) + {}^{-}S_3(2)x + {}^{-}S_3(2)x^2 + \cdots \\ &\quad + {}^{-}S_3(n+1)x^n + \cdots, \ |x| < 1, \ \text{ and} \\ \frac{x^2}{(1-x)^3} &= {}^{-}S_3(1)x + {}^{-}S_3(2)x^2 + {}^{-}S_3(3)x^3 + \cdots \\ &\quad + {}^{-}S_3(n)x^n + \cdots, \ |x| < 1.\end{aligned}$$

As for $m = 4$, the generating function $f(x) = \frac{x(1+x)}{(1-x)^3}$ for the sequence of the generated square numbers with negative indices coincides, clearly, with the generating function for the sequence of the ordinary square numbers.

Now we can get the generating function for the sequence $S_m(0), S_m(1), S_m(-1), S_m(2), S_m(-2), \ldots$ of all generalized m-gonal numbers, written for $n = 0, \pm 1, \pm 2, \ldots$. It can be obtained from the generating functions $f_1(x) = \frac{1+(m-3)x}{(1-x)^3} = S_m(1) + S_m(2)x + \cdots + S_m(n+1)x^n + \cdots$ and $f_2(x) = \frac{(m-3)+x}{(1-x)^3} = S_m(-1) + S_m(-2)x + \cdots + S_m(-(n+1))x^n + \cdots$ using the formula $x^2 f_1(x^2) + x^3 f_2(x^2)$. The direct computation gives $f(x) = \frac{x^2(1+(m-3)x+(m-3)x^2+x^3)}{(1-x^2)^3} = \frac{x^2(1+x)(1+(m-4)x+x^2)}{(1-x^2)^3} = \frac{x^2(1+(m-4)x+x^2)}{(1-x)(1-x^2)^2}$. In other words, it holds

$$\begin{aligned}\frac{x^2(1+(m-4)x+x^2)}{(1-x)(1-x^2)^2} &= S_m(0)x + S_m(1)x^2 + S_m(-1)x^3 \\ &\quad + S_m(2)x^4 + S_m(-2)x^5 + \cdots, \ |x| < 1.\end{aligned}$$

In particular, we have

$$\begin{aligned}\frac{x^2(1-x+x^2)}{(1-x)(1-x^2)^2} &= S_3(0)x + S_3(1)x^2 + S_3(-1)x^3 + S_3(2)x^4 \\ &\quad + S_3(-2)x^5 + \cdots, \ |x| < 1; \\ \frac{x^2(1+x^2)}{(1-x)(1-x^2)^2} &= S_4(0)x + S_4(1)x^2 + S_4(-1)x^3 + S_4(2)x^4 \\ &\quad + S_4(-2)x^5 + \cdots, \ |x| < 1;\end{aligned}$$

$$\frac{x^2(1+x+x^2)}{(1-x)(1-x^2)^2} = S_5(0)x + S_5(1)x^2 + S_5(-1)x^3 + S_5(2)x^4 + S_5(-2)x^5 + \cdots, \ |x| < 1;$$

$$\frac{x^2(1+2x+x^2)}{(1-x)(1-x^2)^2} = S_6(0)x + S_6(1)x^2 + S_6(-1)x^3 + S_6(2)x^4 + S_6(-2)x^5 + \cdots, \ |x| < 1.$$

We can consider some properties of the generated polygonal numbers, similar to the properties of ordinary polygonal numbers.

For example, using the fact $S_4(n) = S_3(n) + S_3(n-1)$ and the identities ${}^{-}S_4(n) = S_4(n)$, ${}^{-}S_3(n) = S_3(n-1)$, we obtain that *any generated square number is a sum of two generalized triangular numbers with opposite indices*:

$$S_4(n) = {}^{-}S_4(n) = S_3(n) + {}^{-}S_3(n).$$

Furthermore, *any generalized square number is a sum of two consecutive generalized triangular numbers*:

$$S_4(n) = {}^{-}S_4(n) = S_3(n-1) + S_3(n) = {}^{-}S_3(n) + {}^{-}S_3(n+1).$$

On the same way, we can rewrite all properties, connecting triangular and square numbers. For example, the property $S_3(2n) = 3S_3(n) + S_3(n-1)$ gets the form

$${}^{-}S_3(2n+1) = 3\,{}^{-}S_3(n+1) + {}^{-}S_3(n).$$

Since $\frac{3n^2+n}{2} = n^2 + \frac{n(n+1)}{2}$, we get the property

$${}^{-}S_5(n) = {}^{-}S_4(n) + {}^{-}S_3(n+1),$$

which is similar to the property $S_5(n) = S_4(n) + S_3(n-1)$.

The reader can easily obtain many new identities, connecting generalized polygonal numbers.

1.8.2. The **generalized centered polygonal numbers**, or, specifically, the **generalized centered m-gonal numbers** are defined as all values of the formula $CS_m(n) = \frac{mn^2-mn+2}{2}$ for any integer value of the index n.

For positive integers n, we get ordinary centered m-gonal numbers. For $n = 0$, it holds $CS_m(0) = 1$. Finally, for negative integers we get $CS_m(-n) = \frac{mn^2+mn+2}{2}$, $n \in \mathbb{N}$.

So, denoting $CS_m(-n)$ by $^{-}CS_m(n)$, we get the formula

$$^{-}CS_m(n) = CS_m(-n) = \frac{mn^2 + mn + 2}{2}, \quad n \in \mathbb{N}.$$

Hence, the **generalized centered triangular numbers** with negative indices have the form

$$^{-}CS_3(n) = CS_3(-n) = \frac{3n^2 + 3n + 2}{2},$$

giving values 4, 10, 19, 31, 46, ... So, the generalized centered triangular numbers are

$$\ldots, 46, 31, 19, 10, 4, 1, 1, 4, 10, 19, 31, \ldots$$

forming sequence $1, 1, 4, 4, 10, 10, 19, 19, 31, \ldots$ for $n = 0, \pm 1, \pm 2, \pm 3, \pm 4, \ldots$ (see Sloane's A005448).

The **generalized centered square numbers** with negative indices have the from

$$^{-}CS_4(n) = CS_4(-n) = 2n^2 + 2n + 1,$$

giving values 5, 13, 25, 41, 61, ... So, the generalized centered square numbers are

$$\ldots, 61, 41, 25, 13, 5, 1, 1, 5, 13, 25, 41, \ldots$$

forming sequence $1, 1, 5, 5, 13, 13, 41, 41, 61, \ldots$ for $n = 0, \pm 1, \pm 2, \pm 3, \pm 4, \ldots$ (see Sloane's A001844).

The **generated centered pentagonal numbers** with negative indices have the form

$$^{-}CS_5(n) = CS_5(-n) = \frac{5n^2 + 5n + 2}{2},$$

giving values $6, 16, 31, 51, 76, \ldots$. So, the generalized centered pentagonal numbers are

$$\ldots, 76, 51, 31, 16, 6, 1, 1, 6, 16, 31, 51, \ldots$$

forming sequence $1, 1, 6, 6, 16, 16, 31, 31, 51, \ldots$ for $n = 0, \pm 1, \pm 2, \pm 3, \pm 4, \ldots$ (see Sloane's A005891).

The **generalized centered hexagonal numbers** (or **generalized hex numbers**) with negative indices have the form

$$^{-}S_6(n) = S_6(-n) = 3n^2 + 3n + 1,$$

giving values $7, 19, 37, 61, 91, \ldots$. So, the generalized hex numbers are

$$\ldots, 91, 61, 37, 19, 7, 1, 1, 7, 19, 37, 61, \ldots,$$

forming the sequence $1, 1, 7, 7, 19, 19, 37, 37, \ldots$ for $n = 0, \pm 1, \pm 2, \pm 3, \pm 4, \ldots$ (see Sloane's A003215).

It is easy to see that in this case the situation is much more simple than in the case of generalized polygonal numbers, since *all generalized centered m-gonal numbers are ordinary centered m-gonal numbers*:

$$^{-}CS_m(n) = CS_m(-n) = CS_m(n+1), \quad n \in \mathbb{N}.$$

In fact, it holds $^{-}CS_m(n) = CS_m(-n) = \frac{mn^2+mn+2}{2} = \frac{(mn^2+2mn+1)-(mn+m)+2}{2} = \frac{m(n+1)^2-m(n+1)+2}{2} = CS_m(n+1)$.

So, the sequence of the generalized centered m-gonal numbers is

$$\ldots, 1+15m, 1+10m, 1+6m, 1+3m, 1+m,$$
$$1, 1, 1+m, 1+3m, 1+6m, 1+10m, \ldots,$$

forming the sequence $1, 1, 1+m, 1+m, 1+3m, 1+3m, 1+6m, 1+6m, 1+10m, \ldots$ for $n = 0, \pm 1, \pm 2, \pm 3, \pm 4, \ldots$.

One can check that the recurrent formula for the sequence of the generalized centered m-gonal numbers with negative indices has the form

$$^{-}CS_m(n+1) = {}^{-}CS_m(n) + (n+1)m, \quad {}^{-}CS_m(1) = 1 + m.$$

The generating function for the sequence $^{-}CS_m(1), {}^{-}CS_m(2), {}^{-}CS_m(3), \ldots$ of generalized centered m-gonal numbers with negative indices, i.e., for the sequence $CS_m(-1), CS_m(-2), CS_m(-3), \ldots$, has the form $f(x) = \frac{x(x^2-2x+(m+1))}{(1-x)^3}$, i.e., it holds

$$\frac{x(x^2-2x+(m+1))}{(1-x)^3} = CS_m(-1)x + CS_m(-2)x^2 + CS_m(-3)x^3 + \cdots + CS_m(-n)x^n + \cdots, \ |x| < 1.$$

In fact, using the identity ${}^{-}CS_m(n) = CS_m(n+1)$ and the decomposition

$$\frac{1+(m-2)x+x^2}{(1-x)^3} = CS_m(1) + CS_m(2)x + \cdots + CS_m(n+1)x^n + \cdots, \quad |x| < 1,$$

one has

$$\begin{aligned}\frac{1+(m-2)x+x^2}{(1-x)^3} - CS_m(1) &= CS_m(2)x + \cdots + CS_m(n+1)x^n + \cdots \\ &= {}^{-}CS_m(1)x + {}^{-}CS_m(2)x^2 + \cdots \\ &\quad + {}^{-}CS_m(n)x^n + \cdots, \quad |x| < 1.\end{aligned}$$

Since $CS_m(1) = 1$, and $\frac{1+(m-2)x+x^2}{(1-x)^3} - 1 = \frac{x(x^2-2x+(m+1))}{(1-x)^3}$, it holds

$$\frac{x(x^2-2x+(m+1))}{(1-x)^3} = {}^{-}CS_m(1)x + {}^{-}CS_m(2)x^2 + \cdots + {}^{-}CS_m(n)x^n + \cdots, \ |x| < 1.$$

The generating function for the sequence $CS_m(0)$, $CS_m(1)$, $CS_m(-1), CS_m(2), CS_m(-2), \ldots$ of all generalized m-gonal numbers, written for $n = 0, \pm1, \pm2, \ldots$, can be obtained from the generating functions $f_1(x) = \frac{1+(m-2)x+x^2}{(1-x)^3} = CS_m(1) + CS_m(2)x + \cdots + CS_m(n+1)x^n + \cdots$ and $f_2(x) = \frac{(m+1)-2x+x^2}{(1-x)^3} = CS_m(-1) + CS_m(-2)$ $x + \cdots + CS_m(-(n+1))x^n + \cdots$ using the formula $x^2 f_1(x^2) + x^3 f_2(x^2)$. The direct computation gives $f(x) = \frac{x^2(1+(m+1)x+(m-2)x^2-2x^3+x^4+x^5)}{(1-x^2)^3}$.

In other words, it holds

$$\frac{x^2(1+(m+1)x+(m-2)x^2-2x^3+x^4+x^5)}{(1-x^2)^3}$$
$$= CS_m(0)x + CS_m(1)x^2 + CS_m(-1)x^3 + CS_m(2)x^4$$
$$+CS_m(-2)x^5 + \cdots, \ |x| < 1.$$

For example, in the case of the generalized hex numbers, we have

$$\frac{x^2(1+7x+4x^2-2x^3+x^4+x^5)}{(1-x^2)^3} = \frac{x^2(1+6x-2x^2+x^4)}{(1-x)(1-x^2)^2}$$
$$= CS_6(0)x + CS_6(1)x^2 + CS_6(-1)x^3$$
$$+ CS_m(2)x^4 + CS_m(-2)x^5 + \cdots, \ |x| < 1.$$

One can obtain many properties of generalized centered polygonal numbers, using known properties for ordinary centered polygonal numbers and the above identity ${}^{-}CS_m(n) = CS_m(n+1)$.

For example, the identity $CS_m(n+1) = 1 + mS_3(n)$ implies that *any generalized centered m-gonal number can be constructed as m copies of generalized triangular numbers, places around a given central point*:

$${}^{-}CS_m(n) = 1 + mS_3(n) = 1 + m({}^{-}S_3(n+1)).$$

1.8.3. Following the similar procedure, one can consider other generalized plane figurate numbers.

For example, the **generalized pronic numbers** are defined as all values of the formula $P(n) = n(n+1)$ for any integer value of the index n.

For positive integer n, we get ordinary pronic numbers. For $n = 0$, one has $P(0) = 0$. Finally, for negative values of the index one has

$$P(-n) = (n-1)n, \quad n \in \mathbb{N}.$$

So, *any generalized pronic number coincides with the ordinary pronic number of smaller size*:

$${}^{-}P(n) = P(-n) = P(n-1), \quad n \in \mathbb{N}.$$

Hence, the generalized pronic numbers $^-P(n) = P(-n)$ with negative integers are 0, 2, 6, 12, 20, ..., while the generalized pronic numbers are

$$\ldots, 20, 12, 6, 2, 0, 0, 2, 6, 12, 20, 30, \ldots,$$

forming the sequence $0, 2, 0, 6, 2, 12, 6, 20, 12, \ldots$ for $n = 0, \pm 1, \pm 2, \pm 3, \pm 4 \ldots$.

The recurrent formula for the generalized pronic numbers with negative indices is

$$^-P(n+1) = {^-P(n)} + 2n, {^-P(1)} = 0,$$

while the generating function for this sequence has the form $f(x) = \frac{2x^2}{(1-x)^3}$, i.e.,

$$\frac{2x^2}{(1-x)^3} = P(-1)x + P(-2)x^2 + P(-3)x^3 + \cdots + P(-n)x^n + \cdots, \quad |x| < 1.$$

The recurrent formula for the sequence $P(0), P(1), P(-1), P(2), P(-2), \ldots$ can be obtained from the generating functions $f_1(x) = \frac{2}{(1-x)^3} = P(1) + P(2)x + \cdots + P(n+1)x^n + \cdots$ and $f_2(x) = \frac{2x}{(1-x)^3} = P(-1) + P(-2)x + \cdots + P(-(n+1))x^n + \cdots$ using the formula $x^2 f_1(x^2) + x^3 f_2(x^2)$. The direct computation gives $f(x) = \frac{2+2x^3}{(1-x^2)^3}$, i.e., it holds

$$\frac{x(2+2x^3)}{(1-x)^3} = P(0)x + P(1)x^2 + P(-1)x^3 + P(2)x^4 + P(-2)x^5 \ldots, \ |x| < 1.$$

Chapter 2

Space figurate numbers

Posing points in some special order in the space, instead of the plane, one obtains **space figurate numbers**. The most known such numbers are **pyramidal numbers**, corresponding to triangular, square, pentagonal, hexagonal, heptagonal, in general, m-gonal pyramid. They can be represented as sums of corresponding polygonal numbers.

Cubic numbers correspond to cubes, which are constructed from balls, and possess many interesting properties. Also well-known are **octahedral numbers**, corresponding to the next, after tetrahedron and cube, Platonic solid, octahedron. They are equal to the sum of two consecutive square pyramidal numbers. **Dodecahedral** and **icosahedral numbers** correspond to the last two Platonic solids, dodecahedron and icosahedron. Their construction is more complicate, and these numbers rarely appear in special literature.

The numbers obtained by adding or subtracting pyramidal numbers of smaller dimension correspond to truncation of corresponding polyhedra or to putting pyramids on their faces, as in constructing of star polyhedra. Examples of such numbers are **truncated tetrahedral numbers**, **rhombic dodecahedral numbers**, **stella octangula numbers**.

Often are considered **centered space figurate numbers**; the idea of their construction is similar to the idea of construction of centered polygonal numbers.

2.1 Pyramidal numbers

2.1.1. The n-th m-**pyramidal number** $S_m^3(n)$ is defined as the sum of the first n m-gonal numbers:

$$S_m^3(n) = S_m(1) + S_m(2) + \cdots + S_m(n).$$

So, it holds the following recurrent formula for the m-pyramidal numbers:

$$S_m^3(n) = S_m^3(n-1) + S_m(n), \quad S_m^3(1) = 1.$$

The general formula for n-th m-gonal pyramidal number has the form

$$S_m^3(n) = \frac{n(n+1)((m-2)n - m + 5)}{6}.$$

Let us prove it by induction. For $n = 1$ one has $S_m^3(1) = S_m(1) = 1 = \frac{1 \cdot 2((m-2) \cdot 1 - m + 5)}{6}$. Going from n to $n+1$, one obtains

$$\begin{aligned}
S_m^3(n+1) &= S_m^3(n) + S_m(n+1) \\
&= \frac{n(n+1)((m-2)n - m + 5)}{6} \\
&\quad + \frac{(n+1)((m-2)(n+1) - m + 4)}{2} \\
&= \frac{(n+1)}{6} \cdot (n^2(m-2) + n(5-m) \\
&\quad + 3(n+1)(m-2) + 3(4-m)) \\
&= \frac{(n+1)}{6} \cdot ((n^2 + 3n + 3)(m-2) + n(5-m) + 3(4-m)) \\
&= \frac{(n+1)}{6} \cdot ((n^2 + 3n + 2)(m-2) + (n+2)(5-m) \\
&\quad + (m-2) - 2(5-m) + 3(4-m)) \\
&= \frac{(n+1)}{6} \cdot ((n+2)(n+1)(m-2) + (n+2)(5-m) \\
&\quad + (m - 2 - 10 + 2m + 12 - 3m)) \\
&= \frac{(n+1)}{6} \cdot ((n+2)(n+1)(m-2) + (n+2)(5-m)) \\
&= \frac{(n+1)(n+2)((m-2)(n+1) - m + 5)}{6}.
\end{aligned}$$

This formula was known to Archimedes (287–212 BC) and was used by him for finding volumes.

2.1.2. A **triangular pyramidal number** (or **tetrahedral number**) is a space figurate number which corresponds to placing dots in the configuration of the tetrahedron. The n-th tetrahedral number $S_3^3(n)$ is the sum of the first n consecutive triangular numbers; it has the form

$$S_3^3(n) = \frac{n(n+1)(n+2)}{6}.$$

So, the sequence of the triangular pyramidal numbers is obtained by consecutive summation of the sequence $1, 3, 6, 10, 15, 21, 28, 36, 45, 55, \ldots$ (Sloane's A000217) of triangular numbers and begins with elements $1, 4, 10, 20, 35, 56, 84, 120, 165, 220, \ldots$ (Sloane's A000292).

Note, that one can check the above formula for $S_3^3(n)$ by packing 6 copies of n-th tetrahedral number into an $n \times (n+1) \times (n+2)$ box. Moreover, the following geometrical construction for $n = 5$

$$\begin{array}{ccccccccc} &&&&1&&&& \\ &&&1&&2&&& \\ &&1&&2&&3&& \\ &1&&2&&3&&4& \\ 1&&2&&3&&4&&5 \end{array} + \begin{array}{ccccccccc} &&&&1&&&& \\ &&&2&&1&&& \\ &&3&&2&&1&& \\ &4&&3&&2&&1& \\ 5&&4&&3&&2&&1 \end{array}$$

$$+ \begin{array}{ccccccccc} &&&&5&&&& \\ &&&4&&4&&& \\ &&3&&3&&3&& \\ &2&&2&&2&&2& \\ 1&&1&&1&&1&&1 \end{array} = \begin{array}{ccccccccc} &&&&7&&&& \\ &&&7&&7&&& \\ &&7&&7&&7&& \\ &7&&7&&7&&7& \\ 7&&7&&7&&7&&7 \end{array}$$

also illustrate the proof (see [CoGu96]): $3(1+3+6+10+15) = (1+2+3+4+5) \cdot 7$, and, in general case, $3(S_3(1)+S_3(2)+\cdots+S_3(n)) = (1+2+\cdots+n)(n+2) = S_3(n)\cdot(n+2) = \frac{n(n+1)}{2}\cdot(n+2)$, i.e., $S_3^3(n) = S_3(1)+S_3(2)+\cdots+S_3(n) = \frac{n(n+1)(n+2)}{6}$.

A **square pyramidal number** (or, simply, **pyramidal number**) corresponds to a configuration of points which form a square pyramid. The n-th square pyramidal number $S_4^3(n)$ is the sum of the first n perfect squares, and has the form

$$S_4^3(n) = \frac{n(n+1)(2n+1)}{6}.$$

The first few square pyramidal numbers, obtained by consecutive summation of the sequence $1, 4, 9, 16, 25, 36, 49, 64, 81, 100, \dots$ (Sloane's A000290) of square numbers, are $1, 5, 14, 30, 55, 91, 140, 204, 285, 385, \dots$ (Sloane's A000330).

One can check the above formula by packing of six copies of $S_4^3(n)$ into an $n \times (n+1) \times (2n+1)$ box. On the other hand, one can use the shallow rectangular box width $2n+1$ and length $S_3(n)$. Pack the layers of two square pyramids (consisting of $*$ and $\bullet$, respectively) into this box. Partitioning the remaining space into strips, as it shown in figure below, we put the square layers of a third square pyramid (constructed from five square layers of one 1, four 2, nine 3, sixteen 4 and fifty five 5) in the box.

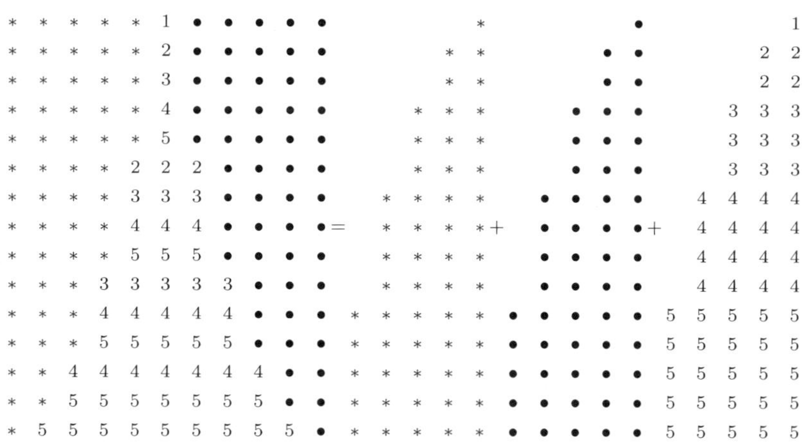

So, for $n = 5$, we have packed three square pyramids into $1 \times 11 \times 15$ box. Therefore, $3S_4^3(5) = S_3(5)(2 \cdot 5 + 1)$. In general, $3S_4^3(n) = S_3(n) \cdot (2n+1)$, and $S_4^3(n) = \frac{n(n+1)(2n+1)}{6}$ (see [CoGu96]).

Moreover, one can use for the proof the following construction.

$$
\begin{array}{ccccccccccccccc}
 & & & & & & & 1 & & & & & & \to & 1^2 \\
 & & & & & & 1 & 2 & 1 & & & & & \to & 2^2 \\
 & & & & & 1 & 2 & 3 & 2 & 1 & & & & \to & 3^2 \\
 & & & & 1 & 2 & 3 & 4 & 3 & 2 & 1 & & & \to & 4^2 \\
 & & & 1 & 2 & 3 & 4 & 5 & 4 & 3 & 2 & 1 & & \to & 5^2 \\
\hline
 & & 1 & 2 & 3 & 4 & 5 & & 5 & 4 & 3 & 2 & 1 & & \\
 & \swarrow & 2 & 3 & 4 & 5 & & & & 5 & 4 & 3 & 2 & \searrow & \\
1^2 & \swarrow & 3 & 4 & 5 & & & & & & 5 & 4 & 3 & \searrow & 1^2 \\
2^2 & \swarrow & 4 & 5 & & & & & & & & 5 & 4 & \searrow & 2^2 \\
3^2 & \swarrow & 5 & & & & & & & & & & 5 & \searrow & 3^2 \\
4^2 & \swarrow & & & & & & & & & & & & \searrow & 4^2 \\
5^2 & & & & & & & & & & & & & & 5^2
\end{array}
$$

In the above figure, there are $(2 \cdot 5) + 1 = 11$ columns, each contains the triangular number $S_3(5) = 1+2+3+4+5$. The five rows above the line add up to 1^2, 2^2, 3^2, 4^2, 5^2 by the *upstairs-downstairs rule* $1+2+3+\cdots+(n-1)+n+(n-1)+\cdots+2+1 = n^2$, while each of the two triangles below the line, when summing diagonally, also contains the first five squares. So, $3(1^2+2^2+3^2+4^2+5^2) = (1+2+3+4+5)\cdot 11$, or $3S_3^4(5) = S_3(5)\cdot 11$. In general, $3(1^2+2^2+\cdots+n^2) = 3S_4(n) = (1+2+\cdots+n)(2n+1) = S_3(n)(2n+1)$, and again $3S_4^3(n) = S_3(n)(2n+1)$, or $S_4^3(n) = \frac{n(n+1)(2n+1)}{6}$ (see [CoGu96]).

A **pentagonal pyramidal number** corresponds to a pentagonal pyramid. The n-th pentagonal pyramidal number $S_5^3(n)$ is the sum of the first n pentagonal numbers and is given by the formula $S_5^3(n) = \frac{n^2(n+1)}{2}$. The first few pentagonal pyramidal numbers, obtained by consecutive summation of the sequence $1, 5, 12, 22, 35, 51, 70, 92, 117, 145, \ldots$ (Sloane's A000326) of pentagonal numbers, are $1, 6, 18, 40, 75, 126, 196, 288, 405, 550, \ldots$ (Sloane's A002411).

A **hexagonal pyramidal number** corresponds to a hexagonal pyramid. The n-th hexagonal pyramidal number $S_6^3(n)$ is the sum of the first n hexagonal numbers and is given by the formula $S_6^3(n) = \frac{n(n+1)(4n-1)}{6}$. The first few hexagonal pyramidal numbers, obtained by consecutive summation of the sequence $1, 6, 15, 28, 45, 66, 91, 120, 153, 190, \ldots$ (Sloane's A000384) of hexagonal numbers, are $1, 7, 22, 50, 95, 161, 252, 372, 525, 715, \ldots$ (Sloane's A002412).

A **heptagonal pyramidal number** corresponds to a heptagonal pyramid. The n-th heptagonal pyramidal number $S_7^3(n)$ is the sum of the first n consecutive heptagonal numbers and is given by the formula $S_7^3(n) = \frac{n(n+1)(5n-2)}{6}$. The first few heptagonal pyramidal numbers, obtained by consecutive summation of the sequence $1, 7, 18, 34, 55, 81, 112, 148, 189, 235, \ldots$ (Sloane's A000566), are $1, 8, 26, 60, 115, 196, 308, 456, 645, 880, \ldots$ (Sloane's A002413).

A **octagonal pyramidal number** corresponds to an octagonal pyramid. The n-th octagonal pyramidal number $S_8^3(n)$ is the sum of the first n consecutive octagonal numbers and is given by the formula $S_8^3(n) = \frac{n(n+1)(6n-1)}{6}$. The first few octagonal pyramidal numbers, obtained by consecutive summation of the sequence $1, 8, 21, 40, 65, 96, 133, 176, 225, 280, \ldots$ (Sloane's A000567) are $1, 9, 30, 70, 135, 231, 364, 540, 765, 1045, \ldots$ (Sloane's A002414).

Similarly, one obtains the sequence $1, 10, 34, 80, 155, 266, 420, 624, 885, 1210, \ldots$ (Sloane's A007584) of **nonagonal pyramidal numbers**; the sequence $1, 11, 38, 90, 175, 301, 476, 708, 1005, 1375, \ldots$ (Sloane's A007585) of **decagonal pyramidal numbers**; the sequence $1, 12, 42, 100, 195, 336, 532, 792, 1125, 1540, \ldots$ (Sloane's A007586) of **hendecagonal pyramidal numbers**; the sequence $1, 13, 46, 110, 215, 371, 588, 876, 1245, 1705, \ldots$ (Sloane's A007587) of **dodecacagonal pyramidal numbers**, and so on.

The formulas for m-gonal pyramidal numbers with $3 \leq m \leq 30$, as well as the first few elements of the corresponding sequences and the numbers of these sequences in the Sloane's *On-Line Encyclopedia of Integer Sequences* (OEIS, [Sloa11]) classification, are given in the table below.

Name	Formula											Sloane
Tetrahedral	$\frac{1}{6}(n(n+1)(n+2))$	1	4	10	20	35	56	84	120	165	220	A000292
Square pyr.	$\frac{1}{6}(n(n+1)(2n+1)$	1	5	14	30	55	91	140	201	285	385	A000330
Pentagonal pyr.	$\frac{1}{6}(n(n+1)(3n+0))$	1	6	18	40	75	126	196	288	405	550	A002411
Hexagonal pyr.	$\frac{1}{6}(n(n+1)(4n-1))$	1	7	22	50	95	161	252	372	525	715	A002412
Heptagonal pyr.	$\frac{1}{6}(n(n+1)(5n-2)$	1	8	26	60	115	196	308	456	645	880	A002413
Octagonal pyr.	$\frac{1}{6}(n(n+1)(6n-3)$	1	9	30	70	135	231	364	540	765	1045	A002413
Nonagonal pyr.	$\frac{1}{6}(n(n+1)(7n-4))$	1	10	34	80	155	266	420	624	885	1210	A007584
Decagonal pyr.	$\frac{1}{6}(n(n+1)(8n-5)$	1	11	38	90	175	301	476	708	1005	1375	A007585
Hendecagonal pyr.	$\frac{1}{6}(n(n+1)(9n-6)$	1	12	42	100	195	336	532	792	1125	1540	A007586
Dodecagonal pyr.	$\frac{1}{6}(n(n+1)(10n-7)$	1	13	46	110	215	371	588	876	1245	1705	A007587
Tridecagonal pyr.	$\frac{1}{6}(n(n+1)(11n-8)$	1	14	50	120	235	406	644	960	1365	1870	
Tetradecagonal pyr.	$\frac{1}{6}(n(n+1)(12n-9)$	1	15	54	130	255	441	700	1044	1485	2035	
Pentadecagonal pyr.	$\frac{1}{6}(n(n+1)(13n-10)$	1	16	58	140	275	476	756	1128	1605	2200	
Hexadecagonal pyr.	$\frac{1}{6}(n(n+1)(14n-11)$	1	17	62	150	295	511	812	1212	1725	2365	
Heptadecagonal pyr.	$\frac{1}{6}(n(n+1)(15n-12)$	1	18	66	160	315	546	868	1296	1845	2530	
Octadecagonal pyr.	$\frac{1}{6}(n(n+1)(16n-13)$	1	19	70	170	335	581	924	1380	1965	2695	
Nonadecagonal pyr.	$\frac{1}{6}(n(n+1)(17n-14)$	1	20	74	180	355	616	980	1464	2085	2860	
Icosagonal pyr.	$\frac{1}{6}(n(n+1)(18n-15)$	1	21	78	190	375	651	1036	1548	2205	3025	
Icosihenagonal pyr.	$\frac{1}{6}(n(n+1)(19n-16)$	1	22	82	200	395	686	1092	1632	2325	3190	
Icosidigonal pyr.	$\frac{1}{6}(n(n+1)(20n-17)$	1	23	86	210	415	721	1148	1716	2445	3355	
Icositrigonal pyr.	$\frac{1}{6}(n(n+1)(21n-18)$	1	24	90	220	435	756	1204	1800	2565	3520	
Icositetragonal pyr.	$\frac{1}{6}(n(n+1)(22n-19)$	1	25	94	230	455	791	1260	1884	2685	3685	
Icosipentagonal pyr.	$\frac{1}{6}(n(n+1)(23n-20)$	1	26	98	240	475	826	1316	1968	2805	3850	
Icosihexagonal pyr.	$\frac{1}{6}(n(n+1)(24n-21)$	1	27	102	250	495	861	1372	2052	2925	4015	
Icosiheptagonal pyr.	$\frac{1}{6}(n(n+1)(25n-22)$	1	28	106	260	515	896	1428	2136	3045	4180	
Icosioctagonal pyr.	$\frac{1}{6}(n(n+1)(26n-23)$	1	29	110	270	535	931	1484	2220	3165	4345	
Icosinonagonal pyr.	$\frac{1}{6}(n(n+1)(27n-24)$	1	30	114	280	555	966	1540	2304	3285	4510	
Triacontagonal pyr.	$\frac{1}{6}(n(n+1)(28n-25)$	1	31	118	290	575	1001	1596	2388	3405	4675	

2.1.3. The generating function for the sequence $S_m^3(1), S_m^3(2), \ldots, S_m^3(n), \ldots$ of the m-pyramidal numbers has the form $f(x) = \frac{x(1+(m-3)x)}{(1-x)^4}$, i.e., it holds

$$\frac{x(1+(m-3)x)}{(1-x)^4} = S_m^3(1)x + S_m^3(2)x^2 + S_m^3(3)x^3 + \cdots + S_m^3(n)x^n + \cdots, \quad |x| < 1.$$

In particular, one gets

$$\frac{x}{(1-x)^4} = x + 4x^2 + 10x^3 + \cdots + S_3^3(n)x^n + \cdots, \quad |x| < 1;$$

$$\frac{x(x+1)}{(x-1)^4} = x + 5x^2 + 14x^3 + \cdots + S_4^3(n)x^n + \cdots, \quad |x| < 1;$$

$$\frac{x(2x+1)}{(x-1)^4} = x + 6x^2 + 18x^3 + \cdots + S_5^3(n)x^n + \cdots, \quad |x| < 1;$$

$$\frac{x(3x+1)}{(x-1)^4} = x + 7x^2 + 22x^3 + \cdots + S_6^3(n)x^n + \cdots, \quad |x| < 1.$$

This result can be obtained by the standard procedure, leading to the linear recurrent equation $S_m^3(n+4) - 4S_m^3(n+3) + 6S_m^3(n+2) - 4S_m^3(n) + S_m^3(n) = 0$ with initial values $S_m^3(1) = 1$, $S_m^3(2) = 1+m$, $S_m^3(3) = 4m-2$, $S_m^3(4) = 10m-10$. This result can be obtained also using the generating function $\frac{1+(m-3)x}{(1-x)^3} = S_m(1) + S_m(2)x + S_m(3)x^2 + \cdots + S_m(n)x^{n-1} + \cdots$ for the m-gonal numbers and the decomposition $\frac{1}{1-x} = 1 + x + x^2 + \cdots + x^n + \cdots$. In fact, one has

$$\begin{aligned}\frac{1+(m-3)x}{(1-x)^4} &= \frac{1+(m-3)x}{(1-x)^3} \cdot \frac{1}{1-x} \\ &= (S_m(1) + S_m(2)x + S_m(3)x^2 \\ &\quad + \cdots + S_m(n)x^{n-1} + \cdots)(1 + x + x^2 + \cdots + x^n + \cdots) \\ &= S_m(1) + (S_m(1) + S_m(2))x + (S_m(1) + S_m(2) \\ &\quad + S_m(3))x^2 + \cdots + (S_m(1) + \cdots + S_m(n))x^{n-1} + \cdots \\ &= S_m^3(1) + S_m^3(2)x + S_m^3(3)x^2 \\ &\quad + \cdots + S_m^3(n)x^{n-1} + \cdots, \quad |x| < 1.\end{aligned}$$

2.1.4. Let us consider now some interesting relations between pyramidal numbers. (see [Weis11], [Wiki11], [CoGu96], [Kim03]).

I. It is easy to show that *four copies of a square pyramidal number form a tetrahedral number*:

$$4S_4^3(n) = S_3^3(2n).$$

In fact, $4 \cdot \frac{n(n+1)(2n+1)}{6} = \frac{2n(2n+1)(2n+2)}{6}$.

II. Moreover, *square pyramidal numbers are sums of consecutive pairs of tetrahedral numbers*:

$$S_4^3(n) = S_3^3(n) + S_3^3(n-1).$$

In fact, one obtains $S_4^3(n) = S_3^3(n) + S_3^3(n-1)$ summing $S_4(k) = S_3(k) + S_3(k-1)$ over all $1 \le k \le n$: $S_4^3(n) = S_4(1) + S_4(2) + \cdots + S_4(n-1) + S_4(n) = S_3(1) + (S_3(2) + (S_3(1)) + \cdots + (S_3(n-1) + S_3(n-2)) + (S_3(n) + S_3(n-1)) = (S_3(1) + \cdots + S_3(n)) + (S_3(1) + \cdots + S_3(n-1)) = S_3^3(n) + S_3^3(n-1)$.

III. The n-th m-pyramidal number satisfies the equality

$$S_m^3(n) = \frac{1}{3}((m-2)n - m + 5)S_3(n),$$

where $S_3(n)$ is n-th triangular number. In particular, it holds

$$S_3^3(n) = \frac{1}{3}(n+2)S_3(n), \quad S_4^3(n) = \frac{1}{3}(2n+1)S_3(n),$$

$$S_5^3(n) = nS_3(n), \quad S_6^3(n) = \frac{1}{3}(4n-1)S_3(n),$$

$$S_7^3(n) = \frac{1}{3}(5n-2)S_3(n), \quad S_8^3(n) = (2n-1)S_3(n).$$

In fact, since $S_3(n) = \frac{n(n+1)}{2}$, this property follows, in an obvious way, from the general formula $S_m^3(n) = \frac{n(n+1)((m-2)n-m+5)}{6}$.

IV. In the Roman Codex Arcerianus ([Cant75]), about 450 AC, occur a number of special cases of the following remarkable formula for pyramidal numbers:

$$S_m^3(n) = \frac{n+1}{6}(2S_m(n) + n).$$

It can be checked by direct computation: $\frac{n+1}{6}(2S_m(n) + n) = \frac{n+1}{6}(n((m-2)n - m + 4) + n) = \frac{n(n+1)}{6}((m-2)n - (m-4) + 1) = \frac{n(n+1)((m-2)n-m+5}{6} = S_m^3(n)$.

V. It is easy to show that *any m-gonal pyramidal number is equal to the sum of the $(m-1)$-gonal pyramidal number, taking in the row the same place, and the tetrahedral number, taking in the row the previous place*:

$$S_m^3(n) = S_{m-1}^3(n) + S_3^3(n-1).$$

This identity is a three-dimensional analogue of *Nicomachus formula* $S^3_m(n) = S^3_{m-1}(n) + S^3_3(n-1)$ and can be obtained by summing this formula over all $1 \leq k \leq n$: $S^3_m(n) = S_m(1) + S_m(2) + \cdots + S_m(n-1) + S_m(n) = S_{m-1}(1) + (S_{m-1}(2) + (m-2)S_3(1)) + \cdots + (S_{m-1}(n-1) + (m-2)S_3(n-2)) + (S_{m-1}(n) + (m-2)S_3(n-1)) = (S_{m-1}(1) + \cdots + S_{m-1}(n)) + (m-2)(S_3(1) + \cdots + S_3(n-1)) = S^3_{m-1}(n) + (m-2)S^3_3(n-1)$.

VI. Moreover, *any m-gonal pyramidal number is a linear combination with non-negative coefficients of the tetrahedral numbers*:

$$S^3_m(n) = S^3_3(n) + (m-3)S^3_3(n-1).$$

It follows from the *Bachet de Méziriac formula* $S_m(n) = S_3(n) + (m-3)S_3(n-1)$:

$$\begin{aligned} S^3_m(n) &= S_m(1) + S_m(2) + \cdots + S_m(n-1) + S_m(n) \\ &= S_3(1) + (S_3(2) + (m-3)S_3(1)) + \cdots + (S_3(n-1) \\ &\quad +(m-3)S_3(n-2)) + (S_3(n) + (m-3)S_3(n-1)) \\ &= (S_3(1) + \cdots + S_3(n)) + (m-3)(S_3(1) + \cdots + S_3(n-1)) \\ &= S^3_3(n) + (m-3)S^3_3(n-1). \end{aligned}$$

VII. Similar arguments allow to show that *any m-pyramidal number is the sum of the triangular number of the same size and m − 2 copies of the tetrahedral numbers of the previous size*:

$$S^3_m(n) = S_3(n) + (m-2)S^3_3(n-1).$$

It follows by the same procedure of summation from the formula $S_m(n) = n + (m-2)S_3(n-1)$.

VIII. It holds also the formula

$$S^3_3(n-1) + S^3_3(n+1) = 2S^3_3(n) + (n+1),$$

which can be obtained from the formula $S_3(n-1) + S_3(n+1) = 2S_3(n) + 1$ for triangular numbers.

2.1.5. Let us consider now some pyramidal numbers, which simultaneously belong to other classes of figurate numbers.

I. A **triangular tetrahedral number** is a number which is simultaneously triangular and tetrahedral. Such numbers correspond to the positive integer solutions of the Diophantine equation

$$\frac{1}{2}u(u+1) = \frac{1}{6}v(v+1)(v+2).$$

The only positive integer solutions of this equation are (u, v) with $u = 1, 3, 8, 20, 34$ and $v = 1, 4, 15, 55, 119$. So, the only triangular tetrahedral numbers are

$$1, 10, 120, 1540, 7140$$

(Sloane's A027568). In fact, $S_3^3(1) = S_3(1) = 1$, $S_3^3(3) = S_3(4) = 10$, $S_3^3(8) = S_3(15) = 120$, $S_3^3(20) = S_3(55) = 1540$, $S_3^3(34) = S_3(119) = 7140$ ([Avan67]).

II. A **triangular square pyramidal number** is a number which is simultaneously triangular and square pyramidal. Such numbers correspond to the positive integer solutions of the Diophantine equation

$$\frac{1}{2}u(u+1) = \frac{1}{6}v(v+1)(2v+1).$$

The only positive integer solutions of this equation are $(u, v) = (1, 1), (10, 5), (13, 6), (645, 85)$. So, the only numbers, which are simultaneously triangular and square pyramidal, are

$$1, 55, 91, 208335$$

(Sloane's A039596). In fact, $S_4^3(1) = S_3(1) = 1$, $S_4^3(5) = S_3(10) = 55$, $S_4^3(6) = S_3(13) = 91$, $S_4^3(85) = S_3(645) = 208335$ ([Beuk88]).

III. A **square tetrahedral number** is a number which are simultaneously square and tetrahedral. Such numbers correspond to the positive integer solutions of the Diophantine equation

$$u^2 = \frac{1}{6}v(v+1)(v+1).$$

The only solutions of this equation are $(u, v) = (1, 1), (2, 2), (140, 48)$. So, the only numbers, which are simultaneously square and tetrahedral, are

$$1, 4, 19600$$

(Sloane's A003556). In fact, $S_3^3(1) = S_4(1) = 1$, $S_3^3(2) = S_4(2) = 4$, and $S_3^3(48) = S_4(140) = 19600$ ([Meyl78]).

IV. A **square square pyramidal number** is a number which is simultaneously square and square pyramidal. Such numbers correspond to the positive integer solutions of the Diophantine equation

$$u^2 = \frac{1}{6}v(v+1)(2v+1).$$

The only solutions of this equations are $(u,v) = (1,1)$, and $(70,24)$. So, the only numbers, which are simultaneously square and square pyramidal, are

$$1, 4900.$$

In fact, $S_4^3(1) = S_4(1) = 1$, and $S_4^3(24) = S_4(70) = 4900$. It was conjectured by Lucas, 1875 ([Luca75]), and proved by Watson, 1918 ([Wats18]). These numbers give the solution of the *Cannon-ball's problem: to construct, from a square on the plane, a square pyramid.*

V. It is known, that no pile of bullets with a triangular or square base contains a number of bullets equal to a cube, biquadrate or fifth power ([Dick05]). In other words, there are no non-trivial **cubic tetrahedral numbers**, **biquadratic tetrahedral numbers**, and five-dimensional **hypercube tetrahedral numbers**, as well as **cubic square pyramidal numbers**, **biquadratic square pyramidal numbers**, and five-dimensional **hypercube square pyramidal numbers**.

VI. A **tetrahedral square pyramidal number** is a number which is simultaneously tetrahedral and square pyramidal. Such numbers correspond to the positive integer solutions of the the Diophantine equation

$$u(u+1)(u+2) = v(v+1)(2v+1).$$

Beukers, 1988, has studied the problem of finding solutions using integral points on an elliptic curve and found that the only solution is the trivial 1 ([Beuk88]).

2.2 Cubic numbers

2.2.1. A **cubic number** (or **perfect cube**) is a space figurate number corresponding to a cube constructed from balls. The n-th cubic number $C(n)$ is the sum of n copies of n-th square number $S_4(n)$, and has the form

$$C(n) = n^3.$$

The series of perfect cubes starts as follows: $1, 8, 27, 64, 125, 216, 343, 512, 729, 1000, \dots$ (Sloane's A000578).

Determination of cube of large numbers was very common in many ancient civilizations. Aryabhatta, the ancient Indian mathematician, in his famous treatise *Aryabhatiya* explains the mathematical meaning of cube as *the continuous product of three equals as also the (rectangular) solid having 12 equal edges are called cube.* Similar definitions can be found in ancient texts such as *Brahmasphuta Siddhanta* (XVIII. 42), *Ganitha Sara Sangraha* (II. 43), and *Siddhanta Sekhara* (XIII. 4). It is interesting that in modern Mathematics too, the term *cube* stands for two mathematical meanings just like in Sanskrit, where the word *Ghhana* means a factor of power with the number, multiplied by itself three times and also a cubical structure ([Wiki11]).

Since $(n+1)^3 = n^3 + 3n^2 + 3n + 1$, one obtains the following recurrent formula for the cubic numbers:

$$C(n+1) = C(n) + 3n^2 + 3n + 1, \quad C(1) = 1.$$

So, the generating function for the sequence of the cubic numbers is $f(x) = \frac{x(x^2+4x+1)}{(x-1)^4}$ ([Plou92]), i.e., it holds

$$\frac{x(x^2+4x+1)}{(x-1)^4} = C(1)x + C(2)x^2 + C(3)x^3 + \cdots + C(n)\dot{x}^n + \cdots, \quad |x| < 1.$$

In fact, let us start with the recurrent equation $C(n+1) = C(n) + 3n^2 + 3n + 1$. Going from n to $n+1$, one gets $C(n+2) = C(n+1) + 9(n+1)^2 + 3(n+1) + 1$. Subtracting first equality from the second one, we have $C(n+2)-C(n+1)=C(n+1)-C(n)+3(2n+1)+3$, i.e., $C(n+2)=2C(n+1)-C(n)+6n+6$. Similarly, one obtains $C(n+3)=2C(n+2)-C(n+1)+6(n+1)+6$, $C(n+3)-C(n+2)=2C(n+2)-2C(n+1)-C(n+1)+C(n)+6$, and $C(n+3)=3C(n+2)-3C(n+1)+C(n)+6$. Finally, writing $C(n+4)=3C(n+3)-3C(n+2)+C(n+1)+6$, one has $C(n+4)-C(n+3)=3C(n+3)-3C(n+2)-3C(n+2)+3C(n+1)+C(n+1)-C(n)$, or $C(n+4)-4C(n+3)+6C(n+2)-4C(n+1)+C(n)=0$. It is a linear recurrent equation of 4-th order. Its initial values are $C(1)=1, C(2)=8, C(3)=27$, and $C(4)=64$. Denoting $C(n+1)$ by c_n, one obtains the linear recurrent equation

$$c_{n+4}-4c_{n+3}+6c_{n+2}-4c_{n+1}+c_n=0, \quad c_0=1,$$
$$c_1=8, \quad c_2=27, \quad c_3=64.$$

Therefore, the generating function for the sequence of the cubic numbers has the form

$$\frac{f(x)}{g(x)}=\frac{a_0+a_1x+a_2x^2+a_3x^3}{b_0+b_1x+b_2x^2+b_3x^3+b_4x^4},$$

where $b_0=1, b_1=-4, b_2=6, b_3=-4, b_4=1$, and $a_0=b_0c_0=1, a_1=b_0c_1+b_1c_0=1\cdot 8+(-4)\cdot 1=4, a_2=b_0c_2+b_1c_1+b_2c_0=1\cdot 27+(-4)\cdot 8+6\cdot 1=1, a_3=b_0c_3+b_1c_2+b_2c_1+b_3c_0=1\cdot 64+(-4)\cdot 27+6\cdot 8+(-4)\cdot 1=0$. Since $g(x)=1-4x+6x^2-4x^3+x^4=(1-x)^4$ has four coincided roots $x_1=x_2=x_3=x_4=1$, the generating function for the sequence of the cubic numbers has the form

$$\frac{1+4x+x^2}{(1-x)^4}=C(1)+C(2)x+C(3)x^2$$
$$+\cdots+C(n)x^{n-1}+\cdots, \quad |x|<1.$$

2.2.2. Let us consider now some interesting properties of the cubic numbers.

I. It is well-known that *the sum of the first n cubic numbers is the squared n-th triangular number*:

$$C(1)+C(2)+\cdots+C(n)=(S_3(n))^2.$$

This property is just a different form of the following beautiful identity:

$$1^3 + 2^3 + \cdots + n^3 = (1 + 2 + \cdots + n)^2.$$

It is easy to prove, checking by induction, that $1^3 + \cdots + n^3 = \frac{n^2(n+1)^2}{4}$. In fact, for $n = 1$ it holds $1^3 = \frac{1^2 \cdot 2^2}{4}$, and, going from n to $n + 1$, we obtain that

$$\begin{aligned} 1^3 + 2^3 + \cdots + n^3 + (n+1)^3 &= \frac{n^2(n+1)^2}{4} + (n+1)^3 \\ &= \frac{(n+1)^2}{4}(n^2 + 4(n+1)) \\ &= \frac{(n+1)^2}{4}(n^2 + 4n + 4) \\ &= \frac{(n+1)^2(n+2)^2}{4}. \end{aligned}$$

Now, reminding that $1 + 2 + \cdots + n = \frac{n(n+1)}{2} = S_3(n)$, we get the above formula.

II. On the other hand, one can observe, that

$$\begin{aligned} &1^3 = 1, \quad 2^3 = 3 + 5, \quad 3^3 = 7 + 9 + 11, \\ &4^3 = 13 + 15 + 17 + 19, \ldots, \end{aligned}$$

i.e., *n-th cubic number $C(n)$ is a sum of n consecutive odd numbers starting from $n^2 - n + 1$.* In other words, it holds

$$\begin{aligned} C(n) = {} & (2S_3(n-1) + 1) + (2S_3(n-1) + 3) + (2S_3(n-1) + 5) \\ & + \cdots + (2S_3(n) - 1), \end{aligned}$$

where $S_3(n)$ is n-th triangular number.

In order to check this property, let us prove by indiction the following identity:

$$n^3 = \sum_{k=\frac{(n-1)n}{2}+1}^{\frac{n(n+1)}{2}} (2k-1) = \sum_{k=S_3(n-1)+1}^{S_3(n)} (2k-1).$$

In fact, for $n = 1$ one has $1^3 = 2 \cdot 1 - 1 = \sum_{k=1}^{1}(2k-1)$. Going from n to $(n+1)$, one obtains

$$
\begin{aligned}
(n+1)^3 = n^3+3n^2+3n+1 &= \sum_{k=\frac{n(n-1)}{2}+1}^{\frac{n(n+1)}{2}} (2k-1)+n \cdot 2n+(n^2+3n+1) \\
&= \sum_{k=\frac{n(n-1)}{2}+1}^{\frac{n(n+1)}{2}} (2k-1) + \sum_{k=\frac{n(n-1)}{2}+1}^{\frac{n(n+1)}{2}} 2n + \left(2\frac{(n+1)(n+2)}{2} - 1\right) \\
&= \sum_{k=\frac{n(n-1)}{2}+1}^{\frac{n(n+1)}{2}} (2(k+n)-1) + \left(2\frac{(n+1)(n+2)}{2} - 1\right) \\
&= \sum_{m=\frac{n(n-1)}{2}+1+n}^{\frac{n(n+1)}{2}+n} (2m-1) + \left(2\frac{(n+1)(n+2)}{2} - 1\right) \\
&= \sum_{k=\frac{n(n+1)}{2}+1}^{\frac{(n+1)(n+2)}{2}-1} (2k-1) + \left(2\frac{(n+1)(n+2)}{2} - 1\right) \\
&= \sum_{k=\frac{n(n+1)}{2}+1}^{\frac{(n+1)(n+2)}{2}} (2k-1).
\end{aligned}
$$

Note, that we can easy obtain an interesting interpretation of the above fact (see [CoGu96]), if consider two following tables:

$$
\begin{array}{c}
1 = 1^2, \\
1+3 = 2^2, \\
1+3+5 = 3^2, \\
1+3+5+7 = 4^2, \\
\dots
\end{array}
\qquad \text{and} \qquad
\begin{array}{c}
1 = 1^3, \\
3+5 = 2^3, \\
7+9+11 = 3^3, \\
13+15+17+19 = 4^3, \\
\dots
\end{array}
$$

So, *adding odd numbers* $1, 3, 5, 7, 9, 11, 13, 15, \dots$ *gives squares or cubes*: if we each time starting from the unity, we get the squares, as in the first figure; if instead of starting with 1 each time we start where we previously left off, as in the second figure, we get the cubes.[1]

III. Obviously, *any cubic number is the difference of two squared consecutive triangular numbers*:

$$C(n) = (S_3(n))^2 - (S_3(n-1))^2.$$

This result can be obtained using the property $C(1) + C(2) + \cdots + C(n) = (S_3(n))^2$:

$$\begin{aligned}(S_3(n))^2 - (S_3(n-1))^2 &= (C(1) + \cdots + C(n-1) + C(n)) \\ &\quad -(C(1) + \cdots + C(n-1)) = C(n).\end{aligned}$$

IV. It was proven already (see Chapter 1), that *the difference of two consecutive cubic numbers is a centered hexagonal number*:

$$C(n) - C(n-1) = CS_6(n).$$

Therefore, the centered hexagonal numbers are the gnomons of the cubic numbers. In other words, the centered hexagonal numbers (i.e., hex numbers) $1+3n+3n^2$ can be seen as projections (or "shadows") of cubes. We obtain $(n+1)$-th cube by starting with a single point and building out three pairwise perpendicular rods of n points each. Then the three spaces between pairs of rods each accommodate a wall of $n \times n$ points and we have a nest that will encase (three adjacent faces of) an $n \times n \times n$ cube, making it up to $(n+1)^3$. This is a very special case $1 + 3n + 3n^2 + n^3 = (1+n)^3$ of the *binomial theorem* ([CoGu96]).

V. Viewed from the opposite perspective, this property yields that *n-th cubic number $C(n)$ is the sum of the first n centered hexagonal numbers*:

$$C(n) = CS_6(1) + CS_6(2) + \cdots + CS_6(n).$$

[1]The second figure illustrates also the property $1^3 + \cdots + n^3 = (1+2+\cdots+n)^2$: the sum $1^3 + 2^3 + \cdots + n^3$ of the first n cubic numbers (right side' entries of the second figure) is equal to the sum of the first $S_3(n)$ odd numbers (left side' entries of the second figure), which, by definition of square numbers, is equal to $(S_3(n))^2$, and, hence, to $(1+\cdots+n)^2$.

It follows immediately from the equation $CS_6(n) = n^3 - (n-1)^3$ using a telescoping sum:

$$\begin{aligned} CS_6(1) + CS_6(2) + \cdots + CS_6(n) &= (1^3 - 0^3) + (2^3 - 1^3) \\ &\quad + \cdots + (n^3 - (n-1)^3) \\ &= n^3. \end{aligned}$$

VI. One can show that *any cubic number can be constructed using only tetrahedral numbers*:

$$C(n) = S_3^3(n) + 4S_3^3(n-1) + S_3^3(n-2), \quad n \geq 2.$$

In fact, we have

$$\begin{aligned} &S_3^3(n) + 4S_3^3(n-1) + S_3^3(n-2) \\ &\quad = \frac{n(n+1)(n+2) + 4(n-1)n(n+1) + (n-2)(n-1)n}{6} \\ &\quad = \frac{n}{6}((n+1)(n+2+4n-4) + (n-2)(n-1)) \\ &\quad = \frac{n}{6}((n+1)(5n-2) + (n-2)(n-1)) \\ &\quad = \frac{n}{6}(5n^2 + 5n - 2n - 2 + n^2 - 2n - n + 2) \\ &\quad = \frac{n}{6}(6n^2) = n^3 = C(n). \end{aligned}$$

2.2.3. Let us consider cubic numbers, which simultaneously belong to other classes of figurate numbers.

I. It is easy to prove, that 1 is the only **triangular cubic number**, i.e., a number that is simultaneously cubic and triangular.

In fact, such a number corresponds to the positive integer solutions of the Diophantine equation

$$\frac{u(u+1)}{2} = v^3.$$

Completing the square and rearranging gives $(2u+1)^2 - (2v)^3 = 1$. Substituting $x = 2u+1$ and $y = 2v$ gives the Diophantine equation $x^2 - y^3 = 1$. According to a special case of the proven *Catalan's conjecture*,[2] it is known that the only pair of a perfect cube and a perfect square differing by 1 is 3^2 and 2^3 (excluding 0 and 1). Hence, the only positive integer solution of the equation $x^2 - y^3 = 1$ is the pair $(x, y) = (3, 2)$. It implies $u = v = 1$. So, the unique cubic triangular number is 1.

II. A **hex cubic number** is a number that is simultaneously a perfect cube and a centered hexagonal number. It is known, that 1 is the only number of such kind ([Weis11]).

2.3 Octahedral numbers

2.3.1. An **octahedral number** is a space figurate number that represents an octahedron, or two pyramids placed together, one upside-down underneath the other. Therefore, n-th octahedral number $O(n)$ is the sum of two consecutive square pyramidal numbers:

$$O(n) = S_4^3(n-1) + S_4^3(n).$$

It implies the following general formula for the octahedral numbers:

$$O(n) = \frac{n(2n^2+1)}{3}.$$

The first few octahedral numbers are $1, 6, 19, 44, 85, 146, 231, 344, 489, 670, \ldots$ (Sloane's A005900).

The recurrent formula $S_4^3(n+1) = S_4^3(n) + (n+1)^2$ for the square pyramidal numbers yields the following recurrent formula for the octahedral numbers:

$$O(n+1) = O(n) + (n+1)^2 + n^2, \quad O(1) = 1.$$

The generating function for the octahedral numbers is $f(x) = \frac{x(x+1)^2}{(x-1)^4}$ ([Wiki11]), i.e., it holds

$$f(x) = \frac{x(x+1)^2}{(x-1)^4} = O(1)x + O(2)x^2 + O(3)x^3 + \cdots + O(n)x^n + \cdots, \quad \text{for } |x| < 1.$$

[2]The conjecture made by Catalan, 1844, that 8 and 9 are the only consecutive powers (excluding 0 and 1) and proven by Mihăilescu in 2002.

It is easy to obtain, using the generating function $\frac{x+1}{(x-1)^4} = S_4^3(1) + S_4^3(2)x + \cdots + S_4^3(n+1)x^n + \cdots$ for the sequence of the square pyramidal numbers. In fact, we have $\frac{x+1}{(x-1)^4} + \frac{x(x+1)}{(x-1)^4} = (S_4^3(1) + S_4^3(2)x + S_4^3(3)x^2 + \cdots + S_4^3(n+1)x^n + \cdots) + (S_4^3(1)x + S_4^3(2)x^2 + S_4^3(3)x^3 + \cdots + S_4^3(n)x^n + \cdots) = S_4^3(1) + (S_4^3(1) + S_4^3(2))x + (S_4^3(2) + S_4^3(3))x^2 + \cdots + (S_4^3(n) + S_4^3(n+1))x^n + \cdots$, $|x| < 1$. Obviously, it holds $O(1) = S_4^3(1) = 1$. Hence, using the definition of the octahedral numbers, we get

$$\frac{(x+1)^2}{(x-1)^4} = O(1) + O(2)x + O(3)x^2 + \cdots + O(n+1)x^n + \cdots, \quad |x| < 1.$$

This result can be obtained also by the standard procedure, leading to the linear recurrent equation $O(n+4) - 4O(n+3) + 6O(n+2) - 4O(n+1) + O(n) = 0$ with initial values $O(1) = 1, O(2) = 6, O(3) = 19$, and $O(4) = 44$.

2.3.2. Let us consider now some interesting properties of the octahedral numbers.

I. It was proven (see Chapter 1), that a sum of two consecutive square numbers is a centered square number: $CS_4(n) = S_4(n) + S_4(n-1)$. Hence, it holds $(n+1)^2 + n^2 = CS_4(n+1)$, and we obtain, that *the difference of two consecutive octahedral numbers is a centered square number*:

$$O(n+1) - O(n) = CS_4(n+1).$$

II. It is easy to see also, that *octahedral numbers can be constructed using only tetrahedral numbers*:

$$O(n) = S_3^3(n) + 2S_3^3(n-1) + S_3^3(n-2), \quad n \geq 2.$$

In fact, $S_3^3(n) + 2S_3^3(n-1) + S_3^3(n-2) = \frac{1}{6}(n(n+1)(n+2) + 2(n-1)n(n+1) + (n-2)(n-1)n) = \frac{n}{6}(n^2 + 3n + 2 + 2n^2 - 2 + n^2 - 3n + 2) = \frac{n}{6}(4n^2 + +2) = \frac{n}{3}(2n^2 + 1) = O(n)$ ([Kim03]).

III. Moreover, *octahedral numbers can be obtained by truncation of tetrahedral numbers*:

$$O(n) = S_3^3(2n-1) - 4S_3^3(n-1).$$

In fact, it holds $S_3^3(2n-1) - 4S_3^3(n-1) = \frac{(2n-1)\cdot 2n\cdot(2n+1)}{6} - 4\frac{(n-1)\cdot n\cdot(n+1)}{6} = \frac{2n((2n-1)(2n+1)-2(n-1)(n+1))}{6} = \frac{n(2n^2+1)}{3} = O(n)$ ([Wiki11]).

2.3.3. A related set of numbers are so-called **Haüy octahedral numbers** ([Weis11]). They give the number of cubes in the *Haüy construction* (i.e., the construction of polyhedra using identical building blocks) of the octahedron.

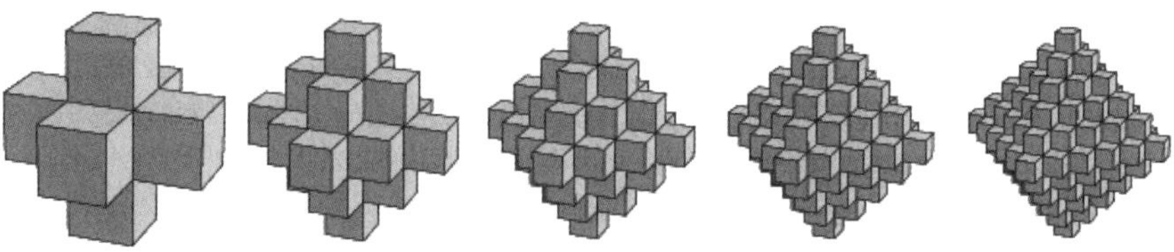

The first few values of this sequence are $1, 7, 25, 63, 129, 231, 377, 575, 833, 1159, \ldots$ (Sloane's A001845). These numbers can be seen as centered octahedral numbers, which will be describe below.

By definition, the number of building blocks in each cross-section of a Haüy octahedral number is a centered square number. Therefore, any Haüy octahedral number can be represented as a sum of centered square numbers. More precisely, for n-th Haüy octahedral number $O_H(n)$ it holds

$$O_H(n) = CS_4(n) + 2(CS_4(n-1) + CS_4(n-2) + \cdots + CS_4(1)).$$

Since $CS_4(n) = 2n^2 - 2n + 1$, we obtain the following general formula for Haüy octahedral numbers:

$$O_H(n) = \frac{(2n-1)(2n^2-2n+3)}{3}.$$

In fact, since $1 + 2 + \cdots + n = \frac{n(n+1)}{2}$, and $1^2 + 2^2 + \cdots + n^2 = \frac{n(n+1)(2n+1)}{6}$, we get $O_H(n) = CS_4(n) + 2\sum_{k=1}^{n-1} CS_4(k) = (2n^2 - 2n + 1) + 2\sum_{k=1}^{n-1}(2k^2 - 2k + 1) = (2n^2 - 2n + 1) + 4\sum_{k=1}^{n-1} k^2 - 4\sum_{k=1}^{n-1} k + 2\sum_{k=1}^{n-1} 1 = (2n^2 - 2n + 1) - 4\cdot\frac{(n-1)\cdot n\cdot(2n-1)}{6} - 4\cdot\frac{(n-1)n}{2} - 2(n-1) = \frac{1}{3}(2n-1)(2n^2 - 2n + 3)$.

The recurrent formula $CS_4(n+1) = CS_4(n)+4n$ for the sequence of the centered square numbers implies the following recurrent formula for the sequence of Haüy octahedral numbers:

$$O_H(n+1) = O_H(n) + 4n^2 + 2, \quad O_H(1) = 1.$$

Really, one has $O_H(n+1) = CS_4(n+1)+2\sum_{k=1}^{n} CS_4(k) = (CS_4(n)+4n) + (2\sum_{k=1}^{n-1} CS_4(k) + 2CS_4(n)) = (CS_4(n) + 2\sum_{k=1}^{n-1} CS_4(k)) + (4n + 2(n^2 - 2n + 1)) = O_H(n) + (4n^2 + 2)$.

Now it is easy to prove, that the generating function for the sequence of Haüy octahedral numbers is $f(x) = \frac{x(1+x)^3}{(1-x)^4}$, i.e., it holds

$$\frac{x(1+x)^3}{(1-x)^4} = O_H(1)x + O_H(2)x^2 + O_H(3)x^3 + \cdots + O_H(n)x^n + \cdots, \quad |x| < 1.$$

In fact, let us start with the recurrent equation $O_H(n+1) = O_H(n)+(4n^2+2)$. Going from n to $n+1$, one gets $O_H(n+2) = O_H(n+1)+4(n+1)^2 + 2$. Subtracting first equality from the second one, we have $O_H(n+2) - O_H(n+1) = O_H(n+1) - O_H(n) + (8n+1)$, i.e., $O_H(n+2) = 2O_H(n+1) - O_H(n) + 8n + 1$. Similarly, one obtains $O_H(n+3) = 2O_H(n+2) - O_H(n+1) + 8(n+1) + 1$, $O_H(n+3) - O_H(n+2) = 2O_H(n+2) - 2O_H(n+1) - O_H(n+1) + O_H(n) + 8$, and $O_H(n+3) = 3O_H(n+2) - 3O_H(n+1) + O_H(n) + 8$. Finally, writing $O_H(n+4) = 3O_H(n+3) - 3O_H(n+2) + O_H(n+1) + 8$, one has $O_H(n+4) - O_H(n+3) = 3O_H(n+3) - 3O_H(n+2) - 3O_H(n+2) + 3O_H(n+1) + O_H(n+1) - O_H(n)$, or

$$O_H(n+4) - 4O_H(n+3) + 6O_H(n+2) - 4O_H(n+1) + O_H(n) = 0.$$

It is a linear recurrent equation of 4-th order. Its initial values are $O_H(1) = 1$, $O_H(2) = 7$, $O_H(3) = 25$, and $O_H(4) = 63$. Denoting $O_H(n+1)$ by c_n, one obtains the linear recurrent equation

$$c_{n+4} - 4c_{n+3} + 6c_{n+2} - 4c_{n+1} + c_n = 0, \quad c_0 = 1, \quad c_1 = 7, \quad c_2 = 25, \quad c_3 = 63.$$

So, the generating function for the sequence of Haüy octahedral numbers has the form

$$\frac{f(x)}{g(x)} = \frac{a_0 + a_1x + a_2x^2 + a_3x^3}{b_0 + b_1x + b_2x^2 + b_3x^3 + b_4x^4},$$

where $b_0 = 1$, $b_1 = -4$, $b_2 = 6$, $b_3 = -4$, $b_4 = 1$, and $a_0 = b_0c_0 = 1$, $a_1 = b_0c_1 + b_1c_0 = 1 \cdot 7 + (-4) \cdot 1 = 3$, $a_2 = b_0c_2 + b_1c_1 + b_2c_0 = 1{\cdot}25+(-4){\cdot}7+6{\cdot}1 = 3$, $a_3 = b_0c_3+b_1c_2+b_2c_1+b_3c_0 = 1{\cdot}63+(-4){\cdot}25+6\cdot 7+(-4)\cdot 1 = 1$. Since $g(x) = 1-4x+6x^2-4x^3+x^4 = (1-x)^4$ has four coincided roots $x_1 = x_2 = x_3 = x_4 = 1$, the generating function for the sequence of Haüy octahedral numbers has the form

$$\frac{1+3x+3x^2+3x^3}{(1-x)^4} = O_H(1) + O_H(2)x + O_H(3)x^2 + \cdots + O_H(n)x^{n-1} + \cdots, \quad |x| < 1.$$

2.4 Other regular polyhedral numbers

2.4.1. The **regular polyhedral numbers** (or **Platonic solid numbers**) are space figurate numbers that correspond to classical convex regular polyhedra, i.e., *Platonic solids.*

A (convex) *regular polyhedron* is a polyhedron with equivalent vertices and faces (faces being congruent convex regular polygons). There are exactly five such solids: *tetrahedron*, *cube*, *octahedron*, *dodecahedron*, and *icosahedron.*

The regular polyhedra can be described by *Schläfli symbol*, which is defined recursively: a convex regular polygon having n sides is denoted by $\{n\}$, while a convex regular polyhedron having faces $\{n\}$ with p faces joining around a vertex is denoted by $\{n, p\}$.

The Schläfli symbols of five Platonic solids are listed in the table below, as well as the number v of the vertices, number s of sides and number f of faces of a given regular polyhedron.

name	Schläfli symbol	v	s	f
tetrahedron	$\{3, 3\}$	4	6	4
octahedron	$\{3, 4\}$	6	12	8
cube	$\{4, 3\}$	8	12	6
icosahedron	$\{3, 5\}$	12	30	20
dodecahedron	$\{5, 3\}$	20	30	12

In the previous sections we have considered space figurate numbers, corresponding to three of five Platonic solids: **tetrahedral numbers**, **cubic numbers**, and **octahedral numbers**. These classes of space figurate numbers were obtained in some natural manner: tetrahedral numbers as sums of consecutive triangular numbers; cubic numbers just as numbers corresponding to cube constructed from balls; octahedral numbers as sums of two consecutive square pyramidal numbers.

However, those three classes of regular polyhedral numbers, as well as **dodecahedral numbers** and **icosahedral numbers**, corresponding to dodecahedron and icosahedron, can be constructed following a standard procedure, which is similar to the procedure of the construction of polygonal numbers, discussed in Chapter 1.

In fact, the general rule for enlarging the regular polyhedron to the next size is *to extend the all edges, containing a given vertex, by one point and then add the required extra faces of the next array between this points.*

In other words, in order to obtain the sequence of tetrahedral numbers, we start from a point, corresponding to $S_3^3(1) = 1$, and then get $S_3^3(2)$, constructing three edges, containing the given point, putting one additional point on each of these edges, and building four triangles three points each, which together form 4-points tetrahedron, corresponding to $S_3^3(2) = 4$. Next we obtain $S_3^3(3)$, using previous procedure: extending the edges, containing the first fixed point, by one new point on each, and building a new six points triangle between these three points, which together with previous construction form 10-points tetrahedron, corresponding to $S_3^3(3) = 10$, etc.

In order to obtain the sequence of cubic numbers, we start from a point, corresponding to $C(1) = 1$, and then get $C(2)$, constructing the three edges containing the given point, putting one additional point on each of these edges, and building "between these points" six squares of four points each, which together form 8-points cube, corresponding to $C(2) = 8$. Next we obtain $C(3)$, using previous procedure: extending the edges, containing the first fixed point, by one new point on each, and building "between these points" 3 new squares of 9 points each, which together with previous construction form 27-points cube, corresponding to $C(3) = 27$, and so on.

In general, for a given Platonic solid, we start from one point, giving the first element of the sequence of corresponding regular polyhedral numbers. Then, assuming that the $(n-1)$-th element of this sequence is obtained, we get n-th element of these sequence, extending each edge, contained the above fixed point, by one new point. Finally, we add the required extra faces, each being n-th polygonal number, in order to obtain again the shape of the given Platonic solid — tetrahedron, octahedron, cube, icosahedron or dodecahedron.

Algebraically, in order to obtain n-th regular polyhedral number, corresponding to a given Platonic solid $\{m, q\}$ with v vertices, s sides and f m-gonal faces, one should add to $(n-1)$-th regular polyhedral number the following values: the value $v-1$ that counts points, corresponding to all new vertices of our construction; the value $(s-q)(n-2)$ that counts points, corresponding to the interior of all new sides of this construction; finally, the value $(s-q)int(S_m(n))$ that counts points, corresponding to the interior of all new faces of this construction; here $int(S_m(n))$ is the number of points in the interior of n-th m-gonal number (see [Kim03]).

2.4.2. For **tetrahedral numbers** we get, by the above construction, the following relation:

$$\begin{aligned} S_3^3(n) - S_3^3(n-1) &= (v-1) + (s-3)\cdot(n-2) + (f-3)\cdot int(S_3(n)) \\ &= (4-1) + (6-3)\cdot(n-2) + (4-3) \\ &\quad \cdot \left(\frac{n(n+1)}{2} - 3(n-1)\right) \\ &= 3 + 3(n-2) + \frac{(n-3)(n-2)}{2} = \frac{n(n+1)}{2}. \end{aligned}$$

So, the classical recurrent equation for tetrahedral numbers is obtained:

$$S_3^3(n) = S_3^3(n-1) + S_3(n), \quad S_3^3(1) = 1.$$

Now we can easily get, for example, by induction, the general formula $S_3^3(n) = \frac{n(n+1)(n+2)}{6}$: it holds for $n=1$, and, going from $n-1$ to n, we have $S_3^3(n) = S_3^3(n-1) + \frac{n(n+1)}{2} = \frac{(n-1)n(n+1)}{6} + \frac{n(n+1)}{2} = \frac{n(n+1)(n+2)}{6}$.

For **cubic numbers** we get, by the above construction, the following relation:

$$\begin{aligned} C(n) - C(n-1) &= (v-1) + (s-3)\cdot(n-2) + (f-3)\cdot int(S_4(n)) \\ &= (8-1) + (12-3)\cdot(n-2) + (6-3) \\ &\quad \cdot(n^2 - 4(n-1)) \\ &= 7 + 9(n-2) + 3(n-2)^2 = 3n^2 - 3n + 1. \end{aligned}$$

Therefore, the classical recurrent equation for tetrahedral numbers is obtained:

$$C(n) = C(n-1) + 3n^2 - 3n + 1, \quad C(1) = 1.$$

Now we can easily get, for example, by induction, the general formula $C(n) = n^3$: it holds for $n = 1$, and, going from $n-1$ to n, we get $C(n) = C(n-1) + (3n^2 - 3n + 1) = (n-1)^3 + (3n^2 - 3n + 1) = n^3$.

For **octahedral numbers** we get, by the above construction, the following relation:

$$\begin{aligned} O(n) - O(n-1) &= (v-1) + (s-4)\cdot(n-2) + (f-4)\cdot int(S_3(n)) \\ &= (6-1) + (12-4)\cdot(n-2) + (8-4) \\ &\quad \cdot\left(\frac{n(n+1)}{2} - 3(n-1)\right) \\ &= 5 + 8(n-2) + 4\cdot\frac{(n-3)(n-2)}{2} = 2n^2 - 2n + 1. \end{aligned}$$

So, the classical recurrent equation for tetrahedral numbers is obtained:

$$O(n) = O(n-1) + 2n^2 - 2n + 1, \quad O(1) = 1.$$

Now we can easily get, for example, by induction, the general formula $O(n) = \frac{n(2n^2+1)}{3}$: it holds for $n = 1$, and, going from $n-1$ to n, we get $O(n) = O(n-1) + (2n^2 - 2n + 1) = \frac{(n-1)(2(n-1)^2+1)}{3} + (2n^2 - 2n + 1) = \frac{n(2n^2+1)}{3}$.

2.4.3. The **icosahedral numbers** are regular polyhedral numbers, obtained by the above constriction, applied to an icosahedron.

Starting with $I(1) = 1$, we get by the above construction the following relation between n-th and $(n-1)$-th icosahedral numbers:

$$
\begin{aligned}
I(n) - I(n-1) &= (v-1) + (s-5)\cdot(n-2) + (f-5)\cdot int(S_3(n)) \\
&= (12-1) + (30-5)\cdot(n-2) + (20-5) \\
&\quad \cdot\left(\frac{n(n+1)}{2} - 3(n-1)\right) \\
&= 11 + 25(n-2) + 15\cdot\frac{(n-3)(n-2)}{2} \\
&= \frac{15n^2 - 25n + 12}{2}.
\end{aligned}
$$

So, the recurrent formula for the sequence of icosahedral numbers has the form

$$I(n) = I(n-1) + \frac{15n^2 - 25n + 12}{2}, \quad I(1) = 1.$$

Now we can easily get, for example by induction, that the general formula for n-th icosahedral number $I(n)$ has the form

$$I(n) = \frac{n(5n^2 - 5n + 2)}{2}.$$

In fact, it holds for $n = 1$, and, going from $n-1$ to n, we get $I(n) = I(n-1) + \frac{15n^2-25n+12}{2} = \frac{(n-1)(5(n-1)^2-5(n-1)+2)}{2} + \frac{15n^2-25n+12}{2} = \frac{n(5n^2-5n+2)}{2}$.

The sequence of the icosahedral numbers starts with the elements $1, 12, 48, 124, 255, 456, 742, 1128, 1629, 2260, \ldots$ (Sloane's A006564).

The generating function for the sequence of icosahedral numbers is $f(x) = \frac{x(1+8x+6x^2)}{(1-x)^4}$ (see [Plou92]), i.e., it holds

$$
\begin{aligned}
\frac{x(1+8x+6x^2)}{(1-x)^4} &= I(1)x + I(2)x^2 + I(3)x^3 \\
&\quad + \cdots + I(n)x^n + \cdots, \quad |x| < 1.
\end{aligned}
$$

This result can be obtained following the standard procedure, leading to the linear recurrent equation $I(n+4)-4I(n+3)+6I(n+2)-4I(n+1)+I(n)=0$ with initial conditions $I(1)=1, I(2)=12, I(3)=48$, and $I(4)=124$.

It is easy to show, that *every icosahedral number can be written as a linear combination of tetrahedral numbers*:

$$I(n)=S_3^3(n)+8S_3^3(n-1)+6S_3^3(n-1).$$

The proof can be obtained by direct computation.

2.4.4. The **dodecahedral numbers** are regular polyhedral numbers, obtained by the above constriction, applied to a dodecahedron.

Starting with $D(1)=1$, we get by the above construction the following relation between n-th and $(n-1)$-th dodecahedral numbers:

$$\begin{aligned} D(n)&-D(n-1)\\ &=(v-1)+(s-3)\cdot(n-2)+(f-3)\cdot int(S_5(n))\\ &=(20-1)+(30-3)\cdot(n-2)+(12-3)\cdot\left(\frac{3n^2-n}{2}-5(n-1)\right)\\ &=19+27(n-2)+9\cdot\frac{3n^2-11n+9}{2}=\frac{27n^2-45n+20}{2}. \end{aligned}$$

So, the recurrent formula for the sequence of dodecahedral numbers has the form

$$D(n)=D(n-1)+\frac{27n^2-45n+20}{2},\quad D(1)=1.$$

Now we can easily get, for example by induction, that the general formula for n-th dodecahedral number $D(n)$ has the form

$$D(n)=\frac{n(9n^2-9n+2)}{2}=\frac{n(3n-1)(3n-2)}{2}.$$

It is true for $n=1$, and, going from $n-1$ to n, we get $D(n)=D(n-1)+\frac{27n^2-45n+20}{2}=\frac{(n-1)(9(n-1)^2-9(n-1)+2)}{2}+\frac{27n^2-45n+20}{2}=\frac{n(9n^2-9n+2)}{2}$.

The sequence of the dodecahedral numbers starts with the elements $1, 20, 84, 220, 455, 816, 1330, 2024, 2925, 4060, \ldots$ (Sloane's A006566).

The generating function for the sequence of dodecahedral numbers is $f(x) = \frac{x(1+16x+10x^2)}{(1-x)^4}$ (see [Plou92]), i.e., it holds

$$\frac{x(1+16x+10x^2)}{(1-z)^4} = D(1)x + D(2)x^2 + D(3)x^3$$
$$+ \cdots + D(n)x^n + \cdots, \quad |x| < 1.$$

This result can be obtained following the standard procedure, leading the linear recurrent equation $D(n+4) - 4D(n+3) + 6D(n+2) - 4D(n+1) + D(n) = 0$ with initial conditions $D(1) = 1, D(2) = 20, D(3) = 84$, and $D(4) = 220$.

It is easy to show that *every dodecahedral number can be written as a linear combination of tetrahedral numbers*:

$$D(n) = S_3^3(n) + 16S_3^3(n-1) + 10S_3^3(n-1).$$

The proof can be obtained by direct computation.

2.5 Some semiregular and star polyhedral numbers

In this section we consider some space figurate numbers corresponding to other three-dimensional polyhedra, in particular, to the *Archimedean solids* and *star solids.*

In fact, the **semiregular polyhedral numbers** correspond to the *semiregular polyhedra* (i.e., polyhedra with equivalent vertices and regular faces), including thirteen *Archimedean solids*, *prism* and *antiprism.*

Among all classes of such space figurate numbers we consider only three classes of **Archimedean solid numbers**, corresponding to the truncated regular polyhedra — the truncated tetrahedral, truncated cubic and truncated octahedral numbers, and discuss possibilities to construct truncated icosahedral and truncated dodecahedral numbers.

The **star polyhedral numbers** correspond to the *star polyhedra*: non-convex polyhedra which contain an arrangement of symmetrically (or nearly symmetrically) arranged spikes giving it the visual appearance of a three-dimensional star.

Among all classes of such space figurate numbers, we consider only **stella octangula numbers**, corresponding to the *stella octangula*, that is the only stellation of the octahedron.

2.5.1. The **truncated tetrahedral numbers** correspond to a truncation of four vertices of tetrahedron. The n-th truncated tetrahedral number $TS_3^3(n)$ is obtained from the tetrahedral number $S_3^3(3n-2)$ by dropping four tetrahedral numbers $S_3^3(n-1)$, and has the form

$$TS_3^3(n) = S_3^3(3n-2) - 4S_3^3(n-1).$$

In other words, it holds $TS_3^3(n) = \frac{(3n-2)(3n-1)3n-4(n-1)n(n+1)}{6} = \frac{n}{6}(27n^2-27n+6-4n^2+4) = \frac{n(23n^2-27n+10)}{6}$, i.e., we obtain the following general formula for n-th truncated tetrahedral number:

$$TS_3^3(n) = \frac{n(23n^2-27n+10)}{6}.$$

The first few truncated tetrahedral numbers are $1, 16, 68, 180, 375, 676, 1106, 1688, 2445, 3400, \ldots$ (Sloane's A005906).

Moreover, since $S_3^3(n+1) = S_3^3(n) + S_3(n+1)$, it follows $TS_3^3(n+1) = S_3^3(3n+1) - 4S_3^3(n) = (S_3^3(3n) + S_3(3n+1)) - 4S_3^3(n) = (S_3^3(3n-1) + S_3(3n) + S_3(3n+1)) - 4S_3^3(n) = (S_3^3(3n-2) + S_3(3n-1) + S_3(3n) + S_3(3n+1)) - 4(S_3^3(n-1) + S_3(n)) = (S_3^3(3n-2) - 4S_3^3(n-1)) + (S_3(3n-1) + S_3(3n) + S_3(3n+1) - 4S_3(n))$. So, the sequence of truncated tetrahedral numbers can be obtained by the following recurrent equation:

$$\begin{aligned} TS_3^3(n+1) = TS_3^3(n) &+ (S_3(3n-1) + S_3(3n) \\ &+ S_3(3n+1) - 4S_3(n)), \quad TS_3^3(1) = 1. \end{aligned}$$

Since

$$\begin{aligned} &S_3(3n-1) + S_3(3n) + S_3(3n+1) - 4S_3(n) \\ &\quad = \frac{(3n-1)3n + 3n(3n+1) + (3n+1)(3n+2) - 4n(n+1)}{2} \\ &\quad = \frac{23n^2+5n+2}{2}, \end{aligned}$$

the above recurrent equation gets the form

$$TS_3^3(n+1) = TS_3^3(n) + \frac{23n^2+5n+2}{2}, \quad TS_3^3(1) = 1.$$

The generating function of this sequence is $f(x) = \frac{x(10x^2+12x+1)}{(x-1)^4}$ (see [Weis11]), i.e., it holds

$$\frac{x(10x^2+12x+1)}{(x-1)^4} = TS_3^3(1)x + TS_3^3(2)x^2 + TS_3^3(3)x^3 + \cdots + TS_3^3(n)x^n + \cdots, \quad |x| < 1.$$

This result can be obtained by the standard procedure, leading to the linear recurrent equation $TS_3^3(n+4) - 4TS_3^3(n+3) + 6TS_3^3(n+2) - 4TS_3^3(n+1) + TS_3^3(n) = 0$ with initial conditions $TS_3^3(1) = 1$, $TS_3^3(2) = 16$, $TS_3^3(3) = 68$, and $TS_3^3(4) = 180$.

2.5.2. The **truncated cubic numbers** correspond to a truncation of eight vertices of cube. The n-th truncated cubic number $TC(n)$ is obtained from the cubic number $C(3n-2)$ by dropping eight tetrahedral numbers $S_3^3(n-1)$, and has the form

$$TC(n) = C(3n-2) - 8S_3^3(n-1).$$

In other words, it holds $TC(n) = (3n-2)^3 - 8\frac{(n-1)n(n+1)}{6} = \frac{n}{3}(77n^3 - 162n^2 + 112n - 24)$, i.e., we obtain the following general formula for n-th truncated cubic number:

$$TC(n) = \frac{n(77n^3 - 162n^2 + 112n - 24)}{3}.$$

The first few truncated cubic numbers are $1, 56, 311, 920, 2037, 3816, 6411, 9976, 14665, 20632, \ldots$ (Sloane's A005912).

It is easy to check that the sequence of truncated cubic numbers can be obtained by the following recurrent equation:

$$TC(n+1) = TC(n) + (77n^2 - 31n + 9), \quad TC(1) = 1.$$

The generating function of this sequence is $f(x) = \frac{x(1+52x+93x^2+8x^3)}{(x-1)^4}$, i.e., it holds

$$\frac{x(1+52x+93x^2+8x^3)}{(x-1)^4} = TC(1)x + TC(2)x^2 + TC(3)x^3 + \cdots + TC(n)x^n + \cdots, \quad |x| < 1.$$

This result can be obtained by the standard procedure, leading to the linear recurrent equation $TC(n+4) - 4TC(n+3) + 6TC(n+2) - 4TC(n+1) + TC(n) = 0$ with initial conditions $TC(1) = 1$, $TC(2) = 56$, $TC(3) = 311$, and $TC(4) = 920$.

2.5.3. A **truncated octahedral number** is obtained by truncation all six vertices of octahedron (see [Weis11]). More precisely, n-th truncated octahedral number $TO(n)$ is obtained from the octahedral number $O(3n-2)$ by dropping six square pyramidal numbers $S_4^3(n-1)$, i.e., has the form

$$TO(n) = O(3n-2) - 6S_4^3(n-1).$$

In other words, it holds $TO(n) = S_4^3(3n-2) + S_4^3(3n-3) - 6S_4^3(n-1) = \frac{(3n-2)(3n-1)(2(3n-2)+1)}{6} + \frac{(3n-3)(3n-2)(2(3n-3)+1)}{6} - 6\frac{(n-1)n(2(n-1)+1)}{6} = \frac{3n-2}{6}(2(3n-1)(3n-2) + (3n-1) + 2(3n-3)^2 + (3n-3)) - (n-1)n(2n-1) = \frac{3n-2}{6}(18n^2 - 36n + 18 + 3n - 3 + 18n^2 - 18n + 4 + 3n - 1) - (n^2-n)(2n-1) = \frac{3n-2}{6}(36n^2 - 48n + 18) - (2n^3 - 3n^2 + n) = (3n-2)(6n^2 - 8n + 3) - (2n^3 - 3n^2 + n) = 18n^3 - 12n^2 - 24n^2 + 16n + 9n - 6 - 2n^3 + 3n^2 - n = 16n^3 - 33n^2 + 24n - 6$, i.e., the general formula for the sequence of truncated octahedral numbers has the form

$$TO(n) = 16n^3 - 33n^2 + 24n - 6.$$

The first few truncated octahedral numbers are $1, 38, 201, 586, 1289, 2406, 4033, 6266, 9201, 12934, \ldots$ (Sloane's A005910).

The recurrent formulas $O(n+1) = O(n) + (n+1)^2 + n^2$ and $S_4^3(n+1) = S_4^3(n) + (n+1)^2$ yield the following recurrent formula for the sequence of the truncated octahedral numbers:

$$TO(n+1) = TO(n) + 48n^2 - 18n + 7, \quad TO(1) = 1.$$

In fact, it holds $TO(n+1) = O(3n+1) - 6S_4^3(n) = (O(3n) + (3n+1)^2 + (3n)^2) - 6S_4^3(n) = ((O(3n-1) + (3n)^2 + (3n-1)^2) + (3n+1)^2 + (3n)^2) - 6S_4^3(n) = (O(3n-1) + (3n+1)^2 + 2(3n)^2 + (3n-1)^2) - 6S_4^3(n) = ((O(3n-2) + (3n-1)^2 + (3n-2)^2) + (3n+1)^2 + 2(3n)^2 + (3n-1)^2) - 6S_4^3(n) = (O(3n-2) + (3n+1)^2 + 2(3n)^2 + 2(3n-1)^2 + (3n-2)^2) - 6(S_4^3(n-1) + n^2) = (O(3n-2) - 6S_4(n-1)) + ((3n+1)^2 + 2(3n)^2 + 2(3n-1)^2 + (3n-2)^2 - 6n^2) = TO(n) + (48n^2 - 18n + 7)$.

This recurrent formula implies, that the generating function for the sequence of truncated octahedral numbers is $f(x) = \frac{x(6x^3+55x^2+34x+1)}{(x-1)^4}$ (see [Weis11]), i.e., it holds

$$\frac{x(6x^3+55x^2+34x+1)}{(x-1)^4} = TO(1)x + TO(2)x^2 + TO(3)x^3 + \cdots + TO(n)x^n + \cdots, \quad |x| < 1.$$

This fact can be obtained following the standard procedure, leading to the linear recurrent equation $TO(n+4) - 4TO(n+3) + 6TO(n+2) - 4TO(n+1) + TO(n) = 0$ with initial values $TO(1) = 1$, $TO(2) = 38$, $TO(3) = 201$, and $TO(4) = 586$.

2.5.4. The **truncated icosahedral numbers** and **truncated dodecahedral numbers** can be defined on the same way by the formulas

$$TI(n) = I(3n-2) - 12S_5^3(n-1), \quad \text{and}$$
$$TD(n) = D(3n-2) - 20S_3^3(n-1),$$

but those formulas lost the practical sense, as in those cases we have no symmetrical properties. For icosahedron the corresponding 5-gonal pyramidal numbers are not symmetrical, since are constructed from non-symmetrical 5-gonal numbers. As for dodecahedron, its faces are 5-gonal numbers which are not symmetrical.

2.5.5. A **stella octangula number** is a space figurate number constructed as an octahedron with a square pyramid appended to each face. More precisely, n-th stella octangula number $SO(n)$ is obtained as the sum of n-th octahedral number $O(n)$ and eight copies of $(n-1)$-th tetrahedral number $S_3^3(n-1)$, i.e., has the form

$$SO(n) = O(n) + 8S_3^3(n-1).$$

In other words, it holds $SO(n) = S_4^3(n) + S_4^3(n-1) - 8S_3^3(n-1) = \frac{n(n+1)(2n+1)}{6} + \frac{(n-1)n(2n-1)}{6} + 8\frac{(n-1)n(n+1)}{6} = \frac{n((n+1)(2n+1)+(n-1)(2n-1))}{6} + \frac{4n(n^2-1)}{3} = \frac{n((2n^2+3n+1)+(2n^2-3n+1))}{6} + \frac{n(4n^2-4)}{3} = \frac{n(4n^2+2)}{6} + \frac{n(4n^2-4)}{3} = \frac{n(2n^2+1)}{3} + \frac{n(4n^2-4)}{3} = \frac{n((2n^2-1)+(4n^2-4))}{3} = \frac{n(6n^2-3)}{3} = n(2n^2-1)$, i.e., we have the following general formula for n-th stella octangula number:

$$SO(n) = n(2n^2 - 1).$$

The first few numbers of this sequence are $1, 14, 51, 124, 245, 426, 679, 1016, 1449, 1990, \ldots$ (Sloane's A007588).

The recurrent formulas $O(n+1) = O(n) + (n+1)^2 + n^2$ and $S_3^3(n+1) = S_3^3(n) + \frac{(n+1)(n+2)}{2}$ yield the following recurrent formula for the sequence of stella octangula numbers:

$$SO(n+1) = SO(n) + 6n^2 + 6n + 1, \quad SO(1) = 1.$$

In fact, it holds $SO(n+1) = O(n+1) + 8S_3^3(n) = (O(n) + (n+1)^2 + n^2) + 8(S_3^3(n-1) + \frac{n(n+1)}{2}) = (O(n) + 8S_3^3(n-1)) + (2n^2 + 2n + 1 + 4n(n+1)) = SO(n) + (6n^2 + 6n + 1)$.

This recurrent formula implies that the generating function for the sequence of the stella octangula numbers is $f(x) = \frac{x(x^2+10x+1)}{(x-1)^4}$ (see [Weis11]), i.e., it holds

$$\frac{x(x^2+10x+1)}{(x-1)^4} = SO(1)x + SO(2)x^2 + SO(3)x^3 + \cdots + SO(n)x^n + \cdots, \quad |x| < 1.$$

This result can be obtained by the standard procedure, leading to the linear recurrent equation $SO(n+4) - 4SO(n+3) + 6SO(n+2) - 4SO(n+1) + SO(n) = 0$ with initial values $SO(1) = 1$, $SO(2) = 14$, $SO(3) = 51$, and $SO(4) = 124$.

The only known **square stella octangula numbers** are 1 and 9653449, which correspond to $SO(u) = S_4(v)$ with $(u, v) = (1, 1)$ and $(169, 3107)$ (see [Weis11]).

2.6 Centered space figurate numbers

In this section we consider space figurate numbers, formed by a central dot, surrounded by successive polyhedral layers, first of all the **centered regular polyhedron numbers**, correspond to the regular polyhedra: **centered tetrahedron numbers**, **centered cube**

numbers, centered octahedron numbers, centered icosahedron numbers, centered dodecahedron numbers. Are considered also general **centered m-gonal pyramid numbers**, and **rhombic dodecahedral numbers.**

2.6.1. A **centered cube number** (or **body-centered cube number**) is a centered space figurate number that represents a "centered cube". Any centered cube number is formed by a central dot, surrounded by successive cubic layers. At n-th step the previous figure is surrounded by a layer of new points, forming a cube, such that each its face contains exactly n^2 points. In other words, n-th layer can be considered as n-th cubic number n^3 without interior, which corresponds to the $(n-2)$-th cubic number $(n-2)^3$.

So, starting from 1, we obtain, on the 2-nd level, the cube, formed by 8 points. It corresponds to the second cubic number $C(2) = 2^3$ without empty interior. The 3-rd level is the cube, formed by 28 points. It corresponds to 3-rd cubic number $C(3) = 3^3$ without 1-point interior: $28 = C(3) - C(1)$. The 4-th level's cube consists from 56 points and corresponds to 4-th cubic number $C(4) = 5^3$ without 8-points interior: $56 = C(4) - C(2)$, etc.

Therefore, centered cube numbers are constructed as consecutive sums of the elements of the series $1^3, 2^3 - 0^3, 3^3 - 1^3, 4^3 - 2^3, \dots, n^3 - (n-2)^3, \dots$. In particular, n-th centered cubic number $\bar{C}(n)$ has the form

$$\bar{C}(n) = 1^3 + (2^3 - 0^3) + (3^3 - 1^3) + (4^3 - 2^3) + \cdots + (n^3 - (n-2)^3).$$

It implies the recurrent formula $\bar{C}(n+1) = \bar{C}(n) + ((n+1)^3 - (n-1)^3)$, $\bar{C}(1) = 1$. Since $(n+1)^3 - (n-1)^3 = 6n^2 + 2$, we obtain the following recurrent equation for the sequence of centered cube numbers:

$$\bar{C}(n+1) = \bar{C}(n) + (6n^2 + 2), \quad \bar{C}(1) = 1.$$

Easy to see, that $1^3 + (2^3 - 0^3) + (3^3 - 1^3) + (4^3 - 2^3) + \cdots + (n^3 - (n-2)^3) = n^3 + (n-1)^3$. So, we obtain, that *any centered cube number is a sum of two consecutive cubic numbers*:

$$\bar{C}(n) = C(n) + C(n-1).$$

Therefore, it holds $\bar{C}(n) = n^3 + (n-1)^3 = n^3 + (n^3 - 3n^2 + 3n - 1) = (2n^3 - n^2) - (2n^2 + n) + (2n - 1) = (2n-1)n^2 - (2n-1)n + (2n-1) = (2n-1)(n^2 - n + 1)$, i.e., the general formula for n-th centered cube number has the form

$$\bar{C}(n) = (2n-1)(n^2 - n + 1).$$

The first few centered cube numbers are $1, 9, 35, 91, 189, 341, 559, 855, 1241, 1729, \ldots$ (Sloane's A005898).

Clearly, we can get the above sequence of centered cube numbers by counting of partial sums of the sequence $1, 8, 26, 56, 98, 152, 218, 296, 386, 488, \ldots$ (Sloane's A005897), giving the number of points on surface of cube. The first element of this sequence is 1, while n-th element, $n \geq 2$, has the form $6(n-1)^2 + 2 = 6n^2 - 12n + 8$.

In fact, it can be obtained by the formula $v + s(n-2) + f \cdot int(S_4(n))$, where $v = 8$, $s = 12$ and $f = 6$ are the numbers of vertices, sides, and faces of cube, while $int(S_4(n))$ is a number of points in the interior of n-th square number. So, for $n \geq 2$ it holds $v + s(n-2) + f \cdot int(S_4(n)) = 8 + 12(n-2) + 6(n^2 - 4(n-1)) = 6n^2 - 12n + 8 = 6(n-1)^2 + 2$.

By this reason, the above recurrent formula for centered cube numbers becomes obvious, while the general formula for $\bar{C}(n)$ can be obtained as the result of the following summation: $\bar{C}(n) = 1 + \sum_{i=1}^{n-1}(6i^2 + 2) = 1 + 6\sum_{i=1}^{n-1} i^2 + 2(n-1) = 6 \cdot \frac{(n-1)n(2n-1)}{6} + (2n-1) = (2n-1)(n^2 - n + 1)$.

The generating function for the centered cube numbers is $f(x) = \frac{x(x^3+5x^2+5x+1)}{(x-1)^4} = \frac{x(1+x)(1+4x+x^2)}{(1-x)^4}$ (see [Weis11]), i.e., it holds

$$\frac{x(1+x)(1+4x+x^2)}{(1-x)^4} = \bar{C}(1)x + \bar{C}(2)x^2 + \bar{C}(3)x^3 + \cdots + \bar{C}(n)x^n + \cdots, \quad |x| < 1.$$

In fact, for the sequence of cubic numbers one has $\frac{x(x^2+4x+1)}{(x-1)^4} = C(1)x + C(2)x^2 + \cdots + C(n)x^n + \cdots$, $|x| < 1$. So, it holds $\frac{x^2(x^2+4x+1)}{(x-1)^4} = C(1)x^2 + C(2)x^3 + C(3)x^4 + \cdots + C(n-1)x^n + \cdots$, $|x| < 1$, and $\frac{x(x^2+4x+1)}{(x-1)^4} + \frac{x^2(x^2+4x+1)}{(x-1)^4} = C(1)x + (C(1) + C(2))x^2 +$

$(C(2)+C(3))x^3+\cdots+(C(n-1)+C(n))x^n+\cdots, \quad |x|<1$. Since $C(n-1)+C(n)=\bar{C}(n)$, the last equation implies

$$\frac{x(x^3+5x^2+5x+1)}{(x-1)^4}=\bar{C}(1)x+\bar{C}(2)x^2+\bar{C}(3)x^3 +\cdots+\bar{C}(n)x^n+\cdots, \quad |x|<1.$$

Of course, one can obtain this result, following the standard procedure, leading to the linear recurrent equation $\bar{C}(n+4)-4\bar{C}(n+3)+6\bar{C}(n+2)-4\bar{C}(n+1)+\bar{C}(n)=0$ with initial values $\bar{C}(1)=1$, $\bar{C}(2)=9$, $\bar{C}(3)=35$, and $\bar{C}(4)=91$.

2.6.2. A **rhombic dodecahedral number** is a space figurate number which is constructed as a centered cube with a square pyramid appended to each face. More precisely, n-th rhombic dodecahedral number $RD(n)$ is obtained from n-th centered cube number $\bar{C}(n)$ by adding six square pyramidal numbers $S_4^3(n-1)$, and has the following form:

$$RD(n)=\bar{C}(n)+6S_4^3(n-1).$$

In other words, it holds $RD(n)=(2n-1)(n^2-n+1)+6\cdot\frac{(n-1)n(2n-1)}{6}=(2n-1)(n^2-n+1)+(2n-1)(n^2-n)=(2n-1)(2n^2-2n+1)$, i.e., the general formula for n-th rhombic dodecahedral number has the form

$$RD(n)=(2n-1)(2n^2-2n+1)=4n^3-6n^2+4n-1.$$

The first few rhombic dodecahedral numbers are $1, 15, 65, 175, 369, 671, 1105, 1695, 2465, 3439, \ldots$ (Sloane's A005917).

The recurrent formula for the sequence of the rhombic dodecahedral numbers has the form

$$RD(n+1)=RD(n)+12n^2+2, \quad RD(1)=1.$$

In fact, we have $RD(n+1)=4(n+1)^3-6(n+1)^2+4(n+1)-1=(4n^4-6n^3+4n-1)+(4(3n^2+3n+1)-6(2n+1)+4)=RD(n)+(12n^2+2)$.

Using this recurrent equation, it is easy to check that the generating function for the sequence of the rhombic dodecahedral numbers is $f(x) = \frac{x(x+1)(x^2+10x+1)}{(x-1)^4} = \frac{x(1+11x+11x^2+x^3)}{(1-x)^4}$ (see [Weis11]), i.e., it holds

$$\frac{x(x+1)(x^2+10x+1)}{(x-1)^4} = RD(1)x + RD(2)x^2 + RD(3)x^3 + \cdots + RD(n)x^n + \cdots, \quad |x| < 1.$$

This result can be obtained by the standard procedure, leading to the linear recurrent equation $RD(n+4) - 4RD(n+3) + 6RD(n+2) - 4RD(n+1) + RD(n) = 0$ with initial values $RD(1) = 1$, $RD(2) = 15$, $RD(3) = 65$, and $RD(4) = 175$.

It is easy to show also that *any rhombic dodecahedral number is the difference of two consecutive bi-squares*:

$$RD(n) = n^4 - (n-1)^4.$$

In fact, the formula $RD(n) = 4n^3 - 6n^2 + 4n - 1$ can be seen as $RD(n) = n^4 - (n^4 - 4n^3 + 6n^2 - 4n + 1) = n^4 - (n-1)^4$.

This property implies that the rhombic dodecahedral numbers are gnomons of bi-squares. Therefore, the rhombic dodecahedral numbers are three-dimensional analogue of the hex numbers, which are gnomons of perfect cubes. [3]

Viewed from the opposite perspective, above property yields that *the sum of the first n rhombic dodecahedral numbers is a bi-square*:

$$RD(1) + RD(2) + \cdots + RD(n) = n^4.$$

It follows immediately from the equation $RD(n) = n^4 - (n-1)^4$, using a telescoping sum: $RD(1) + RD(2) + \cdots + RD(n) = (1^4 - 0^4) + (2^4 - 1^4) + (3^4 - 2^4) + \cdots + (n^4 - (n-1)^4) = n^4 - 0^4 = n^4$.

2.6.3. A related set of numbers consists of so-called **Haüy rhombic dodecahedral numbers**. They give the number of cubes in the *Haüy construction* of the rhombic dodecahedron.

[3] In fact, the hex numbers count the number of cells of honeycomb packing of the plane that are less than n ($n = 0, 1, 2, 3...$) steps away from a central one. Now, among the Platonic and Archimedean solids, the only ones that partition $\mathbb{R}^3$ 3-periodically are the cube and the truncated octahedron. In the packing by truncated octahedra, the nexus of all cells less than n steps away from a given cell has the shape of a rhombic dodecahedron.

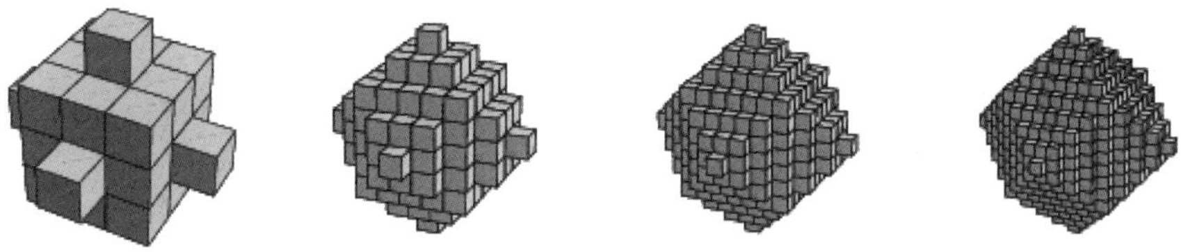

The n-th Haüy rhombic dodecahedral number $RD_H(n)$ is formed from a cube with odd side's length $2n-1$, and six copies of a square pyramid, constructed from consecutive odd square numbers $1^2, 3^2, \ldots, (2n-3)^2$. Therefore, one obtains

$$RD_H(n) = (2n-1)^3 + 6(1^2 + 3^2 + 5^2 + \cdots + (2n-3)^2).$$

In other words, we get the following general formula for n-th Haüy rhombic dodecahedral number:

$$RD_H(n) = (2n-1)(8n^2 - 14n + 7).$$

In fact, it holds $(2n-1)^3+6\sum_{k=1}^{n-1}(2k-1)^2 = (2n-1)^3+24\sum_{k=1}^{n-1}k^2 - 24\sum_{k=1}^{n-1}k + 6\sum_{k=1}^{n-1}1 = (2n-1)^3 + 4n(n-1)(2n-1) - 12n(n-1) + 6(n-1) = (2n-1)((2n-1)^2 + 4n(n-1)) - 6(n-1)(2n-1) = (2n-1)((4n^2-4n+1+4n^2+4n) - (6n-6)) = (2n-1)(8n^2-14n+7)$.

The first few values of the Haüy rhombic dodecahedral numbers are 1, 33, 185, 553, 1233, 2321, 3913, 6105, 8993, 12673, ... (Sloane's A046142).

It holds the following recurrent equation for the sequence of the Haüy rhombic dodecahedral numbers:

$$RD_H(n+1) = RD_H(n) + (48n^2 - 24n + 8), \quad RD_H(1) = 1.$$

In fact, we have $RD_H(n+1) = (2(n+1)-1)((8(n+1)^2 - 14(n+1)+7) = ((2n-1)+2)((8n^2-14n+7)+16n-6) = (2n-1)(8n^2-4n+7) + (2n+1)(16n-6) + 2(8n^2-14n+7) = RD_H(n) + (32n^2 + 16n - 12n - 6 + 16n^2 - 28n + 14) = RD_H(n) + (48n^2 - 24n + 8)$.

Using this recurrent equation, it is easy to check that the generating function for the sequence of Haüy rhombic dodecahedral numbers is $f(x) = \frac{x(1+29x+59x^2-13x^3)}{(x-1)^4}$:

$$\frac{x(1+29x+59x^2-13x^3)}{(x-1)^4} = RD_H(1)x + RD_H(2)x^2 + RD_H(3)x^3 + \cdots + RD_H(n)x^n + \cdots, \quad |x| < 1.$$

This result can be obtained by the standard procedure, leading to the linear recurrent equation $RD_H(n+4) - 4RD_H(n+3) + 6RD_H(n+2) - 4RD_H(n+1) + RD_H(n) = 0$ with initial values $RD_H(1) = 1$, $RD_H(2) = 33$, $RD_H(3) = 185$, and $RD_H(4) = 553$.

2.6.4. The **centered pyramid numbers** are formed by a central dot surrounded by pyramidal layers. Each layer can be considered as corresponding pyramidal number "without interior".

So, **centered tetrahedron numbers** (or, sometimes, **centered tetrahedral numbers**) correspond to centered tetrahedron. Starting from one point, we obtain on the 2-nd level the tetrahedron, formed from 4 points. It corresponds to 2-nd tetrahedral number $S_3^3(2)=4$ without empty interior. The 3-rd level is the tetrahedron, formed from 10 points. It corresponds to 3-rd tetrahedral number $S_3^3(3)=10$ without empty interior. Similarly, the 4-th level corresponds to 4-th tetrahedral number $S_3^3(4)=20$ without empty interior. The 5-th level's tetrahedron is formed from 34 points and corresponds to 5-th tetrahedral number $S_3^3(5)=35$ without 1-point interior: $34=S_3^3(5)-S_3^3(1)$. The 52 points of the 6-th level correspond to the number $S_3^3(6) = 56$ without 4-points interior: $52=S_3^3(6) - S_3^3(2)$, etc.

In general, the centered tetrahedron numbers are consecutive sums of the elements of the series

$$S_3^3(1), S_3^3(2), S_3^3(3), S_3^3(4), S_3^3(5) - S_3^3(1), S_3^3(6) - S_3^3(2), \ldots, S_3^3(n+4) - S_3^3(n), \ldots.$$

So, starting from $n = 4$, n-th centered tetrahedral number $\overline{S}_3^3(n)$ can be obtained by the following formula:

$$\overline{S}_3^3(n) = S_3^3(1) + S_3^3(2) + S_3^3(3) + S_3^3(4) + (S_3^3(5) - S_3^3(1)) + (S_3^3(6) - S_3^3(2)) + \cdots + (S_3^3(n+4) - S_3^3(n)).$$

Since the above telescopic sum is equal to $S_3^3(n)+S_3^3(n-1)+S_3^3(n-2)+S_3^3(n-3)$, we get

$$\overline{S}_3^3(1)=S_3^3(1)=1,\quad \overline{S}_3^3(2)=S_3^3(2)+S_3^3(1)=5,$$

$$\overline{S}_3^3(3)=S_3^3(3)+S_3^3(2)+S_3^3(1)=15,$$

$$\overline{S}_3^3(n)=S_3^3(n)+S_3^3(n-1)+S_3^3(n-2)+S_3^3(n-3),\quad n\geq 4.$$

So, the first few elements of the sequence of centered tetrahedral numbers are 1, 5, 15, 35, 69, 121, 195, 295, 425, 589,... (Sloane's A005894).

Since $S_3^3(n)=S_3^3(n-1)+S_3(n)$, we get, for $n\geq 4$, the recurrent formula $\overline{S}_3^3(n)=\overline{S}_3^3(n-1)+(S_3(n)+S_3(n-1)+S_3(n-2)+S_3(n-3))$. In other words, for $n\geq 4$, it holds $\overline{S}_3^3(n)=\overline{S}_3^3(n-1)+(2n^2-4n+4)$. Moreover, it is not hard to check that these formula works well and for $n=2,3$. In fact, $\overline{S}_3^3(1)+(2\cdot 2^2-4\cdot 2+4)=1+4=5=\overline{S}_3^3(2)$, and $\overline{S}_3^3(2)+(2\cdot 3^2-4\cdot 3+4)=5+10=15=\overline{S}_3^3(3)$. So, noting, that $2(n+1)^2-4(n+1)+4=2n^2+2$, we get the following general recurrent formula for the sequence of centered tetrahedron numbers:

$$\overline{S}_3^3(n+1)=\overline{S}_3^3(n)+(2n^2+2),\quad \overline{S}_3^3(1)=1.$$

In addition, it is easy to obtain for n-th centered tetrahedral number $\overline{S}_3^3(n)$ the general formula of the form

$$\overline{S}_3^3(n)=\frac{(2n-1)(n^2-n+3)}{3}=\frac{2n^3-3n^2+7n-3}{3}.$$

In fact, for $n\geq 4$ it holds $S_3^3(n)+S_3^3(n-1)+S_3^3(n-2)+S_3^3(n-3)=\frac{n(n+1)(n+2)}{6}+\frac{(n-1)n(n+1)}{6}+\frac{(n-2)(n-1)n}{6}+\frac{(n-3)(n-2)(n-1)}{6}=\frac{n(n+1)(2n+1)}{6}+\frac{(n-2)(n-1)(2n-3)}{6}=\frac{(n^2+n)(2n+1)}{6}+\frac{(n^2-3n+2)(2n-3)}{6}=\frac{(2n^3+2n^2+n^2+n)+(2n^3-6n^2+4n-3n^2+9n-6)}{6}=\frac{4n^3-6n^2+14n-6}{6}=\frac{2n^3-3n^2+7n-3}{3}=\frac{(2^3-n^2)-(2n^2-n)+(6n-3)}{3}=\frac{(2n-1)(n^2-n+3)}{3}$. For $n=1,2,3$ one can check it by direct computation: $\frac{(2\cdot 1-1)(1^2-1+3)}{3}=1=\overline{S}_3^3(1)$, $\frac{(2\cdot 2-1)(2^2-2+3)}{3}=5=\overline{S}_3^3(2)$, and $\frac{(2\cdot 3-1)(3^2-3+3)}{3}=35=\overline{S}_3^3(3)$.

Obviously, we can get the above sequence of centered tetrahedron numbers by counting of partial sums of the sequence 1, 4, 10, 20, 34,

52, 74, 100, 130, 164, ... (Sloane's A005893), giving the number of points on surface of tetrahedron. The first element of this sequence is 1, while n-th element, $n \geq 2$, has the form $2(n-1)^2 + 2 = 2n^2 - 4n + 4$.

It can be obtained by the formula $v + s(n-2) + f \cdot int(S_3(n))$, where $v = 4$, $s = 6$ and $f = 4$ are the numbers of vertices, sides, and faces of tetrahedron, while $int(S_3(n))$ is a number of points in the interior of n-th triangular number. So, for $n \geq 2$ it holds $v + s(n-2) + f \cdot int(S_3(n)) = 4 + 6(n-2) + 4(\frac{n(n+1)}{2} - 3(n-1)) = 6n - 8 + 2(n^2 - 5n + 6) = 2n^2 - 4n + 4 = 2(n-1)^2 + 2$.

By this reason, the above recurrent formula for centered tetrahedron numbers becomes obvious, while the general formula for $\overline{S}_3^3(n)$ can be obtained as the result of the following summation: $\overline{S}_3^3(n) = 1 + \sum_{i=1}^{n-1}(2i^2 + 2) = 1 + 2\sum_{i=1}^{n-1} i^2 + 2(n-1) = 2 \cdot \frac{(n-1)n(2n-1)}{6} + (2n-1) = \frac{(2n-1)(n^2-n+1)}{3}$.

The generating function for the sequence of centered tetrahedron numbers is $f(x) = \frac{x(1+x)(1+x^2)}{(x-1)^4} = \frac{x(1+x+x^2+x^3)}{(1-x)^4} = \frac{x(1-x^4)}{(1-x)^5}$ (see [Sloa11]), i.e., it holds

$$\frac{x(1+x)(1+x^2)}{(x-1)^4} = \overline{S}_3^3(1)x + \overline{S}_3^3(2)x^2 + \overline{S}_3^3(3)x^3 + \cdots + \overline{S}_3^3(n)x^4 + \cdots, \quad |x| < 1.$$

This result can be obtained by the standard procedure, leading to the linear recurrent equation $\overline{S}_3^3(n+4) - 4\overline{S}_3^3(n+3) + 6\overline{S}_3^3(n+2) - 4\overline{S}_3^3(n+1) + \overline{S}_3^3(n) = 0$ with initial values $\overline{S}_3^3(1) = 1$, $\overline{S}_3^3(2) = 5$, $\overline{S}_3^3(3) = 15$, and $\overline{S}_3^3(4) = 35$.

A **centered square pyramid number** (or, simply, **centered pyramid number**) corresponds to a centered square pyramid.

Starting from one point, we obtain on the 2-nd level the square pyramid, formed from 5 points. It corresponds to 2-nd square pyramidal number $S_4^3(2) = 5$ without empty interior. The 14 points of the 3-rd level correspond to the number $S_4^3(3) = 14$ without empty interior. The 29 points of the 4-th level correspond to the number $S_4^3(4) = 30$ without 1-point interior: $29 = S_4^3(4) - S_4^3(1)$. The 50 points of the 5-th level correspond to the number $S_4^3(5) = 55$ without 5-points interior: $50 = S_4^3(5) - S_4^3(2)$, etc.

In general, the centered pyramid numbers are consecutive sums of the elements of the series

$$S_4^3(1), S_4^3(2), S_4^3(3), S_3^3(4) - S_4^3(1), S_4^3(5) - S_4^3(2),$$
$$S_4^3(6) - S_4^3(3), \ldots, S_4^3(n+3) - S_4^3(n), \ldots .$$

So, n-th centered pyramid number $\overline{S}_4^3(n), n \geq 3$, has the form

$$\overline{S}_4^3(n) = S_4^3(1) + S_4^3(2) + S_4^3(3) + (S_3^3(4) - S_4^3(1)) + (S_4^3(5) - S_4^3(2)) + \cdots + (S_4^3(n+3) - S_4^3(n)).$$

Since the above telescopic sum is equal to $S_4^3(n)+S_4^3(n-1)+S_4^3(n-2)$, we get

$$\overline{S}_4^3(1) = S_4^3(1) = 1, \quad \overline{S}_4^3(2) = S_4^3(2) + S_4^3(1) = 6,$$
$$\overline{S}_4^3(n) = S_4^3(n) + S_4^3(n-1) + S_4^3(n-2), \quad n \geq 3.$$

Therefore, the first few elements of the sequence of centered square pyramid numbers are 1, 6, 20, 49, 99, 176, 286, 435, 629, 874,... (Sloane's A063488).

Since $S_4^3(n) = S_4^3(n-1) + S_4(n)$, we get, for $n \geq 3$, the recurrent formula $\overline{S}_4^3(n) = \overline{S}_4^3(n-1) + (n^2 + (n-1)^2 + (n-2)^2)$, $\overline{S}_4^3(2) = 6$. Moreover, it is holds also for $n = 2$: $\overline{S}_4^3(1) + (2^2 + 1^2 + 0^2) = 1 + 5 = 6 = \overline{S}_4^3(2)$. Therefore, noting, that $(n+1)^2 + n^2 + (n-1)^2 = 3n^2 + 2$, we get the following recurrent equation for the sequence of centered square pyramid numbers:

$$\overline{S}_4^3(n+1) = \overline{S}_4^3(n) + (3n^2 + 2), \quad \overline{S}_4^3(1) = 1.$$

In addition, it is easy to obtain for n-th centered square pyramid number $\overline{S}_4^3(n)$ the general formula

$$\overline{S}_4^3(n) = \frac{(2n-1)(n^2-n+2)}{2} = \frac{2n^3 - 3n^2 + 5n - 2}{2}.$$

In fact, for $n \geq 3$ it holds

$$S_4^3(n) + S_4^3(n-1) + S_4^3(n-2)$$
$$= \frac{n(n+1)(2n+1)}{6} + \frac{(n-1)n(2n-1)}{6} + \frac{(n-2)(n-1)(2n-3)}{6}$$
$$= \frac{(n^2+n)(2n+1) + (n^2-n)(2n-1) + (n^2-3n+2)(2n-3)}{6}$$

$$= \frac{(2n^3+3n^2+n)+(2n^3-3n^2+n)+(2n^3-9n^2+13n-6)}{6}$$

$$= \frac{6n^3-9n^2+15n-6}{6} = \frac{2n^3-3n^2+5n-2}{2}$$

$$= \frac{(2^3-n^2)-(2n^2-n)+(4n-2)}{2} = \frac{(2n-1)(n^2-n+2)}{2}.$$

For $n = 1, 2$ it can be checked by direct computation: $\frac{(2\cdot1-1)(1^2-1+2)}{2} = 1 = \overline{S}_4^3(1)$, and $\frac{(2\cdot2-1)(2^2-2+2)}{2} = 6 = \overline{S}_4^3(2)$.

Clearly, we can get the above sequence of centered square pyramid numbers by counting of partial sums of the sequence 1, 5, 14, 29, 50, 77, 110, 149, 194, 245, $\ldots$, (Sloane's A005918), giving the number of points on surface of square pyramid. The first element of this sequence is 1, while n-th element, $n \geq 2$, has the form $3(n-1)^2+2 = 3n^2-6n+5$.

It can be obtained by the formula $v + s(n-2) + f_\triangle \cdot int(S_3(n)) + int(S_4(n))$, where $v = 5$, $s = 8$ and $f_\triangle = 4$ are the numbers of vertices, sides, and triangular faces of square pyramid, while $int(S_m(n))$ is a number of points in the interior of n-th m-gonal number, $m = 3, 4$. So, for $n \geq 2$ it holds $v + s(n-2) + f_\triangle \cdot int(S_3(n)) + int(S_4(n)) = 5 + 8(n-2) + 4(\frac{n(n+1)}{2} - 3(n-1)) + (n^2 - 4(n-1)) = 8n - 11 + 2(n^2-5n+6) + (n^2-4n+4) = 3n^2-6n+5 = 3(n-1)^2+2$.

By this reason, the above recurrent formula for centered square pyramid numbers becomes obvious, while the general formula for $\overline{S}_3^3(n)$ can be obtained as the result of the following summation: $\overline{S}_4^3(n) = 1 + \sum_{i=1}^{n-1}(3i^2+2) = 1 + 3\sum_{i=1}^{n-1} i^2 + 2(n-1) = 3 \cdot \frac{(n-1)n(2n-1)}{6} + (2n-1) = \frac{(2n-1)(n^2-n+2)}{2}$.

The generating function for the sequence $\overline{S}_4^3(1), \overline{S}_4^3(2), \overline{S}_4^3(3)x^3 \ldots$ of centered pyramid numbers has the form $f(x) = \frac{x(1+x)(1+x+x^2)}{(x-1)^4} = \frac{x(1+2x+2x^2+x^3)}{(1-x)^4}$, i.e., it holds

$$\frac{x(1+x)(1+x+x^2)}{(x-1)^4} = \overline{S}_4^3(1)x + \overline{S}_4^3(2)x^2 + \overline{S}_4^3(3)x^3 + \cdots + \overline{S}_4^3(n)x^4 + \cdots, \quad |x| < 1.$$

This result can be obtained by the standard procedure, leading to the linear recurrent equation $\overline{S}_4^3(n+4) - 4\overline{S}_4^3(n+3) + 6\overline{S}_4^3(n+2) - 4\overline{S}_4^3(n+1) + \overline{S}_4^3(n) = 0$ with initial values $\overline{S}_4^3(1) = 1$, $\overline{S}_4^3(2) = 6$, $\overline{S}_4^3(3) = 20$, and $\overline{S}_4^3(4) = 49$.

In general, one can construct by the same procedure **centered m-gonal pyramid numbers** for any $m \geq 3$.

More precisely, the sequence $\overline{S}_m^3(1), \overline{S}_m^3(2), \ldots, \overline{S}_m^3(n), \ldots$ of centered m-gonal pyramid numbers can be obtained by counting of partial sums of the sequence 1, $m+1$, $4m-2$, $9m-7$, $16m-14, \ldots$, giving the number of points on surface of m-gonal pyramid. The first element of the last sequence is 1, while n-th element, $n \geq 2$, has the form $(m-1)(n-1)^2 + 2 = (m-1)n^2 - 2(m-1)n + (m+1)$.

It can be obtained by the formula $v + s(n-2) + f_\triangle \cdot int(S_3(n)) + int(S_m(n))$, where $v = m+1$, $s = 2m$ and $f_\triangle = m$ are the numbers of vertices, sides, and triangular faces of m-gonal pyramid, while $int(S_m(n))$ is a number of points in the interior of n-th m-gonal number. So, for $n \geq 2$ it holds $v + s(n-2) + f_\triangle \cdot int(S_3(n)) + int(S_m(n)) = (m+1) + 2m(n-2) + m(\frac{n(n+1)}{2} - 3(n-1)) + (\frac{m-2}{2}(n^2-n) + n - m(n-1)) = (m-1)n^2 - 2(m-1)n + (m+1) = (m-1)(n-1)^2 + 2$.

So, the recurrent formula for the sequence of centered m-gonal pyramid numbers has the form

$$\overline{S}_m^3(n+1) = \overline{S}_m^3(n) + (m-1)n^2 + 2, \quad \overline{S}_m^3(1) = 1.$$

The general formula for n-th centered m-gonal pyramid number $\overline{S}_m^3(n)$ has the form

$$\overline{S}_m^3(n) = \frac{(2n-1)(m-1)n^2 - (m-1)n + 6}{6}.$$

It can be obtained as the result of the following summation: $\overline{S}_m^3(n) = 1 + \sum_{i=1}^{n-1}((m-1)i^2 + 2) = 1 + (m-1)\sum_{i=1}^{n-1} i^2 + 2(n-1) = (m-1) \cdot \frac{(n-1)n(2n-1)}{6} + (2n-1) = \frac{(2n-1)((m-1)n^2-(m-1)n+6)}{6}$.

The generating function for this sequence is $f(x) = \frac{x(1+(m-2)x+(m-2)x^2+x^3)}{(1-x)^4} = \frac{x(1+x)(1+(m-3)x+x^2)}{(1-x)^4}$, i.e., it holds

$$\frac{x(1+x)(1+(m-3)x+x^2)}{(1-x)^4} = \overline{S}_m^3(1)x + \overline{S}_m^3(2)x^2 + \overline{S}_m^3(3)x^3 + \cdots + \overline{S}_m^3(n)x^n + \cdots, \quad |x| < 1.$$

This result can be obtained by the standard procedure, leading to the linear recurrent equation $\overline{S}^3_m(n+4) - 4\overline{S}^3_m(n+3) + 6\overline{S}^3_m(n+2) - 4\overline{S}^3_m(n+1) + \overline{S}^3_m(n) = 0$ with initial values $\overline{S}^3_m(1) = 1$, $\overline{S}^3_m(2) = m+1$, $\overline{S}^3_m(3) = 4m - 2$, and $\overline{S}^3_m(4) = 9m - 7$.

In particular, the **centered pentagonal pyramid numbers** form the sequence 1, 7, 25, 63, 129, 231, 377, 575, 833, ... (Sloane's A001845), following the formula

$$\overline{S}^3_5(n) = \frac{(2n-1)(2n^2-2n+3)}{3}.$$

They are numerically equal to Haüy octahedral numbers, considered before, and to the centered octahedral numbers, that will be considered below.

The **centered hexagonal pyramid numbers** form the sequence 1, 8, 30, 77, 159, 286, 468, 715, 1037, 1444, ... (Sloane's A063489), following the formula

$$\overline{S}^3_6(n) = \frac{(2n-1)(5n^2-5n+6)}{6}.$$

The **centered heptagonal pyramid numbers** form the sequence 1, 9, 35, 91, 189, 341, 559, 855, 1241, 1729, ... (Sloane's A005898), following the formula

$$\overline{S}^3_7(n) = (2n-1)(n^2-n+1) = 2n^3 + 9n^2 + 15n + 9.$$

They are numerically equal to centered cube numbers.

The **centered octagonal pyramid numbers** form the sequence 1, 10, 40, 105, 219, 396, 650, 995, 1445, 2014, ... (Sloane's A063490), following the formula

$$\overline{S}^3_8(n) = \frac{(2n-1)(7n^2-7n+6)}{6}.$$

2.6.5. The **centered octahedron numbers** represent "centered octahedron" and are formed by a central dot surrounded by octahedral layers.

The first centered octahedron numbers are 1, 7, 25, 63, 129, 231, 377, 575, 833, 1159, ... (Sloane's A001845).

They can be obtained as partial sums of the sequence 1, 6, 18, 38, 66, 102, 146, 198, 258, 326, ... (Sloane's A005899), giving number of points on surface of octahedron. The first element of the sequence

A005899 is 1, while n-th element, $n \geq 2$, has the form $4(n-1)^2+2 = 4n^2-8n+6$.

It can be obtained by the formula $v + s(n-2) + f \cdot int(S_3(n))$, where $v = 6$, $s = 12$ and $f = 8$ are the numbers of vertices, sides, and faces of octahedron, while $int(S_3(n))$ is a number of points in the interior of n-th triangular number. So, for $n \geq 2$ it holds $v + s(n-2) + f \cdot int(S_3(n)) = 6 + 12(n-2) + 8(\frac{n(n+1)}{2} - 3(n-1)) = 12n - 18 + 4(n^2 - 5n + 6) = 4n^2 - 8n + 6 = 4(n-1)^2 + 2$.

The recurrent formula for the sequence of centered octahedron numbers has the form

$$\overline{O}(n+1) = \overline{O}(n) + 4n^2 + 2, \quad \overline{O}(1) = 1.$$

The general formula for n-th centered octahedron number $\overline{O}(n)$ has the form

$$\overline{O}(n) = \frac{(2n-1)(2n^2-2n+3)}{3}.$$

It can be obtained as the result of the following summation: $\overline{O}(n) = 1 + \sum_{i=1}^{n-1}(4i^2+2) = 1 + 4\sum_{i=1}^{n-1} i^2 + 2(n-1) = 4 \cdot \frac{(n-1)n(2n-1)}{6} + (2n-1) = \frac{(2n-1)(2n^2-2n+3)}{3}$.

So, the centered octahedron numbers are numerically equal to centered pentagonal pyramid numbers and coinside with the Haüy octahedral numbers, which were obtained using other reasons.

The generating function for the sequence $\overline{O}(1), \overline{O}(2), \dots, \overline{O}(n), \dots$ is $f(x) = \frac{x(1+3x+3x^2+x^3)}{(1-x)^4} = \frac{x(1+x)(1+2x+x^2)}{(1-x)^4}$, i.e., it holds

$$\frac{x(1+x)(1+2x+x^2)}{(1-x)^4} = \overline{O}(1)x + \overline{O}(2)x^2 + \overline{O}(3)x^3 + \cdots + \overline{O}(n)x^n + \cdots, \quad |x| < 1.$$

This result can be obtained by the standard procedure, resulting in the linear recurrent equation $\overline{O}(n+4) - 4\overline{O}(n+3) + 6\overline{O}(n+2) - 4\overline{O}(n+1) + \overline{O}(n) = 0$ with initial values $\overline{O}(1) = 1$, $\overline{O}(2) = 7$, $\overline{O}(3) = 25$, and $\overline{O}(4) = 63$.

2.6.6. The **centered icosahedron numbers** (or **centered cuboctahedron numbers**) represent "centered icosahedron" and are formed by a central dot surrounded by icosahedral layers. The first centered icosahedron numbers are 1, 13, 55, 147, 309, 561, 923, 1415, 2057, 2869, ... (Sloane's A005902).

They can be obtained as partial sums of the sequence 1, 12, 42, 92, 162, 252, 362, 492, 642, 812, ... (Sloane's A005901), giving number of points on surface of icosahedron. The first element of the sequence A005901 is 1, while n-th element, $n \geq 2$, has the form $10(n-1)^2+2 = 10n^2 - 20n + 12$.

It can be obtained by the formula $v + s(n-2) + f \cdot int(S_3(n))$, where $v = 12$, $s = 30$ and $f = 20$ are the numbers of vertices, sides, and faces of icosahedron, while $int(S_3(n))$ is a number of points in the interior of n-th triangular number. So, for $n \geq 2$ it holds $v+s(n-2)+f \cdot int(S_5(n)) = 12+30(n-2)+20(\frac{n(n+1)}{2} - 3(n-1)) = 10n^2 - 20n + 12 = 10(n-1)^2 + 2$.

The recurrent formula for the sequence of centered icosahedron numbers has the form

$$\overline{I}(n+1) = \overline{I}(n) + 10n^2 + 2, \quad \overline{I}(1) = 1.$$

The general formula for n-th centered icosahedron number $\overline{I}(n)$ has the form

$$\overline{I}(n) = \frac{(2n-1)(5n^2-5n+3)}{3}.$$

It can be obtained as the result of the following summation: $\overline{I}(n) = 1 + \sum_{i=1}^{n-1}(10i^2 + 2) = 1 + 10\sum_{i=1}^{n-1} i^2 + 2(n-1) = 10 \cdot \frac{(n-1)n(2n-1)}{6} + (2n-1) = \frac{(2n-1)(5n^2-5n+3)}{3}$.

The generating function for this sequence has the form $f(x) = \frac{x(1+x)(1+8x+x^2)}{(1-x)^4} = \frac{x(1+9x+9x^2+x^3)}{(1-x)^4}$, i.e., it holds

$$\frac{x(1+x)(1+8x+x^2)}{(1-x)^4} = \overline{I}(1)x + \overline{I}(2)x^2 + \overline{I}(3)x^3 + \cdots + \overline{I}(n)x^n + \cdots, \quad |x| < 1.$$

This result can be obtained by the standard procedure, resulting in the linear recurrent equation $\overline{I}(n+4) - 4\overline{I}(n+3) + 6\overline{I}(n+2) - 4\overline{I}(n+1) + \overline{I}(n) = 0$ with initial values $\overline{I}(1) = 1$, $\overline{I}(2) = 13$, $\overline{I}(3) = 55$, and $\overline{I}(4) = 147$.

2.6.7. The **centered dodecahedron numbers** represent "centered dodecahedron" and are formed by a central dot surrounded by dodecahedral layers. The first centered dodecahedron numbers are 1, 21, 95, 259, 549, 1001, 1651, 2535, 3689, 5149, ...

They can be obtained as partial sums of the sequence 1, 20, 74, 164, 290, 452, 650, 884, 1154, 1460, ... (Sloane's A010008; see also Sloane's A005904), giving number of points on surface of dodecahedron. The first element of the sequence A010008 is 1, while n-th element, $n \geq 2$, has the form $18(n-1)^2 + 2 = 18n^2 - 36n + 20$.

It can be obtained by the formula $v + s(n-2) + f \cdot int(S_5(n))$, where $v = 20$, $s = 30$ and $f = 12$ are the numbers of vertices, edges, and faces of dodecahedron, while $int(S_5(n))$ is a number of points in the interior of n-th pentagonal number. So, for $n \geq 2$, it holds $v + s(n-2) + f \cdot int(S_5(n)) = 20 + 30(n-2) + 12(\frac{3n^2-n}{2} - 5(n-1)) = 18n^2 - 36n + 20 = 18(n-1)^2 + 2$.

The recurrent formula for the sequence of centered dodecahedron numbers has the form

$$\overline{D}(n+1) = \overline{D}(n) + 18n^2 + 2, \quad \overline{D}(1) = 1.$$

The general formula for n-th centered dodecahedron number $\overline{D}(n)$ has the form

$$\overline{D}(n) = (2n-1)(3n^2 - 3n + 1).$$

It can be obtained as the result of the following summation: $D(n) = 1 + \sum_{i=1}^{n-1}(18i^2 + 2) = 1 + 18\sum_{i=1}^{n-1} i^2 + 2(n-1) = 18 \cdot \frac{(n-1)n(2n-1)}{6} + (2n-1) = (2n-1)(3n^2 - 3n + 1)$.

The generating function for the sequence $\overline{D}(1)$, $\overline{D}(2), \ldots,$ $\overline{D}(n), \ldots$ of centered dodecahedron numbers has the form $f(x) = \frac{x(1+17x+17x^2+x^3)}{(1-x)^4} = \frac{x(1+x)(1+16x+x^2)}{(1-x)^4}$, i.e., it holds

$$\frac{x(1+x)(1+16x+x^2)}{(1-x)^4} = \overline{D}(1)x + \overline{D}(2)x^2 + \overline{D}(3)x^3 + \cdots + \overline{D}(n)x^n + \cdots, \quad |x| < 1.$$

This result can be obtained by the standard procedure, leading to the linear recurrent equation $\overline{D}(n+4) - 4\overline{D}(n+3) + 6\overline{D}(n+2) - 4\overline{D}(n+1) + \overline{D}(n) = 0$ with initial values $\overline{D}(1) = 1$, $\overline{D}(2) = 21$, $\overline{D}(3) = 95$, and $\overline{D}(4) = 259$.

2.6.8. The **centered truncated tetrahedron numbers** represent "centered truncated tetrahedron" and are formed by a central dot surrounded by corresponding layers. The first centered truncated tetrahedron numbers are 1, 17, 75, 203, 429, 781, 1287, 1975, 2873, 4009, ... (Sloane's A063494).

They can be obtained as partial sums of the sequence 1, 16, 58, 128, 226, 352, 506, 688, 898, ... (Sloane's A005905), giving number of points on surface of truncated tetrahedron. The first element of the sequence A005905 is 1, while n-th element, $n \geq 2$, has the form $14(n-1)^2+2 = 14n^2-28n+16$. This result can be obtained by the standard procedure.

So, the recurrent formula for the sequence of centered truncated tetrahedron numbers has the form

$$\overline{TS}_3^3(n+1) = \overline{TS}_3^3(n) + 14n^2 + 2, \quad \overline{TS}_3^3(1) = 1.$$

The general formula for n-th centered truncated tetrahedron number $\overline{TS}_3^3(n)$ has the form

$$\overline{TS}_3^3(n) = \frac{(2n-1)(7n^2-7n+3)}{3}.$$

It can be obtained as the result of the following summation: $\overline{TS}_3^3(n) = 1 + \sum_{i=1}^{n-1}(14i^2+2) = 1 + 14\sum_{i=1}^{n-1} i^2 + 2(n-1) = 14 \cdot \frac{(n-1)n(2n-1)}{6} + (2n-1) = \frac{(2n-1)(7n^2-7n+3)}{3}$.

The generating function for the sequence $\overline{TS}_3^3(1), \overline{TS}_3^3(2), \ldots, \overline{TS}_3^3(n), \ldots$ of centered truncated tetrahedron numbers has the form $f(x) = \frac{x(1+13x+13x^2+x^3)}{(1-x)^4} = \frac{x(1+x)(1+12x+x^2)}{(1-x)^4}$, i.e., it holds

$$\frac{x(1+x)(1+12x+x^2)}{(1-x)^4} = \overline{TS}_3^3(1)x + \overline{TS}_3^3(2)x^2 + \overline{TS}_3^3(3)x^3 + \cdots + \overline{TS}_3^3(n)x^n + \cdots, \quad |x| < 1.$$

This result can be obtained by the standard procedure, leading to the linear recurrent equation $\overline{TS}_3^3(n+4) - 4\overline{TS}_3^3(n+3) + 6\overline{TS}_3^3(n+2) - 4\overline{TS}_3^3(n+1) + \overline{TS}_3^3(n) = 0$ with initial values $\overline{TS}_3^3(1) = 1$, $\overline{TS}_3^3(2) = 17$, $\overline{TS}_3^3(3) = 75$, and $\overline{TS}_3^3(4) = 203$.

The **centered truncated cube numbers** represent "centered truncated cube" and are formed by a central dot surrounded by corresponding layers. The first centered truncated cube numbers are 1, 49, 235, 651, 1389, 2541, 4199, 6455, 9401, 13129, ...

They can be obtained as partial sums of the sequence 1, 48, 186, 416, 738, 1152, 1658, 2256, 2946, 3728, ... (Sloane's A005911), giving number of points on surface of truncated cube. The first element of the sequence A005911 is 1, while n-th element, $n \geq 2$, has the form $46(n-1)^2+2$. This result can be obtained by the standard procedure.

So, the recurrent formula for the sequence of centered truncated cube numbers has the form

$$\overline{TC}(n+1) = \overline{TC}(n) + 46n^2 + 2, \quad \overline{TS}_3^3(1) = 1.$$

The general formula for n-th centered truncated cube number $\overline{TC}(n)$ has the form

$$\overline{TC}(n) = \frac{(2n-1)(23n^2-23n+3)}{3}.$$

It can be obtained as the result of the following summation: $\overline{TS}_3^3(n) = 1 + \sum_{i=1}^{n-1}(46i^2+2) = 1 + 46\sum_{i=1}^{n-1} i^2 + 2(n-1) = 46 \cdot \frac{(n-1)n(2n-1)}{6} + (2n-1) = \frac{(2n-1)(23n^2-23n+3)}{3}$.

The generating function for the sequence $\overline{TC}(1), \overline{TC}(2), \ldots, \overline{TC}(n), \ldots$ of centered truncated cube numbers has the form $f(x) = \frac{x(1+45x+45x^2+x^3)}{(1-x)^4} = \frac{x(1+x)(1+44x+x^2)}{(1-x)^4}$, i.e., it holds

$$\frac{x(1+x)(1+44x+x^2)}{(1-x)^4} = \overline{TC}(1)x + \overline{TC}(2)x^2 + \overline{TC}(3)x^3 + \cdots + \overline{TC}(n)x^n + \cdots, \quad |x| < 1.$$

This result can be obtained by the standard procedure, resulting in the linear recurrent equation $\overline{TC}(n+4) - 4\overline{TC}(n+3) + 6\overline{TC}(n+2) - 4\overline{TC}(n+1) + \overline{TC}(n) = 0$ with initial values $\overline{TC}(1) = 1$, $\overline{TC}(2) = 49$, $\overline{TC}(3) = 235$, and $\overline{TC}(4) = 651$.

The **centered truncated octahedron numbers** represent "centered truncated octahedron" and are formed by a central dot surrounded by corresponding layers. The first centered truncated octahedron numbers are 1, 33, 155, 427, 909, 1661, 2743, 4215, 6137, 8569, ... (see Sloane's A005904).

They can be obtained as partial sums of the sequence 1, 32, 122, 272, 482, 752, 1082, 1472, 1922, 2432, ... (see Sloane's A005903), giving number of points on surface of truncated octahedron. The first element of the sequence A005903 is 1, while n-th element, $n \geq 2$, has the form $30(n-1)^2+2 = 30n^2-60n+32$. This result can be obtained by the standard procedure.

So, the recurrent formula for the sequence of centered truncated octahedron numbers has the form

$$\overline{TO}(n+1) = \overline{TO}(n) + 30n^2 + 2, \quad \overline{TO}(1) = 1.$$

The general formula for n-th centered truncated octahedron number $\overline{TO}(n)$ has the form

$$\overline{TO}(n) = (2n-1)(5n^2-5n+1).$$

It can be obtained as the result of the following summation: $\overline{TO}(n) = 1 + \sum_{i=1}^{n-1}(30i^2+2) = 1 + 30\sum_{i=1}^{n-1} i^2 + 2(n-1) = 30 \cdot \frac{(n-1)n(2n-1)}{6} + (2n-1) = (2n-1)(5n^2-5n+1)$.

The generating function for the sequence $\overline{TO}(1), \overline{TO}(2), \dots, \overline{TO}(n), \dots$ of centered truncated octahedron numbers has the form $f(x) = \frac{x(1+29x+29x^2+x^3)}{(1-x)^4} = \frac{x(1+x)(1+28x+x^2)}{(1-x)^4}$, i.e., it holds

$$\frac{x(1+x)(1+28x+x^2)}{(1-x)^4} = \overline{TO}(1)x + \overline{TO}(2)x^2 + \overline{TO}(2)x^2 + \dots + \overline{TO}(n)x^n + \cdots, \quad |x| < 1.$$

This result can be obtained by the standard procedure, leading to the linear recurrent equation $\overline{TO}(n+4) - 4\overline{TO}(n+3) + 6\overline{TO}(n+2) - 4\overline{TO}(n+1) + \overline{TO}(n) = 0$ with initial values $\overline{TO}(1) = 1$, $\overline{TO}(2) = 33$, $\overline{TO}(3) = 155$, and $\overline{TO}(4) = 427$.

2.7 Other space figurate numbers

In this section we construct two classes of space figurate numbers, which are used as building blocks centered polygonal numbers. They are **centered pyramidal numbers**, and **prism numbers**.

2.7.1. The **hex pyramidal numbers** (or **centered hexagonal pyramidal numbers**) are the most known objects of such kind. They correspond to a configuration of points which form a hexagonal pyramid with the base a hex (i.e., centered hexagonal) number.

By definition, the hex pyramidal numbers are the consecutive sums of the series 1, 7, 19, 37, 61, 91, 127, 169, 217, 271, ... (Sloane's A003215) of the hex numbers:

$$CS_3(n) = CS_6(1) + CS_6(2) + \cdots + CS_6(n).$$

Since $CS_6(n) = 3n^2 + 3n + 1 = n^3 - (n-1)^3$, one obtains, that the sum of the first n hex numbers is equal to n^3: $CS_6(1) + CS_6(2) + \cdots + CS_6(n) = (1^3 - 0^3) + (2^3 - 1^3) + \cdots + (n^3 - (n-1)^3) = n^3$. Therefore, n-th hex pyramidal number $CS_3(n)$ is equal to n^3:

$$CS_3(n) = n^3.$$

It means that hex pyramidal numbers are equivalent to cubic numbers, but differently arranged in the space. So, the first few hex pyramidal numbers are 1, 8, 27, 64, 125, 216, 343, 512, 729, 1000, ... (Sloane's A000578).

2.7.2. Of course, one can construct similar objects for any type of centered polygonal numbers, though only hex pyramidal numbers are known in the special literature.

Let a **centered m-gonal pyramidal number** be a configuration of points which form an m-gonal pyramid with the base a centered m-gonal number.

By definition, such numbers are the consecutive sums of the series $CS_m(1) = 1, CS_m(2) = 1 + m, CS_m(3) = 1 + 3m, CS_m(4) = 1 + 6m, \ldots, CS_m(n) = 1 + \frac{n(n+1)}{2}, \ldots$ of centered m-gonal numbers. More exactly, n-th centered m-gonal pyramidal number $CS_m^3(n)$ has the form

$$CS_m^3(n) = CS_m(1) + CS_m(2) + \cdots + CS_m(n).$$

In particular, the **centered triangular pyramidal numbers** are obtained as the consecutive sums of the series 1, 4, 10, 19, 31, 46, 64, 85, 109, 136, ... (Sloane's A005448) of centered triangular numbers. The first few centered triangular pyramidal numbers are 1, 5, 15, 34, 65, 111, 175, 260, 369, 505, ...

The **centered square pyramidal numbers** are obtained as the consecutive sums of the series 1, 5, 13, 25, 41, 61, 85, 113, 145, 181, ... (Sloane's A001844) of centered square numbers. The first few centered square pyramidal numbers are 1, 6, 19, 44, 85, 111, 146, 231, 344, 489, ...

The **centered pentagonal pyramidal numbers** are obtained as the consecutive sums of the series 1, 6, 16, 31, 51, 76, 106, 141, 181, 226, ... (Sloane's A005891) of centered pentagonal numbers. The first few centered pentagonal pyramidal numbers are 1, 7, 23, 54, 105, 181, 287, 428, 609, 835, ... They coincide with the Sloane's A004068, giving the number of atoms in dodecahedron with n shells.

As was shown before, the **centered hexagonal pyramidal numbers** are obtained as the consecutive sums of the series 1, 7, 19, 37, 61, 91, 127, 169, 217, 271, ... (Sloane's A003215) of centered pentagonal numbers. The first few centered hexagonal pyramidal numbers are 1, 8, 27, 64, 125, 216, , 512, 729, 1000, ... (Sloane's A000578).

Following this procedure, one can construct **centered heptagonal pyramidal numbers** 1, 9, 31, 74, 145, 251, 399, 596, 849, 1165, ...; **centered octagonal pyramidal numbers** 1, 10, 35, 84, 165, 286, 455, 680, 969, 1330, ...; **centered nonagonal pyramidal numbers** 1, 11, 39, 94, 185, 321, 511, 764, 1089, 1495, ...; **centered decagonal pyramidal numbers** 1, 12, 43, 104, 205, 356, 567, 848, 1209, 1660, ...; **centered hendecagonal pyramidal numbers** 1, 13, 47, 114, 225, 391, 623, 932, 1329, 1825, ...; **centered dodecagonal pyramidal numbers** 1, 14, 51, 124, 245, 426, 679, 1016, 1449, 1990, ... and so on.

By definition, we get the following recurrent formula for the sequence $CS_m^3(1)$, $CS_m^3(2), \ldots$, $CS_m^3(n), \ldots$ of centered m-gonal pyramidal numbers:

$$CS_m^3(n+1) = CS_m^3(n) + CS_m(n+1), \quad CS_m^3(1) = 1.$$

Since $CS_m(n+1) = \frac{mn^2+mn+2}{2}$, it holds

$$CS_m^3(n+1) = CS_m^3(n) + \frac{mn^2+mn+2}{2}, \; CS_m^3(1) = 1.$$

In particular, we get

$$CS_3^3(n+1) = CS_3^3(n) + \frac{3n^2+3n+2}{2},$$
$$CS_4^3(n+1) = CS_4^3(n) + 2n^2 + 2n + 1,$$
$$CS_5^3(n+1) = CS_5^3(n) + \frac{5n^2+5n+2}{2},$$
$$CS_6^3(n+1) = CS_6^3(n) + 3n^2 + 3n + 1.$$

The generating function for the sequence $CS_m^3(1)$, $CS_m^3(2), \ldots,$ $CS_m^3(n), \ldots$ of centered m-pyramidal numbers has the form $f(x) = \frac{x(1+(m-2)x+x^2)}{(1-x)^4}$, i.e., it holds

$$\frac{x(1+(m-2)x+x^2)}{(1-x)^4} = CS_m^3(1)x + CS_m^3(2)x^2 + CS_m^3(3)x^3 + \cdots + CS_m^3(n)x^n + \cdots, \quad |x| < 1.$$

In particular, one gets

$$\frac{x(1+x+x^2)}{(1-x)^4} = x + 4x^2 + 10x^3 + \cdots + CS_3^3(n)x^n + \cdots, \quad |x| < 1;$$
$$\frac{x(1+x)^2}{(x-1)^4} = x + 5x^2 + 13x^3 + \cdots + CS_4^3(n)x^n + \cdots, \quad |x| < 1;$$
$$\frac{x(1+3x+x^2)}{(x-1)^4} = x + 6x^2 + 16x^3 + \cdots + CS_5^3(n)x^n + \cdots, \quad |x| < 1;$$
$$\frac{x(1+4x+x^2)}{(x-1)^4} = x + 8x^2 + 27x^3 + \cdots + CS_6^3(n)x^n + \cdots, \quad |x| < 1.$$

This fact can be obtained by the standard procedure, resulting in the linear recurrent equation $CS_m^3(n+4) - 4CS_m^3(n+3) + 6CS_m^3(n+2) - 4CS_m^3(n) + CS_m^3(n) = 0$ with initial values $CS_m^3(1) = 1$, $CS_m^3(2) = 2 + m$, $CS_m^3(3) = 3 + 4m$, and $CS_m^3(4) = 4 + 10m$.

This result can be obtained also using the generating function $\frac{1+(m-2)x+x^2}{(1-x)^3}$ for centered m-gonal numbers and the decomposition $\frac{1}{1-x} = 1 + x + x^2 + \cdots + x^n + \cdots$. In fact, one has $\frac{1+(m-2)x+x^2}{(1-x)^4} = \frac{1+(m-3)x}{(1-x)^3} \cdot \frac{1}{1-x} = (CS_m(1) + CS_m(2)x + CS_m(3)x^2 + \cdots + CS_m(n)x^{n-1} + \cdots)(1 + x + x^2 + \cdots + x^n + \cdots) = CS_m(1) + (CS_m(1) + CS_m(2))x + (CS_m(1) + CS_m(2) + CS_m(3))x^2 + \cdots + (CS_m(1) + \cdots + CS_m(n))x^{n-1} + \cdots = CS_m^3(1) + CS_m^3(2)x + CS_m^3(3)x^2 + \cdots + CS_m^3(n)x^{n-1} + \cdots$, $|x| < 1$.

Since $CS_m(n) = 1 + mS_3(n-1)$, one obtains that n-th centered m-gonal pyramidal number $CS_m^3(n)$ has the form

$$CS_m^3(n) = n + m(S_3(n-1) + S_3(n-2) + \cdots + S_3(2) + S_3(1)).$$

The sum of the first $(n-1)$ triangular numbers gives the $(n-1)$-th tetrahedral number $S_3^3(n-1)$. So, we get

$$CS_m^3(n) = n + mS_3^3(n-1).$$

Since $S_3^3(n-1) = \frac{(n-1)\cdot n\cdot(n+1)}{6}$, it holds

$$CS_m^3(n) = \frac{mn^3 + n(6-m)}{6}.$$

In particular, we have

$$CS_3^3(n) = \frac{n(n^2+1)}{2}, \quad CS_4^3(n) = \frac{2n^3+n}{3},$$

$$CS_5^3(n) = \frac{5n^3+n}{6}, \quad CS_6^3(n) = n^3.$$

The construction of centered m-gonal pyramidal numbers is similar to the construction of ordinary m-gonal pyramidal numbers. However, any centered m-gonal pyramidal number can be seen as a configuration of points formed by a central dot, surrounded by pyramidal layers. In this sense they form a class of centered figurate numbers and are relative to the centered pyramid numbers. The natural "center" of n-th centered m-gonal pyramidal number is the central dot of n-th centered m-gonal number, forming the base of the corresponding pyramid. This central dot is surrounded by pyramidal layers. Each layer represents k-th m-gonal pyramidal number (with $k = 1, 3, \ldots, n$) without interior; moreover, its base also has no interior.

For instance, the first level of a centered triangular pyramidal number is simply the central dot of its base. The 2-nd level is formed from 4 points. It corresponds to 2-nd tetrahedral number $S_3^3(2) = 4$ without empty interior; its base is 2-nd triangular number without empty interior. The 2-nd level is formed from 4 points. It corresponds to 2-nd tetrahedral number $S_3^3(2) = 4$ without empty interior; its base is 2-nd triangular number 3 without empty interior. Similarly, the 3-rd level is formed from 10 points. It corresponds to 3-rd tetrahedral number $S_3^3(3) = 10$ without empty interior; its base is 3-rd triangular number 6 without empty interior. The 4-th level is formed from 19 points. It corresponds to 4-th tetrahedral number $S_3^3(4) = 20$ without empty interior, which base is 4-th triangular number 10 without 1-points interior. The 5-th level is formed from 31 points. It corresponds to 5-th tetrahedral number $S_3^3(5) = 35$ without 1-point interior, which base is 5-th triangular number 15 without 3-points interior, etc. So, the number of points in the k-th level is equal to $S_3^3(k) - S_3^3(k-3)$, $k \geq 4$.

It implies, that the centered triangular pyramidal numbers are the consecutive sums of the series

$$S_3^3(1), S_3^3(2), S_3^3(3), S_3^3(4) - S_3^3(1),$$
$$S_3^3(5) - S_3^3(2), \ldots, S_3^3(n+3) - S_3^3(n), \ldots.$$

Therefore, we have

$$CS_3^3(1) = CS_3^3(1) = 1, \quad CS_3^3(2) = S_3^3(2) + S_3^3(1) = 5,$$
$$CS_3^3(n) = S_3^3(n) + S_3^3(n-1) + S_3^3(n-2), n \geq 3.$$

So, we proved the following relationship between centered triangular pyramidal numbers and centered tetrahedral numbers:

$$\overline{S}_3^3(1) = CS_3^3(1), \quad \overline{S}_3^3(2) = CS_3^3(2), \quad \overline{S}_3^3(3) = CS_3^3(3),$$
$$\overline{S}_3^3(n) = CS_3^3(n) + S_3^3(n-3), \quad n \geq 4.$$

Similarly, the first level of a centered square pyramidal number is the central dot of its base. The 2-nd level is formed from 5 points.

It corresponds to 2-nd square pyramidal number $S_4^3(2) = 5$ without empty interior; its base is 2-nd square number 4 without empty interior. The 3-rd level is formed from 13 points. It corresponds to 3-rd square pyramidal number $S_4^3(3) = 14$ without empty interior, which base is 3-rd square number 9 without 1-point interior. The 4-th level is formed from 25 points. It corresponds to 4-th square pyramidal number $S_4^3(4) = 30$ without 1-point interior, which base is 4-th square number 16 without 9-points interior. The 5-nd level is formed from 31 points. It corresponds to 5-th tetrahedral number $S_3^3(5) = 35$ without 1-point interior, which base is 5-th triangular number 15 without 3-points interior, etc. So, the number of points in the k-th level is equal to $S_4^3(k) - S_4^3(k-2)$, $k \geq 3$.

It implies that centered square pyramidal numbers are the consecutive sums of the series

$$S_4^3(1), S_4^3(2), S_4^3(3) - S_4^3(1), S_4^3(4) - S_4^3(2),$$
$$S_4^3(5) - S_4^3(3), \ldots, S_4^3(n+2) - S_4^3(n), \ldots .$$

Therefore, we have

$$CS_4^3(1) = CS_4^3(1) = 1, \;\; CS_4^3(n) = S_4^3(n) + S_4^3(n-1), \quad n \geq 2.$$

So, we proved the following relationship between centered square pyramidal numbers and centered pyramid numbers:

$$\overline{S}_4^3(1) = CS_4^3(1), \quad \overline{S}_4^3(2) = CS_4^3(2),$$
$$\overline{S}_4^3(n) = CS_4^3(n) + S_4^3(n-2), \quad n \geq 3.$$

2.7.3. The **prism numbers**, or, more exactly, m**-gonal prism numbers**, correspond to a *regular right* m*-prism*: a polyhedron, possessing two congruent regular m-gonal faces, with all remaining faces being rectangles. They are obtained by adding up several congruent copies of a given centered m-gonal number.

For example, n-th **hexagonal prism number** is obtained as sum of n copies of n-th hex (i.e., centered hexagonal) number $CS_6(n)$, and has the form

$$PCS_6(n) = nCS_6(n) = n(3n^2 - 3n + 1).$$

The first elements of the sequence of hexagonal prism numbers are 1, 14, 57, 148, 305, 546, 889, 1352, 1953, ... (Sloane's A005915). The hexagonal prism numbers can be considered as $\frac{1}{6}(18n^3 - 18n^2 + 6n)$, and, so, coincide with **structured rhombic dodecahedral numbers** (see [Sloa11]).

The recurrent formula for the sequence of hexagonal prism numbers has the form

$$PCS_6(n+1) = PCS_6(n) + (9n^2 + 3n + 1), \quad PCS_6(1) = 1.$$

It follows, for example, from the recurrent formula $CS_6(n+1) = PCS_6(n)+6n$ for hex numbers. In fact, $PCS_6(n+1) = (n+1)CS_6(n+1) = (n+1)(CS_6(n) + 6n) = nCS_6(n) + CS_6(n) + 6n(n+1) = nCS_6(n) + (3n^2 - 3n + 1) + 6n(n+1) = PCS_6(n) + (9n^2 + 3n + 1)$.

The generating function for the sequence $PCS_6(1), PCS_6(2), \dots, PCS_6(n), \dots$ is $f(x) = \frac{x(1+10x+7x^2)}{(1-x)^4}$ (see [Plou92]), i.e., it holds

$$\frac{x(1+10x+7x^2)}{(1-x)^4} = PCS_6(1)x + PCS_6(2)x^2 + PCS_6(3)x^3 + \dots + PCS_6(n)x^n + \cdots, \quad |x| < 1.$$

This fact can be obtained by the standard procedure, leading to he linear recurrent equation $PCS_6(n+4) - 4PCS_6(n+3) + 6PCS_6(n+2) - 4PCS_6(n+1) + PCS_6(n) = 0$ with initial values $PCS_6(1) = 1$, $PCS_6(2) = 14$, $PCS_6(3) = 57$, and $PCS_6(1) = 148$.

2.8 Generalized space figurate numbers

Similarly to the plane case, the **generalized space figurate numbers** are defined as all values of the standard formulas for a given class of space figurate numbers, taken for any integer value of indices.

2.8.1. The **generalized pyramidal numbers**, or, more exactly, **generalized m-gonal pyramidal numbers**, are defined as all values of the formula $S_m^3(n) = \frac{n(n+1)((m-2)n-m+5)}{6}$, taken for each integer value of n.

For positive integers, we get ordinary m-gonal pyramidal numbers. For $n = 0$, it holds $S_m^3(0) = 0$. Finally, for negative indices, we get $S_m^3(-n) = -\frac{n(n-1)((m-2)n+m-5)}{6}$, $n \in \mathbb{N}$.

Let us consider the properties of the above numbers $S_m^3(-n)$, $n \in \mathbb{N}$. For convenience, denote $S_m^3(-n)$ by $^-S_m^3(n)$, i.e., by definition,

$$^-S_m^3(n) = S_m^3(-n) = -\frac{n(n-1)((m-2)n+m-5)}{6}, \quad n \in \mathbb{N}.$$

In particular, it holds

$$^-S_3^3(n) = S_3^3(-n) = -\frac{n(n-1)(n-2)}{6}, \quad ^-S_4^3(n) = S_4^3(-n)$$
$$= -\frac{n(n-1)(2n-1)}{6},$$
$$^-S_5^3(n) = S_5^3(-n) - \frac{n^2(n-1)}{2}, \quad ^-S_6^3(n) = S_6^3(-n)$$
$$= -\frac{n(n-1)(4n+1)}{6}.$$

So, the **generalized tetrahedral numbers** with negative indices have the form $^-S_3^3(n) = S_3^3(-n) = \frac{(n-2)(n-1)n}{6}$, i.e., $^-S_3^3(1) = 0$, $^-S_3^3(2) = 0$, and $^-S_3^3(n) = S_3^3(n-2)$ for $n \geq 3$. They give values $0, 0, -1, -4, -10, \ldots$. Therefore, the generalized tetrahedral numbers are

$$\ldots, -10, -4, -1, 0, 0, 0, 1, 4, 10, 20, 35, \ldots,$$

forming the sequence $0, 1, 0, 4, 0, 10, -1, 15, -4, \ldots$ for $n = 0, \pm 1, \pm 2, \pm 3, \pm 4, \ldots$.

The **generated square pyramidal numbers** with negative indices have the form $^-S_4^3(n) = S_4^3(-n) = -\frac{n(n-1)(2n-1)}{6}$, giving values $0, -1, -5, -14, -30, \ldots$. So, the generated square pyramidal numbers are

$$\ldots, -30, -14, -5, -1, 0, 0, 1, 5, 14, 30, 55, \ldots,$$

forming the sequence $0, 1, 0, 5, -1, 14, -5, 30, -14, \ldots$ for $n = 0, \pm 1, \pm 2, \pm 3, \pm 4, \ldots$. It is easy to see, that $^-S_4^3(n) = -S_4^3(n-1)$, $n \geq 2$, since $\frac{n(n-1)(2n-1)}{6} = \frac{(n-1)n(2(n-1)+1)}{6}$.

The **generated pentagonal pyramidal numbers** with negative indices have the form $^{-}S_5^3(n) = S_5^3(n) = -\frac{n^2(n-1)}{2}$, giving values $0, -2, -9, -24, -50, \ldots$. Therefore, the generalized pentagonal pyramidal numbers are

$$\ldots, -50, -24, -9, -2, 0, 0, 1, 6, 18, 40, 75, \ldots,$$

forming the sequence $0, 1, 0, 6, -2, 18, -9, 40, -24, \ldots$ for $n = 0, \pm1, \pm2, \pm3, \pm4, \ldots$.

The **generalized hexagonal pyramidal numbers** with negative indices have the form $^{-}S_6^3(n) = S_6^3(-n) = -\frac{n(n-1)(4n+1)}{6}$, giving values $0, -3, -13, -34, -70, \ldots$. So, the generated hexagonal pyramidal numbers are

$$\ldots, -70, -34, -13, -3, 0, 0, 1, 7, 22, 50, 95, \ldots,$$

forming the sequence $0, 1, 0, 7, -3, 22, -13, 50, -34, \ldots$ for $n = 0, \pm1, \pm2, \pm3, \pm4, \ldots$.

The recurrent formula for generalized m-gonal pyramidal numbers with negative indices has the form

$$^{-}S_m^3(n+1) = {}^{-}S_m^3(n) - {}^{-}S_m(n), \quad {}^{-}S_m^3(1) = 0,$$

where $^{-}S_m(n) = S_m(-n) = \frac{m-2}{2}(n^2+n) - n$ is n-th generalized m-gonal number with negative index.

In fact, one obtains from the recurrent formula $S_m^3(n+1) = S_m^3(n) + S_3(n+1)$ for the ordinary m-gonal pyramidal numbers that $S_m^3(n) = S_m^3(n+1) - S_m(n+1)$, or $S_m^3(n-1) = S_m^3(n) - S_m(n)$. So, $S_m^0(0) = 0$, and, for negative integers, one has $S_m^3(-(n+1)) = S_m^3(-n) + S_m(-n)$, $n \in \mathbb{N}$.

Hence, starting with $^{-}S_m^3(1) = 0$, we get $^{-}S_m^3(2) = -({}^{-}S_m(1))$, $^{-}S_m(3) = -({}^{-}S_m(1) + {}^{-}S_m(2))$, $^{-}S_m(4) = -({}^{-}S_m(1) + {}^{-}S_m(2) + {}^{-}S_m(3))$, etc., leading to the formula

$$^{-}S_m^3(n+1) = S_m^3(-(n+1)) = -({}^{-}S_m(1) + {}^{-}S_m(2) + \cdots + {}^{-}S_m(n-1) + {}^{-}S_m(n)).$$

The generating function for the sequence $^{-}S_m^3(1), {}^{-}S_m^3(2), {}^{-}S_m^3(3), \ldots$ of generated m-pyramidal numbers with negative indices, i.e., for the sequence $S_m^3(-1)$, $S_3(-2)$, $S_m^3(-3)$, $\ldots$, has the form $f(x) = -\frac{x^2(x+m-3)}{(1-x)^4}$, i.e., it holds

$$-\frac{x^2(x+m-3)}{(1-x)^4} = S_m^3(-1)x + S_m^3(-2)x^2 + S_m^3(-3)x^3 + \cdots + S_m^3(-n)x^n + \cdots, \quad |x| < 1.$$

In particular,

$$-\frac{x^3}{(1-x)^4} = S_3^3(-1)x + S_3^3(-2)x^2 + S_3^3(-3)x^3 + \cdots + S_3^3(-n)x^n + \cdots, \quad |x| < 1.$$

$$-\frac{x^2(x+1)}{(1-x)^4} = S_4^3(-1)x + S_4^3(-2)x^2 + S_4^3(-3)x^3 + \cdots + S_4^3(-n)x^n + \cdots, \quad |x| < 1.$$

$$-\frac{x^2(x+2)}{(1-x)^4} = S_5^3(-1)x + S_5^3(-2)x^2 + S_5^3(-3)x^3 + \cdots + S_5^3(-n)x^n + \cdots, \quad |x| < 1.$$

$$-\frac{x^2(x+3)}{(1-x)^4} = S_6^3(-1)x + S_6^3(-2)x^2 + S_6^3(-3)x^3 + \cdots + S_6^3(-n)x^n + \cdots, \quad |x| < 1.$$

This result can be obtained by the standard procedure, leading to the linear recurrent equation ${}^{-}S_m^3(n+4) - 4\,{}^{-}S_m^3(n+3) + 6\,{}^{-}S_m^3(n+2) - 4\,{}^{-}S_m^3(n+1) + {}^{-}S_m^3(n) = 0$ with initial conditions ${}^{-}S_m^3(1) = 0$, ${}^{-}S_m^3(2) = -(m-3)$, ${}^{-}S_m^3(3) = -(4m-11)$, and ${}^{-}S_m^3(4) = -(10m-26)$.

However, the function $f(x) = \frac{x(1+(m-3)x}{(1-x)^3}$ can be obtained also using the generating function $g(x) = \frac{x(1+(m-3)x)}{(1-x)^3}$ for generalized m-gonal numbers with negative indices. In fact, since it holds ${}^{-}S_m^3(n+1) = S_m^3(-(n+1)) = -({}^{-}S_m(1) + \cdots + {}^{-}S_m(n))$, we get $f(x) = g(x) \cdot \frac{1}{1-x}$.

For $m = 3$ and $m = 4$ the result can be obtained also using corresponding generating functions $\frac{x}{(1-x)4}$ and $\frac{x(x+1)}{(1-x)^4}$ for tetrahedral and square pyramidal numbers, respectively, since ${}^{-}S_3^3(n) = -S_3^3(n-2)$, $n \geq 3$, and ${}^{-}S_4^3(n) = -S_4^3(n-1)$, $n \geq 2$.

Now, we can get the generating function for the sequence $S_m^3(0)$, $S_m^3(1)$, $S_m^3(-1)$, $S_m^3(2)$, $S_m^3(-2), \dots$ of all generalized m-gonal numbers, written for $n = 0, \pm 1, \pm 2, \dots$. It can be obtained from generating functions $f_1(x) = \frac{1+(m-3)x}{(1-x)^4} = S_m^3(1) + S_m^3(2)x + \cdots + S_m^3(n+1)x^n + \cdots$ and $f_2(x) = \frac{(m-3)+x}{(1-x)^4} = S_m^3(-1) + S_m^3(-2)x + \cdots + S_m^3(-(n+1))x^n + \cdots$ using the formula $x^2 f_1(x^2) + x^3 f_2(x^2)$. The direct computation gives $f(x) = \frac{x^2(1+(m-3)x^2-(m-3)x^3-x^5)}{(1-x^2)^4}$. In other words, it holds

$$\frac{x^2(1+(m-3)x^2-(m-3)x^3-x^5)}{(1-x^2)^4} = S_m^3(0)x + S_m^3(1)x^2 + S_m^3(-1)x^3 + S_m^3(2)x^4 + S_m^3(-2)x^5 + \cdots, \quad |x| < 1.$$

In particular,

$$\frac{x^2(1+x^2-x^3-x^5)}{(1-x^2)^4} = S_3^3(0)x + S_3^3(1)x^2 + S_3^3(-1)x^3 + S_3^3(2)x^4 + S_3^3(-2)x^5 + \cdots, \quad |x| < 1;$$

$$\frac{x^2(1+2x^2-2x^3-x^5)}{(1-x^2)^4} = S_4^3(0)x + S_4^3(1)x^2 + S_4^3(-1)x^3 + S_4^3(2)x^4 + S_4^3(-2)x^5 + \cdots, \quad |x| < 1;$$

$$\frac{x^2(1+3x-3x^3-x^5)}{(1-x^2)^4} = S_5^3(0)x + S_5^3(1)x^2 + S_5^3(-1)x^3 + S_5^3(2)x^4 + S_5^3(-2)x^5 + \cdots, \quad |x| < 1;$$

$$\frac{x^2(1+4x-4x^3-x^5)}{(1-x^2)^4} = S_6^3(0)x + S_6^3(1)x^2 + S_6^3(-1)x^3 + S_6^3(2)x^4 + S_6^3(-2)x^5 + \cdots, \quad |x| < 1.$$

2.8.2. The **generalized cubic numbers** $\dots, -125, -64, -27, -8, -1, 0, 1, 8, 27, 64, 125, \dots$ are just signed version of the ordinary cubic numbers: $C(0) = 0$, and $C(-n) = -C(n)$ for $n \in \mathbb{N}$.

So, the recurrent relation for generalized cubic numbers $^-C(n) = C(-n) = -C(n)$ with negative indices is

$$^-C(n+1) = {}^-C(n) - (3n^2+n+1), \quad ^-C(1) = -1,$$

while the generating function is is $f(x) = -\frac{x(1+4x+x^2)}{(1-x)^4}$:

$$-\frac{x(1+4x+x^2)}{(1-x)^4} = C(-1)x + C(-2)x^2 + C(-3)x^3 + \cdots + C(-n)x^n + \cdots, \quad |x| < 1.$$

The generating function for the sequence $C(0), C(1), C(-1), C(2), C(-2), \ldots$ of all generalized cubic numbers, written for $n = 0, \pm1, \pm2, \ldots$, can be obtained from the generating function $g(x) = \frac{1+4x+x^2}{(1-x)^4} = C(1) + C(2)x + \cdots + C(n+1)x^n + \cdots$ using the formula $x^2g(x^2) - x^3g(x^2)$. The direct computation gives $f(x) = \frac{x^2(1-x)(1+4x^2+x^4)}{(1-x^2)^4} = \frac{x^2(1+4x+x^2)}{(1+x)(1-x^2)^3}$. In other words, it holds

$$\frac{x^2(1+4x+x^2)}{(1+x)(1-x^2)^3} = C(0)x + C(1)x^2 + C(-1)x^3 + C(2)x^4 + C(-2)x^5 + \cdots, \quad |x| < 1.$$

Following the standard procedure, we can obtain a lot of properties of generalized cubic numbers. For example, it holds the following identity:

$$^-C(1) + {}^-C(2) + \cdots + {}^-C(n-1) + {}^-C(n) = -({}^-S_3(n+1))^2.$$

A generalization of this fact states that *the sum of several consecutive generalized cubic numbers is a difference of squares of two generalized triangular numbers*:

- $C(k+1) + \cdots + C(k+n) = (S_3(k+n))^2 - (S_3(k))^2 =^- (S_3(k+n-1))^2 - {}^-(S_3(k-1))^2$;
- $^-C(k+1) + \cdots + {}^-C(k+n) = (S_3(k))^2 - (S_3(k+n))^2 = ({}^-S_3(k-1))^2 - ({}^-S_3(k+n-1))^2$;
- $C(-k) + \cdots + C(k) + \cdots + C(k+n) = (S_3(k+n))^2 - (S_3(k))^2 = {}^-(S_3(k+n-1))^2 - {}^-(S_3(k-1))^2$;
- $C(-(k+n)) + \cdots + C(-k) + \cdots + C(k) = (S_3(k))^2 - (S_3(k+n))^2 = ({}^-S_3(k-1))^2 - ({}^-S_3(k+n-1))^2$.

2.8.3. The **generalized octahedral numbers** correspond to all values of the formula $O(n) = \frac{n}{3}(2n^2+1)$ for each integer value of n.

For positive integers n, we get ordinary octahedral numbers. For $n = 0$, it holds $O(0) = 0$. Finally, for negative values of argument, it holds $O(-n) = -\frac{n}{3}(2n^2 + 1)$, $n \in \mathbb{N}$.

Therefore, the generalized octagonal numbers

$$\dots, -85, -44, 19, -6, -1, 0, 1, 6, 19, 44, 85, \dots$$

are just signed analogues of ordinary octagonal numbers.

The recurrent formula for generalized octagonal numbers ${}^{-}O(n) = O(-n)$ with negative indices gets the form

$${}^{-}O(n+1) = {}^{-}O(n) - (n+1)^2 - n^2, \quad {}^{-}O(1) = -1,$$

while the generating function for the sequence ${}^{-}O(1), {}^{-}O(2)$, $\dots$, ${}^{-}O(n), \dots$ is $f(x) = -\frac{x(x+1)^2}{(x-1)^2}$, i.e., it holds

$$-\frac{x(x+1)^2}{(x-1)^4} = {}^{-}O(1)x + {}^{-}O(2)x^2 + {}^{-}O(3)x^3 + \dots + {}^{-}O(n)x^n + \cdots, \quad |x| < 1.$$

The generating function for the sequence $O(0), O(1), O(-1)$, $O(2), O(-2), \dots$ of all generalized octagonal numbers, written for $n = 0, \pm 1, \pm 2, \dots$, can be obtained from the generating function $g(x) = \frac{(x+1)^2}{(1-x)^4} = O(1) + O(2)x + \dots + O(n+1)x^n + \cdots$ using the formula $x^2 g(x^2) - x^3 g(x^2)$. The direct computation gives $f(x) = \frac{x^2(1-x)(1+x^2)^2}{(1-x^2)^4} = \frac{x^2(1+x^2)^2}{(1+x)(1-x^2)^3}$. In other words, it holds

$$\frac{x^2(1+x^2)^2}{(1+x)(1-x^2)^3} = O(0)x + O(1)x^2 + O(-1)x^3 + O(2)x^4 + O(-2)x^5 + \cdots, \quad |x| < 1.$$

2.8.4. The **generalized icosahedral numbers** correspond to all values of the formula $I(n) = \frac{n}{2}(5n^2 - 5n + 2)$ for each integer value of n.

For positive integers n, we get ordinary icosahedral numbers. For $n = 0$, it holds $I(0) = 0$. Finally, for negative values of argument, it holds $I(-n) = -\frac{n}{2}(5n^2 + 5n + 2)$, $n \in \mathbb{N}$.

So, the generalized icosahedral numbers are

$$\dots, -380, -204, -93, -32, -6, 0, 1, 12, 48, 124, 255, \dots,$$

giving the sequence $0, 1, -6, 12, -32, 48, -93, 124, -204, \ldots$ for $n = 0, \pm 1, \pm 2, \pm 3, \pm 4, \ldots$.

The recurrent formula for generalized icosahedral numbers ${}^{-}I(n) = I(-n)$ with negative indices has the form

$$ {}^{-}I(n+1) = {}^{-}I(n) - \frac{15n^2 + 25n + 12}{2}, \quad {}^{-}I(1) = -6. $$

In fact, since $I(n) = I(n-1) + \frac{15n^2 - 25n + 12}{2}$, we get $I(n-1) = I(n) - \frac{15n^2 - 25n + 12}{2}$. So, $I(0) = 0$, and $I(-(n+1)) = I(-n) - \frac{15n^2 + 25n + 12}{2}$.

The generating function for the sequence ${}^{-}I(1), {}^{-}I(2), \ldots {}^{-}I(n), \ldots$ is $f(x) = -\frac{x(6+8x+x^2)}{(x-1)^4}$, i.e., it holds

$$ -\frac{x(6+8x+x^2)}{(x-1)^4} = {}^{-}I(1)x + {}^{-}I(2)x^2 + {}^{-}I(3)x^3 + \cdots + {}^{-}I(n)x^n + \cdots, \quad |x| < 1. $$

It can be obtained following the standard procedure.

The generating function for the sequence $I(0), I(1), I(-1), I(2), I(-2), \ldots$ of all generalized icosahedral numbers, written for $n = 0, \pm 1, \pm 2, \ldots$, can be obtained from the generating functions $f_1(x) = \frac{1+8x+6x^2}{(1-x)^4} = I(1) + I(2)x + \cdots + I(n+1)x^n + \cdots$ and $f_2(x) = -\frac{6+8x+x^2}{(1-x)^4} = I(-1) + I(-2)x + \cdots + I(-(n+1))x^n + \cdots$ using the formula $x^2 f_1(x^2) + x^3 f_2(x^2)$. The direct computation gives $f(x) = \frac{x^2(1-6x+8x^2-8x^3+6x^4-x^5)}{(1-x^2)^4}$. In other words, it holds

$$ \frac{x^2(1-6x+8x^2-8x^3+6x^4-x^5)}{(1-x^2)^4} = I(0)x + I(1)x^2 + I(-1)x^3 + I(2)x^4 + I(-2)x^5 + \cdots, \quad |x| < 1. $$

2.8.5. The **generalized dodecahedral numbers** correspond to all values of the formula $D(n) = \frac{n}{2}(9n^2 - 9n + 2)$ for each integer value of n.

For positive integers n, we get ordinary dodecahedral numbers. For $n = 0$, it holds $D(0) = 0$. Finally, for negative values of argument, it holds $D(-n) = -\frac{n}{2}(9n^2 + 9n + 2)$, $n \in \mathbb{N}$.

Therefore, the generalized dodecahedral numbers are

$$\ldots, -680, -364, -165, -56, -10, 0, 1, 20, 84, 220, 455, \ldots,$$

giving the sequence $0, 1, -10, 20, -56, 84, -165, 220, -364, \ldots$ for $n = 0, \pm 1, \pm 2, \pm 3, \pm 4, \ldots$.

The recurrent formula for generalized dodecahedral numbers ${}^{-}D(n) = D(-n)$ with negative indices gets the form

$$^{-}D(n+1) = {}^{-}D(n) - \frac{27n^2 + 45n + 20}{2}, \quad {}^{-}D(1) = -10,$$

while the generating function for the sequence ${}^{-}D(1)$, ${}^{-}D(2), \ldots$ ${}^{-}D(n), \ldots$ is $f(x) = -\frac{x(10+16x+x^2)}{(x-1)^4}$, i.e., it holds

$$-\frac{x(10 + 16x + x^2)}{(x-1)^4} = {}^{-}D(1)x + {}^{-}D(2)x^2 + {}^{-}D(3)x^3 + \cdots +^{-} D(n)x^n + \cdots, \quad |x| < 1.$$

It can be obtained following the standard procedure.

The generating function for the sequence $D(0), D(1), D(-1), D(2), D(-2), \ldots$ of all generalized dodecahedral numbers, written for $n = 0, \pm 1, \pm 2, \ldots$, can be obtained from the generating functions $f_1(x) = \frac{1+16x+10x^2}{(1-x)^4} = D(1) + D(2)x + \cdots + D(n+1)x^n + \cdots$ and $f_2(x) = -\frac{10+16x+x^2}{(1-x)^4} = D(-1) + D(-2)x + \cdots + D(-(n+1))x^n + \cdots$ using the formula $x^2 f_1(x^2) + x^3 f_2(x^2)$. The direct computation gives $f(x) = \frac{x^2(1-10x+16x^2-16x^3+10x^4-x^5)}{(1-x^2)^4}$. In other words, it holds

$$\frac{x^2(1 - 10x + 16x^2 - 16x^3 + 10x^4 - x^5)}{(1 - x^2)^4} = D(0)x + D(1)x^2 + D(-1)x^3 + D(2)x^4 + D(-2)x^5 + \cdots, \quad |x| < 1.$$

2.8.6. The **generalized centered cube numbers** are defined as all values of the formula $\bar{C}(n) = (2n-1)(n^2 - n + 1)$ for each integer value of n.

For positive integer n, we obtain ordinary centered cube numbers. For $n = 0$, we get $\bar{C}(0) = -1$. Finally, for negative values, one has

$\bar{C}(-n) = -(2n+1)(n^2+n+1)$. So, the generalized centered cube numbers are

$$\dots, -341, -189, -91, -35, -9, -1, 1, 9, 35, 91, 189, \dots,$$

giving the sequence $-1, 1, -9, 9, -35, 35, -91, 91, -189, \dots$ for $n = 0, \pm 1, \pm 2, \pm 3, \pm 4, \dots$.

It is easy to see, that ${}^{-}\bar{C}(n) = \bar{C}(-n) = -\bar{C}(n+1)$, since $(2n+1)(n^2+n+1) = (2(n+1)-1)((n+1)^2-(n+1)+1)$.

So, the recurrent equation for the sequence of generalized centered cube numbers with negative indices has the form

$${}^{-}\bar{C}(n) ={}^{-}\bar{C}(n-1) - (6n^2+2), \quad {}^{-}\bar{C}(1) = -9,$$

while the generating function is $f(x) = \frac{x(x^3-5x^2+x-9)}{(1-x)^4}$:

$$\frac{x(x^3-5x^2+x-9)}{(1-x)^4} = {}^{-}\bar{C}(1)x + {}^{-}\bar{C}(2)x^2 + {}^{-}\bar{C}(3)x^3 + \cdots + {}^{-}\bar{C}(n)x^n + \cdots, \quad |x| < 1.$$

In fact, $\frac{1+5x+5x^2+x^3}{(1-x)^4} = \bar{C}(1) + \bar{C}(2)x + \cdots + \bar{C}(n+1)x^n + \cdots = 1 - {}^{-}\bar{C}(1)x - {}^{-}\bar{C}(1)x - \ldots - {}^{-}\bar{C}(n)x^n - \ldots$. Therefore, $1 - \frac{x^3+5x^2+5x+1}{(1-x)^4} = \frac{x(x^3-5x^2+x-9)}{(1-x)^4} ={}^{-}\bar{C}(1)x + {}^{-}\bar{C}(2)x^2 + \cdots + {}^{-}\bar{C}(n)x^n + \cdots$, $|x| < 1$.

The generating function for the sequence $\bar{C}(0), \bar{C}(1), \bar{C}(-1), \bar{C}(2), \bar{C}(-2), \dots$ of all generalized octagonal numbers, written for $n = 0, \pm 1, \pm 2, \dots$, can be obtained from the generating functions $f_1(x) = \frac{1+5x+5x^2+x^3}{(1-x)^4} = \bar{C}(1) + \bar{C}(2)x + \cdots + \bar{C}(n+1)x^n + \cdots$ and $f_2(x) = \frac{x^3-5x^2+x-9}{(1-x)^4} = \bar{C}(-1) + \bar{C}(-2)x + \cdots + \bar{C}(-(n+1))x^n + \cdots$ using the formula $x^2 f_1(x^2) + x^3 f_2(x^2)$. The direct computation gives $f(x) = \frac{x^2(1-9x+5x^2+x^3+5x^4-5x^5+x^6+x^7)}{(1-x^2)^4}$. In other words, it holds

$$\frac{x^2(1-9x+5x^2+x^3+5x^4-5x^5+x^6+x^7)}{(1-x^2)^4} = \bar{C}(0)x + \bar{C}(1)x^2 + \bar{C}(-1)x^3 + \bar{C}(2)x^4 + \bar{C}(-2)x^5 + \cdots, \quad |x| < 1.$$

2.8.7. The **generalized centered tetrahedron numbers** are defined as all values of the formula $\overline{S}_3^3(n) = \frac{(2n-1)(n^2+n+3)}{3}$ for each integer value of n.

For positive integers n, we obtain the ordinary centered tetrahedron numbers. For $n = 0$, we get $\overline{S}_3^3(0) = -1$. Finally, for negative values, one has $\overline{S}_3^3(-n) = -\frac{(2n+1)(n^2+n+3)}{3}$, $n \in \mathbb{N}$.

So, the generalized centered tetrahedron numbers are

$$\ldots, -121, -69, -35, -15, -5, -1, 1, 5, 15, 35, 69, \ldots,$$

forming the sequence $-1, 1, -5, 5, -15, 15, -35, 35, -69, \ldots$ for $n = 0, \pm 1, \pm 2, \pm 3, \pm 4, \ldots$.

Clearly, ${}^{-}\overline{S}_3^3(n) = \overline{S}_3^3(-n) = -\overline{S}_3^3(n+1)$, since $(2n+1)(n^2+n+3) = (2(n+1)-1)((n+1)^2-(n+1)+3$.

Therefore, the recurrent formula for the sequence of generalized centered tetrahedron numbers with negative indices has the from

$${}^{-}\overline{S}_3^3(n) = {}^{-}\overline{S}_3^3(n-1) - (2n^2+2), \quad {}^{-}\overline{S}_3^3(1) = -5,$$

while he generating function is $f(x) = -\frac{x(5-5x+5x^2-x^3)}{(1-x)^4}$:

$$-\frac{x(5-5x+5x^2-x^3)}{(1-x)^4} = {}^{-}\overline{S}_3^3(1)x + {}^{-}\overline{S}_3^3(2)x^2 + {}^{-}\overline{S}_3^3(3)x^3$$

$$+ \cdots + {}^{-}\overline{S}_3^3(n)x^n + \cdots, \quad |x| < 1.$$

Really, $\frac{1+x+x^2+x^3}{(1-x)^4} = \overline{S}_3^3(1) + \overline{S}_3^3(2)x + \cdots + \overline{S}_3^3(n+1)x^n + \cdots = 1 - {}^{-}\overline{S}_3^3(1)x - {}^{-}\overline{S}_3^3(2)x^2 - \ldots - {}^{-}\overline{S}_3^3(n)x^n + \cdots$. So, we get $1 - \frac{1+x+x^2+x^3}{(1-x)^4} = -\frac{x(5-5x+5x^2-x^3)}{(1-x)^4} = {}^{-}\overline{S}_3^3(1)x + {}^{-}\overline{S}_3^3(2)x^2 + \cdots + {}^{-}\overline{S}_3^3(n)x^n + \cdots$, $|x| < 1$.

The generating function for the sequence $\overline{S}_3^3(0), \overline{S}_3^3(1), \overline{S}_3^3(-1), \overline{S}_3^3(2), \overline{S}_3^3(-2), \ldots$ of all generalized centered tetrahedron numbers, written for $n = 0, \pm 1, \pm 2, \ldots$, can be obtained from the generating functions $f_1(x) = \frac{1+x+x^2+x^3}{(1-x)^4} = \overline{S}_3^3(1) + \overline{S}_3^3(2)x + \cdots + \overline{S}_3^3(n+1)x^n + \cdots$ and $f_2(x) = -\frac{5-5x+5x^2-x^3}{(1-x)^4} = \overline{S}_3^3(-1) + \overline{S}_3^3(-2)x + \cdots + \overline{S}_3^3(-(n+1))x^n + \cdots$ using the formula $x^2 f_1(x^2) + x^3 f_2(x^2)$. The direct computation gives $f(x) = \frac{x^2(1-5x+x^2+5x^3+x^4-5x^5+x^6+x^7)}{(1-x^2)^4}$. In other words, it holds

$$\frac{x^2(1-5x+x^2+5x^3+x^4-5x^5+x^6+x^7)}{(1-x^2)^4}$$
$$=\overline{S}_3^3(0)x+\overline{S}_3^3(1)x^2+\overline{S}_3^3(-1)x^3+\overline{S}_3^3(2)x^4$$
$$+\overline{S}_3^3(-2)x^5+\cdots,\quad |x|<1.$$

2.8.8. The **generalized centered square pyramid numbers** are defined as all values of the formula $\overline{S}_4^3(n)=\frac{(2n-1)(n^2-n+2)^2}{3}$ for each integer value of n.

For positive integers n, we obtain the ordinary centered square pyramid numbers. For $n=0$, we get $\overline{S}_4^3(0)=-1$. Finally, for negative values, one has $\overline{S}_3^3(-n)=-\frac{(2n+1)(n^2+n+3)}{3}$, $n\in\mathbb{N}$.

So, the generalized centered square pyramid numbers are

$$\ldots,-176,-99,-49,-20,-6,-1,1,6,20,49,99,\ldots,$$

forming the sequence $-1,1,-6,6,-20,20,-49,49,-99,\ldots$ for $n=0,\pm1,\pm2,\pm3,\pm4,\ldots$.

Clearly, ${}^{-}\overline{S}_4^3(n)=\overline{S}_4^3(-n)=-\overline{S}_4^3(n+1)$, since $(2n+1)(n^2+n+2)=(2(n+1)-1)((n+1)^2-(n+1)+2)$.

So, the recurrent formula for the sequence of generalized centered square pyramid numbers with negative indices is

$${}^{-}\overline{S}_4^3(n)={}^{-}\overline{S}_4^3(n-1)-(3n^2+2),\quad -\overline{S}_4^3(1)=-6,$$

while the generating function is $f(x)=-\frac{x(x^3-5x^2+4x-6)}{(1-x)^4}$:

$$-\frac{x(x^3-5x^2+4x-6)}{(1-x)^4}={}^{-}\overline{S}_4^3(1)x+{}^{-}\overline{S}_4^3(2)x^2+{}^{-}\overline{S}_4^3(2)x^2$$
$$+\cdots+{}^{-}\overline{S}_4^3(n)x^n+\cdots,\quad |x|<1.$$

In fact, it holds $\frac{1+2x+2x^2+x^3}{(1-x)^4}=\overline{S}_4^3(1)+\overline{S}_4^3(2)x+\cdots+\overline{S}_4^3(n+1)x^n+\cdots=1-{}^{-}\overline{S}_4^3(1)x-{}^{-}\overline{S}_4^3(2)x^2-\ldots-{}^{-}\overline{S}_4^3(n)x^n-\ldots$. So, one obtains $1-\frac{1+2x+2x^2+x^3}{(1-x)^4}=-\frac{x(x^3-5x^2+4x-6)}{(1-x)^4}={}^{-}\overline{S}_4^3(1)x+{}^{-}\overline{S}_4^3(2)x^2+\cdots+{}^{-}\overline{S}_4^3(n)x^n+\cdots$, $|x|<1$.

The generating function for the sequence $\overline{S}_4^3(0), \overline{S}_4^3(1), \overline{S}_4^3(-1), \overline{S}_4^3(2), \overline{S}_4^3(-2), \dots$ of all generalized centered tetrahedron numbers, written for $n = 0, \pm 1, \pm 2, \dots$, can be obtained from the generating functions $f_1(x) = \frac{1+2x+2x^2+x^3}{(1-x)^4} = \overline{S}_4^3(1) + \overline{S}_4^3(2)x + \cdots + \overline{S}_4^3(n+1)x^n + \cdots$ and $f_2(x) = -\frac{x^3-5x^2+4x-6}{(1-x)^4} = \overline{S}_4^3(-1) + \overline{S}_4^3(-2)x + \cdots + \overline{S}_4^3(-(n+1))x^n + \cdots$ using the formula $x^2 f_1(x^2) + x^3 f_2(x^2)$. The direct computation gives $f(x) = \frac{x^2(1+6x+2x^2-4x^3+2x^4+5x^5+x^6-x^7)}{(1-x^2)^4}$. In other words, it holds

$$\frac{x^2(1+6x+2x^2-4x^3+2x^4+5x^5+x^6-x^7)}{(1-x^2)^4}$$
$$= \overline{S}_4^3(0)x + \overline{S}_4^3(1)x^2 + \overline{S}_4^3(-1)x^3 + \overline{S}_4^3(2)x^4$$
$$+ \overline{S}_4^3(-2)x^5 + \cdots, \quad |x| < 1.$$

2.8.9. The **generalized rhombic dodecahedral numbers** are defined as all values of the formula $RD(n) = n^4 - (n-1)^4$ for each integer value of n.

For positive integers n, we obtain the ordinary centered square pyramid numbers. For $n = 0$, we get $RD(0) = -1$. For negative values, one has $RD(-n) = n^4 - (n+1)^4$, $n \in \mathbb{N}$.

So, the generalized rhombic dodecahedral numbers are

$$\dots, -671, -369, -175, -65, -15, -1, 1, 15, 65, 175, 369, \dots,$$

forming the sequence $-1, 1, -15, 15, -65, 65, -175, 175, -369, \dots$ for $n = 0, \pm 1, \pm 2, \pm 3, \pm 4, \dots$.

Clearly, one has ${}^{-}RD(n) = RD(-n) = -RD(n+1)$, since $(-n)^4 - (-n-1)^4 = -((n+1)^4 - n^4)$.

So, the recurrent formula for the sequence of generalized rhombic dodecahedral numbers with negative indices is

$${}^{-}RD(n) = {}^{-}RD(n-1) - (12n^2+2), \ {}^{-}RD(1) = -6,$$

while the generating function is $f(x) = \frac{x(15+5x+5x^2)}{(1-x)^4}$:

$$-\frac{x(15+5x+5x^2)}{(1-x)^4} = {}^{-}RD(1)x + {}^{-}RD(2)x^2 + {}^{-}RD(3)x^3$$
$$+ \cdots + {}^{-}RD(n)x^n + \cdots, \quad |x| < 1.$$

In fact, $\frac{1+11x+11x^2+x^3}{(1-x)^4} = RD(1) + RD(2)x + \cdots + RD(n+1)x^n + \cdots = 1 - {}^{-}RD(1)x - {}^{-}RD(2)x^2 - \ldots - {}^{-}RD(n)x^n - \ldots$. Therefore, it holds $1 - \frac{1+11x+11x^2+x^3}{(1-x)^4} = -\frac{x(15+5x+5x^2)}{(1-x)^4} = {}^{-}RD(1)x + {}^{-}RD(2)x^2 + \cdots + {}^{-}RD(n)x^n + \cdots$, $|x| < 1$.

The generating function for the sequence $RD(0), RD(1), RD(-1), RD(2), RD(-2), \ldots$ of all generalized rhombic dodecahedral numbers, written for $n = 0, \pm 1, \pm 2, \ldots$, can be obtained from the generating functions $f_1(x) = \frac{1+11x+11x^2+x^3}{(1-x)^4} = RD(1) + RD(2)x + \cdots + RD(n+1)x^n + \cdots$ and $f_2(x) = -\frac{15+5x+5x^2}{(1-x)^4} = RD(-1) + RD(-2)x + \cdots + RD(-(n+1))x^n + \cdots$ using the formula $x^2 f_1(x^2) + x^3 f_2(x^2)$. The direct computation gives $f(x) = \frac{x^2(1-15x+11x^2-5x^3+11x^4-5x^5+x^6)}{(1-x^2)^4}$. In other words, it holds

$$\frac{x^2(1 - 15x + 11x^2 - 5x^3 + 11x^4 - 5x^5 + x^6)}{(1 - x^2)^4}$$
$$= RD(0)x + RD(1)x^2 + RD(-1)x^3 + RD(2)x^4 + RD(-2)x^5 + \cdots, \quad |x| < 1.$$

2.8.10. The **generalized centered m-gonal pyramidal numbers** are defined as all values of the formula $CS_m^3(n) = \frac{mn^3+n(6-m)}{6}$ for each integer value of n.

For positive integers n, we obtain the ordinary centered m-gonal pyramidal numbers. For $n = 0$, we get $CS_m^3(0) = 0$. For negative values, one has $CS_m(-n) = -\frac{mn^3+n(6-m)}{6}$, $n \in \mathbb{N}$.

So, generalized centered m-gonal pyramidal numbers

$$\ldots, -(5 + 20m), -(4 + 10m), -(3 + 4), -(2 + m), -1, 0, 1,$$
$$2 + m, 3 + 4m, 4 + 10m, 5 + 20m, \ldots$$

are just signed analogues of ordinary centered m-gonal pyramidal numbers.

Therefore, the recurrent formula for the sequence of generalized centered m-gonal pyramidal numbers with negative indices is

$${}^{-}CS_m^3(n+1) = {}^{-}CS_m^3(n) - CS_m(n+1), \; {}^{-}CS_m^3(-1) = -1,$$

while the generating function is $f(x) = -\frac{x(1+(m-2)x+x^2)}{(1-x)^4}$:

$$-\frac{x(1+(m-2)x+x^2)}{(1-x)^4} = {}^{-}CS_m^3(1)x + {}^{-}CS_m^3(2)x^2 + {}^{-}CS_m^3(3)x^3 + \cdots + {}^{-}CS_m^3(n)x^n + \cdots, \quad |x| < 1.$$

The generating function for the sequence $CS_m^3(0), CS_m^3(1), CS_m^3(-1), CS_m^3(2), CS_m^3(-2), \ldots$ of all generalized centered m-gonal pyramidal numbers, written for $n = 0, \pm 1, \pm 2, \ldots$, can be obtained from the generating function $g(x) = \frac{x(1+(m-2)x+x^2}{(1-x)^4} = CS_m^3(1) + CS_m^3(2)x + \cdots + CS_m^3(n+1)x^n + \cdots$ using the formula $x^2g(x^2) - x^3g(x^2)$. The direct computation gives $f(x) = \frac{x^2(1-x)(1+(m-2)x^2+x^4)}{(1-x^2)^4}$ $= \frac{x^2(1+(m-2)x^2+x^4)}{(1+x)(1-x^2)^3}$. In other words, it holds

$$\frac{x^2(1+(m-2)x^2+x^4)}{(1+x)(1-x^2)^3} = CS_m^3(0)x + CS_m^3(1)x^2 + CS_m^3(-1)x^3 + CS_m^3(2)x^4 + CS_m^3(-2)x^5 + \cdots, \quad |x| < 1.$$

2.8.11. Similarly, one can obtain **generalized prism numbers**, ore, more exactly, **generalized m-gonal prism numbers**.

In particular, **generalized hexagonal prism numbers** are defined as all values of the formula $PCS_6(n) = n(3n^2 - 3n + 1)$ for each integer value of n.

For positive integers n, we obtain the ordinary hexagonal prism numbers. For $n = 0$, we get $PCS_6(0) = 0$. For negative values, one has $PCS_6(-n) = -n(3n^2 + 3n + 1)$, $n \in \mathbb{N}$.

So, the generalized hexagonal prism numbers are

$$\ldots, -455, -244, -111, -38, -7, 0, 1, 14, 57, 148, 305, \ldots,$$

giving the sequence $0, 1, -7, 14, -38, 57, -111, 148, -244, \ldots$ for $n = 0, \pm 1, \pm 2, \pm 3, \pm 4, \ldots$.

The recurrent formula for the sequence of generalized hexagonal prism numbers ${}^{-}PCS_6(n) = PCS_6(-n)$ with negative indices is

$${}^{-}PCS_6(n+1) = {}^{-}PCS_6(n) - (9n^2 + 15n + 7),$$
$${}^{-}PCS_6(1) = -7.$$

In fact, since $PCS_6(n) = PCS_6(n-1) + (9n^2 - 15n + 7)$, we get $PCS_6(n-1) = PCS_6(n) - (9n^2 - 15n + 7)$. So, $PCS_6(0) = 0$, and $PCS_6(-(n+1)) = PCS_6(-n) - (9n^2 + 15n + 7)$.

The generating function for the sequence of generalized hexagonal prism numbers with negative indices is $f(x) = -\frac{x(7+10x+x^2)}{(1-x)^4}$, i.e., it holds

$$-\frac{x(7+10x+x^2)}{(1-x)^4} = {}^{-}PCS_6(1)x + {}^{-}PCS_6(2)x^2 + {}^{-}PCS_6(3)x^3 + \cdots + {}^{-}PCS_6(n)x^n + \cdots, \quad |x| < 1.$$

This result can be obtained following the standard procedure.

The generating function for the sequence of all generalized hexagonal prism numbers, written for $n = 0, \pm 1, \pm 2, \ldots$, can be obtained from the generating functions $f_1(x) = \frac{1+10x+x^2}{(1-x)^4} = PCS_6(1) + PCS_6(2)x + \cdots + PCS_6(n+1)x^n + \cdots$ and $f_2(x) = -\frac{7+10x+x^2}{(1-x)^4} = {}^{-}PCS_6(1) + {}^{-}PCS_6(2)x + \cdots + {}^{-}PCS_6(n+1)x^n + \cdots$ using the formula $x^2 f_1(x^2) - x^3 f_2(x^2)$. The direct computation gives $f(x) = \frac{x^2(1-7x+10x^2-10x^3+7x^4-x^5)}{(1-x^2)^4}$. In other words, it holds

$$\frac{x^2(1-7x+10x^2-10x^3+7x^4-x^5)}{(1-x^2)^4} = PCS_6(0)x + PCS_6(1)x^2 + PCS_6(-1)x^3 + PCS_6(2)x^4 + PCS_6(-2)x^5 + \cdots, \quad |x|<1.$$

Chapter 3

Multidimensional figurate numbers

Formally, figurate numbers can be constructed in any dimension k, but, for $k \geq 4$, such construction loses usual physical sense. In this chapter we consider several classes of such **multidimensional figurate numbers**.

In dimension four, **pentatope numbers** form the first class of figurate numbers, which can be represented as regular discrete geometric patterns. The second natural class of four-dimensional figurate numbers consists of **biquadratic numbers**. The four-dimensional **regular polychoral numbers**, corresponding to the other regular polychora, are **hyperoctahedral numbers**, **hyperdodecahedral numbers**, **hypericosahedral numbers**, and **polyoctahedral numbers**. Their construction is more complicate, and these numbers rarely appear in the special literature.

Among multidimensional figurate numbers, constructed for any dimension k, $k \geq 5$, we consider three classes of such numbers, corresponding to classical multidimensional regular polytopes: **multidimensional hypertetrahedron numbers**, **multidimensional hypercube numbers**, and **multidimensional hyperoctahedron numbers**. Are considered also **multidimensional m-gonal pyramidal numbers**, **centered multidimensional figurate numbers**, and **nexus numbers**.

3.1 Pentatope numbers and their multidimensional analogues

3.1.1. A **pentatope number** (or **hypertetrahedral number**, **triangulo-triangular number**) is a figurate number which represents a four-dimensional hypertetrahedron. It is four-dimensional analogue of three-dimensional tetrahedral numbers and two-dimensional triangular numbers. The n-th pentatope number $S_3^4(n)$ is given as the sum of the first n tetrahedral numbers:

$$S_3^4(n) = S_3^3(1) + S_3^3(1) + \cdots + S_3^3(n).$$

Since $S_3^3(n) = \frac{n(n+1)(n+2)}{6}$, we get the following recurrent formula for the sequence of pentatope numbers:

$$S_3^4(n+1) = S_3^4(n) + \frac{(n+1)(n+2)(n+3)}{6}, \quad S_3^4(1) = 1.$$

It yields the general formula for n-th pentatope number:

$$S_3^4(n) = \frac{n(n+1)(n+2)(n+3)}{24}.$$

Let us check it by induction. For $n = 1$, it holds $\frac{1}{24} \cdot 1 \cdot 2 \cdot 3 \cdot 4 = 1 = S_3^4(1)$. Going from n to $n+1$, one has $S_3^4(n+1) = S_3^4(n) + \frac{(n+1)(n+2)(n+3)}{6} = \frac{1}{24}n(n+1)(n+2)(n+3) + \frac{(n+1)(n+2)(n+3)}{6} = \frac{(n+1)(n+2)(n+3)}{24}(n+4) = \frac{(n+1)(n+2)(n+3)(n+4)}{24}$.

So, the first few pentatope numbers are 1, 5, 15, 35, 70, 126, 210, 330, 495, 715, ... (Sloane's A000332).

Let us give a simple geometrical illustration of the above formula for $n = 5$ (see [CoGu96]).

$$1 + \begin{matrix} 2 \\ 1\ \ 1 \end{matrix} + \begin{matrix} 3 \\ 2\ \ 2 \\ 1\ \ 1\ \ 1 \end{matrix} + \begin{matrix} 4 \\ 3\ \ 3 \\ 2\ \ 2\ \ 2 \\ 1\ \ 1\ \ 1\ \ 1 \end{matrix} + \begin{matrix} 5 \\ 4\ \ 4 \\ 3\ \ 3\ \ 3 \\ 2\ \ 2\ \ 2\ \ 2 \\ 1\ \ 1\ \ 1\ \ 1\ \ 1 \end{matrix} +$$

$$1 + \begin{matrix} 1 \\ 1\ \ 2 \end{matrix} + \begin{matrix} 1 \\ 1\ \ 2 \\ 1\ \ 2\ \ 3 \end{matrix} + \begin{matrix} 1 \\ 1\ \ 2 \\ 1\ \ 2\ \ 3 \\ 1\ \ 2\ \ 3\ \ 4 \end{matrix} + \begin{matrix} 1 \\ 1\ \ 2 \\ 1\ \ 2\ \ 3 \\ 1\ \ 2\ \ 3\ \ 4 \\ 1\ \ 2\ \ 3\ \ 4\ \ 5 \end{matrix} +$$

$$1 + \begin{array}{c} 1 \\ 2\ \ 1 \end{array} + \begin{array}{c} 1 \\ 2\ \ 1 \\ 3\ \ 2\ \ 1 \end{array} + \begin{array}{c} 1 \\ 2\ \ 1 \\ 3\ \ 2\ \ 1 \\ 4\ \ 3\ \ 2\ \ 1 \end{array} + \begin{array}{c} 1 \\ 2\ \ 1 \\ 3\ \ 2\ \ 1 \\ 4\ \ 3\ \ 2\ \ 1 \\ 5\ \ 4\ \ 3\ \ 2\ \ 1 \end{array} +$$

$$5 + \begin{array}{c} 4 \\ 4\ \ 4 \end{array} + \begin{array}{c} 3 \\ 3\ \ 3 \\ 3\ \ 3\ \ 3 \end{array} + \begin{array}{c} 4 \\ 4\ \ 4 \\ 4\ \ 4\ \ 4 \\ 4\ \ 4\ \ 4\ \ 4 \end{array} + \begin{array}{c} 5 \\ 5\ \ 5 \\ 5\ \ 5\ \ 5 \\ 5\ \ 5\ \ 5\ \ 5 \\ 5\ \ 5\ \ 5\ \ 5\ \ 5 \end{array} =$$

$$8 + \begin{array}{c} 8 \\ 8\ \ 8 \end{array} + \begin{array}{c} 8 \\ 8\ \ 8 \\ 8\ \ 8\ \ 8 \end{array} + \begin{array}{c} 8 \\ 8\ \ 8 \\ 8\ \ 8\ \ 8 \\ 8\ \ 8\ \ 8\ \ 8 \end{array} + \begin{array}{c} 8 \\ 8\ 8 \\ 8\ \ 8\ \ 8 \\ 8\ \ 8\ \ 8\ \ 8 \\ 8\ \ 8\ \ 8\ \ 8\ \ 8 \end{array}$$

In the above figure, four copies of the first five tetrahedral numbers (for the last copy we use the representation $S_3^3(1) + \cdots + S_3^3(5) = 5S_3(1)+4S_3(2)+3S_3(3)+2S_3(4)+S_3(5)$) add up to $8 = 5+3$ times the sum of the first five triangular numbers. In other words, the value $4S_3^4(5) = 4(S_3^3(1)+S_3^3(2)+\cdots+S_3^3(5))$ is equal to $8(S_3(1)+S_3(2)+\cdots+S_3(5)) = (5+3)S_3^3(5)$, or, algebraically, $4S_3^4(5) = (5+3)S_3^3(5)$. In general, $4S_3^4(n) = (n+3)S_3^3(n)$, and, since $S_3^3(n) = \frac{n(n+1)(n+2)}{6}$, we get $S_3^4(n) = \frac{(n+3)}{4}S_3^3(n) = \frac{n(n+1)(n+2)(n+3)}{24}$.

3.1.2. Using above recurrent formula, it is easy to obtain the generating function for the sequence of pentatope numbers, which has the form $f(x) = \frac{x}{(1-x)^5}$. In other words, it holds

$$\frac{x}{(1-x)^5} = S_3^4(1)x + S_3^4(2)x^2 + S_3^4(3)x^3 + \cdots + S_3^4(n)x^n + \cdots, \quad |x| < 1.$$

In fact, let us consider the recurrent equation $S_3^4(n+1) = S_3^4(n) + S_3^3(n+1)$. Going from n to $n+1$, one obtains $S_3^4(n+2) = S_3^4(n+1)+S_3^3(n+2)$. Subtracting first equality from second one, we get $S_3^4(n+2)-S_3^4(n+1) = S_3^4(n+1)-S_3^4(n)+(S_3^3(n+2)-S_3^3(n+1))$, i.e., $S_3^4(n+2) = 2S_3^4(n+1) - S_3^4(n) + S_3(n+2)$. Similarly, one has $S_3^4(n+3) = 2S_3^4(n+2) - S_3^4(n+1) + S_3(n+3)$, and

$$S_3^4(n+3) - S_3^4(n+2) = 2S_3^4(n+2) - 2S_3^4(n+1) \\ - S_3^4(n+1) + S_3^4(n) + (S_3(n+3) - S_3(n+2)),$$

i.e., $S_3^4(n+3) = 3S_3^4(n+2) - 3S_3^4(n+1) + S_3^4(n) + (n+3)$. Going from $n+3$ to $n+4$, one obtains

$$S_3^4(n+4) = 3S_3^4(n+3) - 3S_3^4(n+2) + S_3^4(n+1) + (n+4).$$

Hence, $S_3^4(n+4) - S_3^4(n+3) = 3S_3^4(n+3) - 3S_3^4(n+2) - 3S_3^4(n+2) + 3S_3^4(n+1) + S_3^4(n+1) - S_3^4(n) + 1$, i.e., it holds

$$S_3^4(n+4) = 4S_3^4(n+3) - 6S_3^4(n+2) + 4S_3^4(n+1) - S_3^4(n) + 1.$$

Finally, $S_3^4(n+5) = 4S_3^4(n+4) - 6S_3^4(n+3) + 4S_3^4(n+2) - S_3^4(n+1) + 1$, and

$$S_3^4(n+5) - S_3^4(n+4) = 4S_3^4(n+4) - 4S_3^4(n+3) - 6S_3^4(n+3) \\ + 6S_3^4(n+2) + 4S_3^4(n+2) - 4S_3^4(n+1) - S_3^4(n+1) + S_3^4(n),$$

i.e., it holds $S_3^4(n+5) = 5S_3^4(n+4) - 10S_3^4(n+3) + 10S_3^4(n+2) - 5S_3^4(n+1) + S_3^4(n)$. Therefore, we get, for the sequence of pentatope numbers, a linear recurrent equation

$$S_3^4(n+5) - 5S_3^4(n+4) + 10S_3^4(n+3) - 10S_3^4(n+2) + 5S_3^4(n) - S_3^4(n) = 0$$

of 5-th order with initial values $S_3^4(1) = 1$, $S_3^4(2) = 5$, $S_3^4(3) = 15$, $S_3^4(4) = 35$, and $S_3^4(5) = 70$. Denoting $S_3^4(n+1)$ by c_n, one can rewrite above equation as

$$c_{n+4} - 5c_{n+4} + 10c_{n+3} - 10c_{n+2} + 5c_{n+1} - c_n = 0, \\ c_0 = 1, \quad c_1 = 5, \quad c_2 = 15, \quad c_3 = 35, \quad c_4 = 70.$$

Therefore, the generating function for the sequence of pentatope numbers has the form

$$\frac{f(x)}{g(x)} = \frac{a_0 + a_1x + a_2x^2 + a_3x^3 + a_4x^4}{b_0 + b_1x + b_2x^2 + b_3x^3 + b_4x^4 + b_5x^5},$$

where $b_0 = 1$, $b_1 = -5$, $b_2 = 10$, $b_3 = -10$, $b_4 = 5$, $b_5 = -1$, and $a_0 = b_0c_0 = 1$, $a_1 = b_0c_1 + b_1c_0 = 1\cdot 5 + (-5)\cdot 1 = 0$, $a_2 = b_0c_2 + b_1c_1 + b_2c_0 = 1\cdot 15 + (-5)\cdot 5 + 10\cdot 1 = 0$, $a_3 = b_0c_3 + b_1c_2 + b_2c_1 + b_3c_0 = 1\cdot 35 + (-5)\cdot 15 + 10\cdot 5 + (-10)\cdot 1 = 0$, $a_4 = b_0c_4 + b_1c_3 + b_2c_2 + b_3c_1 + b_4c_0 = 1\cdot 70 + (-5)\cdot 35 + 10\cdot 15 + (-10)\cdot 5 + 5\cdot 1 = 0$, Since $g(x) = 1 - 5x + 10x^2 - 10x^3 + 5x^4 - x^5 = (1-x)^5$ has five coinciding roots $x_1 = \cdots = x_5 = 1$, it holds

$$\frac{1}{(1-x)^5} = S_3^4(1) + S_3^4(2)x + S_3^4(3)x^2 + \cdots + S_3^4(n)x^{n-1} + \cdots, \quad |x| < 1.$$

This result can be obtained also using the generating function $\frac{1}{(1-x)^4} = S_3^3(1) + S_3^3(2)x + S_3^3(3)x^2 + \cdots + S_3^3(n)x^{n-1} + \cdots$ for the tetrahedral numbers and the decomposition $\frac{1}{1-x} = 1 + x + x^2 + \cdots + x^n + \cdots$. In fact, one has

$$\begin{aligned}\frac{1}{(1-x)^5} &= \frac{1}{(1-x)^4}\cdot\frac{1}{1-x}\\ &= (S_3^3(1) + S_3^3(2)x + S_3^3(3)x^2 + \cdots + S_3^3(n)x^{n-1} + \cdots)\\ &\quad\times(1 + x + x^2 + \cdots + x^n + \cdots)\\ &= S_3^3(1) + (S_3^3(1) + S_3^3(2))x + (S_3^3(1) + S_3^3(2) + S_3^3(3))x^2\\ &\quad + \ldots + (S_3^3(1) + \cdots + S_3^3(n))x^{n-1} + \cdots\\ &= S_3^4(1) + S_3^4(2)x + S_3^4(3)x^2\\ &\quad + \cdots + S_3^4(n)x^{n-1} + \cdots, \quad |x| < 1.\end{aligned}$$

3.1.3. The formulas $S_3(n) = \frac{n(n+1)}{2}$ and $S_3^3(n) = \frac{n(n+1)(n+2)}{6}$ imply the identities

$$S_3^4(n) = \frac{1}{12}(n+2)(n+3)S_3(n), \quad \text{and} \quad S_3^4(n) = \frac{1}{4}(n+3)S_3^3(n).$$

It is easy to show, that *two of every three consecutive pentatope numbers are also pentagonal numbers*. More exactly, $(3k-2)$-th pentatope number is always $(\frac{3k^2-k}{2})$-th pentagonal number, and $(3k-1)$-th pentatope number is always $(\frac{3k^2+k}{2})$-th pentagonal number. Moreover, $(3k)$-th pentatope number is generalized pentagonal number obtained by taking the negative index $-\frac{3k^2+3k}{2}$:

$$S_3^4(3k-2) = S_5\left(\frac{3k^2-k}{2}\right),$$

$$S_3^4(3k-1) = S_5\left(\frac{3k^2+k}{2}\right), \quad \text{and}$$

$$S_3^4(3k) = S_5\left(-\frac{3k^2+3k}{2}\right).$$

In fact, it holds

$$S_5\left(\frac{3k^2-k}{2}\right) = \frac{\frac{3k^2-k}{2}(3\frac{3k^2-k}{2}-1)}{2} = \frac{k(3k-1)(3k(3k-1)-2)}{8}$$
$$= \frac{k(3k-1)(9k^2-3k-2)}{8}$$
$$= \frac{(3k-2)(3k-1)3k(3k+1)}{24} = S_3^4(3k-2);$$

$$S_5\left(\frac{3k^2+k}{2}\right) = \frac{\frac{3k^2+k}{2}(3\frac{3k^2+k}{2}-1)}{2} = \frac{k(3k+1)(3k(3k+1)-2)}{8}$$
$$= \frac{k(3k+1)(9k^2+3k-2)}{8}$$
$$= \frac{(3k-1)3k(3k+1)(3k+2)}{24} = S_3^4(3k-1);$$

$$S_5\left(-\frac{3k^2+3k}{2}\right) = \frac{-\frac{3k^2+3k}{2}(-3\frac{3k^2+3k}{2}-1)}{2}$$
$$= \frac{k(3k+3)(3k(3k+3)+2)}{8}$$
$$= \frac{k(3k+3)(9k^2+9k+2)}{8}$$
$$= \frac{3k(3k+1)(3k+2)(3k+3)}{24} = S_3^4(3k).$$

3.1.4. In general, for a given dimension k, $k \geq 2$, an k-**dimensional hypertetrahedron number** (or k-**hypertetrahedron number**, S_3^k-**number**, **regular** k-**polytopic number**, **figurate number of**

order k, or, for $k = 5$, **pyramidi-pyramidal number**) is a figurate number which represents an k-dimensional hypertetrahedron.

The n-th k-dimensional hypertetrahedron number $S_3^k(n)$ is given as the sum of the first n $(k-1)$-dimensional hypertetrahedron numbers:

$$S_3^k(n) = S_3^{(k-1)}(1) + S_3^{(k-1)}(2) + \cdots + S_3^{(k-1)}(n).$$

In particular, for $k = 2, 3$, and 4 one obtains triangular, tetrahedral and pentatope numbers, respectively.

The **five-dimensional hypertetrahedron numbers** start with elements 1, 6, 21, 56, 126, 252, 462, 792, 1287, 2002, ... (Sloane's A000389). They are obtained as partial sums of the sequence 1, 5, 15, 35, 70, 126, 210, 330, 495, 715, ... (Sloane's A000332) of pentatope numbers.

The **six-dimensional hypertetrahedron numbers** start with elements 1, 7, 28, 84, 210, 462, 924, 1716, 3003, 5005, ... (Sloane's A000579). They are obtained as partial sums of the sequence 1, 7, 28, 84, 210, 462, 924, 1716, 3003, 5005, ... (Sloane's A000579) of five-dimensional hypertetrahedron numbers.

By definition, we get the following recurrent formula for the sequence of k-dimensional hypertetrahedron numbers:

$$S_3^k(n+1) = S_3^k(n) + S_3^{(k-1)}(n+1), \quad S_3^k(1) = 1.$$

Now it is easy to obtain the general formula for n-th k-dimensional hypertetrahedron number, which has the form

$$S_3^k(n) = \frac{n(n+1)(n+2)\cdots(n+(k-1))}{k!}, \quad \text{or} \quad S_3^k(n) = \frac{n^{\overline{k}}}{k!},$$

where $n^{\overline{k}} = n(n+1)(n+2)\cdots(n+(k-1))$ is a *rising factorial.*

Let us prove it by induction. In fact, the formula holds for $k = 2$, 3 and 4, i.e., for triangular, tetrahedral and pentatope numbers. Supposing that the formula holds for $k-1$, we get for k the following considerations. If $n = 1$, one has $S_3^k(1) = 1$, i.e., the formula holds: $S_3^k(1) = \frac{1\cdot(1+1)\ldots(1+(k-1))}{k!}$. Going from n to $n+1$, we obtain

$$S_3^k(n+1) = S_3^k(n) + S_3^{k-1}(n+1) = \frac{n(n+1)(n+2)\cdots(n+(k-1))}{k!} + \frac{(n+1)(n+2)(n+3)\cdots((n+1)+(k-2))}{(k-1)!} = \frac{(n+1)(n+2)\cdots(n+(k-1))(n+k)}{k!} = \frac{(n+1)(n+2)\cdots((n+1)+(k-1))}{k!},$$

which concludes the proof.

3.1.5. Using above recurrent formula, it is easy to obtain the generating function for the sequence of k-dimensional hypertetrahedron numbers, which has the form $f(x) = \frac{x}{(1-x)^{k+1}}$. In other words, it holds

$$\frac{x}{(1-x)^{k+1}} = S_3^k(1)x + S_3^k(2)x^2 + S_3^k(3)x^3 + \cdots + S_3^k(n)x^n + \cdots, \quad |x| < 1.$$

This result can be proved by induction on k. For $k = 2, 3, 4$ it was obtained before, as generating functions for triangular, tetrahedral and pentatope numbers. Let the result be true for k, i.e., let the generating function for k-dimensional hypertetrahedron numbers be of the form

$$\frac{1}{(1-x)^k} = S_3^k(1) + S_3^k(2)x + S_3^k(3)x^2 + \cdots + S_3^k(n)x^{n-1} + \cdots, \quad |x| < 1.$$

Then, using the decomposition $\frac{1}{1-x} = 1 + x + x^2 + \cdots + x^n + \cdots$, we obtain

$$\frac{1}{(1-x)^{k+1}} = \frac{1}{(1-x)^k}\cdot\frac{1}{1-x} = (S_3^k(1) + S_3^k(2)x + S_3^k(3)x^2 + \cdots + S_3^k(n)x^{n-1} + \cdots) \times (1 + x + x^2 + \cdots + x^n + \cdots)$$

$$= S_3^k(1) + (S_3^k(1) + S_3^k(2))x + (S_3^k(1) + S_3^k(2)$$
$$+S_3^k(3))x^2 + \cdots + (S_3^k(1) + \cdots + S_3^k(n))x^{n-1} + \cdots$$
$$= S_3^{k+1}(1) + S_3^{k+1}(2)x + S_3^{k+1}(3)x^2$$
$$+ \cdots + S_3^{k+1}(n)x^{n-1} + \cdots, \quad |x| < 1.$$

3.1.6. Clearly, one get that $nS_3^k(n+1) = (k+1)S_3^{k+1}(n)$. So, the following identities hold:

$$S_3^k(n) = \frac{1}{k}(n + (k-1))S_3^{(k-1)}(n),$$

$$S_3^k(n) = \frac{1}{(k-1)k}(n + (k-2))(n + (k-3))S_3^{(k-2)}(n), \ldots,$$

$$S_3^k(n) = \frac{1}{4 \cdot \cdots \cdot k}(n+3) \cdots (n + (k-1))S_3^3(n) \quad \text{and}$$

$$S_3^k(n) = \frac{1}{3 \cdot \cdots \cdot k}(n+2) \cdots (n + (k-1))S_3(n).$$

3.2 Biquadratic numbers and their multidimensional analogues

3.2.1. A **biquadratic number** (or **hypercubic number**) is a fourth power of a positive integer. So, n-th biquadratic number $BC(n)$ can be obtained as the sum of n copies of n-th cubic number $C(n)$ and is given by the formula

$$BC(n) = n^4.$$

The first few biquadratic numbers are 1, 16, 81, 256, 625, 1296, 2401, 4096, 6561, 10000, ... (Sloane's A000583).

Since $(n+1)^4 = n^4 + 4n^3 + 6n^2 + 4n + 1$, we obtain the following recurrent formula for the sequence of biquadratic numbers:

$$BC(n+1) = BC(n) + 4n^3 + 6n^2 + 4n + 1, \quad BC(1) = 1.$$

As was shown before, *any biquadratic number is a sum of the first n rhombic dodecahedral numbers*:

$$BC(n) = RD(1) + RD(2) + \cdots + RD(n).$$

So, it holds also the identity

$$BC(n) - BC(n-1) = RD(n),$$

i.e., the rhombic dodecahedral numbers are gnomons of biquadratic numbers.

Geometrically, $(n+1)$-th rhombic dodecahedral number $RD(n+1) = 1 + 4n + 6n^2 + 4n^3$ is a projection of $(n+1)^4$. In fact, starting from one point, build out four rods of n cells; then add six walls, each of n^2 cells, one between each pair of rods; finally, insert four blocks of n^3 cells, one in each solid angle formed by three walls. We get a nest that will enclose $n \times n \times n \times n$ bi-square, making it up to $(n+1)^4$. This is a special case $1 + 4n + 6n^2 + 4n^3 = (1+n)^3$ of the binomial theorem (see [CoGu96]).

3.2.2. The standard procedure gives now the generating function for the sequence of biquadratic numbers, which has the form $f(x) = \frac{x(1+11x+11x^2+x^3)}{(1-x)^5}$. In other words, it holds

$$\frac{x(1+11x+11x^2+x^3)}{(1-x)^5} = BC(1)x + BC(2)x^2 + BC(3)x^3 + \cdots + BC(n)x^n + \cdots, \quad |x| < 1.$$

In fact, let us consider the recurrent equation $BC(n+1) = BC(n) + 4n^3 + 6n^2 + 4n + 1$. Going from n to $n+1$, one obtains $BC(n+2) = BC(n+1) + 4(n+1)^3 + 6(n+1)^2 + 4(n+1) + 1$. Subtracting first equality from second one, we get

$$BC(n+2) - BC(n+1) = BC(n+1) - BC(n) + (12n^2 + 24n + 14),$$

i.e., $BC(n+2) = 2BC(n+1) - BC(n) + (12n^2 + 4n + 14)$. Similarly, one has $BC(n+3) = 2BC(n+2) - BC(n+1) + (12(n+1)^2 + 4(n+1) + 14)$, and

$$BC(n+3) - BC(n+2) = 2BC(n+2) - 2BC(n+1) - BC(n+1) + BC(n) + (24n + 16),$$

i.e., $BC(n+3) = 3BC(n+2) - 3BC(n+1) + BC(n) + (24n+16)$. Going from $n+3$ to $n+4$, one obtains

$$BC(n+4) = 3BC(n+3) - 3BC(n+2) + BC(n+1) + (24(n+1) + 16).$$

Hence,

$$BC(n+4)-BC(n+3)=3BC(n+3)-3BC(n+2)-3BC(n+2)$$
$$+3BC(n+1)+BC(n+1)-BC(n)+24,$$

i.e., it holds $BC(n+4) = 4BC(n+3) - 6BC(n+2) + 4BC(n+1) - BC(n) + 24$. Finally, $BC(n+5) = 4BC(n+4) - 6BC(n+3) + 4BC(n+2) - BC(n+1) + 1$, and

$$BC(n+5)-BC(n+4)=4BC(n+4)-4BC(n+3)-6BC(n+3)$$
$$+6BC(n+2)+4BC(n+2)$$
$$-4BC(n+1)-BC(n+1)+BC(n),$$

i.e., it holds

$$BC(n+5)=5BC(n+4)-10BC(n+3)+10BC(n+2)$$
$$-5BC(n+1)+BC(n).$$

So, we get for the sequence of biquadratic numbers the linear recurrent equation

$$BC(n+5)-5BC(n+4)$$
$$+10BC(n+3)-10BC(n+2)+5BC(n)-BC(n)=0$$

of 5-th order with initial values $BC(1) = 1$, $BC(2) = 16$, $BC(3) = 81$, $BC(4) = 256$, and $BC(5) = 625$. Denoting $BC(n+1)$ by c_n, one can rewrite above equation as

$$c_{n+4}-5c_{n+4}+10c_{n+3}-10c_{n+2}+5c_{n+1}-c_n=0,$$
$$c_0=1,\quad c_1=16,\quad c_2=81,\quad c_3=256,\quad c_4=625.$$

Therefore, the generating function for the sequence of biquadratic numbers has the form

$$\frac{f(x)}{g(x)}=\frac{a_0+a_1x+a_2x^2+a_3x^3+a_4x^4}{b_0+b_1x+b_2x^2+b_3x^3+b_4x^4+b_5x^5},$$

where $b_0 = 1$, $b_1 = -5$, $b_2 = 10$, $b_3 = -10$, $b_4 = 5$, $b_5 = -1$, and $a_0 = b_0c_0 = 1$, $a_1 = b_0c_1 + b_1c_0 = 1 \cdot 16 + (-5) \cdot 1 = 11$,

$a_2 = b_0c_2 + b_1c_1 + b_2c_0 = 1 \cdot 81 + (-5) \cdot 16 + 10 \cdot 1 = 11$, $a_3 = b_0c_3 + b_1c_2 + b_2c_1 + b_3c_0 = 1 \cdot 256 + (-5) \cdot 81 + 10 \cdot 16 + (-10) \cdot 1 = 1$, $a_4 = b_0c_4 + b_1c_3 + b_2c_2 + b_3c_1 + b_4c_0 = 1 \cdot 625 + (-5) \cdot 256 + 10 \cdot 81 + (-10) \cdot 16 + 5 \cdot 1 = 0$, Since $g(x) = 1 - 5x + 10x^2 - 10x^3 + 5x^4 - x^5 = (1-x)^5$ has five coinciding roots $x_1 = \cdots = x_5 = 1$, it holds

$$\frac{1 + 11x + 11x^2 + x^3}{(1-x)^5} = BC(1) + BC(2)x + BC(3)x^2 + \cdots + BC(n)x^{n-1} + \cdots, \quad |x| < 1.$$

This result can be obtained also using the generating function $\frac{1+11x+11x^2+x^3}{(1-x)^4} = RD(1) + RD(2)x + RD(3)x^2 + \cdots + RD(n)x^{n-1} + \cdots$ and the decomposition $\frac{1}{1-x} = 1 + x + x^2 + \cdots + x^n + \cdots$. In fact, one has

$$\begin{aligned}
&\frac{1 + 11x + 11x^2 + x^3}{(1-x)^5} \\
&= \frac{1 + 11x + 11x^2 + x^3}{(1-x)^4} \cdot \frac{1}{1-x} \\
&= (RD(1) + RD(2)x + RD(3)x^2 + \cdots + RD(n)x^{n-1} + \cdots) \\
&\quad \times (1 + x + x^2 + \cdots + x^n + \cdots) \\
&= RD(1) + (RD(1) + RD(2))x + (RD(1) + RD(2) \\
&\quad + RD(3))x^2 + \cdots + (RD(1) + \cdots + RD(n))x^{n-1} + \cdots \\
&= BC(1) + BC(2)x + BC(3)x^2 + \cdots + BC(n)x^{n-1} + \cdots, \\
&\qquad |x| < 1.
\end{aligned}$$

3.2.3. Obviously, the following identities hold:

$$BC(n) = nC(n), \quad \text{and} \quad BC(n) = (S_4(n))^2.$$

It holds also that *any biquadratic number is a linear combination with non-negative coefficients of polytope numbers*:

$$BC(n) = S_3^4(n) + 11S_3^4(n-1) + 11S_3^4(n-2) + S_3^4(n-3).$$

The proof can be obtained by direct computation (see [Kim03]).

3.2.4. Given a dimension k, $k \geq 5$, one obtains **k-dimensional hypercube numbers** (or **k-hypercube numbers**, **C^k-numbers**) as an k-dimensional analogue of four-dimensional biquadratic numbers, three-dimensional cubic numbers, and two-dimensional square numbers, putting

$$C^k(n) = n^k.$$

So, it holds $S_4(n) = C^2(n)$ for square numbers, $C(n) = C^3(n)$ for cubic numbers, and $BC(n) = C^4(n)$ for biquadratic numbers.

The sequence of **five-dimensional hypercube numbers**, i.e., 5-th powers n^5, starts with the elements 1, 32, 243, 1024, 3125, 7776, 16807, 32768, 59049, 100000, ... (Sloane's A000584).

The sequence of **six-dimensional hypercube numbers**, i.e., 6-th powers n^6, starts with elements 1, 64, 729, 4096, 15625, 46656, 117649, 262144, 531441, 1000000, ... (Sloane's A001014).

Since $(n+1)^k = n^k + \binom{k}{1}n^{k-1} + \binom{k}{2}n^{k-2} + \cdots + \binom{k}{k-1}n + 1$, one obtains the following recurrent formula for the sequence of k-dimensional hypercube numbers:

$$C^k(n+1) = C^k(n) + \binom{k}{1}n^{k-1} + \binom{k}{2}n^{k-2} + \cdots + \binom{k}{k-1}n + 1, \quad C^k(1) = 1.$$

3.2.5. The generation function for the sequence of k-dimensional hypercube numbers has the form

$$\frac{x(1 + a_1^k x^2 + \cdots + a_{k-2}^k x^{k-2} + x^{k-1})}{(1-x)^{k+1}} = C^k(1)x + C^k(2)x^2 + C^k(3)x^3 + \cdots + C^k(n)x^n + \cdots, \quad |x| < 1,$$

where $a_1^k = a_{k-2} = 2^k - (k+1)$, $a_2^k = a_{k-3}^k = 3^k - (k+1)2^k + \frac{k(k+1)}{2}, \ldots, a^k_{\lfloor \frac{k-1}{2} \rfloor} = a^k_{k-\lfloor \frac{k-1}{2} \rfloor - 1} = (\lfloor \frac{k-1}{2} \rfloor + 1)^k - (k+1)(\lfloor \frac{k-1}{2} \rfloor)^k + \frac{k(k+1)}{2}(\lfloor \frac{k-1}{2} \rfloor - 1)^k + \cdots + (-1)^{\lfloor \frac{k-1}{2} \rfloor}\binom{k+1}{\lfloor \frac{k-1}{2} \rfloor}$.

In particular, we have

$$\frac{x(1+x)}{(1-x)^3} = x + 4x^2 + 9x^3 + \cdots + n^2x^n + \cdots, \quad |x| < 1,$$

$$\frac{x(1+4x+x^2)}{(1-x)^4} = x + 8x^2 + 27x^3 + \cdots + n^3x^n + \cdots, \quad |x| < 1,$$

$$\frac{x(1+11x+11x^2+x^3)}{(1-x)^5} = \frac{x(x+1)(1+10x+x^2)}{(1-x)^5}$$
$$= x + 16x^2 + 81x^3 + \cdots + n^4x^n + \cdots, \quad |x| < 1,$$

$$\frac{x(1+26z+66x^2+26x^3+x^4)}{(1-x)^6}$$
$$= x + 32x^2 + 243x^3 + \cdots + n^5x^n + \cdots, \quad |x| < 1,$$

$$\frac{x(1+57x+302x^2+302x^3+57x+x^5)}{(1-x)^7}$$
$$= \frac{x(x+1)(1+56x+246x^2+56x^3+x^4)}{(1-x)^7}$$
$$= x + 64x^2 + 729x^3 + \cdots + n^6x^n + \cdots, \quad |x| < 1,$$

$$\frac{x(1+120x+1191x^2+2416x^3+1191x^4+120x^5+x^6)}{(1-x)^8}$$
$$= x + 128x^2 + 2187x^3 + \cdots + n^7x^n + \cdots, \quad |x| < 1,$$

$$\frac{x(1+247x+4293x^2+15619x^3+15619x^4+4293x^5+247x^6+x^7)}{(1-x)^9}$$
$$= \frac{x(1+x)(1+246x+4047x^2+11572x^3+4047x^4+246x^5+x^6)}{(1-x)^9}$$
$$= x + 256x^2 + 6561x^3 + \cdots + n^8x^n + \cdots, \quad |x| < 1.$$

This result can be obtained by the standard procedure. In fact, let us consider the recurrent equation $C^k(n+1) = C^k(n) + f_1(n^{k-1})$, where $f_1(n^{k-1}) = \binom{k}{1}n^{k-1} + \binom{k}{2}n^{k-2} + \cdots + \binom{k}{k-1}n + 1$ is a polynomial in n of degree $k-1$. Going from n to $n+1$, one obtains

$C^k(n+2) = C^k(n+1) + f_1((n+1)^{k-1})$. Subtracting first equality from second one, we get

$$C^k(n+2) - C^k(n+1) = C^k(n+1) - C^k(n) + f_2(n^{k-2}),$$

where $f_2(n^{k-2}) = f_1((n+1)^{k-1}) - f_1(n^{k-1})$ is a polynomial in n of degree $k-2$. In other words, $C^k(n+2) = 2C^k(n+1) - C^k(n) + f_2(n^{k-2})$. Similarly, one has

$$C^k(n+3) = 2C^k(n+2) - C^k(n+1) + f_2((n+1)^{k-2}),$$

and

$$\begin{aligned} &C^k(n+3) - C^k(n+2) \\ &\quad = 2C^k(n+2) - 2C^k(n+1) - C^k(n+1) + C^k(n) + f_3(n^{k-3}), \end{aligned}$$

where $f_3(n^{k-3}) = f_2((n+1)^{k-2}) - f_2(n^{k-2})$ is a polynomial in n of the degree $k-3$. In other words, $C^k(n+3) = 3C^k(n+2) - 3C^k(n+1) + C^k(n) + f_3(n^{k-3})$. Going from $n+3$ to $n+4$, one obtains

$$C^k(n+4) = 3C^k(n+3) - 3C^k(n+2) + C^k(n+1) + f_3((n+1)^{k-3}),$$

and

$$\begin{aligned} &C^k(n+4) - C^k(n+3) \\ &\quad = 3C^k(n+3) - 3C^k(n+2) - 3C^k(n+2) + 3C^k(n+1) \\ &\qquad + C^k(n+1) - C^k(n) + f_4(n^{k-4}), \end{aligned}$$

where $f_4(n^{k-4}) = f_3((n+1)^{k-3}) - f_3(n^{k-3})$ is a polynomial in n of degree $k-4$. Hence, it holds

$$C^k(n+4) = 4C^k(n+3) - 6C^k(n+2) + 4C^k(n+1) - C^k(n) + f_4(n^{k-4}).$$

So, on the i-th step of this procedure we get

$$\begin{aligned} B_0^i C^k(n+i) = {} & B_1^i C^k(n+i-1) - B_2^i C^k(n+i-2) \\ & + B_3^i C^k(n+i-3) - \cdots + (-1)^{i-1} B_i^i C^k(n) + f_i(n^{k-i}), \end{aligned}$$

where $B_0^i = B_i^i = 1$. Going from $n+i$ to $n+i+1$, we obtain

$$\begin{aligned} B_0^i C^k(n+i+1) = B_1^i C^k(n+i) - B_2^i C^k(n+i-1) \\ + B_3^i C^k(n+i-2) - \cdots + (-1)^{i-1} B_i^i C^k(n+1) \\ + f_i((n+1)^{k-i}). \end{aligned}$$

So, subtracting first equality from the second one, we obtain

$$\begin{aligned} &B_0^i C^k(n+i+1) \\ &\quad = (B_1^i + B_0^i) C^k(n+i) - (B_2^i + B_1^i) C^k(n+i-1) \\ &\qquad + (B_3^i + B_2^i) C^k(n+i-2) - \cdots + (-1)^{i-1}(B_i^i + B_{i-1}^i) C^k(n+1) \\ &\qquad + (-1)^i B_i^i C^k(n) + (f_i((n+1)^{k-i}) - f_i(n^{k-i})). \end{aligned}$$

In other words,

$$\begin{aligned} &B_0^{i+1} C^k(n+i+1) \\ &\quad = B_1^{i+1} C^k(n+i) - B_2^{i+1} C^k(n+i-1) \\ &\qquad + B_3^{i+1} C^k(n+i-2) - \cdots + (-1)^{i-1} B_i^{i+1} C^k(n+1) \\ &\qquad + (-1)^i B_{i+1}^{i+1} C^k(n) + f_{i+1}(n^{k-i-1}), \end{aligned}$$

where

$$B_0^{i+1} = B_0^i = \cdots = B_0^1 = 1, \quad B_{i+1}^{i+1} = B_i^i = \cdots = B_1^1 = 1,$$
$$1 \le m \le i, \quad B_m^{i+1} = B_m^i + B_{m-1}^i,$$

and $f_{i+1}(n^{k-i-1}) = f_i((n+1)^{k-i}) - f_i(n^{k-i})$ is a polynomial in n of degree $k-i-1$.

It is easy to see that the coefficients B_m^{i+1}, $0 \le m \le i+1$, have the same recurrent equation and the same initial conditions as the binomial coefficients $\binom{i+1}{m}$, $0 \le m \le i+1$. In fact, $\binom{i+1}{0} = \binom{i}{0} = \cdots = \binom{1}{0} = 1$, $\binom{i+1}{i+1} = \binom{i}{i} = \cdots = \binom{1}{1} = 1$, and $\binom{i+1}{m} = \binom{i}{m} + \binom{i}{m-1}$,

$1 \le m \le i$. So, we get, $B_m^{i+1} = \binom{i+1}{m}$, $0 \le m \le i+1$, and, hence,

$$\begin{aligned}\binom{i+1}{0}&C^k(n+i+1)\\&= \binom{i+1}{1}C^k(n+i) - \binom{i+1}{2}C^k(n+i-1)\\&\quad+\binom{i+1}{3}C^k(n+i-1) - \cdots + (-1)^{i-1}\binom{i+1}{i}C^k(n+1)\\&\quad+(-1)^i\binom{i+1}{i+1}C^k(n) + f_{i+1}(n^{k-i-1}),\end{aligned}$$

where $f_{i+1}(n^{k-i-1})$ is a polynomial in n of degree $k-i-1$.

Taking $i = k-1$, we obtain the corresponding recurrent equation with coefficients $\binom{k}{m}$, $0 \le m \le k$, and the polynomial $f_k(n^{k-k})$ of degree 0, which is some constant. So, on the last step, for $i = k$, we get $f_{k+1} \equiv 0$, and, hence, the following recurrent equation:

$$\begin{aligned}\binom{k+1}{0}&C^k(n+k+1)\\&= \binom{k+1}{1}C^k(n+k) - \binom{k+1}{2}C^k(n+k-1)\\&\quad+\binom{k+1}{3}C^k(n+k-1) - \cdots + (-1)^{k-1}\binom{k+1}{k}C^k(n+1)\\&\quad+(-1)^k\binom{k+1}{k+1}C^k(n).\end{aligned}$$

In other words, we have

$$\begin{aligned}&\binom{k+1}{0}C^k(n+k+1) - \binom{k+1}{1}C^k(n+k)\\&\quad+\binom{k+1}{2}C^k(n+k-1) - \binom{k+1}{3}C^k(n+k-1)\\&\quad+\cdots+(-1)^k\binom{k+1}{k}C^k(n+1) + (-1)^{k+1}\binom{k+1}{k+1}C^k(n) = 0.\end{aligned}$$

So, we get for the sequence of k-dimensional hypercube numbers a linear recurrent equation of $(k+1)$-th order. Its initial values are $C^k(1) = 1^k$, $C^k(2) = 2^k$, $C^k(3) = 3^k$, $C^k(4) = 4^k, \ldots, C^k(k+1) = (k+1)^k$. Denoting $C^k(n+1)$ by c_n, one can rewrite above recurrent equation as

$$\binom{k+1}{0} c_{n+k} - \binom{k+1}{1} c_{n+k-1} + \binom{k+1}{2} c_{n+k-2} - \binom{k+1}{3} c_{n+k-2} + \cdots + (-1)^k \binom{k+1}{k} c_n + (-1)^{k+1} \binom{k+1}{k+1} c_{n-1} = 0,$$

$$c_0 = 1, \quad c_1 = 2^k, \quad c_2 = 3^k, \ldots, c_k = (k+1)^k.$$

So, the generating function for the sequence of k-dimensional hypercube numbers has the form

$$\frac{f(x)}{g(x)} = \frac{a_0 + a_1 x + a_2 x^2 + a_3 x^3 + \cdots + a_k x^k}{b_0 + b_1 x + b_2 x^2 + b_3 x^3 + b_4 x^4 + \cdots + b_{k+1} x^{k+1}},$$

where $b_0 = \binom{k+1}{0} = 1$, $b_1 = -\binom{k+1}{1}$, $b_2 = \binom{k+1}{2}$, $b_3 = -\binom{k+1}{3}, \ldots$, $b_{k+1} = (-1)^{k+1}\binom{k+1}{k+1} = (-1)^{k+1}$, and $a_0 = b_0 c_0 = 1$, $a_1 = b_0 c_1 + b_1 c_0 = 1 \cdot 2^k + (-k-1) \cdot 1 = 2^k - k - 1$, $a_2 = b_0 c_2 + b_1 c_1 + b_2 c_0 = 1 \cdot 3^k + (-k-1) \cdot 2^k + \frac{k(k+1)}{2} \cdot 1 = 3^k - (k+1) \cdot 2^k + \frac{k(k+1)}{2}, \ldots, a_m = b_0 c_m + b_1 c_{m-1} + \cdots + b_{m-2} c_2 + b_{m-1} c_1 + b_m c_0 = \sum_{i=0}^{m} (-1)^i \binom{k+1}{i} (m+1-i)^k, \ldots, a_k = b_0 c_k + b_1 c_{k-1} + \cdots + b_{k-2} c_2 + b_{k-1} c_1 + b_k c_0 = \sum_{i=0}^{k} (-1)^k \binom{k+1}{i} (k+1-i)^k$. Since $g(x) = 1 - \binom{k+1}{1} x + \binom{k+1}{2} x^2 - \cdots + (-1)^{k+1} x^{k+1} = (1-x)^{k+1}$ has $k+1$ coinciding roots $x_1 = \cdots = x_{k+1} = 1$, it holds

$$\frac{a_0 + a_1 x + a_2 x^2 + a_3 x^3 + \cdots + a_k x^k}{(1-x)^{k+1}} = C^k(1) + C^k(2) x + C^k(3) x^2 + \cdots + C^k(n) x^{n-1} + \cdots, \quad |x| < 1,$$

where $a_m = \sum_{i=0}^{m} (-1)^i \binom{k+1}{i} (m+1-i)^k$, $0 \le m \le k$. It is easy to see that

$$a_k = 0.$$

In fact,

$$a_k = \binom{k+1}{0}(k+1)^k - \binom{k+1}{1}k^k + \binom{k+1}{2}(k-1)^k - \cdots + (-1)^k\binom{k+1}{k}\cdot 1^k$$
$$= \binom{k+1}{0}C^k(k+1) - \binom{k+1}{1}C^k(k) + \binom{k+1}{2}C^k(k-1) - \cdots + (-1)^k\binom{k+1}{k}\cdot C^k(1).$$

Going back to the above recurrent equation

$$\binom{k+1}{0}C^k(n+k+1) - \binom{k+1}{1}C^k(n+k) + \cdots + (-1)^{k+1}\binom{k+1}{k+1}C^k(n) = 0,$$

we can note, that it holds for any integer n. So, considering it for $n = 0$, we get the equality $\binom{k+1}{0}C^k(k+1) - \binom{k+1}{1}C^k(k) + \cdots + (-1)^k\binom{k+1}{k}C^k(1) + (-1)^{k+1}\binom{k+1}{k+1}C^k(0) = 0$. Since $C^k(0) = 0^k = 0$, we get

$$\binom{k+1}{0}C^k(k+1) - \binom{k+1}{1}C^k(n+k) + \cdots + (-1)^k\binom{k+1}{k}C^k(1) = 0,$$

and, hence, $a_k = 0$. Similarly, one can prove, that

$$a_{k-1} = 1.$$

In fact,

$$a_{k-1} = \binom{k+1}{0}k^k - \binom{k+1}{1}(k-1)^k + \binom{k+1}{2}(k-2)^k - \cdots + (-1)^{k-1}\binom{k+1}{k-1}\cdot 1^k$$

$$= \binom{k+1}{0} C^k(k) - \binom{k+1}{1} C^k(k-1) + \binom{k+1}{2} C^k(k-2)$$
$$- \cdots + (-1)^{k-1} \binom{k+1}{k-1} \cdot C^k(1).$$

But considering the above recurrent equation

$$\binom{k+1}{0} C^k(n+k+1) - \binom{k+1}{1} C^k(n+k)$$
$$+ \cdots (-1)^{k+1} \binom{k+1}{k+1} C^k(n) = 0$$

for $n = -1$, we get the equality

$$\binom{k+1}{0} C^k(k) - \binom{k+1}{1} C^k(k-1) + \cdots + (-1)^{k-1} \binom{k+1}{k-1} C^k(1)$$
$$+ (-1)^k \binom{k+1}{k} C^k(0) + (-1)^{k+1} \binom{k+1}{k+1} C^k(-1) = 0.$$

Since $C^k(0) = 0^k = 0$ and $C^k(-1) = (-1)^k$, we get that

$$\binom{k+1}{0} C^k(k) - \binom{k+1}{1} C^k(k-1) + \cdots + (-1)^{k-1} \binom{k+1}{k-1} C^k(1)$$
$$= -(-1)^{k+1} \binom{k+1}{k+1} (-1)^k = (-1)^{2k+2} \cdot 1,$$

and, hence, $a_{k-1} = 1$. In general, it is easy to see that

$$a_m = a_{k-m-1}, \quad 0 \le m \le k-1.$$

In fact,

$$a_m = \binom{k+1}{0} (m+1)^k - \binom{k+1}{1} (m)^k + \binom{k+1}{2} (m-1)^k$$
$$- \cdots + (-1)^{k-m} \binom{k+1}{m} \cdot 1^k$$

$$= \binom{k+1}{0}C^k(m+1) - \binom{k+1}{1}C^k(m) + \binom{k+1}{2}C^k(m-1)$$
$$- \cdots + (-1)^{k-m}\binom{k+1}{m} \cdot C^k(1).$$

But considering the recurrent equation

$$\binom{k+1}{0}C^k(n+k+1) - \binom{k+1}{1}C^k(n+k)$$
$$+ \cdots + (-1)^{k+1}\binom{k+1}{k+1}C^k(n) = 0$$

for $n = m - k$, we get the equality

$$\binom{k+1}{0}C^k(m+1) - \binom{k+1}{1}C^k(m)$$
$$+ \cdots + (-1)^{k-m}\binom{k+1}{m}C^k(1) + (-1)^{m+1}\binom{k+1}{m+1}C^k(0)$$
$$+(-1)^{m+2}\binom{k+1}{m+2}C^k(-1) + \cdots + (-1)^k\binom{k+1}{k}C^k(m-k+1)$$
$$+(-1)^{k+1}\binom{k+1}{k+1}C^k(m-k) = 0.$$

Since $C^k(0) = 0^k = 0$ and $C^k(-i) = (-i)^k = (-1)^k C^k(i)$, it follows

$$\binom{k+1}{0}C^k(m+1) - \binom{k+1}{1}C^k(m)$$
$$+ \cdots + (-1)^{k-m}\binom{k+1}{k-m}C^k(1)$$
$$= -((-1)^{m+2}\binom{k+1}{m+2}(-1)^k C^k(1)$$
$$+ \cdots + (-1)^k\binom{k+1}{k}(-1)^k C^k(k-m-1)$$
$$+(-1)^{k+1}\binom{k+1}{k+1}(-1)^k C^k(k-m)).$$

Since $\binom{k+1}{i} = \binom{k+1}{k+1-i}$, $i = 0, 1, \ldots, k+1$, it follows

$$\begin{aligned}
&-((-1)^{m+2}\binom{k+1}{m+2}(-1)^k C^k(1) \\
&\quad + \cdots + (-1)^k\binom{k+1}{k}(-1)^k C^k(k-m-1) \\
&\quad + (-1)^{k+1}\binom{k+1}{k+1}(-1)^k C^k(k-m)) \\
&= \binom{k+1}{0}C^k(k-m) - \binom{k+1}{1}C^k(k-m-1) \\
&\quad + \cdots + (-1)^{k-m-1}\binom{k+1}{k-m-1}\cdot C^k(1) \\
&= \binom{k+1}{0}(k-m)^k - \binom{k+1}{1}(k-m-1)^k \\
&\quad + \cdots + (-1)^{k-m-1}\binom{k+1}{k-m-1}\cdot 1^k,
\end{aligned}$$

and, hence, $a_m = a_{k-m-1}$.

So, we proved that the generation function for the sequence of k-dimensional hypercube numbers has the form

$$\begin{aligned}
&\frac{x(1 + a_1^k x^2 + \cdots + a_{k-2}^k x^{k-2} + x^{k-1})}{(1-x)^{k+1}} \\
&\quad = C^k(1)x + C^k(2)x^2 + C^k(3)x^3 + \cdots + C^k(n)x^n + \cdots, \quad |x| < 1,
\end{aligned}$$

where $a_1^k = a_{k-2} = 2^k - (k+1)$, $a_2^k = a_{k-3}^k = 3^k - (k+1)2^k + \frac{k(k+1)}{2}, \ldots, a^k_{\lfloor\frac{k-1}{2}\rfloor} = a^k_{k-\lfloor\frac{k-1}{2}\rfloor-1} = (\lfloor\frac{k-1}{2}\rfloor + 1)^k - (k+1)(\lfloor\frac{k-1}{2}\rfloor)^k + \frac{k(k+1)}{2}(\lfloor\frac{k-1}{2}\rfloor - 1)^k + \cdots + (-1)^{\lfloor\frac{k-1}{2}\rfloor}\binom{k+1}{\lfloor\frac{k-1}{2}\rfloor}$.

3.3 Other regular polytope numbers

3.3.1. The **regular polytope numbers** are figurate numbers that correspond to classical regular polytopes for a given dimension k.

For $k \geq 4$ they give multidimensional analogues of two-dimensional polygonal and three-dimensional regular polyhedral numbers.

An *k-dimensional polytope* is defined as a finite region of k-dimensional space enclosed by a finite number of hyperplanes. A *regular polytope* is a polytope whose group of symmetry is transitive on its *flags*: the sequences of faces of a polytope, each contained in the next, with just one face from each dimension. Thus, regular polytopes have the highest degree of symmetry. A regular k-dimensional polytope may be defined as having regular *facets* (i.e., $(n-1)$-dimensional faces) and regular *vertex figures* (broadly speaking, the figure exposed when a corner of a polyhedron or polytope is sliced off). These two conditions are sufficient to ensure that all faces are alike and all vertices are alike.

So, the convex regular polytopes represent a generalized analog of two-dimensional convex regular polygons and three-dimensional convex regular polyhedra (*Platonic solids*) in an arbitrary dimension.

The regular polytopes were discovered by Schläfli, 1852. He showed that there are six regular convex four-dimensional polytopes, five of these correspond to the Platonic solids and the other one is the *polyoctahedron*. For dimensions k with $k \geq 5$, there are only three regular convex polytopes: the *hypercube*, *cross polytope*, and *regular simplex*, which are multidimensional analogs of the *cube*, *octahedron*, and *tetrahedron*, respectively ([Coxe73], [Well91]).

The regular polytopes can be described by *Schläfli symbol*, which is defined recursively. A convex regular polygon having n edges is denoted by $\{n\}$. There exists a convex regular polygon for any positive integer n, $n \geq 3$.

A convex regular polyhedron, having faces $\{n\}$ with p faces joining around a vertex, is denoted by $\{n, p\}$. The Schläfli symbols of five Platonic solids are listed in the table below, as well as the number N_0 of the vertices, N_1 of edges, and N_2 of faces of a given regular polyhedron.

name	Schläfli symbol	N_0	N_1	N_2
tetrahedron	$\{3, 3\}$	4	6	4
octahedron	$\{3, 4\}$	6	12	8
cube	$\{4, 3\}$	8	12	6
icosahedron	$\{3, 5\}$	12	30	20
dodecahedron	$\{5, 3\}$	20	30	12

A regular four-dimensional polytope, called *polychoron* (or *polycell, polyhedroid*) having cells (i.e., three-dimensional faces) $\{n, p\}$ with q cells joining around an edge is denoted by $\{n, p, q\}$. The Schläfli symbols of six polychora are listed in the table below, as well as the number N_0 of the vertices, N_1 of the edges, N_2 of faces, and N_3 of cells of a given polychoron.

name	Schäfli symbol	N_0	N_1	N_2	N_3
hypertetrahedron (5-cell)	$\{3,3,3\}$	5	10	10	5
hyperoctahedron (16-cell)	$\{3,3,4\}$	8	24	32	16
hypercube (8-cell)	$\{4,3,3\}$	16	32	24	8
polyoctahedron (24-cell)	$\{3,4,3\}$	24	96	96	24
hypericosahedron (600-cell)	$\{3,3,5\}$	120	720	1200	600
hyperdodecahedron (120-cell)	$\{5,3,3\}$	600	1200	720	120

A five-dimensional regular polytope is described as $\{n, p, q, r\}$, while Schläfli symbol for an k-dimensional regular polytope is $\{p_1, p_2, \ldots, p_{k-1}\}$. The Schläfli symbols of three k-dimensional regular polytopes, $k \geq 5$, are listed in the table below, as well as the numbers N_j, $0 \leq j \leq k-1$, of i-dimensional faces of a given k-dimensional regular polytope.

name	Schäfli symbol	$N_j (0 \leq j \leq k-1)$
regular simplex α^k	$\{3,3,\ldots,3,3\}$	$\binom{d+1}{j+1}$
cross polytope β^k	$\{3,3,\ldots,3,4\}$	$2^{j+1}\binom{d}{j+1}$
hypercube γ^k	$\{4,3,\ldots,3\}$	$2^{d-j}\binom{d}{j}$

In the previous sections we considered multidimensional figurate numbers corresponding to the two classes of regular polytopes: **k-dimensional hypertetrahedron numbers**, related to k-dimensional *regular simplex*, and **k-dimensional hypercube numbers**, realted to k-dimensional *hypercube*.

These classes of multidimensional figurate numbers were obtained in some natural inductive manner: k-dimensional hypertetrahedron numbers as sums of consecutive $(k-1)$-dimensional hypertetrahedron numbers, and k-dimensional hypercube numbers just as layers of equal $(k-1)$-dimensional hypercube numbers.

However, two above classes of regular polytope numbers, as well as **k-dimensional hyperoctahedron numbers**, corresponding to k-dimensional *cross polytope*, and, for $k = 4$, three addition classes of **regular polychoral numbers (hyperdodecahedral numbers,**

hypericosahedral numbers, and **polyoctahedral numbers**), can be obtained following a standard procedure, similar to the procedure of the construction of regular polyhedral numbers.

In fact, the general rule for enlarging the regular polytope to the next size is *to extend all edges, containing a given vertex, by one point and then add the required extra faces of the next array between these points in order to obtain similar shape.*

More exactly, for a given k-dimensional regular polytope V^k in the Euclidean space, one can construct the sequence $V^k(1), V^k(2), \ldots, V^k(n), \ldots$ of regular polytope numbers, associated to V^k, using induction on the dimension k of V^k.

For $k = 0, 1, 2$ and 3 we will use, as k-dimensional regular polytope numbers, the number $S^0(n) = 1$, the linear numbers $S^1(n) = n$, the polygonal numbers $S_m(n) = S^2_m(n)$, $m = 3, 4, \ldots$, and the regular polyhedral numbers (i.e., tetrahedral, cubic, octahedral, icosahedral and dodecaherdal numbers), respectively.

Suppose now that the sequences of regular polytope numbers have been constructed for any regular polytope of dimension less than k, and let V^k be a regular polytope in $\mathbb{R}^k$. By convention, we put $V^k(1) = 1$ and consider a fixed point $v \in \mathbb{R}^k$ as a regular polygon V^k_1. Then we define the sequence $V^k(n)$, $n \geq 2$, using induction on n. So, we assume that $V^k(n-1)$ has been constructed on a regular polytope V^k_{n-1} of the same shape as V^k, and this polytope is contained the initial point $v = V^k_1$. We extend the edges of V^k_{n-1}, containing the initial vertex v to form a lager regular polytope V^k_n containing V^k_{n-1} and having the same shape. Next we make n-th array of points associated to V^k on V^k_n, placing on it all $(n-1)$-th array's points of V^k_{n-1} (remind that V^k_{n-1} is contained in V^k_n), and then put to each new j-dimensional face of V^k_n, $0 \leq j \leq k-1$, n-th array of points associated to the corresponding j-dimensional regular polytope. Finally, we count all the points in V^k_n to define $V^k(n)$ ([Kim03]).

Algebraically, in order to obtain n-th regular polytope number $V^k(n)$, corresponding to a given k-dimensional regular polytope V^k with N_j j-dimensional faces and N^v_j j-dimensional faces contained the initial vertex v, $0 \leq j \leq k-1$, one should add to $(n-1)$-th regular polyhedral number $V^k(n-1)$ the values $(N_j - N^v_j)int(V^j)$,

$0 \le j \le k-1$, that count points, corresponding to the interiors of all new j-dimensional faces of this construction. Here $int(V^j)$ is the number of points in the interior of n-th j-dimensional regular polytope number $V^j(n)$, corresponding to the regular polytope V^j, forming j-dimensional faces of V^k.

In particular, for $n = 4$, the general calculation has the following form. In order to obtain n-th regular polytope number, corresponding to a given polychoron V^4 with N_0 vertices, N_1 edges, N_2 faces, and N_3 cells, as well as N_0^v vertices, N_1^v edges, N_2^v faces, and N_3^v cells containing the initial vertex v, one should add to $(n-1)$-th regular polytope number the following values:

- the value $N_0 - 1$ that counts points, corresponding to all new vertices of this construction (let $int(V^0)$ be equal to 1);
- the value $(N_1 - N_1^v)(n-2)$ counting points, corresponding to the interior of all new edges of this construction (here $int(V^1) = int(S^1(n)) = int(n) = n-2$);
- the value $(N_2 - N_2^v)int(V^2)$ counting points, corresponding to the interior of all new faces of this construction, each being n-th polygonal number;
- finally, the value $(N_3 - N_3^v)int(V^3)$ counting points, corresponding to the interior of all new cells of our construction, each being n-th regular polyhedral number.

3.3.2. The **polytope numbers** correspond to a regular simplex in $\mathbb{R}^4$, i.e., are four-dimensional **hypertetrahedron numbers** (or **pentachoron numbers**, **hyperpyramid numbers**, **4-simplex numbers**, **5-cell numbers**, α^4**-numbers**, $\{3,3,3\}$**-numbers**).

For a regular four-dimensional simplex, the number of vertices, edges, triangles and tetrahedrons (i.e., one- two- and three-dimensional faces) is 5, 10, 10 and 5, while the number of edges, triangles and tetrahedrons containing a given vertex, is 4, 6 and 4, respectively.

So, for polytope numbers we get by the above construction the following relation:

$$
\begin{aligned}
S_3^4(n) - S_3^4(n-1) &= (N_0 - N_0^v) + (N_1 - N_1^v) \cdot (n-2) \\
&\quad + (N_2 - N_2^v) \cdot int(S_3(n)) \\
&\quad + (N_3 - N_3^v) \cdot int(S_3^3(n)) \\
&= (5-1) + (10-4) \cdot (n-2) + (10-6) \\
&\quad \cdot \left(\frac{n(n+1)}{2} - 3(n-1)\right) + (5-4) \\
&\quad \cdot \left(\frac{n(n+1)(n+2)}{6} - 4 - 6(n-2)\right. \\
&\quad \left. -4\left(\frac{n(n+1)}{2} - 3(n-1)\right)\right) \\
&= 2n^2 - 4n - 4 + \frac{n(n+1)(n+2)}{6} - 2n^2 + 4n - 4 \\
&= \frac{n(n+1)(n+2)}{6}.
\end{aligned}
$$

Hence, the classical recurrent equation for polytope numbers is obtained:

$$S_3^4(n) = S_3^4(n-1) + \frac{n(n+1)(n+2)}{6}, \quad S_3^4(1) = 1.$$

Now we can easily get, for example by induction, the general formula $S_3^4(n) = \frac{n(n+1)(n+2)(n+3)}{24}$. It holds for $n = 1$, and, going from $n-1$ to n, we have $S_3^4(n) = S_3^4(n-1) + \frac{n(n+1)(n+2)}{6} = \frac{(n-1)n(n+1)(n+2)}{24} + \frac{n(n+1)(n+2)}{6} = \frac{n(n+1)(n+2)}{24}(n-1+4) = \frac{n(n+1)(n+2)(n+3)}{24}$.

3.3.3. The biquadratic numbers correspond to a hypercube in $\mathbb{R}^4$, i.e., are four-dimensional **hypercube numbers** (or **octahoron numbers**, **tessaract numbers**, **4-measure-polytope numbers**, **8-cell numbers**, γ^4**-numbers**, $\{4,3,3\}$**-numbers**).

For a four-dimensional hypercube, the number of vertices, edges, squares and cubes (i.e., one- two- and three-dimensional faces) is 16,

32, 24 and 8, while the number of edges, squares and cubes containing a given vertex, is 4, 6 and 4, respectively.

So, for biquadratic numbers we get, by the above construction, the following relation:

$$\begin{aligned} BC(n) - BC(n-1) &= (N_0 - 1) + (N_1 - N_1^v) \cdot (n-2) \\ &\quad +(N_2 - N_2^v) \cdot int(S_4(n)) \\ &\quad +(N_3 - N_3^v) \cdot int(C(n)) \\ &= (16-1) + (32-4)(n-2) \\ &\quad +(24-6)(n^2 - 4(n-1)) \\ &\quad +(8-4)(n^3 - 8 - 12(n-2) \\ &\quad -6(n^2 - 4(n-1))) \\ &= 4n^3 - 6n^2 + 4n - 1. \end{aligned}$$

Therefore, the classical recurrent equation for biquadratic numbers is obtained:

$$BC(n) = BC(n-1) + (4n^3 - 6n^2 + 4n - 1), \quad BC(1) = 1.$$

Now we can easily get, for example by induction, the general formula $BC(n) = n^4$: it holds for $n = 1$, and, going from $n-1$ to n, we have $BC(n) = BC(n-1) + (4n^3 - 6n^2 + 4n - 1) = (n-1)^4 + (4n^3 - 6n^2 + 4n - 1) = n^4$.

3.3.4. The **hyperoctahedral numbers** correspond to a hyperoctahedron in $\mathbb{R}^4$, i.e., are four-dimensional **hyperoctahedron numbers** (or **hexadecahoron numbers**, **4-cross-polytope numbers**, **4-orthoplex numbers**, **16-cell numbers**, β^4**-numbers**, $\{3,3,4\}$ **numbers**).

For a four-dimensional cross polytope the number of vertices, edges, triangles and tetrahedrons (i.e., one- two- and three-dimensional faces) is 8, 24, 32 and 16, while the number of edges, triangles and tetrahedrons containing a given vertex, is 6, 12 and 8, respectively.

So, for hyperoctahedral numbers we get, by the above construction, the following relation:

$$\begin{aligned} HO(n) - HO(n-1) &= (N_0 - 1) + (N_1 - N_1^v) \cdot (n-2) \\ &\quad + (N_2 - N_2^v) \cdot int(S_3(n)) \\ &\quad + (N_3 - N_3^v) \cdot int(S_3^3(n)) \\ &= (8-1) + (24-6)(n-2) + (32-12) \\ &\quad \times \left(\frac{n(n+1)}{2} - 3(n-1) \right) \\ &\quad + (16-8) \left(\frac{n(n+1)(n+2)}{2} - 4 \right. \\ &\quad \left. -6(n-2) - 4 \left(\frac{n(n+1)}{2} - 3(n-1) \right) \right) \\ &= \frac{4n^3 - 6n^2 + 8n - 3}{3}. \end{aligned}$$

Hence, the recurrent equation for hyperoctahedral numbers has the form

$$HO(n) = HO(n-1) + \frac{4n^3 - 6n^2 + 8n - 3}{3}, \quad HO(1) = 1.$$

The general formula for hyperoctahedral numbers has the form

$$HO(n) = \frac{n^2(n^2+2)}{3}.$$

It can be proved by induction. In fact, it holds for $n = 1$. Going from $n-1$ to n, we get $HO(n) = HO(n-1) + \frac{4n^3-6n^2+8n-3}{3} = \frac{(n-1)^2((n-1)^2+2)}{3} + \frac{4n^3-6n^2+8n-3}{3} = \frac{n^2(n^2+2)}{3}$.

The sequence of the hyperoctahedral numbers starts with the elements 1, 8, 33, 96, 225, 456, 833, 1408, 2241, 3400, … (Sloane's A014820).

The generating function for the sequence $HO(1), HO(2), \ldots,$ $HO(n), \ldots$ of hyperoctahedral numbers is $f(x) = \frac{x(1+x)^3}{(1-x)^5}$. In other words, it holds

$$\frac{x(1+x)^3}{(1-x)^5} = HO(1)x + HO(2)x^2 + HO(3)x^3 + \cdots + HO(n)x^n + \cdots, \quad |x| < 1.$$

This result can be obtained by the standard procedure, resulting in the linear recurrent equation $HO(n+5) - 5HO(n+4) + 10HO(n+3) - 10HO(n+2) + 5HO(n+1) - HO(n) = 0$ with initial conditions $HO(1) = 1$, $HO(2) = 8$, $HO(3) = 33$, $HO(4) = 96$, and $HO(5) = 225$.

It holds that *every hyperoctahedral number can be written as a linear combination with non-negative integer coefficients of polytope numbers*:

$$HO(n) = S_3^4(n) + 3S_3^4(n-1) + 3S_3^4(n-2) + S_3^4(n-3).$$

The proof can be obtained by direct computation (see [Kim03]).

3.3.5. The **hypericosahedral numbers** correspond to a hypericosahedron in $\mathbb{R}^4$, i.e., are **hexacisihoron numbers** (or **tetraplex numbers**, **polytetrahedron numbers**, **600-cell numbers**, **$\{3,3,5\}$-numbers**).

Using the standard procedure, we get for them the following relation:

$$HI(n) - HI(n-1) = \frac{290n^3 - 855n^2 + 889n - 321}{3}.$$

So, the recurrent equation for hypericosahedral numbers has the form

$$HI(n) = HI(n-1) + \frac{290n^3 - 855n^2 + 889n - 321}{3}, \quad HI(1) = 1.$$

The general formula for hypericosahedral numbers has the form

$$HI(n) = \frac{n(145n^3 - 280n^2 + 179n - 38)}{6}.$$

It can be proved by indiction. In fact, it holds for $n=1$. Going from $n-1$ to n, we get $HI(n) = HI(n-1) + \frac{290n^3-855n^2+889n-321}{3} = \frac{(n-1)(145(n-1)^3-280(n-1)^2+179(n-1)-38)}{6} + \frac{290n^3-855n^2+889n-321}{3} = \frac{n(145n^3-280n^2+179n-38)}{6}$.

The sequence of the hypericosahedral numbers starts with the elements 1, 120, 947, 3652, 9985, 22276, 43435, 76952, 126897, 197920, ... (Sloane's A092182).

The generating function for the sequence $HI(1), HI(2), \ldots, HI(n), \ldots$ of hypericosahedral numbers is

$$f(x) = \frac{x(1+115x+357x^2+107x^3)}{(1-x)^5}.$$

In other words, it holds

$$\frac{x(1+115x+357x^2+107x^3)}{(1-x)^5} = HI(1)x + HI(2)x^2 + HI(3)x^3 + \cdots + HI(n)x^n + \cdots, \quad |x|<1.$$

This result can be obtained by the standard procedure, leading to the linear recurrent equation $HI(n+5) - 5HI(n+4) + 10HI(n+3) - 10HI(n+2) + 5HI(n+1) - HI(n) = 0$ with initial conditions $HI(1) = 1$, $HI(2) = 120$, $HI(3) = 947$, $HI(4) = 3652$, and $HI(5) = 9985$.

It holds that *every hypericosahedral number can be written as a linear combination with non-negative integer coefficients of polytope numbers*:

$$HI(n) = S_3^4(n) + 115S_3^4(n-1) + 357S_3^4(n-2) + 107S_3^4(n-3).$$

The proof can be obtained by direct computation (see [Kim03]).

3.3.6. The **hyperdodecahedral numbers** correspond to a hyperdodecahedron in $\mathbb{R}^4$, i.e., are **hecatonicosachoron numbers** (or **dodecaplex numbers**, **polydodecahedron numbers**, **120-cell numbers**, $\{5,3,3\}$**-numbers**).

Using the standard procedure, we get for them the following relation:

$$HD(n) - HD(n-1) = 522n^3 - 1539n^2 + 1561n - 543.$$

Therefore, the recurrent equation for hyperdodecahedral numbers has the form

$$HD(n) = HD(n-1) + 522n^3 - 1539n^2 + 1561n - 543,$$
$$HD(1) = 1.$$

The general formula for hyperdodecahedral numbers has the form

$$HD(n) = \frac{n(261n^3 - 504n^2 + 283n - 38)}{2}.$$

It can be proved by induction. In fact, it holds for $n = 1$. Going from $n-1$ to n, we get $HD(n) = HD(n-1) + 522n^3 - 1539n^2 + 1561n - 543 = \frac{(n-1)(261(n-1)^3 - 504(n-1)^2 + 283(n-1) - 38)}{2} + 522n^3 - 1539n^2 + 1561n - 543 = \frac{n(261n^3 - 504n^2 + 283n - 38)}{2}$.

The sequence of the hyperdodecahedral numbers starts with the elements 1, 600, 4983, 19468, 53505, 119676, 233695, 414408, 683793, 1066960, ... (Sloane's A092183).

The generating function for the sequence $HD(1), HD(2), \dots, HD(n), \dots$ of hyperdodecahedral numbers is

$$f(x) = \frac{x(1 + 595x + 1993x^2 + 543x^3)}{(1-x)^5}.$$

In other words, it holds

$$\frac{x(1 + 595x + 1993x^2 + 543x^3)}{(1-x)^5} = HD(1)x + HD(2)x^2 + HD(3)x^3 + \cdots + HD(n)x^n + \cdots, \quad |x| < 1.$$

This result can be obtained by the standard procedure, leading to the linear recurrent equation $HD(n+5) - 5HD(n+4) + 10HD(n+3) - 10HD(n+2) + 5HD(n+1) - HD(n) = 0$ with initial conditions $HD(1) = 1$, $HD(2) = 600$, $HD(3) = 4983$, $HD(4) = 19468$, and $HD(5) = 53505$.

It holds that *every hyperdodecahedral number can be written as a linear combination with non-negative integer coefficients of polytope numbers*:

$$HD(n) = S_3^4(n) + 45S_3^4(n-1) + 1993S_3^4(n-2) + 543S_3^4 4(n-3).$$

The proof can be obtained by direct computation (see [Kim03]).

3.3.7. The **polyoctahedral numbers** correspond to a polyoctahedron in $\mathbb{R}^4$, i.e., are **icositetrachoron numbers** (or **octaplex numbers, hyperdiamond numbers, 24-cell numbers, $\{3,4,3\}$-numbers**).

Using the standard procedure, we get for them the following relation:

$$PO(n) - PO(n-1) = 12n^3 - 30n^2 + 28n - 9.$$

So, the recurrent equation for polyoctahedral numbers has the form

$$PO(n) = PO(n-1) + 12n^3 - 30n^2 + 28n - 9, \quad PO(1) = 1.$$

The general formula for polyoctahedral numbers has the form

$$PO(n) = n^2(3n^2 - 4n + 2).$$

It can be proved by induction. In fact, it holds for $n = 1$. Going from $n-1$ to n, one get $PO(n) = PO(n-1) + 12n^3 - 30n^2 + 28n - 9 = (n-1)^2(3(n-1)^2 - 4(n-1) + 2) + 12n^3 - 30n^2 + 28n - 9 = n(3n^2 - n + 2)$.

The sequence of the polyoctahedral numbers starts with the elements 1, 24, 153, 544, 1425, 3096, 5929, 10368, 16929, 26200, ... (Sloane's A092181).

The generating function for the sequence $PO(1), PO(2), \ldots, PO(n), \ldots$ of polyoctahedral numbers is $f(x) = \frac{x(1+19x+43x^2+9x^3)}{(1-x)^5}$. In other words, it holds

$$\frac{x(1 + 19x + 43x^2 + 9x^3)}{(1-x)^5} = PO(1)x + PO(2)x^2 + PO(3)x^3 + \cdots + PO(n)x^n + \cdots, \quad |x| < 1.$$

This result can be obtained by the standard procedure, leading to the linear recurrent equation $PO(n+5) - 5PO(n+4) + 10PO(n+3) - 10PO(n+2) + 5PO(n+1) - PO(n) = 0$ with initial conditions $PO(1) = 1$, $PO(2) = 24$, $PO(3) = 153$, $PO(4) = 544$, and $PO(5) = 1425$.

It holds that *every polyoctahedral number can be written as a linear combination with non-negative integer coefficients of polytope numbers*:

$$PO(n) = S_3^4(n) + 19S_3^4(n-1) + 43S_3^4(n-2) + 9S_3^4(n-3).$$

The proof can be obtained by direct computation (see [Kim03]).

3.3.8. The **k-dimensional hypertetrahedron numbers** correspond to a regular simplex in $\mathbb{R}^k$, $k \geq 5$, i.e., are k-dimensional **simplex numbers** (or **k-simplex numbers**, **α^k-numbers**, **$\{3^{k-1}\}$-numbers**).

For a regular k-dimensional simplex α^k the number of j-dimensional faces, all being also j-dimensional regular simplexes α^j, is $N_j = \binom{k+1}{j+1}$, $0 \leq j \leq k-1$. The number of j-dimensional faces of α^k containing a given vertex v is $N_j^v = \binom{k}{j}, 1 \leq j \leq k-1$.

So, for k-dimensional hypertetrahedron numbers $S_3^k(n)$, we get the following relation:

$$\begin{aligned} S_3^k(n) - S_3^k(n-1) &= \sum_{j=0}^{k-1} (N_j - N_j^v) int(S_3^j(n)) \\ &= \sum_{j=0}^{k-1} \left(\binom{k+1}{j+1} - \binom{k}{j} \right) int(S_3^j(n)), \end{aligned}$$

where $int(S_3^0(n)) = 1$ by convention. Since $\binom{k+1}{j+1} - \binom{k}{j} = \binom{k}{j+1}$, it holds

$$S_3^k(n) - S_3^k(n-1) = \sum_{j=0}^{k-1} \binom{k}{j+1} int(S_3^j(n)).$$

On the other hand, each face of a regular simplex is again a regular simplex of lower dimension, and it follows easily from this fact and our construction that

$$S_3^r(n) = \sum_{i=0}^{r} \binom{r+1}{i+1} int(S_3^i(n)), \quad 1 \leq r \leq k-1.$$

In particular, $S_3^{k-1}(n) = \sum_{i=0}^{k-1} \binom{k}{i+1} int(S_3^i(n))$. Hence, we have $S_3^k(n) - S_3^k(n-1) = S_3^{k-1}(n)$. In other words, we obtain the classical recurrent formula for k-dimensional hypertetrahedron numbers:

$$S_3^k(n) = S_3^k(n-1) + S_3^{k-1}(n-1), \quad S_3^k(1) = 1.$$

So, the general formula for the k-dimensional hypertetrahedron numbers is

$$S_3^k(n) = \frac{n(n+1)\cdots(n+k-1)}{k!}.$$

It can be obtained by induction. In fact, this formula holds for $k = 2, 3$ and 4, i.e., for triangular, tetrahedral and polytope numbers, respectively. Let $S_3^{k-1}(n) = \frac{n(n+1)\cdots(n+k-2)}{(k-1)!}$. Then $S_3^k(n) = S_3^k(n-1) + \frac{n(n+1)\cdots(n+k-2)}{(k-1)!}$. So, the formula $S_3^k(n) = \frac{n(n+1)\cdots(n+k-1)}{k!}$ holds for $n = 1$, and, going from $n-1$ to n, we get $S_3^k(n) = S_3^k(n-1) + \frac{n(n+1)\cdots(n+k-2)}{(k-1)!} = \frac{(n-1)n(n+1)\cdots(n+k-2)}{k!} + \frac{n(n+1)\cdots(n+k-2)}{(k-1)!} = \frac{n(n+1)\cdots(n+k-1)}{k!}$. From this it follows that $S_3^k(n) = \frac{n(n+1)\cdots(n+k-1)}{k!}$, $n = 1, 2, 3, \ldots$.

3.3.9. The **k-dimensional hypercube numbers** correspond to a hypercube in $\mathbb{R}^k$, $k \geq 5$, i.e., are k-dimensional **measure-polytope numbers** (or **k-measure-polytope numbers**, **γ^k-numbers**, **$\{4, 3^{k-2}\}$-numbers**).

For a k-dimensional hypercube γ^k the number of j-dimensional faces, all being also j-dimensional hypercubes γ^j, is $N_j = 2^{k-j}\binom{k}{j}$, $0 \leq j \leq k-1$. The number of j-dimensional faces of γ^k containing a given vertex v, is $N_j^v = \binom{k}{j}, 1 \leq j \leq k-1$.

So, for k-dimensional hypercube numbers $C^k(n)$, we get the following relation:

$$\begin{aligned} C^k(n) - C^k(n-1) &= \sum_{j=0}^{k-1}(N_j - N_j^v)int(C^j(n)) \\ &= \sum_{j=0}^{k-1}\left(2^{k-j}\binom{k}{j} - \binom{k}{j}\right) int(C^j(n)) \\ &= \sum_{j=0}^{k-1}\left(2^{k-j} - 1\right)\binom{k}{j} int(C^j(n)) \end{aligned}$$

$$= \sum_{j=0}^{k-2} \left(2^{k-j} - 1\right) \binom{k}{j} int(C^j(n))$$
$$+ \binom{k}{k-1} int(C^{k-1}(n)).$$

On the other hand, each face of a hypercube is again a hypercube of lower dimension, and from this fact it follows easily that

$$C^r(n) = \sum_{i=0}^{r} 2^{r-i} \binom{r}{i} int(C^i(n)), \quad 1 \le r \le k-1. \text{ In particular,}$$

$$C^{k-1}(n) = \sum_{i=0}^{k-1} 2^{k-1-i} \binom{k-1}{i} int(C^i(n))$$
$$= \sum_{i=0}^{k-2} 2^{k-1-i} \binom{k-1}{i} int(C^i(n)) + int(C^{k-1}(n)).$$

Therefore, $int(C^{k-1}(n)) = C^{k-1}(n) - \sum_{j=0}^{k-2} 2^{k-1-j} \binom{k-1}{j} int(C^j(n))$. Thus, we have

$$C^k(n) - C^k(n-1) = \binom{k}{k-1} C^{k-1}(n) + \sum_{j=0}^{k-2} \left(\left(2^{k-j} - 1\right) \binom{k}{j} \right.$$
$$\left. - 2^{k-1-j} \binom{k-1}{j} \binom{k}{k-1} \right) int(C^j(n))$$
$$= \binom{k}{k-1} C^{k-1}(n) - \binom{k}{k-2} int(C^{k-2}(n))$$
$$+ \sum_{j=0}^{k-3} \left(\left(2^{k-j} - \binom{k-j}{1} 2^{k-1-j} - 1 \right) \right.$$
$$\left. \times \binom{k}{j} \right) int(C^j(n))$$

$$= \binom{k}{k-1} C^{k-1}(n) - \binom{k}{k-2} C^{k-2}(n)$$
$$+ \sum_{j=0}^{k-3} \left(\left(2^{k-j} - \binom{k-j}{1} 2^{k-1-j} \right.\right.$$
$$\left.\left. + \binom{k-j}{2} 2^{k-j-2} - 1 \right) \binom{k}{j} \right) int(C^j(n)).$$

Continuing this process, we finally have $C^k(n) - C^k(n-1) = \sum_{i=1}^{k}(-1)^{i-1}\binom{k}{i}C^{k-i}(n)$. Hence, we get a variation of the classical recurrent formula for k-dimensional hypercube numbers:

$$C^k(n) = C^k(n-1) + \sum_{i=1}^{k}(-1)^{i-1}\binom{k}{i}C^{k-i}(n)$$
$$= C^k(n-1) + \binom{k}{1}C^{k-1}(n) - \binom{k}{2}C^{k-2}(n)$$
$$+ \cdots + (-1)^{k-1}, \quad C^k(1) = 1.$$

Therefore, the general formula for the k-dimensional hypercube numbers is

$$C^k(n) = n^k.$$

It can be obtained by induction. In fact, this formula holds for $k = 2, 3$ and 4, i.e., for square, cubic and biquadratic numbers, respectively. Let $C^r(n) = n^r$, $0 \le r \le k-1$. Then $C^k(n) = C^k(n-1) + \sum_{i=1}^{k}(-1)^{i-1}\binom{k}{i}C^{k-i}(n) = C^k(n-1) + \sum_{i=1}^{k}(-1)^{i-1}\binom{k}{i}n^{k-i}$. So, the formula $C^k(n) = n^k$ holds for $n = 1$, and, going from $n-1$ to n, we get $C^k(n) = C^k(n-1) + \sum_{i=1}^{k}(-1)^{i-1}\binom{k}{i}n^{k-i} = (n-1)^k + \sum_{i=1}^{k}(-1)^{i-1}\binom{k}{i}n^{k-i} = n^k$ due to the binomial theorem. From this it follows $C^k(n) = n^k$, $n = 1, 2, 3, \ldots$.

It holds (see [Kim03]) that *every k-dimensional hypercube number can be written as a linear combination with non-negative integer coefficients of k-dimensional hypertetrahedron numbers:*

$$C^k(n) = \sum_{i=0}^{k-1} \left\langle {k \atop i} \right\rangle S_3^k(n-i), \quad \textit{where } \left\langle {k \atop i} \right\rangle \textit{ is an Eulerian number.}$$

The **Eulerian number** $\left\langle {k \atop i} \right\rangle$ gives the number of permutations of $\{1, 2, \ldots, k\}$ having exactly i *permutation ascents*: for a given permutation $\{a_1, a_2, \ldots, a_k\}$ the number j is a permutation ascent if $a_j < a_{j+1}$. The number $\left\langle {k \atop i} \right\rangle$ can be given explicitly by the sum $\left\langle {k \atop i} \right\rangle = \sum_{j=0}^{i+1}(-1)^j \binom{k+1}{j}(i-j+1)^k$. Special cases are given by $\left\langle {k \atop 0} \right\rangle = 1$, $\left\langle {k \atop 1} \right\rangle = 2^k - k - 1$, $\left\langle {k \atop 2} \right\rangle = 3^k - 2^k(k+1) + \frac{n(n+1)}{2}$.

The Eulerian numbers share similar properties with the binomial coefficients (see [GKP94]). For example, we have identities

$$\left\langle {k \atop i} \right\rangle = \left\langle {k \atop d-1-i} \right\rangle, \quad \left\langle {d \atop i} \right\rangle = (i+1)\left\langle {d-1 \atop i} \right\rangle + (d-i)\left\langle {d-1 \atop i-1} \right\rangle,$$

$$\sum_{i=0}^{k} \left\langle {k \atop i} \right\rangle = n!, \quad \sum_{i=0}^{k-1} \left\langle {k \atop i} \right\rangle \binom{n+i}{k} = n^k.$$

The last two identities are called *sum identity* and *Worpitzky's identity*, respectively. Since $\binom{n+k-1}{k} = \frac{(n+k-1)(n+k-2)\cdots n}{k!} = S_3^k(n)$, the Worpitzky's identity can be reformulated as follow:

$$C^k(n) = \sum_{i=0}^{k-1} \left\langle {k \atop i} \right\rangle \binom{n+i}{k} = \sum_{i=0}^{k-1} \left\langle {k \atop k-i-1} \right\rangle \binom{n+i}{k}$$
$$= \sum_{j=0}^{k-1} \left\langle {k \atop j} \right\rangle \binom{n+k-j-1}{k} = \sum_{j=0}^{k-1} \left\langle {k \atop j} \right\rangle S_3^k(n-j).$$

Thus, we have $C^k(n) = \sum_{i=0}^{k-1} \left\langle {k \atop i} \right\rangle S_3^k(n-i)$.

3.3.10. The k**-dimensional hyperoctahedron numbers** correspond to a hyperoctahedron in $\mathbb{R}^k$, $k \geq 5$, i.e., are k-dimensional **cross-polytope numbers** (or k**-cross-polytope numbers**, β^k**-numbers**, $\{3^{k-2}, 4\}$**-numbers**).

For an k-dimensional hyperoctahedron β^k, the number of j-dimensional faces, all being j-dimensional regular simplexes α^j, is $N_j = 2^{j+1}\binom{k}{j+1}$, $0 \leq j \leq k-1$. The number of j-dimensional faces of β^k, containing a given vertex v, is $N_j^v = 2^j\binom{k-1}{j}, 1 \leq j \leq k-1$.

So, for k-dimensional hypertetrahedron numbers $O^k(n)$, we get the following relation:

$$O^k(n) - O^k(n-1) = \sum_{j=0}^{k-1}(N_j - N_j^v)int(S_3^j(n))$$
$$= \sum_{j=0}^{k-1}\left(2^{j+1}\binom{k}{j+1} - 2^j\binom{k-1}{j}\right) int(S_3^j(n)).$$

Replacing j by $k-1-j$, we get

$$O^k(n) - O^k(n-1) = \sum_{j=0}^{k-1}\left(2^{k-j}\binom{k}{j} - 2^{k-1-j}\binom{k-1}{j}\right)$$
$$\times\, int(S_3^{k-1-r}(n)).$$

By a simple manipulation on binomial coefficients, we can verify that

$$2^{k-j}\binom{k}{j} - 2^{k-1-j}\binom{k-1}{j} = \sum_{u=0}^{j}(-1)^u\binom{k-1}{u}2^{k-1-u}\binom{k-u}{j-u}.$$

Hence, we have

$$O^k(n) - O^k(n-1)$$
$$= \sum_{j=0}^{k-1}\left(\sum_{u=0}^{j}(-1)^u\binom{k-1}{u}2^{k-1-u}\binom{k-u}{j-u}\right) int(S_3^{k-1-j}(n)$$
$$= \sum_{u=0}^{k-1}(-1)^u\binom{k-1}{u}2^{k-1-u}\left(\sum_{j=u}^{k-1}\binom{k-u}{j-u}int(S_3^{k-1-j}(n))\right)$$
$$= \sum_{u=0}^{k-1}(-1)^u\binom{k-1}{u}2^{k-1-u}\sum_{s=0}^{k-u-1}\binom{k-u}{s+1}int(S_3^s(n))$$
$$= \sum_{u=0}^{k-1}(-1)^u\binom{k-1}{u}2^{k-u-1}S_3^{k-u-1}(n).$$

Thus, the recurrent formula for the sequence of k-dimensional hyperoctahedron numbers has the from

$$O^k(n) = O^k(n-1) + \sum_{j=0}^{k-1}(-1)^j \binom{k-1}{j} 2^{k-j-1} S_3^{k-j-1}(n),$$

$$O^k(1) = 1.$$

Now we can obtain a general formula for n-th k-dimensional hyperoctahedron number:

$$O^k(n) = \sum_{j=0}^{k-1}(-1)^j \binom{k-1}{j} 2^{k-j-1} S_3^{k-j}(n).$$

In fact, since $S_3^{k-1}(n) = S_3^k(n) - S_3^k(n-1)$, we get

$$\begin{aligned}
O^k(n) - O^k(n-1) &= \sum_{j=0}^{k-1}(-1)^j \binom{k-1}{j} 2^{k-j-1} S_3^{k-j-1}(n) \\
&= \sum_{j=0}^{k-1}(-1)^j \binom{k-1}{j} 2^{k-j-1} (S_3^{k-j}(n) - S_3^{k-j}(n-1)) \\
&= \sum_{j=0}^{k-1}(-1)^j \binom{k-1}{j} 2^{k-j-1} S_3^{k-j}(n) \\
&\quad - \sum_{j=0}^{k-1}(-1)^j \binom{k-1}{j} 2^{k-j-1} S_3^{k-j}(n-1).
\end{aligned}$$

This proves that $O^k(n) = \sum_{j=0}^{k-1}(-1)^j \binom{k-1}{j} 2^{k-j-1} S_3^{k-j}(n)$.

For $k = 5$ we get the formula $O^5(n) = \frac{n(2n^4+10n^2+3)}{15}$, and the first values of the sequence of five-dimensional hyperoctahedron numbers are 1, 10, 51, 180, 501, 1182, 2471, 4712, 8361, 14002, ... (Sloane's A069038).

For $k = 6$ we get the formula $O^6(n) = \frac{n^2(2n^4+20n^2+23)}{45}$, and the first values of the sequence of six-dimensional hyperoctahedron numbers are 1, 12, 73, 304, 985, 2668, 6321, 13504, 26577, 48940, ... (Sloane's A069039).

The formulas for k-dimensional hyperoctahedron numbers of dimension k, $2 \leq k \leq 10$, as well as the first few values of the corresponding sequences, are given in the table below (see [Kim03]).

k	$O^k(n)$									*Sloane*
2	n^2	1	4	9	16	25	36	49	64	A000290
3	$\frac{1}{3}n(2n^2+1)$	1	6	19	44	85	146	231	344	A005900
4	$\frac{1}{3}n^2(n^2+2)$	1	8	33	96	225	456	833	1408	A014820
5	$\frac{n}{15}(2n^4+10n^2+3)$	1	10	51	180	501	1182	2471	4712	A069038
6	$\frac{1}{45}n^2(2n^4+20n^2+23)$	1	12	73	304	985	2668	6321	13504	A069039
7	$\frac{n}{315}(4n^6+70n^4+196n^2+45)$	1	14	99	476	1765	5418	14407	34232	A099193
8	$\frac{n^2}{315}(n^6+28n^4+154n^2+132)$	1	16	129	704	2945	10128	29953	78592	A099195
9	$\frac{n}{2835}(2n^8+84n^6+798n^4+1636n^2+315)$	1	18	163	996	4645	17718	57799	166344	A099196
10	$\frac{n^2}{14175}(2n^8+120n^6+1806n^4+7180n^2+5067)$	1	20	201	1360	7001	29364	104881	329024	A099197

The generating function for this sequence is $f(x) = \frac{x(1+x)^{k-1}}{(1-x)^k}$, i.e., it holds

$$\frac{x(1+x)^{k-1}}{(1-x)^k} = O^k(1)x + O^k(2)x^2 + O^k(3)x^3 + \cdots + O^k(n)x^n + \cdots, \quad |x| < 1.$$

In particular,

$$\frac{x(1+x)^4}{(1-x)^5} = x + 10x^2 + 51x^3 + \cdots + O^5(n)x^n + \cdots, \quad |x| < 1;$$

$$\frac{x(1+x)^5}{(1-x)^7} = x + 12x^2 + 73x^3 + \cdots + O^6(n)x^n + \cdots, \quad |x| < 1.$$

This result can be obtained by the standard procedure, leading to the linear recurrent equation $O^k(n+k+1) - \binom{k+1}{1}O^k(n+k) + \binom{k+1}{2}O^k(n+k-1) - \cdots + (-1)^{k+1}\binom{k+1}{k+1}O^k(n) = 0$, $O^k(1) = 1$, $O^k(2) = \sum_{j=0}^{k-1}(-1)^j\binom{k-1}{j}2^{k-j-1}S_3^{k-j}(2), \ldots, O^k(k+1) = \sum_{j=0}^{k-1}(-1)^j\binom{k-1}{j}2^{k-j-1}S_3^{k-j}(k+1)$.

However, much more simple proof can be obtained using the above decomposition $O^k(n) = \sum_{j=0}^{k-1}(-1)^j\binom{k-1}{j}2^{k-j-1}S_3^{k-j}(n)$ and the generating function $\frac{1}{(1-x)^{k+1}} = S_3^k(1) + S_3^k(2)x + \cdots + S_3^k(n)x^{n-1} + \cdots$ for the sequence of k-dimensional hypertetrahedron numbers. In fact,

$$\sum_{i=1}^{\infty} O^k(n)x^{n-1} = \sum_{i=0}^{\infty}\left(\sum_{j=0}^{k-1}(-1)^j\binom{k-1}{j}2^{k-j-1}S_3^{k-j}(n)\right)x^{n-1}$$
$$= \sum_{j=0}^{k-1}(-1)^j\binom{k-1}{j}2^{k-j-1}\left(\sum_{i=0}^{\infty}S_3^{k-j}(n)x^{n-1}\right)$$
$$= \sum_{j=0}^{k-1}(-1)^j\binom{k-1}{j}2^{k-j-1}\cdot\frac{1}{(1-x)^{k-j+1}}$$
$$= \frac{1}{(1-x)^{k+1}}\cdot\sum_{j=0}^{k-1}(-1)^j\binom{k-1}{j}2^{k-j-1}(1-x)^j$$
$$= \frac{1}{(1-x)^{k+1}}\cdot\sum_{j=0}^{k-1}\binom{k-1}{j}2^{k-j-1}(x-1)^j.$$

But using binomial theorem, we get that $\sum_{j=0}^{k-1}\binom{k-1}{j}2^{k-j-1}(x-1)^j = (2+(x-1))^{k-1} = (x+1)^{k-1}$, and, so, it holds $\sum_{i=1}^{\infty} O^k(n)x^{n-1} = \frac{(1+x)^{k-1}}{(1-x)^{k+1}}$, $|x| < 1$.

It is easy to check that *every k-dimensional hyperoctahedron number can be written as a linear combination with non-negative integer coefficients of k-dimensional hypertetrahedron numbers*:

$$O^k(n) = \sum_{i=0}^{k-1}\binom{k-1}{i}S_3^k(n-i).$$

In fact, we have $O^k(n) = \sum_{r=0}^{k-1}(-1)^r\binom{k-1}{r}2^{k-1-r}S_3^{k-r}(n)$. Note that $S_3^{k-1}(n) = S_3^k(n) - S_3^k(n-1)$. By a successive application of this relation, we have $S_3^{k-r}(n) = \sum_{i=0}^{r}(-1)^i\binom{r}{i}S_3^k(n-i)$. It implies

$$O^k(n) = \sum_{r=0}^{k-1}(-1)^r\binom{k-1}{r}2^{k-1-r}\sum_{i=0}^{r}(-1)^i\binom{r}{i}S_3^k(n-i)$$

$$= \sum_{i=0}^{k-1} \sum_{r=i}^{k-1} (-1)^i \binom{k-1}{r} 2^{k-1-r} (-1)^r \binom{r}{i} S_3^k(n-i)$$

$$= \sum_{i=0}^{k-1} a_i S_3^k(n-i),$$

where $a_i = (-1)^i \sum_{r=i}^{k-1} (-1)^r \binom{k-1}{r}\binom{r}{i} 2^{k-1-r}$. Since $\binom{k}{r} \cdot \binom{r}{i} = \binom{k}{i} \cdot \binom{k-i}{k-r}$, we get, replacing $k-1-r$ by s, that

$$a_i = (-1)^i \sum_{r=i}^{k-1} (-1)^r \binom{k-1}{r} \binom{r}{i} 2^{k-1-r}$$

$$= (-1)^i \sum_{r=i}^{k-1} (-1)^r \binom{k-1}{i} \binom{k-1-i}{k-1-r} 2^{k-1-r}$$

$$= \binom{k-1}{i} \sum_{s=0}^{k-1-i} (-1)^{k-1-i-s} \binom{k-1-i}{s} 2^s = \binom{k-1}{i}.$$

This proves that $O^k(n) = \sum_{i=0}^{k-1} \binom{k-1}{i} S_3^k(n-1)$ (see [Kim03]).

3.3.11. The above results imply that *every k-dimensional regular polytope number can be written as a linear combination with non-negative integer coefficients of k-dimensional hypertetrahedron numbers.*

In fact, for $k = 2$ it holds due to *Bachet de Méziriac formula*:

$$S_m(n) = S_3(n) + (m-3)S_3(n-1), \quad m \geq 3.$$

For $k = 3$, it was shown, that any n-th regular polyhedral number can be written in the form $a_1 S_3^3(n) + a_2 S_3^3(n-1) + a_3 S_3^3(n-2)$, $a_i \geq 0$, $i = 1, 2, 3$. The corresponding coefficients are given below.

	$S_3^3(n)$	$C(n)$	$O(n)$	$D(n)$	$I(n)$
a_1	1	1	1	1	1
a_2	0	4	2	16	8
a_3	0	1	1	10	6

For $k = 4$, it was shown, that any n-th regular polychoral number can be written in the form $a_1 S_3^4(n) + a_2 S_3^4(n-1) + a_3 S_3^4(n-2) + a_4 S_3^3(n-3)$, $a_i \geq 0$, $i = 1, 2, 3, 4$. The corresponding coefficients are given below.

	$S_3^4(n)$	$BC(n)$	$HO(n)$	$HD(n)$	$HI(n)$	$RO(n)$
a_1	1	1	1	1	1	1
a_2	0	11	3	45	115	19
a_3	0	11	3	1993	357	43
a_4	0	1	1	543	107	9

For $k \geq 5$, it was shown, that any n-th k-dimensional regular polytope number can be written in the form $a_1 S_3^k(n) + a_2 S_3^k(n-1) + a_3 S_3^k(n-2) + \cdots + a_k S_3^k(n-k+1)$, $a_i \geq 0$, $i = 1, \ldots, k$. The corresponding coefficients are given below.

	$S_3^k(n)$	$O^k(n)$	$C^k(n)$
a_1	1	$\binom{k-1}{0} = 1$	$\left\langle {k \atop 0} \right\rangle = 1$
a_2	0	$\binom{k-1}{1} = k-1$	$\left\langle {k \atop 1} \right\rangle = 2^k - k - 1$
a_3	0	$\binom{k-1}{2} = \frac{(k-1)(k-2)}{2}$	$\left\langle {k \atop 2} \right\rangle = 3^k - 2^k(k+1) + \frac{n(n+1)}{2}$
...	...	...	...
a_2	0	$\binom{k-1}{k-2} = \frac{(k-1)(k-2)}{2}$	$\left\langle {k \atop k-2} \right\rangle = 2^k - k - 1$
a_k	0	$\binom{k-1}{k-1} = 1$	$\left\langle {k \atop k-1} \right\rangle = 1$

3.4 Nexus numbers

3.4.1. The **nexus numbers** are figurate numbers built up of the nexus of cells less than a fixed number of steps away from a given cell. So, n-th k-dimensional nexus number $N^k(n)$ is given by the formula

$$N^k(n) = 1 + \binom{k+1}{1} n + \binom{k+1}{2} n^2 + \cdots + \binom{k+1}{k} n^k.$$

The formula $(n+1)^{k+1} = 1 + \binom{k+1}{1} n + \binom{k+1}{2} n^2 + \cdots + \binom{k+1}{k} n^k + n^{k+1}$ implies

$$N^k(n) = (n+1)^{k+1} - n^{k+1}, \quad \text{i.e.,}$$
$$N^k(n) = C^{k+1}(n+1) - C^{k+1}(n).$$

It means that *k-dimensional nexus numbers form gnomons for $(k+1)$-dimensional hypercube numbers.*

For $k = 1$, n-th one-dimensional nexus number $N^1(n)$ have the form $2n+1$, i.e., coincides with $(n+1)$-th gnomonic number $Gn(n+1)$, giving gnomons for the sequence of square numbers. The first few elements of this sequence, starting from $n = 0$, are 1, 3, 5, 7, 9, 11, 13, 15, 17, 19, ... (Sloane's A005408).

For $k = 2$, n-th two-dimensional nexus number $N^2(n)$ have the form $3n^2 + 3n + 1$, i.e., coincides with $(n + 1)$-th hex number $CS_6(n + 1)$, giving gnomons for the sequence of cubic numbers. The first few elements of this sequence, starting from $n = 0$, are 1, 7, 19, 37, 61, 91, 127, 169, 217, 271, ... (Sloane's A003215).

For $k = 3$, n-th three-dimensional nexus number $N^3(n)$ have the form $4n^3 + 6n^2 + 4n + 1$, i.e., coincides with $(n + 1)$-th rhombic dodecahedral number $RD(n+1)$, giving gnomons for the sequence of biquadratic numbers. The first few elements of this sequence, starting from $n = 0$, are 1, 15, 65, 175, 369, 671, 1105, 1695, 2465, 3439, ... (Sloane's A005917).

For $k = 4$, n-th four-dimensional nexus number has the form $N^4(n) = 5n^4 + 10n^3 + 10n^2 + 5n + 1$. Such numbers are gnomons for the sequence of five-dimensional hypercube numbers. Usually the term *nexus number* is applied precisely to entries of the sequence $N^4(0), N^4(1), N^4(2), \ldots, N^4(n), \ldots$, starting with elements 1, 31, 211, 781, 2101, 4651, 9031, 15961, 26281, 40951, ... (Sloane's A022521).

For $k = 5$, the numbers $N^5(n) = (n+1)^6 - n^6$ are 1, 63, 665, 3367, 11529, 31031, 70993, 144495, 269297, 468559, ... (Sloane's A022522).

For $k = 6$, we obtain the numbers $N^6(n) = (n+1)^7 - n^7$, starting with elements 1, 127, 2059, 14197, 61741, 201811, 543607, 1273609, 2685817, 5217031, ... (Sloane's A002523).

Formally, for $k = 0$, we define $N^0(n) = 1$, and the unities can be considered as gnomons for the sequence of linear numbers, which, in turn, can be seen as one-dimensional hypercube numbers.

3.4.2. The symbolic representations for first k-dimensional nexus numbers are given in the table below.

k	$N^k(n)$	*name*
0	1	unity
1	$2n+1$	gnomonic number
2	$3n^2+3n+1$	hex number
3	$4n^3+6n^2+4n+1$	rhombic dodecahedral number
4	$5n^4+10n^3+10n^2+5n+1$	nexus number

The next table gives first few k-dimensional nexus numbers for $0 \le k \le 23$.

k	$N^k(0)$	$N^k(1)$	$N^k(2)$	$N^k(3)$	$N^k(4)$	$N^k(5)$	*Sloane*
0	1	1	1	1	1	1	
1	1	3	5	7	9	11	A005408
2	1	7	19	37	61	91	A003215
3	1	15	65	175	369	671	A005917
4	1	31	211	781	2101	4651	A022521
5	1	63	665	3367	11529	31031	A022522
6	1	127	2059	14197	61741	201811	A022523
7	1	255	6305	58975	325089	1288991	A002524
8	1	511	19171	242461	1690981	8124571	A002525
9	1	1023	58025	989527	8717049	50700551	A002526
10	1	2047	175099	4017157	44633821	313968931	A002527
11	1	4095	527345	16245775	227363409	1932641711	A002528
12	1	8191	1586131	65514541	1153594261	11839990891	A002529
13	1	16383	4766585	263652487	5835080169	72260648471	A002530
14	1	32767	14316139	1059392917	29443836301	439667406451	A002531
15	1	65535	42981185	4251920575	148292923329	2668522016831	A002532
16	1	131071	129009091	17050729021	745759583941	16163719991611	A0022533
17	1	262143	387158345	68332056247	3745977788889	97745259402791	A0022534
18	1	524287	1161737179	273715645477	18798608421181	590286253682371	A0022535
19	1	1048575	3485735825	1096024843375	94267920012849	3560791008422351	A0022536
20	1	2097151	10458256051	4387586157901	472439111692021	21460113482174731	A0022537
21	1	4194303	31376865305	17560804984807	2366593604971209	129237518051251511	A0022538
22	1	8388607	94134790219	70274600998837	11850560210900461	777809294098524691	A022539
23	1	16777215	282412759265	281192547174175	59323169798679969	4678776693546226271	A022540

3.4.3. It is easy to see, that the formula

$$N^k(n) + C^{k+1}(n) = C^{k+1}(n+1),$$

having for $k = 1, 2, 3$ the form

$$\begin{aligned} 1 + 2n + n^2 &= (n+1)^2, \\ 1 + 3n + 3n^2 + n^3 &= (n+1)^3, \\ 1 + 4n + 6n^2 + 4n^3 + n^4 &= (n+1)^4, \end{aligned}$$

respectively, corresponds to the general case of *binomial theorem*:

$$1 + \binom{k+1}{1} n + \binom{k+1}{2} n^2 + \cdots + \binom{k+1}{k} n^k + n^{k+1} = (n+1)^{k+1}.$$

Viewing from the opposite perspective, one can say, that *the sum of the first n k-dimensional nexus numbers is equal to $(n+1)$-th $(k+1)$-dimensional hypercube number*:

$$N^k(0) + N^k(1) + \cdots + N^k(n) = C^{k+1}(n+1).$$

In fact, $N^k(0) + N^k(1) + \cdots + N^k(n) = (C^{k+1}(1) - C^{k+1}(0)) + (C^{k+1}(2)(C^{k+1}(2) - C^{k+1}(1)) + \cdots + (C^{k+1}(n+1) - C^{k+1}(n)) = C^{k+1}(n+1) - C^{k+1}(0) = C^{k+1}(n+1)$.

Moreover, since $\binom{n}{k} = S_3^k(n-k+1)$, and $n^i = C^i(n)$, $i = 0, 1, 2, \ldots, k$, we get

$$\begin{aligned} N^k(n) &= S_3^0(k+2)C^0(n) + S_3^1(k+1)C^1(n) + S_3^2(k)C^2(n) \\ &\quad + \cdots + S_3^{k-1}(3)C^{k-1}(n) + S_3^k(2)C^k(n). \end{aligned}$$

3.4.5. The generating function for the sequence of k-dimensional nexus numbers has the form

$$\begin{aligned} \frac{x(1+\overline{a}_1^k x + \cdots + \overline{a}_{k-1}^k x^{k-1} + x^k)}{(1-x)^{k+1}} &= N^k(0)x + N^k(1)x^2 + N^k(2)x^3 \\ &\quad + \cdots + N^k(n-1)x^n \\ &\quad + \cdots, \quad |x| < 1, \end{aligned}$$

where $\overline{a}_1^k = \overline{a}_{k-1}^k = 2^{k+1} - (k+2)$, $\overline{a}_2^k = \overline{a}_{k-2}^k = 3^{k+1} - (k+2)2^{k+1} + \frac{(k+1)(k+2)}{2}, \ldots, \overline{a}_{\lfloor \frac{k}{2} \rfloor}^k = \overline{a}_{k-\lfloor \frac{k}{2} \rfloor}^k = (\lfloor \frac{k}{2} \rfloor + 1)^{k+1} - (k+2)(\lfloor \frac{k}{2} \rfloor)^{k+1} + \frac{(k+1)(k+2)}{2}(\lfloor \frac{k}{2} \rfloor - 1)^{k+1} + \cdots + (-1)^{\lfloor \frac{k}{2} \rfloor}\binom{k+2}{\lfloor \frac{k}{2} \rfloor}$. In particular, we have

$$\frac{x(1+x)}{(1-x)^2} = x + 3x^2 + 5x^3 + \cdots + N^1(n-1)x^n + \cdots, \quad |x| < 1;$$

$$\begin{aligned} \frac{x(1+4x+x^2)}{(1-x)^3} &= x + 7x^2 + 19x^3 \\ &\quad + \cdots + N^2(n-1)x^n + \cdots, \quad |x| < 1; \end{aligned}$$

$$\frac{x(1+11x+11x^2+x^3)}{(1-x)^4} = \frac{x(x+1)(1+10x+x^2)}{(1-x)^5}$$

$$= x + 15x^2 + 65x^3 + \cdots + N^3(n-1)x^n + \cdots, \quad |x| < 1;$$

$$\frac{x(1 + 26x + 66x^2 + 26x^3 + x^4)}{(1-x)^5} = x + 31x^2 + 211x^3$$

$$+ \cdots + N^4(n-1)x^n + \cdots, \quad |x| < 1;$$

$$\frac{x(1 + 57x + 302x^2 + 302x^3 + 57x + x^5)}{(1-x)^6}$$

$$= \frac{x(x+1)(1 + 56x + 246x^2 + 56x^3 + x^4)}{(1-x)^6}$$

$$= x + 63x^2 + 665x^3 + \cdots + N^5(n-1)x^n + \cdots, \quad |x| < 1;$$

$$\frac{x(1 + 120x + 1191x^2 + 2416x^3 + 1191x^4 + 120x^5 + x^6)}{(1-x)^7}$$

$$= x + 127x^2 + 2059x^3 + \cdots + N^6(n-1)x^n + \cdots, \quad |x| < 1;$$

$$\frac{x(1+247x+4293x^2+15619x^3+15619x^4+4293x^5+247x^6+x^7)}{(1-x)^8}$$

$$= \frac{x(1 + 246x + 4047x^2 + 11572x^3 + 4047x^4 + 246x^5 + x^6)}{(1-x)^8}$$

$$= x + 255x^2 + 6305x^3 + \cdots + N^7(n-1)x^n + \cdots, \quad |x| < 1.$$

In other words, the coefficients $\overline{a}_i^k$ of the numerator of the generating function $f_{N^k}(x)$ for the sequence of k-dimensional nexus numbers are equal to the coefficients of the numerator of the generating function $f_{C^{k+1}}(x)$ for the sequence of $(k+1)$-dimensional hypercube numbers: $\overline{a}_i^k = a_i^{k+1}$, $i = 0, 1, \ldots, k+1$. Moreover, those two generating functions are connected by the equation $f_{N^k}(x) = (1-x)f_{C^{k+1}}(x)$.

This result can be obtained using the generating function $\frac{(1+a_1^{k+1}x+\cdots+a_{k-1}^{k+1}x^{k-1}+x^k)}{(1-x)^{k+2}}$ for the sequence of $(k+1)$-dimensional hypercube numbers and the decomposition $\frac{1}{1-x} = 1 + x + x^2 + \cdots + x^n + \cdots$. In fact, considering the generating function $f_{N^k}(x) = N^k(0) + N^k(1)x + N^k(2)x^2 + \cdots + N^k(n)x^n + \cdots$, one has

$$f_{N^k}(x) \cdot \frac{1}{1-x} = (N^k(0) + N^k(1)x + N^k(2)x^2$$

$$+ \cdots + N^k(n-1)x^{n-1} + \cdots)$$

$$\begin{aligned}
&\times (1 + x + x^2 + \cdots + x^n + \cdots) \\
&= N^k(0) + (N^k(0) + N^k(1))x + (N^k(0) + N^k(1) \\
&\quad + N^k(2))x^2 + \cdots + (N^k(0) + \cdots + N^k(n-1))x^{n-1} + \cdots \\
&= C^{k+1}(1) + C^{k+1}(2)x + C^{k+1}(3)x^2 \\
&\quad + \cdots + C^{k+1}(n)x^{n-1} + \cdots \\
&= \frac{(1 + a_1^{k+1}x + \cdots + a_{k-1}^{k+1}x^{k-1} + x^k)}{(1-x)^{k+2}}, \quad |x| < 1.
\end{aligned}$$

Therefore, $f_{N^k}(x) \cdot \frac{1}{1-x} = \frac{(1+a_1x+\cdots+a_{k-1}x^{k-1}+x^k)}{(1-x)^{k+2}}$, and $f_{N^k}(x) = \frac{(1+a_1x+\cdots+a_{k-1}x^{k-1}+x^k)}{(1-x)^{k+1}}$, i.e.,

$$\frac{(1 + a_1^{k+1}x + \cdots + a_{k-1}^{k+1}x^{k-1} + x^k)}{(1-x)^{k+1}} = N^k(0) + N^k(1)x + N^k(2)x^2 + \cdots + N^k(n)x^n + \cdots, \quad |x|<1.$$

So, denoting a_i^{k+1} by $\overline{a}_i^k$, $i = 0, 1, 2, \ldots, k+1$, we get

$$\frac{x(1 + \overline{a}_1^k x + \cdots + \overline{a}_{k-1}^k x^{k-1} + x^k)}{(1-x)^{k+1}} = N^k(0)x + N^k(1)x^2 + N^k(2)x^2 + \cdots + N^k(n-1)x^n + \cdots, \quad |x|<1.$$

Of course, one can easily get this result by the standard procedure, using the linear recurrent equation $N^k(n+k+1) - \binom{k+1}{1} N^k(n+k) + \binom{k+1}{2} N^k(n+k-1) + \cdots + (-1)^{k+1}\binom{k+1}{k+1} N^k(n) = 0$ with initial values $N^k(1) = 1$, $N^k(1) = 2^{k+1} - 1$, $N^k(2) = 3^{k+1} - 2^{k+1}, \ldots,$ $N^k(k) = (k+1)^{k+1} - k^{k+1}$.

It can be obtained, for example, using the corresponding linear recurrent equation for $(k+1)$-dimensional hypercube numbers:

$$C^{k+1}(n+k+2) - \binom{k+2}{1} C^{k+1}(n+k+1) + \binom{k+2}{2} C^{k+1}(n+k) + \cdots + (-1)^{k+2}\binom{k+2}{k+2} C^{k+1}(n) = 0.$$

In fact, rewriting the above equation as $C^{k+1}(n+k+3) - \binom{k+2}{1}C^{k+1}(n+k+2) + \binom{k+2}{2}C^{k+1}(n+k+1) + \cdots (-1)^{k+2}\binom{k+2}{k+2}$ $C^{k+1}(n+1) = 0$ and subtracting first equation from the second one, we get $(C^{k+1}(n+k+3) - C^{k+1}(n+k+2)) - \binom{k+2}{1}$ $(C^{k+1}(n+k+2) - C^{k+1}(n+k+1)) + \binom{k+2}{2}(C^{k+1}(n+k+1) - C^{k+1}(n+k+3)) + \cdots + (-1)^{k+2}\binom{k+2}{k+2}(C^{k+1}(n+1) - C^{k+1}(n)) = 0$, i.e., $N^k(n+k+1) - \binom{k+1}{1}N^k(n+k) + \binom{k+1}{2}N^k(n+k-1) + \cdots + (-1)^{k+1}\binom{k+1}{k+1}N^k(n) = 0$.

The third way to obtain this result is just use the decompositions $\frac{1+a_1^{k+1}x+a_2^{k+1}x^2+\cdots+x^k}{(1-x)^{k+2}} = C^{k+1}(1) + C^{k+1}(2)x + C^{k+1}(3)x^2 + \cdots + C^{k+1}(n+1)x^n + \cdots$ and $\frac{x(1+a_1^{k+1}x+a_2^{k+1}x^2+\cdots+x^k)}{(1-x)^{k+2}} = C^{k+1}(1)x + C^{k+1}(2)x^2 + C^{k+1}(3)x^3 + \cdots + C^{k+1}(n)x^n + \cdots$ in order to get the decomposition

$$\frac{(1-x)(1+a_1^{k+1}x+a_2^{k+1}x^2+\cdots+x^k)}{(1-x)^{k+2}} = (C^{k+1}(1) - C^{k+1}(0))$$

$$+ (C^{k+1}(2) - C^{k+1}(1))x + \cdots + (C^{k+1}(n+1)$$

$$- C^{k+1}(n))x^n + \cdots, \quad |x| < 1, \text{ i.e., the decomposition}$$

$$\frac{1+a_1^{k+1}x+a_2^{k+1}x^2+\cdots+x^k}{(1-x)^{k+1}} = N^k(0) + N^k(1)x + N^k(2)x^2$$

$$+ N^k(3)x^3 + \cdots + N^k(n)x^n + \cdots, \quad |x| < 1.$$

3.5 Pyramidal numbers of the second order and their multidimensional analogues

3.5.1. An m-**gonal pyramidal number of the second order** (or **four-dimensional** m-**gonal pyramidal number**) can be constructing, similarly to polytope numbers, as the sum of consecutive m-gonal pyramidal numbers. So, the n-th m-gonal pyramidal number $S_m^4(n)$ of the second order is defined as

$$S_m^4(n) = S_m^3(1) + S_m^3(2) + \cdots + S_m^3(n).$$

In particular, the **four-dimensional square pyramidal numbers**, obtained by consecutive summation of the sequence 1, 5, 14, 30, 55, 91, 140, 204, 285, 385,... (Sloane's A000330) of square pyramidal numbers, are 1, 6, 20, 50, 105, 196, 336, 540, 825, 1210,... (Sloane's A002415).

The **four-dimensional pentagonal pyramidal numbers**, obtained by consecutive summation of the sequence 1, 6, 18, 40, 75, 126, 196, 288, 405, 550,... (Sloane's A002411) of pentagonal pyramidal numbers, are 1, 7, 25, 65, 140, 266, 462, 750, 1155, 1705,... (Sloane's A001296).

The **four-dimensional hexagonal pyramidal numbers**, obtained by consecutive summation of the sequence 1, 7, 22, 50, 95, 161, 252, 372, 525, 715,... (Sloane's A002412) of hexagonal pyramidal numbers, are 1, 8, 30, 80, 175, 336, 588, 960, 1485, 2200,... (Sloane's A002417).

Similarly, we obtain **four-dimensional heptagonal pyramidal numbers** 1, 9, 35, 95, 210, 406, 714, 1170, 1815, 2695,... (Sloane's A002418); **four-dimensional octagonal pyramidal numbers** 1, 10, 40, 110, 245, 476, 840, 1380, 2145, 3190,... (Sloane's A002419); **four-dimensional nonagonal pyramidal numbers** 1, 11, 45, 125, 280, 546, 966, 1590, 2475, 3685,... (Sloane's A051740); **four-dimensional decagonal pyramidal numbers** 1, 12, 50, 140, 315, 616, 1092, 1800, 2805, 4180,... (Sloane's A051797); **four-dimensional hendecagonal pyramidal numbers** 1, 13, 55, 155, 350, 686, 1218, 2010, 3135, 4675,... (Sloane's A051798); **four-dimensional dodecagonal pyramidal numbers** 1, 14, 60, 170, 385, 756, 1344, 2220, 3465, 5170, 7436, 10374,... (Sloane's A051799) and so on.

By definition, we obtain the following recurrent formula for the sequence of m-gonal pyramidal numbers of the second order:

$$S_m^4(n) = S_m^4(n-1) + S_m^3(n), \quad S_m^4(1) = 1.$$

The general formula for n-th m-gonal pyramidal number of the second order has the form

$$S_m^4(n) = \frac{n(n+1)(n+2)((m-2)n - m + 6)}{24}.$$

In particular, we have

$$S_4^4(n) = \frac{n(n+1)^2(n+2)}{12}, \quad S_5^4(n) = \frac{n(n+1)(n+2)(3n+1)}{24},$$
$$S_6^4(n) = \frac{n^2(n+1)(n+2)}{6}, \quad S_7^4(n) = \frac{n(n+1)(n+2)(5n-1)}{24},$$
$$S_8^4(n) = \frac{n(n+1)(n+2)(3n-1)}{12}.$$

Let us prove it by induction. For $n = 1$ one has $S_m^4(1) = S_m^3(1) = 1 = \frac{1\cdot 2\cdot 3((m-2)\cdot 1-m+6)}{24}$. Going from n to $n+1$, one obtains

$$\begin{aligned} S_m^4(n+1) &= S_m^4(n) + S_m^3(n+1) \\ &= \frac{n(n+1)(n+2)((m-2)n-m+6)}{24} \\ &\quad + \frac{(n+1)(n+2)((m-2)(n+1)-m+5)}{6} \\ &= \frac{(n+1)(n+2)(n+3)((m-2)(n+1)-m+6)}{24}. \end{aligned}$$

3.5.2. The generating function for the sequence $S_m^4(1), S_m^4(2), \ldots, S_m^4(n), \ldots$ of m-pyramidal numbers of the second order has the form $f(x) = \frac{x(1+(m-3)x)}{(1-x)^5}$, i.e., it holds

$$\begin{aligned} \frac{x(1+(m-3)x)}{(1-x)^5} &= S_m^4(1)x + S_m^4(2)x^2 + S_m^4(3)x^3 \\ &\quad + \cdots + S_m^4(n)x^n + \cdots, \quad |x| < 1. \end{aligned}$$

In particular, one gets

$$\frac{x(x+1)}{(x-1)^5} = x + 6x^2 + 20x^3 + \cdots + S_4^4(n)x^n + \cdots, \quad |x| < 1;$$
$$\frac{x(2x+1)}{(x-1)^5} = x + 7x^2 + 25x^3 + \cdots + S_5^4(n)x^n + \cdots, \quad |x| < 1;$$
$$\frac{x(3x+1)}{(x-1)^5} = x + 8x^2 + 30x^3 + \cdots + S_6^4(n)x^n + \cdots, \quad |x| < 1.$$

This result can be obtained by the standard procedure, leading to the linear recurrent equation $S_m^4(n+5)-5S_m^4(n+4)+10S_m^4(n+3)-10S_m^4(n+2)+5S_m^4(n)-S_m^4(n)=0$ with initial values $S_m^4(1)=1$, $S_m^4(2)=m+2$, $S_m^4(3)=5m$, $S_m^4(4)=15m-10$, and $S_m^4(5)=35m-35$.

This result can be obtained also using the generating function $\frac{1+(m-3)x}{(1-x)^4}=S_m^3(1)+S_m^3(2)x+S_m^3(3)x^2+\cdots+S_m^3(n)x^{n-1}+\cdots$ for the sequence of m-gonal pyramidal numbers and the decomposition $\frac{1}{1-x}=1+x+x^2+\cdots+x^n+\cdots$. In fact, one has

$$\begin{aligned}\frac{1+(m-3)x}{(1-x)^5} &= \frac{1+(m-3)x}{(1-x)^4}\cdot\frac{1}{1-x}\\ &= (S_m^3(1)+S_m^3(2)x+S_m^3(3)x^2+\cdots\\ &\quad +S_m^3(n)x^{n-1}+\cdots)(1+x+x^2+\cdots+x^n+\cdots)\\ &= S_m^3(1)+(S_m^3(1)+S_m^3(2))x+(S_m^3(1)+S_m^3(2)\\ &\quad +S_m^3(3))x^2+\cdots+(S_m^3(1)+\cdots+S_m^3(n))x^{n-1}+\cdots\\ &= S_m^4(1)+S_m^4(2)x+S_m^4(3)x^2\\ &\quad +\cdots+S_m^4(n)x^{n-1}+\cdots,\quad |x|<1.\end{aligned}$$

3.5.3. Let us consider now some interesting relations between pyramidal numbers of the second order.

I. It holds that n-th m-gonal pyramidal number of the second order satisfies the equality

$$S_m^3(n)=\frac{1}{4}((m-2)n-m+6)S_3^3(n),$$

where $S_3^3(n)$ is n-th tetrahedral number. In particular, it holds

$$S_4^4(n)=\frac{1}{2}(n+1)S_3^3(n),\quad S_5^3(n)=\frac{1}{4}(3n+1)S_3^3(n),$$

$$S_6^3(n)=nS_3^3(n),\quad S_7^3(n)=\frac{1}{4}(5n-1)S_3^3(n),$$

$$S_8^3(n)=\frac{1}{5}(3n-1)S_3(n).$$

Really, since $S_3^3(n) = \frac{n(n+1)(n+2)}{6}$, this property follows in an obvious way from the general formula for $S_m^4(n)$.

II. It is easy to show that *four-dimensional square pyramidal numbers are sums of consecutive pairs of polytope numbers*:

$$S_4^4(n) = S_3^4(n) + S_3^4(n-1).$$

In fact, one obtains this identity summing the similar three-dimensional identity $S_4^3(k) = S_3^3(k) + S_3^3(k-1)$ over all $1 \le k \le n$.

III. It is easy to check also that *any m-gonal pyramidal number of the second order is equal to the sum of the $(m-1)$-gonal pyramidal number of the second order, taking in the row the same place, and the pentatope number, taking in the row the previous place*:

$$S_m^4(n) = S_{m-1}^4(n) + S_3^4(n-1).$$

This identity is a four-dimensional analogue of the identity $S_m^3(k) = S_{m-1}^3(k) + S_3^3(k-1)$ and can be obtained by summing it over all $1 \le k \le n$.

IV. Moreover, *any m-gonal pyramidal number of the second order is a linear combination with non-negative integer coefficients of pentatope numbers*:

$$S_m^4(n) = S_3^4(n) + (m-3)S_3^4(n-1).$$

It follows by the summation over all $1 \le k \le n$ the following three-dimensional analogue of *Bachet de Méziriac formula*: $S_m^3(k) = S_3^3(k) + (m-3)S_3^3(k-1)$.

V. Similar arguments allow us to show that *any m-gonal pyramidal number of the second order is the sum of the tetrahedral number of the same size and $m-2$ copies of pentatope number of the previous size*:

$$S_m^4(n) = S_3^3(n) + (m-2)S_3^4(n-1).$$

It follows by the same procedure from the formula $S_m^3(k) = S_3(k) + (m-2)S_3^3(k-1)$.

3.5.4. In general, for a given dimension k, $k \ge 2$, an m-**gonal pyramidal number of the $(k-2)$-th order** (or k-**dimensional**

m**-gonal pyramidal number**) is given as the consecutive sum of the first m-gonal pyramidal numbers of the $(k-3)$-th order:

$$S_m^k(n) = S_m^{(k-1)}(1) + S_m^{(k-1)}(2) + \cdots + S_m^{(k-1)}(n).$$

In particular, for $k = 2, 3$, and 4 one obtains m-gonal, m-gonal pyramidal and m-gonal pyramidal of the second order numbers, respectively.

By definition, we obtain the following recurrent formula for the m-pyramidal numbers of the $(k-2)$-th order:

$$S_m^k(n) = S_m^k(n-1) + S_m^{k-1}(n), \quad S_m^k(1) = 1.$$

The general formula for n-th m-gonal pyramidal number of the $(k-2)$-th order has the form

$$S_m^k(n) = \frac{n(n+1)\cdots(n+k-2)((m-2)n - m + k + 2)}{k!}.$$

It can be checked in the usual way by induction.

In particular, for $k = 5$ we get the formula

$$S_m^5(n) = \frac{n(n+1)(n+2)(n+3)((m-2)n - m + 7)}{120}.$$

So, **five-dimensional square pyramidal numbers** $S_4^5(n) = \frac{n(n+1)(n+2)(n+3)(2n+3)}{5!}$ form the sequence 1, 7, 27, 77, 182, 378, 714, 1254, 2079, 3289, ... (Sloane's A005585); **five-dimensional pentagonal pyramidal numbers** $S_5^5(n) = \frac{n(n+1)(n+2)(n+3)(3n+2)}{5!}$ form the sequence 1, 8, 33, 98, 238, 504, 966, 1716, 2871, 4576, ... (Sloane's A051836); **five-dimensional hexagonal pyramidal numbers** $S_6^5(n) = \frac{n(n+1)(n+2)(n+3)(4n+1)}{5!}$ form the sequence 1, 9, 39, 119, 294, 630, 1218, 2178, 3663, 5863, ... (Sloane's A034263); **five-dimensional heptagonal pyramidal numbers** $S_7^5(n) = \frac{n^2(n+1)(n+2)(n+3)}{4!}$ form the sequence 1, 10, 45, 140, 350, 756, 1470, 2640, 4455, 7150, ... (Sloane's A027800); **five-dimensional octagonal pyramidal numbers** $S_8^5(n) = \frac{n(n+1)(n+2)(n+3)(6n-1)}{5!}$ form the sequence 1, 11, 51, 161, 406, 882, 1722, 3102, 5247, 8437, ... (Sloane's A051843) and so on.

For $k = 6$ we have the formula

$$S_m^6(n) = \frac{n(n+1)(n+2)(n+3)(n+4)((m-2)n-m+8)}{6!}.$$

So, **six-dimensional square pyramidal numbers** $S_4^6(n) = \frac{n(n+1)(n+2)^2(n+3)(n+4)}{360}$ form the sequence 1, 8, 35, 112, 294, 672, 1386, 2640, 4719, 8008, ... (Sloane's A040977); **six-dimensional pentagonal pyramidal numbers** $S_5^6(n) = \frac{n(n+1)^2(n+2)(n+3)(n+4)}{240}$ form the sequence 1, 9, 42, 140, 378, 882, 1848, 3564, 6435, 11011, ... (Sloane's A051923); **six-dimensional hexagonal pyramidal numbers** $S_6^6(n) = \frac{n(n+1)(n+2)(n+3)(2n+1)}{360}$ form the sequence 1, 10, 49, 168, 462, 1092, 2310, 4488, 8151, 14014, ... (Sloane's A051947); **six-dimensional heptagonal pyramidal numbers** $S_7^6(n) = \frac{n(n+1)(n+2)(n+3)(n+4)(5n+1)}{720}$ form the sequence 1, 11, 56, 196, 546, 1302, 2772, 5412, 9867, 17017, ... (Sloane's A051946); **six-dimensional octagonal pyramidal numbers** $S_8^6(n) = \frac{n^2(n+1)(n+2)(n+3)(n+4)}{120}$ form the sequence 1, 12, 63, 224, 630, 1512, 3234, 6336, 11583, 20020, ... (Sloane's A027810) and so on.

3.5.5. The generating function for the sequence $S_m^k(1), S_m^k(2), \ldots, S_m^k(n), \ldots$ of m-gonal pyramidal numbers of the $(k-2)$-th order has the form $f(x) = \frac{x(1+(m-3)x)}{(1-x)^{k+1}}$, i.e., it holds

$$\frac{x(1+(m-3)x)}{(1-x)^{k+1}} = S_m^k(1)x + S_m^k(2)x^2 + S_m^k(3)x^3 + \cdots + S_m^k(n)x^n + \cdots, \quad |x| < 1.$$

In particular one gets

$$\frac{x(x+1)}{(x-1)^{k+1}} = S_4^k(1)x + S_4^k(2)x^2 + S_4^k(3)x^3 + \cdots + S_4^k(n)x^n + \cdots, \quad |x| < 1;$$

$$\frac{x(2x+1)}{(x-1)^{k+1}} = S_5^k(1)x + S_5^k(2)x^2 + S_5^k(3)x^3 + \cdots + S_5^k(n)x^n + \cdots, \quad |x| < 1;$$

$$\frac{x(3x+1)}{(x-1)^{k+1}} = S_6^k(1)x + S_6^k(2)x^2 + S_6^k(3)x^3 + \cdots + S_6^k(n)x^n + \cdots, \quad |x| < 1.$$

This result can be obtained by the standard procedure, leading to the linear recurrent equation $S_m^k(n+k+1) - \binom{k+1}{1}S_m^k(n+k) + \binom{k+1}{2}S_m^k(n+k-1) + \cdots + (-1)^{k+1}\binom{k+1}{k+1}S_m^k(n) = 0$ with initial values $S_m^k(1) = 1$, $S_m^k(2) = m+k-2$, $S_m^k(3) = \frac{(k+1)(2m+k-4)}{2}, \ldots, S_m^k(k+1) = \frac{(k+1)(k+2)\cdots(2k+1)(m-1)}{(k-1)!}$.

This result can be obtained also by induction using the generating function $\frac{1+(m-3)x}{(1-x)^k} = S_m^{k-1}(1) + S_m^{k-1}(2)x + S_m^{k-1}(3)x^2 + \cdots + S_m^{k-1}(n)x^{n-1} + \cdots$ for m-gonal pyramidal numbers of the $(k-3)$-th order. In fact, it holds for $k = 2, 3, 4$, and, going from $k-1$ to k, one gets

$$\begin{aligned}\frac{1+(m-3)x}{(1-x)^k} &= \frac{1+(m-3)x}{(1-x)^{k-1}} \cdot \frac{1}{1-x} = (S_m^{k-1}(1) + S_m^{k-1}(2)x \\ &\quad + S_m^{k-1}(3)x^2 + \cdots + S_m^{k-1}(n)x^{n-1} + \cdots) \\ &\quad \times (1 + x + x^2 + \cdots + x^n + \cdots) \\ &= S_m^{k-1}(1) + (S_m^{k-1}(1) + S_m^{k-1}(2))x + (S_m^{k-1}(1) \\ &\quad + S_m^{k-1}(2) + S_m^{k-1}(3))x^2 + \cdots + (S_m^{k-1}(1) \\ &\quad + \cdots + S_m^{k-1}(n))x^{n-1} + \cdots = S_m^k(1) + S_m^k(2)x \\ &\quad + S_m^k(3)x^2 + \cdots + S_m^k(n)x^{n-1} + \cdots, \quad |x| < 1.\end{aligned}$$

3.5.6. Let us consider now some interesting relations between pyramidal numbers of the $(k-2)$-th order.

I. It is easy to see that n-th m-pyramidal number of the $(k-2)$-th order satisfies the equality

$$S_m^k(n) = \frac{1}{k}((m-2)n - m + k + 2)S_3^{k-1}(n),$$

where $S_3^{k-1}(n)$ is n-th $(k-1)$-dimensional hypertetrahedron number. In particular, it holds

$$S_4^k(n) = \frac{1}{k}(2n + k - 2)S_3^{k-1}(n),$$

$$S_5^k(n) = \frac{1}{k}(3n + k - 3)S_3^{k-1}(n),$$

$$S_6^k(n) = \frac{1}{k}(4n + k - 4)S_3^{k-1}(n).$$

In fact, since $S_3^{k-1}(n) = \frac{n(n+1)(n+2)\cdots(n+k-2)}{(k-1)!}$, this property follows in an obvious way from the general formula for $S_m^k(n)$.

II. It is easy to show also that *k-dimensional square pyramidal numbers are sums of consecutive pairs of k-dimensional hypertetrahedron numbers*:

$$S_4^k(n) = S_3^k(n) + S_3^k(n-1).$$

In fact, one can prove this identity by induction on k, using the summation of the similar identity $S_4^{k-1}(i) = S_3^{k-1}(i) + S_3^{k-1}(i-1)$ over all $1 \leq i \leq n$.

III. Similarly, it holds that *any m-gonal pyramidal number of the $(k-2)$-th order is equal to the sum of the $(m-1)$-gonal pyramidal number of the $(k-2)$-th order, taking in the row the same place, and the k-dimensional hypertetrahedron number, taking in the row the previous place*:

$$S_m^k(n) = S_{m-1}^k(n) + S_3^k(n-1).$$

This identity is an k-dimensional analogue of the identity $S_m^3(i) = S_{m-1}^3(i) + S_3^3(i-1)$ and can be obtained by induction on k using the procedure of summation over all $1 \leq i \leq n$.

IV. Moreover, *any m-gonal pyramidal number of the $(k-2)$-th order is a linear combination with non-negative integer coefficients of k-dimensional hypertetrahedron numbers*:

$$S_m^k(n) = S_3^k(n) + (m-3)S_3^k(n-1).$$

It follows by induction on k and the procedure of consecutive summation from the three-dimensional analogue of *Bachet de Méziriac formula* $S_m^3(i) = S_3^3(i) + (m-3)S_3^3(i-1)$.

V. Also, *any m-pyramidal number of the $(k-2)$-th order is the sum of the $(k-1)$-dimensional hypertetrahedron number of the same size and $m-2$ copies of k-dimensional hypertetrahedron number of the previous size*:

$$S_m^k(n) = S_3^{k-1}(n) + (m-2)S_3^k(n-1).$$

It follows by the same procedure from the formula $S_m^3(n) = S_3(n) + (m-2)S_3^3(n-1)$.

3.6 Centered multidimensional figurate numbers

It this section we consider multidimensional analogues of some centered space figurate numbers. Precisely, we will construct **centered regular polytope numbers**, i.e., **centered hypertetrahedron numbers**, **centered hypercube numbers**, and **centered hyperoctahedron numbers**, for any dimension k, $k \geq 4$.

3.6.1. The **centered biquadratic numbers** represent a four-dimensional "centered hypercube". The first elements of this sequence are 1, 17, 97, 337, 881, 1921, 3697, 6497, 10657, 16561, ... (Sloane's A008514).

They can be obtained as partial sums of the sequence 1, 16, 80, 240, 544, 1040, 1776, 2800, 4160, 5904, ... (Sloane's A008511), giving number of points on surface of four-dimensional hypercube. The first element of the sequence A008511 is 1, while n-th element, $n \geq 2$, has the form $n^4 - (n-2)^4$.

This fact can be obtained by the formula $N_0 + N_1(n-2) + N_2 \cdot int(S_4(n)) + N_3 int(C(n))$, where $N_0 = 16$, $N_1 = 32$, $N_2 = 24$, $N_3 = 8$ are numbers of vertices, edges, squares and cubes of a four-dimensional hypercube. So, for $n \geq 2$, it holds $N_0 + N_1(n-2) + N_2 \cdot int(S_4(n)) + N_3 int(C(n)) = 16 + 32(n-2) + 24(n^2 - 4(n-1)) + 8(n^3 - (8 + 12(n-2) + 6(n^2 - 4(n-1)))) = 16 + 32(n-2) + 24(n-2)^2 + 8(n-2)^3 = ((n+2)-2)^4 - (n-2)^4 = n^4 - (n-2)^4$.

Hence, the recurrent formula for the sequence of centered biquadratic numbers is

$$\overline{BC}(n+1) = \overline{BC}(n) + (n+1)^4 - (n-1)^4, \quad \overline{BC}(1) = 1.$$

It is easy to see, that the general formula for n-th centered biquadratic number $\overline{BC}(n)$ is

$$\overline{BC}(n) = n^4 + (n-1)^4,$$

i.e., *centered biquadratic numbers are sums of two concecutice ordinary biquadratic numbers.*

In fact, it can be obtained as the result of the following telescoping summation: $\overline{BC}(n) = 1^4 + (2^4 - 0^4) + (3^4 - 1^4) + \cdots + (n^4 - (n-2)^4) = n^4 + (n-1)^4$.

The generation function for the sequence of centered biquadratic numbers has the form $f(x) = \frac{x(x^2+10x+1)(x+1)^2}{(x-1)^5}$ (see [Weis11]), i.e., it holds

$$\frac{x(x^2+10x+1)(x+1)^2}{(x-1)^5} = \overline{BC}(1)x + \overline{BC}(2)x^2 + \overline{BC}(3)x^3 + \cdots + \overline{BC}(n)x^4 + \cdots, \quad |x| < 1.$$

This result can be obtained from the generating function $\frac{(1+x)(1+10x+x^2)}{(1-x)^5} = BC(1) + BC(2)x + \cdots + BC(n+1)x^n + \cdots$ for biquadratic numbers:

$$\begin{aligned}\frac{(1+x)^2(1+10x+x^2)}{(1-x)^5} &= \frac{(1+x)(1+10x+x^2)}{(1-x)^5} \\ &\quad + \frac{x(1+x)(1+10x+x^2)}{(1-x)^5} = BC(1) + BC(2)x \\ &\quad + \cdots + BC(n+1)x^n + \cdots + BC(1)x + BC(2)x^2 \\ &\quad + \cdots + BC(n)x^n + \cdots \\ &= BC(1) + (BC(2) + BC(1))x + (BC(3) + BC(2))x^2 \\ &\quad + \cdots + (BC(n+1) + BC(n))x^n + \cdots \\ &= \overline{BC}(1) + \overline{BC}(2)x + \cdots + \overline{BC}(n+1)x^n + \cdots, \quad |x| < 1.\end{aligned}$$

3.6.2. The **k-dimensional centered hypercube numbers** represent k-dimensional centered cube.

They can be obtained as partial sums of the sequence, giving number of points on the surface of k-dimensional hypercube. The first element of the last sequence is 1, while n-th element, $n \geq 2$, has the form $n^k - (n-2)^k$.

It follows from the fact that *the number $int(C^k(n))$ of the interior's points of k-dimensional hypercube of size n is equal to the* *number of points in k-dimensional hypercube of size $n-2$:*

$$int(C^k(n)) = C^k(n-2), \quad n \geq 2.$$

This result can be obtained by induction on k, using the formula $\sum_{i=0}^{k-1} N_i int(C^i(n))$, where $N_i = 2^{k-i}\binom{k}{i}$, $0 \leq i \leq k-1$, is the number of i-dimensional faces of an k-dimensional hypercube γ^k.

In fact, it holds for $k = 2, 3$ and 4 (see corresponding formulas for centered square, centered cubic and centered biquadratic numbers). Let it be true for all dimensions lower k. In other words, for a given $n \geq 2$, let

$$int(C^i(n)) = C^i(n-2) = (n-2)^i, \quad 0 \leq i \leq k-1.$$

Then

$$\sum_{i=0}^{k-1} N_i int(C^i(n)) = \sum_{i=0}^{k-1} 2^{k-i}\binom{k}{i}(n-2)^i = \sum_{i=0}^{k-1}\binom{k}{i}2^{k-i}(n-2)^i$$

$$= \sum_{i=0}^{k}\binom{k}{i}2^{k-i}(n-2)^i - (n-2)^k$$

$$= ((n-2)+2)^k - (n-2)^k = n^k - (n-2)^k.$$

So, the recurrent formula for the sequence $\overline{C}^k(1)$, $\overline{C}^k(2), \ldots$, $\overline{C}^k(n), \ldots$ of k-dimensional centered hypercube numbers has the form

$$\overline{C}^k(n+1) = \overline{C}^k(n) + (n+1)^k - (n-1)^k, \quad \overline{C}^k(1) = 1.$$

Then the general formula for n-th k-dimensional centered cube number $\overline{C}^k(n)$ is

$$\overline{C}^k(n) = n^k + (n-1)^k,$$

i.e., *k-dimensional centered hypercube numbers are sums of two consecutive ordinary k-dimensional hypercube numbers.*

This result can be obtained by the following telescoping summation: $\overline{C}^k(n) = 1^k + (2^k - 0^k) + (3^k - 1^k) + \cdots + (n^k - (n-2)^k) = n^k + (n-1)^k$.

So, the **five-dimensional centered hypercube numbers** have the form $\overline{C}^5(n) = n^5 + (n-1)^5$. These numbers can be obtained as partial sums of the sequence 1, 32, 242, 992, 2882, 6752, 13682, 24992,

42242, 67232, ... (Sloane's A008512), giving the number of points on surface of five-dimensional hypercube, and are 1, 33, 275, 1267, 4149, 10901, 24583, 49575, 91817, 159049, ... (Sloane's A008515).

The **six-dimensional centered hypercube numbers** have the form $\overline{C}^6(n) = n^6 + (n-1)^6$. These numbers can be obtained as partial sums of the sequence 1, 64, 728, 4032, 14896, 42560, 102024, 215488, 413792, 737856, ... (Sloane's A008513), giving the number of points on surface of six-dimensional hypercube, and are 1, 65, 793, 4825, 19721, 62281, 164305, 379793, 793585, 1531441, ... (Sloane's A008516).

The generation function for the sequence of k-dimensional centered hypercube numbers has the form $f(x) = (1+x)g(x)$, where $g(x)$ is the generated function for the sequence of k-dimensional hypercube numbers, i.e., it holds

$$\frac{x(1+x)(1+a_1^k x^2 + \cdots + a_{k-2}^k x^{k-2} + x^{k-1})}{(1-x)^{k+1}}$$
$$= \overline{C}^k(1)x + \overline{C}^k(2)x^2 + \overline{C}^k(3)x^3 + \cdots + \overline{C}^k(n)x^n + \cdots, \quad |x|<1,$$

where $a_1^k = a_{k-2} = 2^k - (k+1)$, $a_2^k = a_{k-3}^k = 3^k - (k+1)2^k + \frac{k(k+1)}{2}, \ldots, a^k_{\lfloor\frac{k-1}{2}\rfloor} = a^k_{k-\lfloor\frac{k-1}{2}\rfloor-1} = (\lfloor\frac{k-1}{2}\rfloor+1)^k - (k+1)(\lfloor\frac{k-1}{2}\rfloor)^k + \frac{k(k+1)}{2}(\lfloor\frac{k-1}{2}\rfloor-1)^k + \cdots + (-1)^{\lfloor\frac{k-1}{2}\rfloor}\binom{k+1}{\lfloor\frac{k-1}{2}\rfloor}$.

In particular, we get

$$\frac{x(1+x)(1+26z+66x^2+26x^3+x^4)}{(1-x)^6}$$
$$= x + 32x^2 + 242x^3 + \cdots + \overline{C}^5(n)x^n + \cdots, \quad |x| < 1;$$
$$\frac{x(1+x)(1+57x+302x^2+302x^3+57x+x^5)}{(1-x)^7}$$
$$= \frac{x(x+1)^2(1+56x+246x^2+56x^3+x^4)}{(1-x)^7}$$
$$= x + 33x^2 + 275x^3 + \cdots + \overline{C}^6(n)x^n + \cdots, \quad |x| < 1.$$

3.6.3. The **centered polytope numbers** represents a four-dimensional "centered hypertetrahedron". The first elements of the sequence of the centered polytope numbers are 1, 6, 21, 56, 126, 251, 456, 771, 1231, 1876, ... (Sloane's A008498).

They can be obtained as partial sums of the sequence 1, 5, 15, 35, 70, 125, 205, 315, 460, 645, ... (Sloane's A008487), giving the number of points on surface of four-dimensional hypertetrahedron. The first elements of the sequence A008487 are $S_3^4(1), S_3^4(2), S_3^4(3), S_3^4(4)$, while n-th element, $n \geq 5$, has the form $S_3^4(n) - S_3^4(n-5)$.

In fact, it can be obtained by the formula $N_0 + N_1(n-2) + N_2 \cdot int(S_3(n)) + N_3 int(S_3^3(n))$, where $N_0 = 5$, $N_1 = 10$, $N_2 = 10$, $N_3 = 5$ are the numbers of vertices, edges, triangles and tetrahedrons of a four-dimensional hypertetrahedron. So, for $n \geq 5$ it holds[1]

$$N_0 + N_1(n-2) + N_2 \cdot int(S_3(n)) + N_3 int(S_3^3(n)) = 5 + 10(n-2)$$
$$+10\left(\frac{n(n+1)}{2} - 3(n-1)\right) + 5\left(\frac{n(n+1)(n+2)}{6}\right.$$
$$\left.-\left(4 + 6(n-2) + 4\left(\frac{n(n+1)}{2} - 3(n-1)\right)\right)\right) = 5 + 10(n-2)$$
$$+10\frac{(n-3)(n-2)}{2} + 5\frac{(n-4)(n-3)(n-2)}{6}$$
$$= \left(\binom{5}{4}\binom{n-2}{0} + \binom{5}{3}\binom{n-2}{1} + \binom{5}{2}\binom{n-2}{2}\right.$$
$$\left. + \binom{5}{1}\binom{n-2}{3} + \binom{5}{0}\binom{n-2}{4}\right) - \binom{5}{0}\binom{n-2}{4}$$
$$= \binom{n+3}{4} - \binom{n-2}{4} = S_3^4(n) - S_3^4(n-5).$$

Therefore, we get

$$\overline{S}_3^4(1) = S_3^4(1) = 1, \quad \overline{S}_3^4(2) = \overline{S}_3^4(1) + S_3^4(2),$$
$$S_3^4(3) = \overline{S}_3^4(2) + S_3^4(3), \quad S_3^4(4) = \overline{S}_3^4(3) + S_3^4(4),$$
$$\overline{S}_3^4(n+1) = \overline{S}_3^4(n) + S_3^4(n+1) - S_3^4(n-4), \quad n \geq 5.$$

[1]Here we use the formula $\binom{n}{r}\binom{m}{0} + \binom{n}{r-1}\binom{m}{1} + \cdots + \binom{n}{0}\binom{m}{r} = \binom{n+m}{r}$.

So, starting from $n = 5$, *centered polytope numbers are sums of five consecutive ordinary polytope numbers*:

$$\overline{S}_3^4(1) = S_3^4(1), \quad \overline{S}_3^4(2) = S_3^4(2) + S_3^4(1),$$

$$\overline{S}_3^4(3) = S_3^4(3) + S_3^4(2) + S_3^4(1),$$

$$\overline{S}_3^4(4) = S_3^4(4) + S_3^4(3) + S_3^4(2) + S_3^4(1),$$

$$\overline{S}_3^4(n) = S_3^4(n) + S_3^4(n-1) + S_3^4(n-2) + S_3^4(n-3) + S_3^4(n-4), \quad n \geq 5.$$

This fact can be obtained as a result of the following telescopic summation: $\overline{S}_3^4(n) = S_3^4(1)+S_3^4(2)+S_3^4(3)+S_3^4(4)+(S_3^4(5)-S_3^4(0))+(S_3^4(6) - S_3^4(1)) + \cdots + (S_3^4(n) - S_3^4(n-5)) = S_3^4(n) + S_3^4(n-4)$.

It implies the following general formula for n-th centered polytope number:

$$\overline{S}_3^4(n) = \frac{5n^4 - 10n^3 + 55n^2 - 50n + 24}{24}.$$

Moreover, since $S_3^4(n)+S_3^4(n-1)+\cdots+S_3^4(n-4) = \binom{n+3}{4} + \binom{n+2}{4} + \binom{n+1}{4} + \binom{n}{4} + \binom{n-1}{4} = \binom{n+4}{5} - \binom{n-1}{5} = S_3^5(n) - S_3^5(n-5)$, we get also that

$$\overline{S}_3^4(n) = S_3^5(n) - S_3^5(n-5), \quad n \geq 5.$$

The generation function for this sequence is

$$f(x) = \frac{x(1 + x + x^2 + x^3 + x^4)}{(1-x)^5} = \frac{x(1-x^5)}{(x-1)^6}$$

(see [Weis11]), i.e., it holds

$$\frac{x(1-x^5)}{(x-1)^6} = \overline{S}_3^4(1)x + \overline{S}_3^4(2)x^2 + \overline{S}_3^4(3)x^3 + \cdots + \overline{S}_3^4(n)x^4 + \cdots, \quad |x| < 1.$$

It can be obtained using the generating function $\frac{1}{(1-x)^5} = S_3^4(1) + S_3^4(2)x + \cdots + S_3^4(n+1)x^n + \cdots$ for polytope numbers:

$$\frac{x(1-x^5)}{(x-1)^6} = \frac{1 + x + \cdots + x^4}{(1-x)^5} = \frac{1}{(1-x)^5} + \frac{x}{(1-x)^5} + \cdots + \frac{x^4}{(1-x)^5}$$

$$= (S_3^4(1) + S_3^4(2)x + \cdots + S_3^4(n+1)x^n + \cdots)$$

$$+(S_3^4(1)x + S_3^4(2)x^2 + \cdots + S_3^4(n)x^n + \cdots)$$

$$+\cdots+(S_3^4(1)x^4+S_3^4(2)x^5+\cdots+S_3^4(n-3)x^n+\cdots)$$
$$=S_3^4(1)+(S_3^4(1)+S_3^4(2))x+\cdots+(S_3^4(1)$$
$$+\cdots+S_3^4(4))x^4+(S_3^4(2)+\cdots+S_3^4(5))x^5$$
$$+\cdots+(S_3^4(n-3)+\cdots+S_3^4(n+1))x^n+\cdots$$
$$=\overline{S}_3^4(1)+\overline{S}_3^4(2)x+\cdots+\overline{S}_3^4(n+1)x^n+\cdots,\quad |x|<1.$$

3.6.4. The **k-dimensional centered hypertetrahedron numbers** represent k-dimensional centered hypertetrahedron.

They can be obtained as partial sums of the sequence, giving number of points on the surface of k-dimensional hypertetrahedron α^k. The first k elements of the last sequence are $S_3^k(1)=1$, $S_3^k(2),\ldots$, $S_3^k(k)$, while n-th element, $n\geq k+1$, has the form $S_3^k(n)-S_3^k(n-k-1)$.

It follows from the fact that *the number $int(S_3^k(n))$ of the interior's points of k-dimensional hypertetrahedron of size n is zero for $1\leq n\leq k$, and is equal to the number of points in k-dimensional hypertetrahedron of size $n-k-1$, otherwise*:

$$int(S_3^k(n))=S_3^k(n-k-1),\quad n\geq k+1.$$

It can be obtained by induction on k, using the formula $\sum_{i=0}^{k-1}N_i int(S_3^i(n))$, where $N_i=\binom{k+1}{i+1}$, $0\leq i\leq k-1$, is the number of i-dimensional faces of an k-dimensional hypertetrahedron α^k.

In fact, it holds for $k=2,3$ and 4 (see corresponding formulas for centered triangular, centered tetrahedral, and centered polytope numbers). Let it be true for all dimensions lower k:

$$int(S_3^i(n))=S_3^i(n-i-1),\quad n\geq i+1,\ 0\leq i\leq k-1.$$

Then, for $n\geq k+1$, it holds

$$\sum_{i=0}^{k-1}N_i int(S_3^i(n))=\sum_{i=0}^{k-1}\binom{k+1}{i+1}S_3^i(n-i-1)$$
$$=\sum_{i=0}^{k-1}\binom{k+1}{k-i}\binom{n-2}{i}$$

$$= \sum_{i=0}^{k} \binom{k+1}{k-i}\binom{n-2}{i} - \binom{n-2}{k}$$
$$= \binom{(n-2)+k+1}{k} - \binom{n-2}{k}$$
$$= S_3^k(n) - S_3^k(n-k-1).$$

Therefore, we have

$$\overline{S}_3^k(1) = S_3^k(1) = 1, \quad \overline{S}_3^k(2) = \overline{S}_3^k(1) + S_3^k(2), \ldots,$$
$$\overline{S}_3^k(k) = \overline{S}_3^k(k-1) + S_3^k(k),$$
$$\overline{S}_3^k(n+1) = \overline{S}_3^k(n) + S_3^k(n+1) - S_3^k(n-k), \quad n \geq k.$$

So, starting from $n = k+1$, *n-th k-dimensional centered hypertetrahedron number is the sum of* $k+1$ *consecutive ordinary k-dimensional hypertetrahedron numbers*:

$$\overline{S}_3^k(n) = S_3^k(n) + S_3^k(n-1) + \cdots + S_3^k(1), \quad 1 \leq n \leq k,$$
$$\overline{S}_3^k(n) = S_3^k(n) + S_3^k(n-1) + \cdots + S_3^k(n-k), \quad n \geq k+1.$$

This result can be obtained by the following telescoping summation: $\overline{S}_3^k(n) = S_3^k(1) + S_3^k(2) + \cdots + S_3^k(k) + (S_3^k(k+1) - S_3^k(0)) + (S_3^k(k+2) - S_3^k(1)) + \cdots + (S_3^k(n) - S_3^k(n-k-1)) = S_3^k(n) + S_3^k(n-1) + \cdots + S_3^k(n-k)$.

For a given k, the above formulas allow tho obtain the general formula for n-th k-dimensional centered hypertetrahedron number $\overline{S}_3^k(n)$.

So, the **five-dimensional centered hypertetrahedron numbers** 1, 7, 28, 84, 210, 462, 923, 1709, 2975, 4921, ... (Sloane's A008499) have the form $\overline{S}_3^5(n) = \frac{n^5+15n^4+160n^3+225n^2+314n+120}{120}$. These numbers can be obtained as partial sums of the sequence 1, 6, 21, 56, 126, 252, 461, 786, 1266, 1946, ... (Sloane's A008488), giving the number of points on surface of five-dimensional hypertatrahedron.

The **six-dimensional centered hypertetrahedron numbers** $1, 8, 36, 120, 330, 792, 1716, 3431, 6427, 11404, \ldots$ (Sloane's A008500) can be obtained as partial sums of the sequence 1, 7, 28, 84, 210, 462, 924, 1715, 2996, 4977, ... (Sloane's A008489), giving the number of points on surface of six-dimensional hypertatrahedron.

The generation function for the sequence of k-dimensional centered hypertetrahedron numbers has the form $f(x) = \frac{x(1+x+\cdots+x^k)}{(1-x)^{k+1}} = \frac{x(1-x^{k+1})}{(1-x)^{k+2}}$, i.e., it holds

$$\frac{x(1-x^{k+1})}{(1-x)^{k+2}} = \overline{S}_3^k(1)x + \overline{S}_3^k(2)x^2 + \overline{S}_3^k(3)x^3 + \cdots + \overline{S}_3^k(n)x^n + \cdots, \quad |x| < 1.$$

In particular, we have

$$\frac{x(1-x^6)}{(1-x)^7} = x + 7x^2 + 28x^3 + \cdots + \overline{S}_3^5(n)x^n + \cdots, \quad |x| < 1;$$

$$\frac{x(1-x^7)}{(1-x)^8} = x + 8x^2 + 36x^3 + \cdots + \overline{S}_3^6(n)x^n + \cdots, \quad |x| < 1.$$

This result can be obtained from the generating function $\frac{1}{(1-x)^{k+1}} = S_3^k(1) + S_3^k(2)x + \cdots + S_3^k(n+1)x^n + \cdots$ for k-dimensional hypertetrahedron numbers:

$$\begin{aligned}\frac{x(1-x^{k+1})}{(1-x)^{k+2}} &= \frac{1+x+\cdots+x^k}{(1-x)^{k+1}}\\ &= \frac{1}{(1-x)^{k+1}} + \frac{x}{(1-x)^{k+1}} + \cdots + \frac{x^k}{(1-x)^{k+1}}\\ &= (S_3^k(1) + S_3^k(2)x + \cdots + S_3^k(n+1)x^n + \cdots)\\ &\quad + (S_3^k(1)x + S_3^k(2)x^2 + \cdots + S_3^k(n)x^n + \cdots)\\ &\quad + \cdots + (S_3^k(1)x^k + S_3^k(2)x^{k+1}\\ &\quad + \cdots + S_3^k(n-k+1)x^n + \cdots)\end{aligned}$$

$$= S_3^k(1) + (S_3^k(1) + S_3^k(2))x + \cdots + (S_3^k(1) + \cdots + S_3^k(k))x^k + (S_3^k(2) + \cdots + S_3^k(k+1))x^5 + \cdots + (S_3^k(n-k+1) + \cdots + S_3^k(n+1))x^n + \cdots$$
$$= \overline{S}_3^k(1) + \overline{S}_3^k(2)x + \cdots + \overline{S}_3^k(n+1)x^n + \cdots, \quad |x| < 1.$$

3.6.5. In the dimension four, the **centered hyperoctahedral numbers** (or **orthoplex numbers**) represent a four-dimensional "centered hyperoctahedron". The first elements of the sequence of centered hyperoctahedral numbers are 1, 9, 41, 129, 321, 681, 1289, 2241, 3649, 5641, ... (Sloane's A001846).

They can be obtained as partial sums of the sequence 1, 8, 32, 88, 192, 360, 608, 952, 1408, 1992, ... (Sloane's A008412), giving the number of points on surface of four-dimensional cross-polytope. The first elements of the sequence A008412 is 1, while n-th element, $n \geq 2$, has the form $\frac{8}{3}(n^3 - 3n^2 + 5n - 3) = \frac{8}{3}((n-1)^3 + 2(n-1))$.

In fact, it can be obtained by the formula $N_0 + N_1(n-2) + N_2 \cdot int(S_3(n)) + N_3 int(S_3^3(n))$, where $N_0 = 8$, $N_1 = 24$, $N_2 = 32$, $N_3 = 16$ are numbers of vertices, edges, triangles and tetrahedrons of a four-dimensional hyperoctahedron. So, for $n \geq 2$, it holds

$$N_0 + N_1(n-2) + N_2 \cdot int(S_3(n)) + N_3 int(S_3^3(n))$$
$$= 8 + 24(n-2) + 32\left(\frac{n(n+1)}{2} - 3(n-1)\right)$$
$$+16\left(\frac{n(n+1)(n+2)}{6} - \left(4 + 6(n-2) + 4\left(\frac{n(n+1)}{2} - 3(n-1)\right)\right)\right)$$
$$= 8 + 24(n-2) + 32\frac{(n-3)(n-2)}{2} + 16\frac{(n-4)(n-3)(n-2)}{6}$$
$$= \frac{8}{3}(n^3 - 3n^2 + 5n - 3).$$

Hence, the recurrent formula for the sequence of centered hyperoctahedral numbers is

$$\overline{HO}(n+1) = \overline{HO}(n) + \frac{8}{3}(n^3 + 2n), \quad \overline{HO}(1) = 1.$$

The general formula for n-th centered hyperoctahedral number is

$$\overline{HO}(n) = \frac{2n^4 - 4n^3 + 10n^2 - 8n + 3}{3}.$$

In fact, it can be obtained as the result of the following summation: $\overline{HO}(n) = 1+\sum_{i=1}^{n-1} n\frac{8}{3}(i^3+2i) = 1+\frac{8}{3}\sum_{i=1}^{n-1} ni^3+\frac{16}{3}\sum_{i=0}^{n-1} i = 1+\frac{8}{3}\cdot \frac{n^2(n-1)^2}{4}+\frac{16}{3}\cdot\frac{n(n-1)}{2} = 1+\frac{2n(n-1)(1+8(n^2-n))}{3} = \frac{2n^4-4n^3+10n^2-8n+3}{3}$.

The generation function for the sequence of centered hyperoctahedral numbers is $f(x) = \frac{x(1+x)^4}{(1-x)^5}$ (see [Weis11]), i.e., it holds

$$\frac{x(1+x)^4}{(x-1)^5} = \overline{HO}(1)x + \overline{HO}(2)x^2 + \overline{HO}(3)x^3 + \cdots + \overline{HO}(n)x^n + \cdots, \quad |x| < 1.$$

This result can be obtained by the formula $\frac{g(x)}{1-x}$, where $g(x) = \frac{(1+x)^4}{(1-x)^4}$ is the generating function for the sequence A008412. In turn, the function $g(x)$ can be obtained by the standard procedure. However, using the decomposition $\frac{1}{(1-x)^4} = S_3^4(1) + S_3^4(2)x + \cdots + S_3^4(n+1)x^n + \cdots$ we can obtain the above result for $g(x)$ by the following arguments:

$$\begin{aligned}
&1+\sum_{n=1}^{\infty}\left(8+24(n-1)+32\frac{(n-2)(n-1)}{2}\right.\\
&\qquad\left.+16\frac{(n-2)(n-3)(n-1)}{6}\right)x^n\\
&\quad = 1+8\sum_{n=1}^{\infty}x^n+24\sum_{i=1}^{\infty}S_3^1(n-1)x^n+32\sum_{i=1}^{\infty}S_3^2(n-2)x^n\\
&\qquad +16\sum_{n=1}^{\infty}S_3^3(n-3)x^n\\
&\quad = 1+8\frac{x}{1-x}+24\frac{x^2}{(1-x)^2}+32\frac{x^3}{(1-x)^3}+16\frac{x^4}{(1-x)^4}\\
&\quad = \frac{1}{(1-x)^4}\left(\binom{4}{0}(2x)^0(1+x)^4\right.
\end{aligned}$$

$$+\binom{4}{1}(2x)^1(1+x)^3+\binom{4}{2}(2x)^2(1+x)^2$$

$$+\binom{4}{3}(2x)^3(1+x)^1+\binom{4}{4}(2x)^4(1+x)^0)$$

$$=\frac{(2x+(1-x))^4}{(1-x)^4}=\frac{(1+x)^4}{(1-x)^4}.$$

3.6.6. The **k-dimensional centered hypeoctahedron numbers** represent k-dimensional centered hyperoctahedron.

They can be obtained as partial sums of the sequence, giving number of points on surface of k-dimensional hyperoctahedron β^k. The first element of the last sequence is 1, while n-th element, $n \geq 2$, has the form $\sum_{i=0}^{k-1} 2^{i+1}\binom{k}{i+1}\binom{n-2}{i}$.

It can be obtained using the formula $\sum_{i=0}^{k-1} N_i int(S_3^i(n))$, where $N_i = 2^{i+1}\binom{k}{i+1}$, $0 \leq i \leq k-1$, is the number of i-dimensional faces of an k-dimensional hyperoctahedron β^k.

In fact, it was proven that $int(S_3^i(n)) = S_3^i(n-i-1)$ for $n \geq i+1$, and is zero otherwise. Therefore, it holds $int(S_3^i(n)) = \binom{n-2}{i}$, $n \geq 2$. Then we get $\sum_{i=0}^{k-1} N_i int(S_3^i(n)) = \sum_{i=0}^{k-1} 2^{i+1}\binom{k}{i+1}\binom{n-2}{i}$.

So, the recurrent formula for the sequence of k-dimensional centered hyperoctahedron numbers has the form

$$\overline{O}^k(n+1)=\overline{O}^k(n)+\sum_{i=0}^{k-1} 2^{i+1}\binom{k}{i+1}\binom{n-1}{i}, \quad \overline{O}^k(1)=1.$$

Then the general formula for n-th k-dimensional centered hyperoctahedron number can be obtained using the following identity:

$$\overline{O}^k(n)=1+\sum_{j=1}^{n-1}\sum_{i=0}^{k-1} 2^{i+1}\binom{k}{i+1}\binom{n-1}{i}.$$

In particular, the **five-dimensional centered hyperoctahedron numbers** can be calculated by the formula $\overline{O}^5(n) = \frac{4n^5+10n^4+40n^3+50n^2+46n+15}{15}$. These numbers can be obtained as partial sums of the sequence 1, 10, 50, 170, 450, 1002, 1970, 3530, 5890, 9290, ... (Sloane's A008413) of the form $\frac{1}{3}(4(n-1)^4+20(n-1)^2)+2$, giving the number of points on surface of five-dimensional hyperoctahedron, and are 1, 11, 61, 231, 681, 1683, 3653, 7183, 13073, 22363, ... (Sloane's A001847).

Similarly, the **six-dimensional centered hyperoctahedron numbers** can be calculated by the formula $\overline{O}^6(n) = \frac{4n^6-12n^5+70n^4-120n^3+196n^2-138n+45}{45}$. These numbers can be obtained as partial sums of the sequence 1, 12, 72, 292, 912, 2364, 5336, 10836, 20256, 35436, ... (Sloane's A008414) of the form $\frac{4(n-1)}{15}((2(n-1)^4 + 20(n-1)^2+23)$, giving the number of points on surface of six-dimensional hyperoctahedron, and are 1, 13, 85, 377, 1289, 3653, 8989, 19825, 40081, 75517, ... (Sloane's A001848).

The generation function for the sequence of k-dimensional centered hyperoctahedron numbers has the form $f(x) = \frac{x(1+x)^k}{(1-x)^{k+1}}$, i.e., it holds

$$\frac{x(1+x)^k}{(1-x)^{k+1}} = \overline{O}^k(1)x + \overline{O}^k(2)x^2 + \overline{O}^k(3)x^3 + \cdots + \overline{O}^k(n)x^n + \cdots, \quad |x| < 1.$$

In particular, it follows

$$\frac{x(1+x)^5}{(1-x)^6} = x + 11x^2 + 61x^3 + \cdots + \overline{O}^5(n)x^n + \cdots, \quad |x| < 1;$$

$$\frac{x(1+x)^6}{(1-x)^7} = x + 13x^2 + 85x^3 + \cdots + \overline{O}^6(n)x^n + \cdots, \quad |x| < 1.$$

This result can be obtained by the formula $f(x) = \frac{g(x)}{1-x}$, where $g(x) = \frac{(1+x)^k}{(1-x)^k}$ is the generating function for the sequence, giving number of points on surface of β^k. In turn, the above formula for $g(x)$ can be obtained using the following arguments:

$$1 + \sum_{n=1}^{\infty} \left(\sum_{i=0}^{k-1} 2^{i+1} \binom{k}{i+1} \binom{n-1}{i} \right) x^n$$

$$= 1 + \sum_{i=0}^{k-1} 2^{i+1} \binom{k}{i+1} \left(\sum_{n=1}^{\infty} \binom{n-1}{i} x^n \right)$$

$$= 1 + \sum_{i=0}^{k-1} 2^{i+1} \binom{k}{i+1} \left(\sum_{n=i+1}^{\infty} S_3^k(n-i)x^n \right)$$

$$= 1 + \sum_{i=0}^{k-1} 2^{i+1} \binom{k}{i+1} x^{i+1} \left(\sum_{n=1}^{\infty} S_3^i(n)x^{n-1} \right)$$

$$= 1 + \sum_{i=0}^{k-1} (2x)^{i+1} \binom{k}{i+1} \frac{1}{(1-x)^{i+1}} = \frac{1}{(1-x)^k}$$

$$\times \left(\sum_{i=0}^{k} \binom{k}{i} (2x)^i (1-x)^{k-i} \right)$$

$$= \frac{((2x)+(1-x))^k}{(1-x)^k} = \frac{(1+x)^k}{(1-x)^k}.$$

3.7 Generalized multidimensional figurate numbers

Similar to the plane and space cases, the **generalized multidimensional figurate numbers** are defined as all values of the standard formulas for a given class of multidimensional figurate numbers, taken for any integer value of the index.

3.7.1. The **generalized pentatope numbers** are defined as all values of the formula $S_3^4(n) = \frac{n(n+1)(n+2)(n+3)}{24}$, taken for any integer n.

For positive integers n, we get ordinary pentatope numbers. For $n = 0$, it holds $S_3^4(0) = 0$. Finally, for negative indices, we get $S_m^3(-n) = \frac{n(n-1)(n-2)(n-3)}{24}$, $n \in \mathbb{N}$.

So, the generalized pentatope numbers with negative indices ${}^{-}S_3^4(n) = S_3^4(-n)$ have the form ${}^{-}S_3^4(n) = \frac{(n-3)(n-2)(n-1)n}{24}$, giving ${}^{-}S_3^4(1) = 0$, ${}^{-}S_3^4(2) = 0$, ${}^{-}S_3^4(3) = 0$, and ${}^{-}S_3^3(n) = S_3^3(n-3)$, $n \geq 4$. They start with elements $0, 0, 0, 1, 5, \ldots$. Hence, generated pentatope numbers are

$$\ldots, 5, 1, 0, 0, 0, 0, 1, 5, 15, 35, 70, \ldots,$$

forming the sequence $0, 1, 0, 5, 0, 15, 0, 35, 1, \ldots$ for $n = 0, \pm 1, \pm 2, \pm 3, \pm 4, \ldots$.

The recurrent formula for generalized pentatope numbers with negative indices has the form

$$^{-}S_3^4(n+1) = {}^{-}S_3^4(n) - {}^{-}S_3^3(n), \quad {}^{-}S_3^4(1) = 0,$$

where ${}^{-}S_3^3(n) = S_3^3(-n)$ is n-th generalized tetrahedral number with negative index.

In fact, one obtains from the recurrent formula $S_3^4(n+1) = S_3^4(n) + S_3^3(n+1)$ for ordinary pentatope numbers, that $S_3^4(n) = S_3^4(n+1) - S_3^3(n+1)$, or $S_3^4(n-1) = S_3^4(n) - S_3^3(n)$. So, $S_3^4(0) = 0$, and for negative integers one has $S_3^4(-(n+1)) = S_3^4(-n) - S_3^3(-n)$, $n \in \mathbb{N}$. In other words, it holds

$$\begin{aligned} {}^{-}S_3^4(n+1) &= {}^{-}S_3^4(n) + \frac{(n-2)(n-1)n}{6}, \\ {}^{-}S_3^4(1) &= 0. \end{aligned}$$

So, starting with ${}^{-}S_3^4(1) = 0$, we get ${}^{-}S_3^3(2) = -({}^{-}S_3^3(1))$, ${}^{-}S_3^4(3) = -({}^{-}S_3^3(1) + {}^{-}S_3^3(2))$, ${}^{-}S_3^4(4) = -({}^{-}S_3^3(1) + {}^{-}S_3^3(2) + {}^{-}S_3^3(3))$, etc., leading to the formula

$$\begin{aligned} {}^{-}S_3^4(n+1) = S_3^4(-(n+1)) = -({}^{-}S_3^3(1) + {}^{-}S_3^3(2) \\ + \cdots + {}^{-}S_3^3(n)). \end{aligned}$$

The generating function for the sequence ${}^{-}S_3^4(1), {}^{-}S_3^4(2), \ldots, {}^{-}S_3^4(n), \ldots$ of generalized pentatope numbers with negative indices, i.e., for the sequence $S_3^4(-1), S_3^4(-2), \ldots, S_3^4(-n), \ldots$, has the form $f(x) = \frac{x^3}{(1-x)^5}$, i.e., it holds

$$\begin{aligned} \frac{x^3}{(1-x)^5} &= S_3^3(-1)x + S_3^4(-2)x^2 + S_3^4(-3)x^3 \\ &\quad + \cdots + S_3^3(-n)x^n + \cdots, \quad |x| < 1. \end{aligned}$$

This result can be obtained by the standard procedure, leading to the linear recurrent equation ${}^{-}S_3^4(n+5) - 5{}^{-}S_3^4(n+4) + 10{}^{-}S_3^4(n+3) - 10{}^{-}S_3^4(n+2) + 5{}^{-}S_m^3(n+1) - {}^{-}S_m^3(n) = 0$ with initial conditions ${}^{-}S_3^4(1) = 0$, ${}^{-}S_3^4(2) = 0$, ${}^{-}S_3^4(3) = 0$, ${}^{-}S_3^4(4) = 1$, and ${}^{-}S_3^4(5) = 5$.

However, it can be obtained much more easily using the generating function $\frac{1}{(1-x)^5}$ for pentatope numbers. In fact, it holds

$$\begin{aligned}\frac{1}{(1-x)^5} &= S_3^4(1) + S_3^4(2)x + \cdots + S_3^4(n+1)x^n + \cdots \\ &= {}^-S_3^3(4)x + {}^-S_3^4(5)x^2 \\ &\quad + \cdots + {}^-S_3^4(n+3)x^n + \cdots, \quad |x| < 1.\end{aligned}$$

Since ${}^-S_3^4(1) = {}^-S_3^3(2) = {}^-S_3^4(3) = 0$, we get

$$\begin{aligned}\frac{x^3}{(1-x)^5} &= {}^-S_3^4(1)x + {}^-S_3^4(2)x^2 + {}^-S_3^4(3)x^3 \\ &\quad + \cdots + {}^-S_3^3(n)x^n + \cdots, \quad |x| < 1.\end{aligned}$$

The generating function for the sequence $S_3^4(0)$, $S_3^4(1)$, $S_3^4(-1)$, $S_3^4(2)$, $S_3^4(-2), \ldots$ of all generalized pentatope numbers, written for $n = 0, \pm 1, \pm 2, \ldots$, can be obtained from the generating functions $f_1(x) = \frac{1}{(1-x)^5} = S_3^4(1) + S_3^4(2)x + \cdots + S_3^4(n+1)x^n + \cdots$ and $f_2(x) = \frac{x^2}{(1-x)^5} = S_3^4(-1) + S_3^4(-2)x + \cdots + S_3^4(-(n+1))x^n + \cdots$ using the formula $x^2 f_1(x^2) + x^3 f_2(x^2)$. The direct computation gives $f(x) = \frac{x^2(1+x^5)}{(1-x^2)^5}$. In other words, it holds

$$\begin{aligned}\frac{x^2(1+x^5)}{(1-x^2)^5} &= S_3^4(0)x + S_3^4(1)x^2 + S_3^4(-1)x^3 \\ &\quad + S_3^4(2)x^4 + S_3^4(-2)x^5 + \cdots, \quad |x| < 1.\end{aligned}$$

3.7.2. For any dimension k, $k \geq 5$, one can obtain **generalized k-dimensional hypertetrahedron numbers** as all values of the formula $S_3^k(n) = \frac{n(n+1)\cdots(n+k-1)}{k!}$ for each integer value of n.

For positive integers n, we get ordinary k-dimensional hypertetrahedron numbers. For $n = 0$, it holds $S_3^k(0) = 0$. Finally, for negative indices, we get $S_3^k(-n) = (-1)^k \frac{n(n-1)(n-2)\ldots(n-k+1)}{k!}$, $n \in \mathbb{N}$.

So, the generalized k-dimensional hypertetrahedron numbers with negative indices ${}^-S_3^k(n) = S_3^k(-n)$ have the form ${}^-S_3^k(n) = (-1)^k \frac{(n-k+1)\cdots(n-3)(n-2)(n-1)n}{k!}$, giving ${}^-S_3^4(1) = 0$, ${}^-S_3^4(2) = 0, \ldots$, ${}^-S_3^4(k-1) = 0$, and ${}^-S_3^k(n) = (-1)^k S_3^3(n-k+1)$, $n \geq k$. Hence, absolute values of generalized k-dimensional hypertetrahedron numbers with negative indices coincide with ordinary k-dimensional hypertetrahedron numbers of a smaller size; all such numbers are non-negative integers for even k, and non-positive integers for odd k.

In particular, the generalized five-dimensional hypertetrahedron numbers with negative indices are $0, 0, 0, 0, -1, -6, -21, \ldots$. So, **generated five-dimensional hypertetrahedron numbers** are

$$\ldots, -21, -6, -1, 0, 0, 0, 0, 0, 1, 6, 21, 56, 126, \ldots,$$

forming the sequence $0, 1, 0, 6, 0, 21, 0, 56, 0, \ldots$ for $n = 0, \pm 1, \pm 2, \pm 3, \pm 4, \ldots$.

The generalized six-dimensional hypertetrahedron numbers with negative indices are $0, 0, 0, 0, 0, 1, 7, 28, \ldots$. So, **generalized six-dimensional hypertetrahedron numbers** are

$$\ldots, 28, 7, 1, 0, 0, 0, 0, 0, 0, 1, 7, 28, 84, 210, \ldots,$$

forming the sequence $0, 1, 0, 7, 0, 28, 0, 84, 0, \ldots$ for $n = 0, \pm 1, \pm 2, \pm 3, \pm 4, \ldots$.

The recurrent formula for generalized k-dimensional hypertetrahedron numbers with negative indices has the form

$$^{-}S_3^k(n+1) = {}^{-}S_3^k(n) - {}^{-}S_3^{k-1}(n), \quad {}^{-}S_3^k(1) = 0,$$

where $^{-}S_m^{k-1}(n) = S_m^{k-1}(-n)$ is n-th generalized $(k-1)$-dimensional hypertetrahedron number with negative index.

In fact, one obtains from the recurrent formula $S_3^k(n+1) = S_3^k(n) + S_3^{k-1}(n+1)$ for ordinary k-dimensional hypertetrahedron numbers, that $S_3^k(n) = S_3^k(n+1) - S_3^{k-1}(n+1)$, or $S_3^k(n-1) = S_3^k(n) - S_3^{k-1}(n)$. So, $S_3^k(0) = 0$, and, for negative integers, one has $S_3^k(-(n+1)) = S_3^k(-n) - S_3^{k-1}(-n)$, $n \in \mathbb{N}$. In other words, it holds

$$^{-}S_3^4(n+1) = {}^{-}S_3^4(n) + (-1)^k \frac{(n-k+2)\ldots(n-2)(n-1)n}{(k-1)!},$$
$$^{-}S_3^k(1) = 0.$$

So, starting with $^{-}S_3^k(1) = 0$, we get $^{-}S_3^k(2) = -({}^{-}S_3^{k-1}(1))$, $^{-}S_3^k(3) = -({}^{-}S_3^{k-1}(1) + {}^{-}S_3^{k-1}(2))$, $^{-}S_3^k(4) = -({}^{-}S_3^{k-1}(1) + {}^{-}S_3^{k-1}(2) + {}^{-}S_3^{k-1}(3))$, etc., leading to the formula

$$\begin{aligned} {}^{-}S_3^k(n+1) &= S_3^k(-(n+1)) \\ &= -({}^{-}S_3^{k-1}(1) + {}^{-}S_3^{k-1}(2) + \cdots + {}^{-}S_3^{k-1}(n)). \end{aligned}$$

The generating function for the sequence ${}^{-}S_3^k(1), {}^{-}S_3^k(2), \dots$ ${}^{-}S_3^k(n), \dots$ of generalized k-dimensional hypertetrahedron numbers with negative indices, i.e., for the sequence $S_3^3(-1)$, $S_3^k(-2), \dots,$ $S_3^k(-n), \dots$, has the form $f(x) = (-1)^k \frac{x^{k-1}}{(1-x)^{k+1}}$, i.e., it holds

$$(-1)^k \frac{x^{k-1}}{(1-x)^{k+1}} = S_3^k(-1)x + S_3^k(-2)x^2 + S_3^k(-3)x^3 + \cdots + S_3^k(-n)x^n + \cdots, \quad |x| < 1.$$

In particular, it holds

$$-\frac{x^4}{(1-x)^6} = S_3^5(-1)x + S_3^5(-2)x^2 + S_3^5(-3)x^3 + \cdots + S_3^5(-n)x^n + \cdots, \quad |x| < 1,$$

$$\frac{x^5}{(1-x)^7} = S_3^6(-1)x + S_3^6(-2)x^2 + S_3^6(-3)x^3 + \cdots + S_3^6(-n)x^n + \cdots, \quad |x| < 1.$$

This result can be obtained by the standard procedure, leading to the linear recurrent equation ${}^{-}S_3^k(n+k+1) - \binom{k+1}{1}{}^{-}S_3^4(n+k) + \binom{k+1}{2}{}^{-}S_3^4(n+k-1) + \cdots + (-1)^{k-1}\binom{k+1}{k-1}{}^{-}S_3^k(n+1) + (-1)^k S_3^k(n) = 0$ with initial conditions ${}^{-}S_3^k(1) = 0$, ${}^{-}S_3^k(2) = 0, \dots,$ ${}^{-}S_3^k(k-1) = 0$, ${}^{-}S_3^k(k) = 1$, ${}^{-}S_3^k(k+1) = k+1$.

However, it can be obtained much more easy using the generating function $\frac{1}{(1-x)^5}$ for k-dimensional hypertetrahedron numbers. In fact, it holds

$$\begin{aligned}(-1)^k \frac{1}{(1-x)^{k+1}} &= (-1)^k S_3^k(1) + (-1)^k S_3^k(2)x \\ &\quad + \cdots + (-1)^k S_3^k(n+1)x^n + \cdots \\ &= {}^{-}S_3^k(k)x + {}^{-}S_3^k(k+1)x^2 \\ &\quad + \cdots + {}^{-}S_3^k(n+k-1)x^n + \cdots \quad |x| < 1.\end{aligned}$$

Since ${}^{-}S_3^k(1) = {}^{-}S_3^k(2) = \cdots = {}^{-}S_3^k(k-1) = 0$, we get

$$\frac{x^{k-1}}{(1-x)^5} = {}^{-}S_3^k(1)x + \cdots + {}^{-}S_3^k(k-1)x^{k-1} + {}^{-}S_3^k(k)x^k + \cdots + {}^{-}S_3^k(n)x^n + \cdots, \quad |x| < 1.$$

The generating function for the sequence $S_3^k(0)$, $S_3^k(1)$, $S_3^k(-1)$, $S_3^k(2)$, $S_3^4(-2)$, ... of all generalized k-dimensional hypertetrahedron numbers, written for $n = 0, \pm1, \pm2, \ldots$, can be obtained from the generating functions $f_1(x) = \frac{1}{(1-x)^{k+1}} = S_3^k(1)+S_3^k(2)x+\cdots+S_3^k(n+1)x^n + \cdots$ and $f_2(x) = (-1)^k \frac{x^{k-2}}{(1-x)^{k+1}} = S_3^k(-1) + S_3^k(-2)x + \cdots + S_3^k(-(n+1))x^n + \cdots$, using the formula $x^2 f_1(x^2) + x^3 f_2(x^2)$. The direct computation gives $f(x) = \frac{x^2(1+(-1)^k x^{2k-3})}{(1-x^2)^{k+1}}$, i.e.,

$$\frac{x^2(1+(-1)^k x^{2k-3})}{(1-x^2)^{k+1}} = S_3^k(0)x + S_3^k(1)x^2 + S_3^k(-1)x^3 + S_3^k(2)x^4 + S_3^k(-2)x^5 + \cdots, \quad |x| < 1.$$

In particular,

$$\frac{x^2(1-x^7)}{(1-x^2)^6} = S_3^5(0)x + S_3^5(1)x^2 + S_3^5(-1)x^3 + S_3^5(2)x^4 + S_3^5(-2)x^5 + \cdots, \quad |x| < 1;$$

$$\frac{x^2(1+x^9)}{(1-x^2)^7} = S_3^6(0)x + S_3^6(1)x^2 + S_3^6(-1)x^3 + S_3^6(2)x^4 + S_3^6(-2)x^5 + \cdots, \quad |x| < 1.$$

3.7.3. The **generalized biquadratic numbers** $\ldots, 625, 256, 81, 16, 1, 0, 1, 16, 81, 256, 625, \ldots$ are just doubled version of ordinary biquadratic numbers: $BC(0) = 0$, and $BC(-n) = BC(n) = n^4$, $n \in \mathbb{N}$.

The sequence $0, 1, 1, 16, 16, 81, 81, 256, 256, \ldots$ of the generalized biquadratic numbers, written for $n = 0, \pm1, \pm2, \ldots$, has the generating function

$$f(x) = \frac{x^2(1+x)(1+11x^2+11x^4+x^6)}{(1-x^2)^5} = \frac{x^2(1+x+11x^2+11x^3+11x^4+11x^5+x^6+x^7)}{(1-x^2)^5},$$

which is equal to $(x^2+x^3)g(x^2)$, where $g(x) = \frac{1+11x+11x^2+x^3}{(1-x)^5}$ is the generating function for the sequence of ordinary biquadratic numbers.

3.7.4. Given a dimension k, $k \geq 5$, we obtain the **generalized k-dimensional hypercube numbers** as all values of the formula $C^k(n) = n^k$ for each integer value of n.

For positive integers n, we get ordinary k-dimensional hypercube numbers. For $n = 0$, it holds $C^k(0) = 0$. Finally, for negative values, we get the formula $C^k(-n) = (-1)^k C^k(n)$. So, absolute values of generalized k-dimensional hypercube numbers with negative indices $^-C^k(n) = C^k(-n)$ coincide with ordinary k-dimensional hypercube numbers of the same size; all such numbers are positive integers for even k, and negative integers for odd k.

In particular, the generalized five-dimensional hypercube numbers with negative indices are $-1, -32, -243, -1024, -3125, \ldots$. Hence, **generalized five-dimensional hypercube numbers** are

$$\ldots, -3125, -1024, -243, -32, -1, 0, 1, 32, 243, 1024, 3125, \ldots,$$

forming the sequence $0, 1, -1, 32, -32, 243, -243, 1024, -1024, \ldots$ for $n = 0, \pm 1, \pm 2, \pm 3, \pm 4, \ldots$.

The generalized six-dimensional hypercube numbers with negative indices are 1, 64, 729, 4096, 15625, ... So, **generalized six-dimensional hypercube numbers** are

$$\ldots, 15625, 4096, 729, 64, 1, 0, 1, 64, 729, 4096, 15625, \ldots,$$

forming the sequence $0, 1, 1, 64, 64, 729, 729, 4096, 4096, \ldots$ for $n = 0, \pm 1, \pm 2, \pm 3, \pm 4, \ldots$.

It was proven, that the generating function for the sequence $C^k(1)$, $C^k(2), \ldots C^k(n), \ldots$ has the form $g(x) = \frac{1 + a_1^k x^2 + \cdots + a_{k-2}^k x^{k-2} + x^{k-1}}{(1-x)^{k+1}} = C^k(1) + C^k(2)x + C^k(3)x^2 + \cdots$, where $a_1^k = a_{k-2} = 2^k - (k+1)$, $a_2^k = a_{k-3}^k = 3^k - (k+1)2^k + \frac{k(k+1)}{2}, \ldots, a^k_{\lfloor \frac{k-1}{2} \rfloor} = a^k_{k - \lfloor \frac{k-1}{2} \rfloor - 1} = (\lfloor \frac{k-1}{2} \rfloor + 1)^k - (k+1)(\lfloor \frac{k-1}{2} \rfloor)^k + \frac{k(k+1)}{2}(\lfloor \frac{k-1}{2} \rfloor - 1)^k + \cdots + (-1)^{\lfloor \frac{k-1}{2} \rfloor} \binom{k+1}{\lfloor \frac{k-1}{2} \rfloor}$. Hence, the generating function for the sequence $^-C^k(1), {}^-C^k(2), \ldots, {}^-C^k(n), \ldots$ of generalized k-dimensional hypercube numbers with negative indices, i.e., for the sequence $C^k(-1)$, $C^k(-2), \ldots, C^k(-n), \ldots$, has the form $f(x) = (-1)^k g(x)$, giving

$$(-1)^k x g(x) = C^k(-1)x + C^k(-2)x^2 + C^k(-3)x^3 + \cdots + C^k(-n)x^n + \cdots, \quad |x| < 1.$$

Similarly, the generating function for the sequence $C^k(0)$, $C^k(1)$, $C^k(-1)$, $C^k(2)$, $C^k(-2)$, ... of all generalized k-dimensional hypercube numbers, written for $n = 0, \pm 1, \pm 2, \ldots$, is $x^2(1+(-1)^k x)g(x^2)$, giving

$$x^2(1+(-1)^k x)g(x^2) = C^k(0)x + C^k(1)x^2 + C^k(-1)x^3 \\ +C^k(2)x^4 + C^k(-2)x^5 + \cdots, \quad |x| < 1.$$

In particular,

$$\frac{x^2(1-x)(1+26x^2+66x^4+26x^6+x^8)}{(1-x^2)^6}$$
$$= C^5(0)x + C^5(1)x^2 + C^5(-1)x^3 \cdots, \quad |x| < 1$$
$$= \frac{x^2(x+1)^2(1+56x^2+246x^4+56x^6+x^8)}{(1-x^2)^7}$$
$$= C^6(0)x + C^6(1)x^2 + C^6(-1)x^3 + \cdots, \quad |x| < 1.$$

3.7.5. For $k = 4$, the **generalized hyperoctahedral numbers** correspond to all values of the formula $HO(n) = \frac{n^2(n^2+2)}{3}$ for any integer value of n.

Similarly to the case of biquadratic numbers, generalized hyperoctahedral numbers $\ldots, 225, 96, 33, 8, 1, 0, 1, 8, 33, 96, 225, \ldots$ are just doubled version of ordinary hyperoctahedral numbers: $HO(0) = 0$, and $HO(-n) = HO(n) = \frac{n^2(n^2+2)}{3}$, $n \in \mathbb{N}$.

The sequence $0, 1, 1, 8, 8, 33, 33, 96, 96, \ldots$ of the generalized four-dimensional hyperoctahedral numbers, written for $n = 0, \pm 1, \pm 2, \ldots$, has the generating function $f(x) = \frac{x^2(1+x)(1+x^2)^2}{(1-x^2)^5}$, which can be obtained by the formula $(x^2+x^3)g(x^2)$ from the generating function $g(x) = \frac{(1+x)^3}{(1-x)^5}$ for the sequence of ordinary hyperoctahedral numbers.

3.7.6. Given a dimension k, $k \geq 5$, we obtain **generalized k-dimensional hyperoctahedron numbers** as all values of the formula $O^k(n) = \sum_{i=0}^{k-1}(-1)^i\binom{k-1}{i}2^{k-1-i}S_3^{k-i}(n)$ for each integer value of n.

For positive integers n, we get ordinary k-dimensional hyperoctahedron numbers. For $n = 0$, it holds $O^k(0) = 0$. Finally, for negative values, we get the formula $O^k(-n) = (-1)^k O^k(n)$, $n \in \mathbb{N}$. Thus, absolute values of generalized k-dimensional hyperoctahedron numbers with negative indices ${}^-O^k(n) = O^k(-n)$ coincide with ordinary

k-dimensional hyperoctahedron numbers of the same size; all such numbers are positive integers for even k, and are negative integers for odd k.

So, the generalized five-dimensional hyperoctahedron numbers with negative indices are $-1, -10, -51, -180, -501, \ldots$. Hence, **generated five-dimensional hyperoctahedron numbers** are

$$\ldots, -501, -180, -51, -10, -1, 0, 1, 10, 51, 180, 501, \ldots,$$

forming the sequence $0, 1, -1, 10, -10, 51, -51, 180, -180, \ldots$ for $n = 0, \pm 1, \pm 2, \pm 3, \pm 4, \ldots$.

The generalized six-dimensional hyperoctahedron numbers with negative indices are 1, 12, 73, 304, 985, ... Hence, **generalized six-dimensional hyperoctahedron numbers** are

$$\ldots, 985, 304, 73, 12, 1, 0, 1, 12, 73, 304, 985, \ldots,$$

forming the sequence $0, 1, 1, 12, 12, 73, 73, 304, 304, \ldots$ for $n = 0, \pm 1, \pm 2, \pm 3, \pm 4, \ldots$.

It was proven that the generating function for the sequence $O^k(1)$, $O^k(2), \ldots, O^k(n), \ldots$ is $g(x) = \frac{(1+x)^{k-1}}{(1-x)^{k+1}} = O^k(1) + O^k(2)x + O^k(3)x^2 + \cdots + O^k(n+1)x^n + \cdots$. So, the generating function for the sequence ${}^{-}O^k(1), {}^{-}O^k(2), \ldots, {}^{-}O^k(n), \ldots$ of generalized k-dimensional hyperoctahedron numbers with negative indices, i.e., for the sequence $O^k(-1)$, $O^k(-2), \ldots, O^k(-n), \ldots$, has the form $f(x) = (-1)^k g(x)$, giving

$$(-1)^k \frac{x(1+x)^{k-1}}{(1-x)^{k+1}} = O^k(-1)x + O^k(-2)x^2 + O^k(-3)x^3 + \cdots + {}^{-}O^k(n)x^n + \cdots, \quad |x| < 1.$$

Similarly, the generating function for the sequence $O^k(0)$, $O^k(1)$, $O^k(-1)$, $O^k(2)$, $O^k(-2), \ldots$ of all generalized k-dimensional hyperoctahedron numbers, written for $n = 0, \pm 1, \pm 2, \ldots$, has the form $x^2(1 + (-1)^k x)g(x^2)$, giving

$$\frac{x^2(1+x^2)^{k-1}(1+(-1)^k x)}{(1-x^2)^{k+1}} = O^k(0)x + O^k(1)x^2 + O^k(-1)x^3 + O^k(2)x^4 + xO^k(-2)x^5 + \cdots, \quad |x| < 1.$$

In particular, we have

$$\frac{x^2(1-x)(1+x^2)^4}{(1-x^2)^5} = O^5(0)x + O^5(1)x^2 + O^5(-1)x^3 + O^5(2)x^4 + O^5(-2)x^5 + \cdots, \quad |x| < 1;$$

$$\frac{x^2(x+1)(1+x^2)^5}{(1-x^2)^6} = O^6(0)x + O^6(1)x^2 + O^6(-1)x^3 + O^6(2)x^4 + O^k(-2)x^5 + \cdots, \quad |x| < 1.$$

3.7.7. The reader can obtain similar results on other classes of generalized multidimensional figurate numbers: three additional classes of four-dimensional **generalized polychoral numbers**, including **generalized hyperdodecahedral numbers**, **generalized hypericosahedral numbers**, and **generalized polyoctahedral numbers**; **generalized k-dimensional m-gonal pyramidal numbers**; **generalized centered regular polytope numbers**, including **generalized k-dimensional centered hypercube numbers**, **generalized k-dimensional centered hypertetrahedron numbers**, and **generalized k-dimensional centered hyperoctahedron numbers**, etc.

For example, the **generalized nexus numbers** are defined as all values of the formula $N^k(n) = (n+1)^{k+1} - n^{k+1}$ for any integer value of n.

For positive integers n, we obtain ordinary nexus numbers. For $n = 0$, we get $N^k(0) = 1$. Finally, for negative values of the index, one has

$$N_k(-n) = (-1)^k(n^{k+1} - (n-1)^{k+1}) = (-1)^k N_k(n-1), \quad n \in \mathbb{N}.$$

Hence, absolute values of the generalized nexus numbers with negative indices coincide with ordinary nexus numbers of smaller size; all these numbers are positive for even k, and are negative for odd k.

Chapter 4

Areas of number theory including figurate numbers

The theory of figurate numbers does not belong to the central domains of Mathematics, but the beauty of these numbers attracted the attention of many scientists during thousands years. Moreover, many mathematical facts have deep connections with figurate numbers, and a lot of well-known theorems can be formulated in terms of these numbers. In this chapter we give a large list of areas of Number Theory, related to figurate numbers.

4.1 Addition and multiplication tables

Let us consider the simplest mathematical structures, addition and multiplication tables, and try to find in them some connections with figurate numbers.

4.1.1. For example, in the addition table below it is possible to find differences of triangular numbers.

+	1	2	3	4	5	6	7	8	9	...	n
1	2	3	4	5	6	7	8	9	10	...	$n+1$
2	3	4	5	6	7	8	9	10	11	...	$n+2$
3	4	5	6	7	8	9	10	11	12	...	$n+3$
4	5	6	7	8	9	10	11	12	13	...	$n+4$
5	6	7	8	9	10	11	12	13	14	..	$n+5$
6	7	8	9	10	11	12	13	14	15	...	$n+6$
7	8	9	10	11	12	13	14	15	16	...	$n+7$
8	9	10	11	12	13	14	15	16	17	...	$n+8$
9	10	11	12	13	14	15	16	17	18	...	$n+9$
...	...	...	...	...	...	...	...	...	...	...	...
n	$n+1$	$n+2$	$n+3$	$n+4$	$n+5$	$n+6$	$n+7$	$n+8$	$n+9$	...	$2n$

More exactly, *the sum of elements in k-th column of $n \times n$ addition table is equal to the difference between $(n+k)$-th and k-th triangular numbers.*

In fact, for first column we get $2+3+\cdots+(n+1) = S_3(n+1) - S_3(1)$, for second column we get $3+4+\cdots+(n+2) = S_3(n+2) - S_3(2), \ldots$, for n-th column we get $(n+1)+\cdots+2n = S_3(2n) - S_3(n)$.

It implies that *the sum of all entries of $n \times n$ addition table is equal to $(2n)$-th tetrahedral number without two n-th tetrahedral numbers.*

In fact, due to the previous observation, this sum can be written as

$$\begin{aligned}(S_3(n+1) - S_3(1)) + (S_3(n+2) - S_3(1)) + \cdots + (S_3(2n) - S_3(n)) \\ = (S_3(n+1) + \cdots + S_3(2n)) - (S_3(1) + \cdots + S_3(n)) = (S_3(1) \\ + \cdots + S_3(2n)) - 2(S_3(1) + \ldots S_3(n)) = S_3^3(2n) - 2S_3^3(n).\end{aligned}$$

On the other hand, *the sum of all elements of $n \times n$ addition table is $2nS_3(n)$.*

Indeed, the summation of the form

$$\begin{array}{ccccccc} (1+1) & + & (1+2) & +\cdots+ & (1+n) & + \\ (2+1) & + & (2+2) & +\cdots+ & (2+n) & + \\ & & & \cdots & & + \\ (n+1) & + & (n+2) & +\cdots+ & (n+n) & = \\ (S_3(n)+n) & + & (S_3(n)+2n) & +\cdots+ & (S_3(n)+n\cdot n) & \end{array}$$

gives for this sum the value $nS_3(n) + n(1+2+\cdots+n) = nS_3(n) + nS_3(n) = 2nS_3(n)$.

Furthermore, *the sums of elements in the rising diagonals "2", "3, 3", "4, 4, 4",... of the addition table are consecutive pronic numbers:* $2 \cdot 1 = P(1)$, $3 \cdot 2 = P(2)$, $4 \cdot 3 = P(3)$, etc.

It holds also, that *the sum of elements in all rising diagonals of $n \times n$ addition table is equal to the sum of n-th square pyramidal number and n-th triangular number.* Moreover, *the sum of elements in any upper-left isosceles triangle of the table is the sum of square pyramidal and triangular numbers of the same size.*

In fact, any such sum can be written as $2 \cdot 1 + \cdots + (k+1) \cdot k = \sum_{i=1}^{k}(i+1)i = \sum_{i=1}^{k} i^2 + \sum_{i=1}^{k} i = S_4^3(k) + S_3(k)$.

It is easy to see that *the sums of elements in the falling diagonals of the addition table can be represented as differences between pronic numbers for "even" diagonals "2, 4,..., 2n", "4, 6,..., 2n-2",..., and as differences between square numbers for "odd" diagonals "3, 5,..., 2n-1", "5, 7,...,2n-3",...*.

In fact, $2+4+6+\cdots+2n = 2(1+2+\cdots+n) = n(n+1) = P(n)-P(0)$, $3+5+7+\cdots+(2n-1) = S_4(2n-1)-S_4(1)$, $4+6+\cdots+(2n-2) = 2(2+3+\cdots+(n-1)) = 2(S_3(n-1)-S_3(1)) = P(n-1)-P(1)$, $5+7+\cdots+(2n-3) = S_4(2n-3)-S_4(2)$, etc.

Also, *the sums of elements in the gnomons*

$$\begin{array}{ccc} & & 3 \\ & 2 & 4 \\ 2\ , & 2\ 3\ , & 3\ 4\ 5\ ,\ldots \end{array}$$

of the addition table can be represented as differences between square and pronic numbers.

Indeed, it holds $2 = (1+2+1)-2\cdot 1 = S_4(2)-P(1)$, $3+4+3 = (1+2+3+4+3+2+1)-2(1+2) = S_4(4)-P(2)$, etc. In general, $(k+1)+\cdots+2k+\cdots+(k+1) = (1+\cdots k+(k+1)+\cdots+2k+(k+1)+k+\cdots+1)-2(1+\cdots+k) = S_4(2k)-P(k)$.

4.1.2. The multiplication table also has many hidden connections with figurate numbers (see, for example, [CoGu96]).

$\times$	1	2	3	4	5	6	7	8	9	...	n
1	1	2	3	4	5	6	7	8	9	...	n
2	2	4	6	8	10	12	14	16	18	...	$2n$
3	3	6	9	12	15	18	21	24	27	...	$3n$
4	4	8	12	16	20	24	28	32	36	...	$4n$
5	5	10	15	20	25	30	35	40	45	..	$5n$
6	6	12	18	24	30	36	42	48	54	...	$6n$
7	7	14	21	28	35	42	49	56	63	...	$7n$
8	8	16	24	32	40	48	56	64	72	...	$8n$
9	9	18	27	36	45	54	63	72	81	...	$9n$
...	...	...	...	...	...	...	...	...	...	...	...
n	n	$2n$	$3n$	$4n$	$5n$	$6n$	$7n$	$8n$	$9n$	...	n^2

Obviously, *on the main diagonal of $n \times n$ multiplication table we have square numbers 1, 4, 9, 16, 25, ...* .

It implies that *the sum of all elements of the main diagonal in $n \times n$ multiplication table is equal to the square pyramidal number $S_4^3(n)$.*

It is easy to see that *elements of the two falling diagonals nearest to the main diagonal are consecutive pronic numbers*: $2 = 2\cdot 1 = P(1)$,$6 = 3\cdot 2 = P(2),\ldots,n\cdot(n-1) = P(n-1)$.

Moreover, *the sums of elements in the rising diagonals "1", "2, 2", "3, 4, 3",... give consecutive tetrahedral numbers 1, 4, 10, 20, 35,...*.

In fact, any such sum has the form $k \cdot 1 + (k-1) \cdot 2 + \cdots + 1 \cdot k = (k + (k-1) + \cdots + 1) + ((k-1) + (k-2) + \cdots + 1) + \cdots + (2+1) + 1 = S_3(k) + S_3(k-1) + \cdots + S_3(2) + S_3(1) = S_3^3(k)$.

Furthermore, *the sums of elements in the gnomons "1", "2, 4, 2", "3, 6, 9, 6, 3",... of the multiplication table give consecutive cubic numbers 1, 8, 27, 64, 125, ...*.

In fact, $1 = 1^3$; $2 + 4 + 2 = 2(1 + 2 + 1) = 2^3$; $3 + 6 + 9 + 6 + 3 = 3(1 + 2 + 3 + 2 + 1) = 3 \cdot 3^2 = 3^3$, etc. In general, we have $k \cdot 1 + k \cdot 2 + \cdots + k \cdot (k-1) + k \cdot k + (k-1) \cdot k + \cdots + 2 \cdot k + 1 \cdot k = k \cdot (1 + 2 + \cdots + (k-1) + k + (k-1) + \cdots + 2 + 1) = k \cdot k^2 = k^3$.

So, *the sum of all entries of $n \times n$ multiplication table is the sum $1^3 + 2^3 + \cdots + n^3$ of the first n cubic numbers.*

On the other hand, *the sum of all elements of $n \times n$ multiplication table is $(S_3(n))^2$.*

In fact, a summation of the form

$$\begin{array}{ccccccccc}
1 \cdot 1 & + & 1 \cdot 2 & + \cdots + & 1 \cdot n & + \\
2 \cdot 1 & + & 2 \cdot 2 & + \cdots + & 2 \cdot n & + \\
 & & & \cdots & & + \\
n \cdot 1 & + & n \cdot 2 & + \cdots + & n \cdot n & = \\
\multicolumn{6}{c}{1(1 + 2 + \cdots + n) + 2(1 + 2 + \cdots + n) + \cdots + n(1 + 2 + \cdots + n)}
\end{array}$$

gives for this sum value $(1 + 2 + \cdots + n)(1 + 2 + \cdots + n) = (1 + 2 + \cdots + n)^2 = (S_3(n))^2$.

Hence, we get another proof of the formula

$$1^3 + 2^3 + \cdots + n^3 = (1 + 2 + \cdots + n)^2.$$

Its geometrical illustration for $n = 5$ is given below: we construct a square with the side $S_3(5) = 15$, using as constructed blocks one element 1, corresponding to the cube 1^3, eight elements 2, corresponding to the cube 2^3,...,125 elements 5, corresponding to the cube 5^3.

$$\begin{array}{ccccccccccccccc}
1 & 2 & 2 & 3 & 3 & 3 & 4 & 4 & 4 & 4 & 5 & 5 & 5 & 5 & 5 \\
2 & 2 & 2 & 3 & 3 & 3 & 4 & 4 & 4 & 4 & 5 & 5 & 5 & 5 & 5 \\
2 & 2 & 2 & 3 & 3 & 3 & 4 & 4 & 4 & 4 & 5 & 5 & 5 & 5 & 5 \\
3 & 3 & 3 & 3 & 3 & 3 & 4 & 4 & 4 & 4 & 5 & 5 & 5 & 5 & 5 \\
3 & 3 & 3 & 3 & 3 & 3 & 4 & 4 & 4 & 4 & 5 & 5 & 5 & 5 & 5 \\
3 & 3 & 3 & 3 & 3 & 3 & 4 & 4 & 4 & 4 & 5 & 5 & 5 & 5 & 5 \\
4 & 4 & 4 & 4 & 4 & 4 & 4 & 4 & 4 & 4 & 5 & 5 & 5 & 5 & 5 \\
4 & 4 & 4 & 4 & 4 & 4 & 4 & 4 & 4 & 4 & 5 & 5 & 5 & 5 & 5 \\
4 & 4 & 4 & 4 & 4 & 4 & 4 & 4 & 4 & 4 & 5 & 5 & 5 & 5 & 5 \\
4 & 4 & 4 & 4 & 4 & 4 & 4 & 4 & 4 & 4 & 5 & 5 & 5 & 5 & 5 \\
5 & 5 & 5 & 5 & 5 & 5 & 5 & 5 & 5 & 5 & 5 & 5 & 5 & 5 & 5 \\
5 & 5 & 5 & 5 & 5 & 5 & 5 & 5 & 5 & 5 & 5 & 5 & 5 & 5 & 5 \\
5 & 5 & 5 & 5 & 5 & 5 & 5 & 5 & 5 & 5 & 5 & 5 & 5 & 5 & 5 \\
5 & 5 & 5 & 5 & 5 & 5 & 5 & 5 & 5 & 5 & 5 & 5 & 5 & 5 & 5 \\
5 & 5 & 5 & 5 & 5 & 5 & 5 & 5 & 5 & 5 & 5 & 5 & 5 & 5 & 5 \\
5 & 5 & 5 & 5 & 5 & 5 & 5 & 5 & 5 & 5 & 5 & 5 & 5 & 5 & 5
\end{array}$$

4.2 Pascal's triangle and binomial theorem

4.2.1. The *Pascal's triangle* is an arithmretic triangle, the sides of which are formed by 1, and any inner entry is obtained by adding the two entries diagonally above. So, the 0-th row of the Plascal's triangle contains only the number 1, the 0-th and the first entries in the first row are equal to 1, the 0-th and the second entries in the second row are equal to 1, while the first one is $2 = 1 + 1$, the third row has the form $1, 3 = 1 + 2, 3 = 2 + 1, 1$, etc. Moving step by step, we obtain the following figure:

$$\begin{array}{ccccccccc}
 & & & & 1 & & & & \\
 & & & 1 & & 1 & & & \\
 & & 1 & & 2 & & 1 & & \\
 & 1 & & 3 & & 3 & & 1 & \\
1 & & 4 & & 6 & & 4 & & 1 \\
\vdots & \vdots & \vdots & \vdots & \vdots & \vdots & \vdots & \vdots & \vdots
\end{array}$$

Denoting k-the element of n-th row of the Pascal's triangle by T_n^k, $n \geq 0$, $0 \leq k \leq n$, one can write the above definition algebraically: $T_n^k = T_{n-1}^k + T_{n-1}^{k-1}$, $1 \leq k \leq n-1$, and $T_0^0 = T_n^0 = T_n^n = 1$.

It is easy to check that elements T_n^k of the Pascal's triangle are *binomial coefficients* $\binom{n}{k}$, i.e., coefficients of the following decomposition:

$$(1+x)^n = \binom{n}{0} + \binom{n}{1}x + \cdots + \binom{n}{n}x^n.$$

In fact, for $n = 0$, it holds $(1+x)^0 = 1$, i.e., $\binom{0}{0} = 1$. For $n = 1$, we get $(1+x)^1 = 1 + x$, i.e., $\binom{1}{0} = 1$, $\binom{1}{1} = 1$. Moreover,

$$\begin{aligned}(1+x)^n &= (1+x)^{n-1}(1+x) = \left(\binom{n-1}{0} + \binom{n-1}{1}x\right. \\ &\quad \left.+ \cdots + \binom{n-1}{n-1}x^{n-1}\right)(1+x) = \binom{n-1}{0} \\ &\quad + \left(\binom{n-1}{1} + \binom{n-1}{0}\right)x + \left(\binom{n-1}{2}\right. \\ &\quad \left.+ \binom{n-1}{1}\right)x^2 + \cdots\cdots + \left(\binom{n-1}{n-1}\right. \\ &\quad \left.+ \binom{n-1}{n-2}\right)x^{n-1} + \binom{n-1}{n-1}x^n.\end{aligned}$$

So, $\binom{n-1}{0} = \binom{n}{0}$, $\binom{n-1}{n-1} = \binom{n}{n}$, and $\binom{n-1}{k} + \binom{n-1}{k-1} = \binom{n}{k}, 1 \leq k \leq n-1$. Therefore, binomial coefficients $\binom{n}{k}$ have the same recurrent equation $\binom{n}{k} = \binom{n-1}{k-1} + \binom{n-1}{k}, 1 \leq k \leq n-1$, and the same initial conditions $\binom{n}{n} = \binom{n-1}{n-1} = \cdots = \binom{1}{1} = \binom{0}{0} = 1$, $\binom{n}{0} = \binom{n-1}{0} = \cdots = \binom{1}{0} = \binom{0}{0} = 1$ as elements T_n^k of Pascal's triangle. Hence, it holds $T_n^k = \binom{n}{k}$ for all $n \geq 0$, $0 \leq k \leq n$.

Since the function $f(n,k) = \frac{n!}{k!(n-k)!}$, containing **factorial numbers** $n!, k!$ and $(n-k)!$, also has the same recurrent equation $f(n,k) = f(n-1,k) + f(n-1,k-1)$ with the same initial conditions $f(0,0) = f(0,n) = f(n,n) = 1$, we obtain the formula

$$\binom{n}{k} = \frac{n!}{k!(n-k)!}.$$

This arithmetic triangle was studied in the Pascal's treatise *Traité du triangle arithmétique*, 1654 ([Pasc54]), although it had been described centuries earlier by Chinese mathematician Yanghui and the Persian astronomer-poet Omar Khayyam (see [Wiki11]). In this treatise the constriction of the triangle was based on the following rule: starting from first row and first column, constracted from 1, any other entrie A of the table is equal to the sum of previous number in the same row and the previous number in the same column:

1	1	1	1	1	1
1	2	3	4	5	6
1	3	6	10	15	21
1	4	10	20	35	56
1	5	15	35	70	126
1	6	21	56	126	252

Therefore, our triangle can be obtained from the original triangle of Pascal by rotation on 45^0.

4.2.2. One can easily find the family of figurate numbers in the Pascal's triangle. In fact, any triangular number can be written in the form

$$S_3(n) = \binom{n+1}{2},$$

i.e., it is a *binomial coefficient.* Therefore, *the triangular numbers form 3-th diagonal 1, 3, 6, 10, 15, ... of the Pascal's triangle.* In the other words, triangular numbers are found in the third position either from left to right or from right to left in the Pascal's triangle. Similarly,

$$S_3^3(n) = \binom{n+2}{3},$$

i.e., any tetrahedral number is a binomial coefficient. The tetrahedral numbers form 4-th diagonal 1, 4, 10, 20, 35, ... of the Pascal's triangle.

Pentatope numbers, which are four-dimensional analogues of three-dimensional tetrahedral numbers and two-dimensional triangular numbers, also coincide with binomial coefficients:

$$S_3^4(n) = \binom{n+3}{4}.$$

They form 5-th diagonal 1, 5, 15, 35, 70, ... of the Pascal's triangle.

In general, for a given dimension k, $k \geq 2$, n-th k-dimensional hypertetrahedron number $S_3^k(n)$, representing an k-dimensional simplex, has the form

$$S_3^k(n) = \frac{n^{\overline{k}}}{k!} = \binom{n+k-1}{k}.$$

Hence, *k-dimensional hypertetrahedron numbers form $(k+1)$-th diagonal of the Pascal's triangle*, i.e., are found in the $(k+1)$-th position either from left to right or from right to left in the Pascal's triangle.

Moreover, the linear numbers 1, 2, 3, 4, 5, ... can be considered as an one-dimensional analog of the two-dimensional triangular and three-dimensional tetrahedral numbers. The n-th linear number $S_3^1(n) = n$ is equal to the binomial coefficient $\binom{n}{1}$, and such numbers form the second diagonal 1, 2, 3, 4, 5, ... of the Pascal's triangle.

The sequence 1, 1, 1, 1, 1,... can be seen as a zero-dimenzional analogue of linear, triangular, tetrahedral etc. numbers. Thus, n-th element $S_3^0(n) = 1$ of this sequence coincides with the binomial coefficient $\binom{n-1}{0}$. These zero-dimensional figurate numbers form the first diagonal 1, 1, 1, 1, 1, ... of the Pascal's triangle.

So, viewing from the opposite perspective, any element $\binom{n}{k}, n \geq 0, 0 \leq k \leq n$, of the Pascal's triangle is a figurate number, representing an k-dimensional analog of triangular numbers. In fact,

$$\binom{n}{k} = S_3^k(n-k+1),$$

and we can rewrite the Pascal's triangle in the form

$$\begin{array}{ccccccccc}
 & & & & S_3^0(1) & & & & \\
 & & & S_3^0(2) & & S_3^1(1) & & & \\
 & & S_3^0(3) & & S_3^1(2) & & S_3^2(1) & & \\
 & S_3^0(4) & & S_3^1(3) & & S_3^2(2) & & S_3^3(1) & \\
S_3^0(5) & & S_3^1(4) & & S_3^2(3) & & S_3^3(2) & & S_3^4(1) \\
\vdots & \vdots & \vdots & \vdots & \vdots & \vdots & \vdots & \vdots & \vdots
\end{array}$$

In the other words, one can rewrite binomial theorem in the form

$$(1+x)^n = S_3^0(n+1) + S_3^1(n)x + S_3^2(n-1)x^2 + \cdots + S_3^{n-1}(2)x^{n-2} + S_3^n(1)x^n.$$

Moreover, using classical properties of the Pascal's triangle, we can formulate some additional properties of figurate numbers. For example, the identity $\binom{n}{k} = \binom{n}{n-k}$ (check, for example, that $\frac{n!}{k!(n-k)!} = \frac{n!}{(n-k)!(n-(n-k))!}$) implies the identity

$$S_3^k(n-k+1) = S_3^{n-k}(k+1).$$

In particular, we have

$$S_3^n(1) = S_3^0(n+1) = 1, \quad S_3^{n-1}(2) = S_3^1(n) = n,$$

$$S_3^{n-2}(3) = S_3^2(n-1) = \frac{(n-1)n}{2},$$

$$S_3^{n-3}(4) = S_3^3(n-2) = \frac{(n-2)(n-1)n}{6}.$$

So, it is proven that *n-th triangular number $S_3(n)$ is equal to the third $(n-1)$-dimensional simplex number $S_3^{n-1}(3)$, n-th tetrahedral number $S_3^3(n)$ is equal to the fourth $(n-1)$-dimensional hypertetrahedron number $S_3^{n-1}(4)$, n-th pentatope number $S_3^4(n)$ is equal to the 5-th $(n-1)$-dimensional hypertetrahedron number $S_3^{n-1}(5)$*, etc.

Using the identity $\binom{n}{0} + \binom{n}{1} + \cdots + \binom{n}{n} = 2^n$ (take $x = 1$ in the binomial theorem) we obtain the identity

$$S_3^0(n+1) + S_3^1(n) + S_3^2(n-1) + \cdots + S_3^{n-1}(2) + S_3^n(1) = 2^n.$$

Similarly, since $\binom{n}{0} - \binom{n}{1} + \cdots + (-1)^n\binom{n}{n} = 0$, it holds

$$S_3^0(n+1) - S_3^1(n) + S_3^2(n-1) + \cdots (-1)^{n-1} S_3^{n-1}(2)$$
$$+(-1)^n S_3^n(1) = 0.$$

On the other hand, we can obtain some properties of the Pascal's triangle, using the corresponding properties of figurate numbers.

Thus, the classical recurrent equations $S_3^k(n) = S_3^k(n-1) + S_3^{k-1}(n)$, $k \geq 0$, $n \geq 0$, for k-dimensional hypertetrahedron numbers allows to state, that any triangular number $S_3(n)$ is the sum of the previous triangular number $S_3(n-1)$, placed above to the right in the Pascal's triangle, and the number n, placed above to the left in the Pascal's triangle; any pyramidal number $S_3^3(n)$ is the sum of the previous pyramidal number $S_3^3(n-1)$, placed above to the right in the Pascal's triangle, and n-th triangular number $S_3(n)$, placed above to the left in the Pascal's triangle, etc. It gives the above recurrent rule for the construction of the Pascal's triangle: *each inner element of the Pascal's triangle can be obtained by adding two elements, placed above to the right and above to the left in the triangle, while all outside elements are equal to 1.*

Since n-th k-dimensional hypertetrahedron number is the sum of the first n $(k-1)$-dimensional hypertetrahedron numbers, we get a proof of an other well-known fact: *any element of the Pascal's triangle is the sum of all elements placing up and to the right in the above diagonal.*

4.2.3. The above consideration shows, that the Pascal's triangle consists only of figurate numbers. Moreover, the Pascal's triangle, written as a square grid and padded with zeros (as it was considered by Jacob Bernoulli, [Smit84]) is called the **figurate number triangle**.

$$\begin{matrix} 1 & 0 & 0 & 0 & 0 & \cdots \\ 1 & 1 & 0 & 0 & 0 & \cdots \\ 1 & 2 & 1 & 0 & 0 & \cdots \\ 1 & 3 & 3 & 1 & 0 & \cdots \\ 1 & 4 & 6 & 4 & 1 & \cdots \\ \vdots & \vdots & \vdots & \vdots & \vdots & \ddots \end{matrix}$$

This triangle has therefore entries $a_{ij} = \binom{i}{j}$, where $i = 0, 1, 2, 3, \ldots$ is the row number, and $j = 0, 1, 2, 3, \ldots$ is the column number. In this construction, starting with 0-th column, consisting of 1, we obtain the linear numbers in the first column, the triangular numbers in the second column, the tetrahedral numbers in the third column, etc. In general, the k-dimensional hypertetrahedron numbers form k-th column of the figurate number triangle.

4.2.4. The square pyramidal numbers $S_4^3(n)$ can be expressed as sums of binomial coefficients:

$$S_4^3(n) = \binom{n+2}{3} + \binom{n+1}{3}.$$

So, one obtains the sequence of these numbers by adding each two consecutive elements of the 4-th diagonal $1, 4, 10, 20, 35, \ldots$ of the Pascal's triangle: $1 = 1$, $5 = 1 + 4$, $14 = 10 + 4$, $30 = 10 + 20$, etc.

In general, it was proven that n-th k-dimensional square pyramidal number $S_4^k(n)$ can be written as $S_4^k(n) = S_3^k(n) + S_3^k(n-1)$, i.e., it holds

$$S_4^k(n) = \binom{n+k-1}{k} + \binom{n+k-2}{k}.$$

So, one obtains the sequence of these numbers by adding each two consecutive elements of the $(k+1)$-th diagonal of the Pascal's triangle.

4.2.5. The **tritriangular numbers** are defined in terms of binomial coefficients in the following way:

$$TT(n) = \binom{\binom{n+2}{2}}{2} = \frac{n(n+1)(n+2)(n+3)}{8}.$$

So, they also are entries of the Pascal's triangle. The first few values of such numbers are 3, 15, 45, 105, 210, 378, 630, 990, 1485, 2145, ... (Sloane's A050534).

By the above formula, n-th tritriangular number $TT(n)$ is three times n-th pentatope number:

$$TT(n) = 3S_3^4(n).$$

Thus, the sequence $TT(1), TT(2), \dots, TT(n), \dots$ can be seen as 5-th diagonal of the Pascal's triangle, multiplied by 3.

4.2.6. The consideration of Pascal's triangle modulo 2 gives the sequence 1; 1, 1; 1, 0, 1; 1, 1, 1, 1; 1, 0, 0, 0, 1; ... (Sloane's A047999) and leads to an interpretation of the *Sierpiński sieve*, i.e., a *fractal*, obtained by consecutive deleting of the "central part triangle" in a given equilateral triangle, in each of three remaining triangles, etc. In other words, coloring all odd numbers black and all even numbers white in Pascal's triangle with 2^n rows produces a figure, which is an approximation to the *Sierpiński triangle*: the limit of this parity-colored 2^n-row Pascal's triangle for n approaches infinity is the Sierpiński sieve.

Obviously, the central white triangles, as well as other white triangles, arising in this construction, corresponds to some triangular numbers: the first white triangle has one point, giving $S_3(1) = 1$, the second one has 6 points, giving $S_3(3) = 6$, the third one has 28 points, giving $S_3(7) = 28$, etc. In fact, n-th central white triangle has $2^{n-1}(2^n - 1)$ points, giving the triangular number $S_3(2^n - 1)$.

There are many other hidden connections with special numbers, which we will consider in corresponding sections.

4.3 Pythagorean triples and other Diophantine equations

The figurate numbers are closely associated with many types of Diophantine equations. For instance, the square numbers are involved in a natural way in the theory of Pythagorean triples, i.e., are closely connected with the Diophantine equation of the form $x^2 + y^2 = z^2$. The cubic numbers and their multidimensional analogues can be used to formulate *Last Fermat's theorem.* A lot of Diophantine equations, in particular, many Pell-like equations, arise in the theory of highly figurate numbers. In this section we consider some of the most interesting facts of such kind.

4.3.1. Since $S_4(n) = n^2$, one can formulate many well-known facts of the theory of **Pythagorean numbers** (see, for example, [Sier64], [Sier03]) in the language of square numbers.

Consider the *Pythagorean equation* $x^2+y^2 = z^2$. Its positive integer solution (x, y, z) is called *Pythagorean triple* and corresponds geometrically to the catheti and hypotenuse of some right triangle. The famous *Egyptian triangle* is obtained for $x = 3, y = 4, z = 5$.

A Pythagorean triple (x, y, z) is called *primitive*, if the numbers x, y and z are relatively prime, i.e., if $gcd(x, y, z) = 1$. It is easy to see that in a primitive Pythagorean triple (x, y, z) the numbers x, y and z are pairwise relatively prime: $gcd(x, y) = gcd(x, z) = gcd(y, z) = 1$. Really, if $gcd(x, y) = d, d > 1$, then one has $x = dx_1$, $y = dy_1$, and $z^2 = x^2+y^2 = d^2(x_1^2+y_1^2)$. Hence, $d|z$, and $gcd(x, y, z) = d > 1$, that gives contradiction.

Since the numbers x and y are relatively primes, they can not be both even. But they also could not be both odd. In fact, if $x = 2k+1$ and $y = 2t + 1$, then $x^2 = 4(k^2 + k) + 1$, $y^2 = 4(t^2 + t) + 1$, and $x^2+y^2 = 4m+2$. On the other hand, any integer z can be represented in the form $z = 4l + r$ with $r \in \{0, 1, 2, 3\}$, hence, $z^2 = 4(l^2 + 2lr) + r^2$, and possible reminders by dividing by 4 of the square of an integer can be only 0 and 1, in other words, $z^2 \neq 4m + 2$, that concludes the proof.

Therefore, in any primitive Pythagorean triple one cathetus is even, and other is odd. Obviously, the hypotenuse z is an odd number. Without loss of generality, let x be even, and y be odd. Now we can prove, that *all primitive Pythagorean triples* (x, y, z), *in which* y *is even, can be obtained by the formulas*

$$x = m^2 - n^2, y = 2mn, z = m^2 + n^2, m > n,$$

where m *and* n *are relative prime positive integers of different parity. Every primitive Pythagorean triple with even* y *is defined on this way unique.*

At first, let as show that the triple $(m^2 - n^2, 2mn, m^2 + n^2)$ is a primitive Pythagorean triple. In fact, direct checking shows, that

$$(m^2 - n^2)^2 + (2mn)^2 = (m^2 + n^2)^2.$$

If $gcd(x, y, z) = d > 1$, then d is odd, as, for example, z is odd. Then $d|z, d|x$, i.e., $d|(m^2 - n^2), d|(m^2 + n^2)$, and we get that $d|2m^2$ (as d divides the sum $z + x$) and $d|2n^2$ (as d divides the difference $z - x$). Since $gcd(d, 2) = 1$, then $d|m^2, d|n^2$, and we get the contradiction with the condition $gcd(m, n) = 1$. Therefore, we have proven, that $(m^2 - n^2, 2mn, m^2 + n^2)$ is a primitive Pythagorean triple.

On the other hand, consider any primitive Pythagorean triple (x, y, z) with even y. Since z and x are odd, then $z - x$ and $z + x$ are even, i.e., $y = 2c$, $z - x = 2a, z + x = 2b$, where $a, b, c \in \mathbb{N}$. Moreover, a and b are relatively prime: if $gcd(a, b) = d > 1$, then $d|(a + b) = z$ and $d|(b - a) = x$ - a contradiction with the primitivity of the triple (x, y, z). Since $x^2 + y^2 = z^2$, i.e., $y^2 = z^2 - x^2 = (z - x)(z + x)$, then $c^2 = ab$. Since $gcd(a, b) = 1$, then $a = n^2$, and $b = m^2$, where m and n are some relatively prime positive integers. Therefore, $z = a + b = n^2 + m^2$, $x = b - a = m^2 - n^2$, and $y = 2c = 2mn$. For odd x and z we need m and n with different parity.

Finally, let us check, that different pairs of m and n correspond to different primitive Pythagorean triples (x, y, z). In fact, for a given pairwise relatively prime numbers x, y, z we have the equalities $x + z = 2m^2, z - x = 2n^2$, which define the numbers n and m uniquely.

Now we can state, that *all solutions* (x, y, z) *of the Pythagorean equation* $x^2 + y^2 = z^2$ *have the form*

$$x = k(m^2 - n^2), \quad y = 2kmn, \quad z = k(m^2 + n^2),$$
$$k, m, n \in \mathbb{N}, \quad m > n,$$

where m *and* n *are relative prime positive integers of different parity.*

Therefore, we can establish, that *there exist infinitely many square numbers, which are equal to the sum of two other square numbers*:

$$S_4(k(m^2 - n^2)) + S_4(2kmn) = S_4(k(m^2 + n^2)).$$

Viewed from the opposite perspective, *there are infinitely many square numbers, which are equal to the difference of two other square numbers*:

$$S_4(2kmn) = S_4(k(n^2 + m^2)) - S_4(k(m^2 - n^2)).$$

It is known that *there are infinitely many pairs of consecutive square numbers, the sum of which is equal to a square number*:

$$S_4(x) + S_4(x + 1) = S_4(z).$$

In fact, the above equation holds for $x = \frac{(\sqrt{2}+1)^{2n+1}-(\sqrt{2}-1)^{2n+1}}{4} - \frac{1}{2}$ and $z = \frac{(\sqrt{2}+1)^{2n+1}+(\sqrt{2}-1)^{2n+1}}{2\sqrt{2}}$ for any positive integer n (see also **1.3.10.**).

It is easy to show, that *there are no square numbers* $S_4(x)$ *and* $S_4(y)$*, such that the sum and the difference of them are square numbers.* In other words, the system

$$\begin{cases} S_4(x) + S_4(y) = S_4(z) \\ S_4(x) - S_4(y) = S_4(t) \end{cases}$$

has no solutions in positive integers. It is so-called *Fermat's theorem.*

In fact, let us suppose that this system has positive integer solutions. Let (x, y, z, t) be such solution with the least x^2+y^2, i.e., which the least z.

Obviously, $gcd(x, y) = 1$, and $z^2 = x^2 + y^2 > x^2 - y^2 = t^2$, i.e., $z > t$. Since $2x^2 = z^2 + t^2$, we have $x^2 = (\frac{z+t}{2})^2 + (\frac{z-t}{2})^2$. Since $2|z^2+t^2$, we obtain that the numbers z and t have the same parity, i.e., $2|z+t$, $2|z-t$. So, the numbers $\frac{z-t}{2}$ and $\frac{z+t}{2}$ are positive integers. It means that $(\frac{z-t}{2}, \frac{z+t}{2}, x)$ is a Pythagorean triple.

It is easy to see that it is, moreover, a primitive triple. In fact, if $gcd(\frac{z-t}{2}, \frac{z+t}{2}, x) = d$, then $d|z$ and $d|x$. Since $d|z$ and $d|x$, then $d|y$. Hence, $d|gcd(x, y)$, i.e., $d|1$. Therefore, we get $d = 1$.

In this case, we can find positive integers m and n (with the conditions $gcd(m, n) = 1$, $m > n$, m and n have different parity) such that $\frac{z-t}{2} = m^2 - n^2$, and $\frac{z+t}{2} = 2mn$.

Since $2y^2 = z^2 - t^2$, we obtain

$$2y^2 = (z-t)(z+t) = 4mn \cdot 2(m^2 - n^2)$$
$$= 8(m^2 - n^2)mn, \quad \text{i.e., } y^2 = 4(m^2 - n^2)mn.$$

It means that $2|y$, i.e., $y = 2k$, $k \in \mathbb{N}$. Therefore, one gets

$$k^2 = (m^2 - n^2)mn = (m-n)(m+n)mn.$$

Since $gcd(m, n) = 1$ and the numbers m, n have different parity, one obtains that the numbers $m-n$, $m+n$, m and n are pairwise relatively primes. Their product is a perfect square k^2. Hence, each of them is a perfect square itself:

$$m - n = a^2, \quad m + n = b^2, \quad m = c^2, \quad n = d^2,$$

where a, b, c and d are some positive integers. Hence, we obtain the equalities

$$c^2 + d^2 = b^2, \text{ and } c^2 - d^2 = a^2,$$

i.e., a new solution (c, d, b, a) of above system. Clearly, in this case one has

$$b^2 = c^2 + d^2 = m + n < 2m \le 2mn = \frac{z+t}{2} < z \le z^2,$$

i.e., $b < z$. This fact shows that the first solution (x, y, z, t) is not a solution with minimal z.

Now it is easy to show that *there are no square numbers, so that the difference of their squares is a square number*:

$$(S_4(x))^2 - (S_4(y))^2 \neq S_4(z).$$

It implies, that *the sum of two squared square numbers is not a squared square number*:

$$(S_4(x))^2 + (S_4(y))^2 \neq (S_4(z))^2.$$

This fact is equivalent to the special case $n = 4$ of the *Fermat's last theorem*: for any $n > 2$ the equality $x^n + y^n = z^n$ has no positive integer solutions.

The only known Pythagorean triangle, the sides of which are triangular numbers, is given by the triple $(S_3(132), S_3(143), S_3(164)) = (8778, 10293, 13530)$. In the other words, *the only one positive integer solution is known for the equation*

$$S_4(S_3(x)) + S_4(S_3(y)) = S_4(S_3(z)).$$

The situation becomes more simple, if only catheti of a Pythagorean triangle should be triangular numbers. In this case, one gets

$$S_4(S_3(2x)) + S_4(S_3(2x+1)) = S_4(z(2x+1)),$$

where (x, z) is a solution of the equation $S_4(x) + S_4(1+x) = S_4(z)$. For example, $S_4(S_3(6)) + S_4(S_3(7)) = S_4(35)$, and $S_4(S_3(40)) + S_4(S_3(41)) = S_4(1189)$ (see [Sier64]).

It is known also, that *the sum of two squared square numbers is not a triangular number*:

$$(S_4(x))^2 + (S_4(y))^2 \neq S_3(z).$$

4.3.2. As a natural generalization of the above constructions, one can formulate, in the terms of figurate numbers, the *Fermat's last theorem*: *for any $n > 2$ the equality*

$$x^n + y^n = z^n$$

has no positive integer solutions (see, for example, [Ribe99], [Edva77]).

So we get, for $n = 3$, that *a sum of two cubic numbers can not be a cubic number*:

$$C(x) + C(y) \neq C(z).$$

Similarly, one obtains, for $n = 4$, that *a sum of two biquadratic numbers can not be a biquadratic number*:

$$BC(x) + BC(y) \neq BC(z).$$

In general, *for $k \geq 3$, a sum of two k-dimensional hypercube numbers can not be a k-dimensional hypercube number*:

$$C^k(x) + C^k(y) \neq C^k(z), \quad k \geq 3.$$

Euler, 1769, conjectured that for any integers $k, m > 1$, if $C^k(x_1) + \cdots + C^k(x_m) = C^k(z)$, then $m \geq k$. For $m = 2$, due to the Fermat's last theorem, it holds. But for $m = 3, 4$ counterexamples are known; $95800^4 + 217514^4 + 414560^4 = 422481^4$ (Frye, 1988), and $27^5 + 84^5 + 110^5 + 133^5 = 144^5$ (Lander-Parkin, 1966).

4.3.3. The square numbers are related to **Brown numbers**, which are pairs (m, n) of integers satisfying the conditions of the *Brocard's problem* (1876, [Broc76]) i.e., such that

$$n! + 1 = m^2.$$

Only three such pairs are known: $(4, 5), (5, 11), (7, 71)$. There are no other pairs with $n < 10^7$ ([Well86]). Erdös ([ErOb37]) conjectured, that there are no other solutions of the Brocard's problem.

However, there are many solutions of the Diophantine equation

$$n! + k^2 = m^2.$$

In fact, the least k, such that $n! + k^2$ is a square, are 1, 1, 3, 1, 9, 27, 15, 18, 288, 288, ... (Sloane's A038202) for $n = 4, 5, 6, 7, 8, 9, 10, 11, 12, 13, \ldots$, respectively (see [Weis11]).

Using two classical number-theoretical results, stating that every odd prime divisor of x^2+1 is of the form $4n+1$, and there are roughly as many primes $4n + 1$ as $4n + 3$, we can give a general observation, that a similar Diophantine equation

$$n! + 1 = x^8$$

has only a finite number of solutions.

In fact, $n! + 1 = x^8$ gives $n! = x^8 - 1 = (x^4 + 1)(x^2 + 1)(x^2 - 1)$; now, on the right side the contribution of primes $4k + 1$ and $4k - 1$ is about the same, while on the left side all the odd prime factors of $(x^4 + 1)(x^2 + 1)$ i.e., about $(n!)^{3/4}$ of the product, are of the form $4n + 1$ ([Moze09]).

4.3.4. It is proven (see [Weis11]), that *the only consecutive positive integers, one of which is a square number, and other is a cubic number, are* $8 = 2^3$ *and* $9 = 3^2$. This is the only known (except trivial 0 and 1) solution of the more general *Catalan's Diophantine problem* of finding two consecutive integers which are some powers, i.e., of solving the Diophantine equation

$$x^m - y^n = \pm 1$$

for positive integers x, y, n and m.

Catalan's conjecture (1844, [Cata44]) is, that there are no other solutions, i.e., *there are no other consecutive positive integers, one of which is an m-dimensional hypercube number, and other is an n-dimensional hypercube number.* This conjecture has been proved in 2002 by Mihăilescu.

4.3.5. One of the *Diophantus' problem* (see [Dick05]) consists in finding three numbers which, if multiplied in turn by their sum, give a triangular number, a square, and a cube.

A possible solution, with an additional requirement that the sum of the desired numbers is a square, is as follows. Let the sum be x^2. Then the desired numbers are $\frac{\alpha(\alpha+1)}{2x^2}$, $\frac{\beta^2}{x^2}$ and $\frac{\gamma^3}{x^2}$.Therefore, it holds $\frac{\alpha(\alpha+1)}{2}+\beta^2+\gamma^3=x^4$. Let $\beta=x^2-1$. Then $\frac{\alpha(\alpha+1)}{2}=2x^2-\gamma^3-1$. For $x=\frac{8\gamma^3+\delta^2+7}{8\delta}$ we get that $8\frac{a(a+1)}{2}+1=(2\alpha+1)^2=16x^2-8\gamma^3-7=(4x-\delta)^2$. Let $\gamma=2$, and $\delta=1$; then $x=9$, $\alpha=\frac{4x-\delta-1}{2}=14$, $\beta=x^2-1=80$, and the desired numbers are $\frac{153}{81}$, $\frac{6400}{81}$, and $\frac{8}{81}$.

On the other hand, if the sum of the desired numbers is x^2, then the simplest solution of the problem is given by the three numbers $\frac{x^2+1}{2}$, θ^2, and x, giving $\alpha=x^2$, $\beta=x\theta$, and $\gamma=x$. In fact, $\frac{x^2+1}{2}\cdot x^2=\frac{x^2(x^2+1)}{2}=S_3(x^2)$, $\theta^2\cdot x^2=(x\theta)^2=S_4(x\theta)$, and $x\cdot x^2=x^3=C(x)$. Since $\frac{1}{2}(x^2+1)+\theta^2+x=x^2$, then for $x=2y+1$ we get the Pell's equation $\theta^2-2y^2=-1$. Its solutions $(y,\theta)=(1,1),(5,7),(29,41),\dots$ produces the desired numbers $(5,1,3);(61,49,11);(1741,1681,59),\dots$.

However, the original problem did not require that the sum of the numbers be a square. In these general case let the sum of the three desired numbers in the Diophantus' problem be x, the obtained triangular number be Δ, the square be β^2 and the cube be γ^3. Then $\Delta+\beta^2+\gamma^3=x^2$. In other words, it holds that $\Delta+\beta^2+\gamma^3=(\beta+n)^2$, or $\beta=\frac{\Delta+\gamma^3-n^2}{2n}$, where $x=\beta+n$. Now we may assign any admissible values to Δ, γ^3, and n in order to find β and x. For example, for $n=1$, $\gamma=1$, and $\Delta=6$, we get $\beta=3$, and $x=4$. Hence, the desired numbers are $\frac{6}{4}=\frac{3}{2}$, $\frac{9}{4}$, and $\frac{1}{4}$. In fact, it holds $\frac{6}{4}+\frac{9}{4}+\frac{1}{4}=4$, $\frac{6}{4}\cdot 4=6=\Delta$, $\frac{9}{4}\cdot 4=3^2$, $\frac{1}{4}\cdot 4=1^3$.

4.3.6. Bachet de Méziriac (see [Dick05]) proposed the problem to find five numbers which, if multiplied in turn by their sum, give, a triangular number, a square, a cube, a pentagonal number, and a biquadrate.

A possible solution of this problem with an additional requirement that the sum of the desired numbers is a square, is thus. Let the sum of these five numbers be x^2, the square be $(x^2-1)^2$, the cube be 8, the pentagonal number be 5, and the biquadrate be 1. Since the sum of the corresponding triangular number Δ, square, cube, pentagonal number, and biquadrate will be x^4, we get that the triangular number Δ is $2x^2-15$. Thus, $8\Delta+1=16x^2-119$ is a square, say $(4x-1)^2$. Hence, $x=15$, and the desired numbers are $\frac{\Delta}{x^2}=\frac{535}{225}=\frac{107}{45}$, $\frac{(x^2-1)^2}{x^2}=\frac{224^2}{225}=\frac{50176}{225}$, $\frac{8}{225}$, $\frac{5}{225}=\frac{1}{45}$, and $\frac{1}{225}$.

However, the Bachet's generalization of the Diophantus' problem also did not require that the sum of the numbers be a square. If we let the sum of the desired numbers be x, the obtained triangular number be Δ, the square be β^2, the cube be γ^3, the pentagonal number be Υ, and the biquadrate be δ^4, then $\Delta+\beta^2+\gamma^3+\Upsilon+\delta^4=x^2$, and it holds that $\Delta+\beta^2+\gamma^3+\Upsilon+\delta^4=(\beta+n)^2$, or $\beta=\frac{\Delta+\gamma^3+\Upsilon+\delta^4-n^2}{2n}$, where $x=b+n$. Now we may assign any admissible values to Δ, γ^3, Υ, δ^4 and n, in order to find β and x.

4.3.7. In 1638, Sainte-Croix proposed to Descartes the problem: *trouver un trigone qui, plus un trigone tétragone, fasse un tétragone, et de rechef, et que de la somme des côtés des tétragones résulte le premier des trigones et de la multiplication d'elle par son milieu le second. J'ai donné 15 et 120. J'attends que quelqu'un y satisfasse par d'autres nombres ou qu'il montre que la chose est impossible.* The same problem, without the example, he proposed to Fermat in 1636, who did not solve it (see [Dick05]).

Descartes understood a trigone tétragone to be the square of a triangular number, and proved that $(15,120)$ is the only solution if the problem is understood as follows: *to require two triangular numbers such that, if either be added to the same $(S_3(n))^2$, the sum is a square; further, the sum of the square roots of these squares must equal the first required triangular number and must also be the first factor n used in forming the second triangular number.*

However, if one is permitted to add both $(S_3(n))^2$ and a new $(S_3(k))^2$ to the second required triangular number, the two triangular numbers may be taken to be 45 and 1035, since

$$45 + 6^2 = 9^2, \quad 1035 + 6^2 + 15^2 = 36^2,$$
$$36 + 9 = 45, \quad 45 \cdot \frac{46}{2} = 1035.$$

Sainte-Croix did not admit the validity of Descartes' solution, and, probably, meant a *trigone tétragone* to be a number both triangular and square, like 1, 36. The question would then be to find two numbers of the form $\frac{n(n+1)}{2}$ such that, if a number both triangular and square be added to each, there result two squares, such that the sum of the square roots of these squares is simultaneously the first required triangular number, and also the first factor n used in forming the second triangular number:

$$\frac{m(m+1)}{2} + x^2 = y^2; \quad \frac{n(n+1)}{2} + x^2 = z^2;$$
$$y + z = \frac{m(m+1)}{2}, \quad y + z = n,$$

where $x^2 = \frac{k(k+1)}{2}$. If, as it seems intended, the numbers to be added to the triangular numbers are to be identical, the only solution is 15, 120:

$$15 + 1 = 4^2, \ 120 + 1 = 11^2, \ 4 + 11 = 15, \quad \text{and}$$
$$\frac{(4+11) \cdot 16}{2} = \frac{15 \cdot 16}{2} = 120.$$

4.3.8. Square numbers appear in the *Ramanujan-Nagell equation*, which is an exponential Diophantine equation of the form

$$2^n - 7 = x^2.$$

The solutions of this equation in positive integers n and x exist just when $n = 3, 4, 5, 7$ and 15 (Sloane's A060728) with corresponding $x = 1, 3, 5, 11$, and 181 (Sloane's A038198). It was conjectured in 1913 by Ramanujan (see [Rama00]), and proved in 1948 by Nagell (see [Nage61]).

4.3.9. Exactly two cubic numbers have the form $S_4(n) + 4$, i.e., the Diophantine equation $x^3 - 4 = y^2$ has exactly two positive integer solutions: $4 = 2^3 - 4$, and $121 = 5^3 - 4$ ([LeLi83]).

4.3.10. The only triple of consecutive positive integers, whose cubes sum to a cube, is $(3, 4, 5)$:

$$3^3 + 4^3 + 5^3 = 6^3.$$

In other words, the Diophantine equation $x^3 + y^3 + z^3 = u^3$ has exactly one positive integer solution (see [Weis11]).

4.4 Perfect numbers

4.4.1. A positive integer is called **perfect number**, if it is the sum of its positive divisors excluding the number itself. The first perfect numbers are 6, 28, 496, 8128, 33550336, ... (Sloane's A000396). Perfect numbers were deemed to have important numerological properties by the ancients, and were extensively studied by the Greeks, including Euclid.

So, Euclid proved (Proposition IX.36 of *Elements*), that *any even positive integer of the form* $2^{k-1}(2^k - 1)$ *with prime* $2^k - 1$ *is perfect* (see [Weis11]).

In order to prove this fact, it is sufficient to show that for such n it holds $\sigma(n) = 2n$:

$$\sigma(n) = \sigma(2^{k-1}(2^k - 1)) = \frac{2^k - 1}{2 - 1} \cdot \frac{(2^k - 1)^2 - 1}{(2^k - 1) - 1}$$
$$= (2^k - 1)((2^k - 1) + 1) = (2^k - 1)2^k = 2n.$$

In 1849, Euler provided the first proof that *Euclid's construction gives all possible even perfect numbers* (see [Dick05]).

Let us consider this proof. Let n be an even perfect number, having the form $n = 2^\alpha m$, where m is odd. Since $gcd(2^\alpha, m) = 1$, it holds $\sigma(n) = \sigma(2^\alpha)\sigma(m)$. Since n is perfect, it holds $\sigma(n) = 2n$. Then $2n = \sigma(2^\alpha)\sigma(m)$, or $2 \cdot 2^\alpha m = (2^{\alpha+1} - 1)\sigma(m)$. So, we get $2^{\alpha+1}m = (2^{\alpha+1} - 1)\sigma(m)$. Since $gcd(2^{\alpha+1}, 2^{\alpha+1} - 1) = 1$, it holds that $2^{\alpha+1}$ divides $\sigma(m)$, and $2^{\alpha+1} - 1$ divides m. Hence, we have $\sigma(m) = 2^{\alpha+1}t$, and $m = (2^{\alpha+1} - 1)t$, $t \in \mathbb{N}$. If $t > 1$, we get that

$\sigma(m) \geq 1+t+m = 1+t+(2^{\alpha+1}-1)t = 2^{\alpha+1}t+1 > 2^{\alpha+1}t = \sigma(m)$, a contradiction. So, $t = 1$, and $m = 2^{\alpha+1} - 1$, while $\sigma(m) = 2^{\alpha+1}$. It means that $\sigma(m) = m + 1$, and $m = 2^{\alpha+1} - 1$ is a prime number.

However, it is not known if any odd perfect numbers exist, although numbers up to 10^{300} have been checked without success (see [Guy94a]).

4.4.2. It is easy to show that *all even perfect numbers are triangular.*

In fact, the Euclid-Euler's theorem implies, that any even perfect number has the form $2^{k-1}(2^k-1)$, where 2^k-1 is a *Mersenne prime.* Since $2^{k-1}(2^k-1) = \frac{(2^k-1)2^k}{2}$, it holds

$$2^{k-1}(2^k-1) = S_3(2^k-1),$$

i.e., *any even perfect number is a triangular number index of which is a Mersenne prime.*

The largest known perfect number is

$$2^{43112608}(2^{43112609}-1) = S_3(2^{43112609}-1).$$

It has 25956377 decimal digits and corresponds to the largest known (in fact, 47-known) Mersenne prime $2^{43112609} - 1$ having 12978189 decimal digits (see [PrPe11]).

4.4.3. Moreover, if an even perfect number $2^{k-1}(2^k-1)$ is greater than 6, then k is an odd number greater than 1. Hence, we have $2^k - 1 \equiv 2^{2m+1} - 1 = 2(4^m) - 1 \equiv 2 - 1 \equiv 1(mod\,3)$. Therefore, $2^k - 1 = 3n+1$. Noting that $S_3(3n+1) = 1+9S_3(n)$, we obtain that *any even perfect number, except 6, can be represented as*

$$2^{k-1}(2^k-1) = S_3(3n+1) = 1 + 9S_3(n).$$

On the other hand, if $k \geq 3$, then it holds $2^k - 1 \equiv -1(mod\,8)$, i.e., $3n+1 \equiv 7(mod\ 8)$, or $3n \equiv 6(mod\,8)$. Since the numbers 3 and 8 are relatively prime, one can divide the last congruence by 3 in order to get $n \equiv 2(mod\ 8)$. Therefore, $n = 8t + 2$, and we obtain, that *any even perfect number, except 6, can be represented as*

$$2^{k-1}(2^k-1) = S_3(24t+7) = 1 + 9S_3(8t+2).$$

In particular, we have

$$28 = S_3(7) = S_3(3 \cdot 2 + 1) = 1 + 9S_3(2), \quad 2 \equiv 2(mod\ 8);$$
$$496 = S_3(31) = S_3(3 \cdot 10 + 1) = 1 + 9S_3(10), \quad 10 \equiv 2(mod\ 8);$$
$$8128 = S_3(127) = S_3(3 \cdot 42 + 1) = 1 + 9S_3(42), \quad 42 \equiv 2(mod\ 8).$$

4.4.4. Since any even perfect number is a triangular number, then, like all triangular numbers, it is the sum of all natural numbers up to a certain point, in this case, up to $2^k - 1$. In particular,

$$6 = 2^1(2^1 - 1) = 1 + 2 + 3;$$
$$28 = 2^2(2^3 - 1) = 1 + 2 + 3 + 4 + 5 + 6 + 7;$$
$$496 = 2^4(2^5 - 1) = 1 + 2 + 3 + \cdots + 31;$$
$$8128 = 2^6(2^7 - 1) = 1 + 2 + 3 + \cdots + 127.$$

Moreover, any even perfect number, except 6, is the sum of the first $2^{(k-1)/2}$ odd perfect cubes. In particular,

$$28 = 1^3 + 3^3;$$
$$496 = 1^3 + 3^3 + 5^3 + 7^3;$$
$$8128 = 1^3 + 2^3 + 3^3 + \cdots + 15^3.$$

4.4.5. Any triangular number with odd index is hexagonal: $S_3(2n - 1) = S_6(n)$. Therefore, *any even perfect number is a hexagonal number with index a power of two*:

$$2^{k-1}(2^k - 1) = S_6(2^{k-1}).$$

4.4.6. On the other hand, *a perfect number can not be a square number, neither it can be cubic or biquadratic number.* In general, *a perfect number can not be an k-dimensional hypercube number for any dimension k, $k \geq 2$.*

In fact, a perfect number can not be a non-trivial integer power, since it contains the prime factor $2^k - 1$ only once.

4.4.7. The perfect numbers arise, in an unexpected way, in Pascal's triangle. If we consider Pascal's triangle modulo 2, replacing each odd element of the triangle by 1, and each even element by 0, we obtain the chain of central triangles, consisted only from 0.

It is easy to check that any 2^n-th row of Pascal's triangle for $n = 1, 2, 3, 4, \ldots$ is *divisible by 2*, i.e., all its inner entries are even and, hence, give 0 in the modulo 2 representation of the triangle. In turn, it implies that n-th central triangle in the above modulo 2 representation has $2^{n-1}(2^n - 1)$ zeros (see [Uspe76]).

In fact, it holds for $n = 1$: in the second row $1, 2, 1$ the only inner element 2 is even. So, two inner elements of the third row $1, 3, 3, 1$ are odd as sums of 1 and an even number 2. So, we obtain the first one-pointed central triangle.

By the same reason, all three inner elements of the fourth row are even as sums of two odd numbers. So, the 2^2-th row is divisible by 2. Now, two central inner elements of the 5-th row are even as sums of two even numbers, while two inner elements on the right and on the left are odd, as sums of 1 and an even number. The inner elements of the 6-th row have now alternate parity, with one central zero, and the 7-th row contains only odd numbers. So, the second central triangle contains $3 + 2 + 1 = 6$ zeros.

Furthermore, all inner elements of the 8-th row are even, and it is checked that the 2^3-th row is divisible by 2. Therefore, the 8-th row has 7 central zeros, the 9-th row has 6 central zeros, ..., the 14-th row has one central zero, while the 15-th row contains only odd numbers. So, the third central triangle contains $7 + 6 + \cdots + 1 = 28$ zeros.

Since all elements of the 15-th row are odd, it holds that 2^4-th row is divisible by 2, etc. In general, the $(2^n - 1)$-th row is odd, while the 2^n-th row is divisible by 2, having $2^n - 1$ inner zeros. Then next rows have $2^n - 2$, $2^n - 3, \ldots, 1$ central zeros, respectively, while the $(2^{n+1} - 1)$-th row is odd. So, we obtain, that n-th central triangle contains $(2^n - 1) + (2^n - 2) + \cdots + 1 = 2^{n-1}(2^n - 1)$ zeros.

As it was shown before, if the number $2^n - 1$ is prime, then the number $2^{n-1}(2^n - 1)$ is perfect, and all even perfect numbers can be obtained by this formula. Therefore, all even perfect numbers are represented in Pascal's triangle, considered modulo 2.

For a given prime number p, one can obtain a generalization of the above construction, considering Pascal's triangle modulo p and using the following property: *the k-th row of Pascal's triangle is divisible by p if and only if $k = p^n, n \in \mathbb{N}$.*

4.4.8. A positive integer n is called **almost perfect number** (or **least deficient number**, or **slightly defective number**) if $\sigma(n) = 2n - 1$.

The only known almost perfect numbers are the powers of 2, namely $1, 2, 4, 8, 16, 32, \ldots$ (Sloane's A000079). In the other words, *the only known almost perfect numbers are the second k-dimensional hypercube number for any dimension $k = 0, 1, 2, 3, \ldots$.*

4.5 Mersenne and Fermat numbers

4.5.1. A positive integer of the form $M_n = 2^n - 1$, $n \in \mathbb{N}$, is called **Mersenne number**. The first few Mersenne numbers are 1, 3, 7, 15, 31, 63, 127, 255, 511, 1023, ... (Sloane's A000225).

This class of numbers has a long and reach history, which go back to early study of perfect numbers. The main questions of the theory of Mersenne numbers are related the problem of their primality (see [Weis11], [Wiki11], [PrPe11], [Dick05]).

It is easy to see that in order for the Mersenne number $2^n - 1$ to be prime, n must be prime. This is true since for composite n it holds $n = ab$, $1 < a \le b < n$, and, hence, $2^n - 1 = 2^{ab} - 1 = (2^a - 1)(2^{(b-1)a} + 2^{(b-2)a} + \cdots + 1)$, where $1 < 2^a - 1 < 2^n - 1$. Therefore, the number $2^n - 1$ can be factored and is composite.

Therefore, for any *Mersenne prime* $M_p = 2^p - 1$ the exponent p itself must be a prime number. However, often the number $M_p = 2^p - 1$ is not prime even for a prime exponent p. For instance, the Mersenne number $M_{11} = 2^{11} - 1 = 2047 = 23 \cdot 89$ is not prime, even though 11 is a prime number. The lack of an obvious rule to determine whether a given Mersenne number is prime makes the search for Mersenne primes an interesting task leading to many records of prime numbers. For example, the greatest known prime is the Mersenne prime $2^{43112609} - 1$, obtained in 2008 and having about 13 millions (exactly, 12978189) decimal digits (see [PrPe11]).

A positive integer of the form $F_n = 2^{2^n} + 1$, $n = 0, 1, 2, 3, \dots$, is called **Fermat number**. The first few Fermat numbers are 3, 5, 17, 257, 65537, 4294967297, ... (Sloane's A000215).

Being a Fermat number is the necessary form a number $2^n + 1$ must have, in order to be prime. In fact, if $2^n + 1$ is a prime, then n can not have any odd factor $a > 1$: for $n = ab$ with odd $a > 1$ it holds $2^n+1 = 2^{ab}+1 = (2^b)^a+1 = (2^b+1)(2^{(a-1)b}-2^{(a-2)b}+2^{(a-3)b}-\cdots+1)$, where $1 < 2^a+1 < 2^n+1$, so 2^n+1 can be factored and is composite. Hence, for prime $2^n + 1$ the exponent n should be a power of 2.

The first five Fermat numbers $F_0 = 3$, $F_1 = 5$, $F_2 = 17$, $F_3 = 257$, and $F_4 = 65537$ are primes (Sloane's A019434). In 1650 Fermat conjectured that every Fermat number is prime. However, Euler (1732) proved, that already next Fermat number F_5 is composite. In fact,

$$F_5 = 4294967297 = 641 \cdot 6700417$$

is a product of two prime numbers. Moreover, at present only composite Fermat numbers are known for $n \geq 5$ (see [Ribe96a]).

Like Mersenne numbers, which are connected with the theory of perfect numbers, Fermat numbers also are related to an old and beautiful arithmetical problem of finding *constructible polygons*, i.e., regular polygons that can be constructed with ruler and compass. It was stated by Gauss (see [Gaus01], [Want36]) that in order for a m-gon to be circumscribed about a circle, i.e., be a constructible polygon, it must have a number of sides given by

$$m = 2^r \cdot p_1 \cdot \ldots \cdot p_k,$$

where $p_1, \dots, p_k$ are distinct Fermat primes. So, the regular m-gon is constructible with ruler and compass for $m = 1$, 2, 3, 4, 5, 6, 8, 10, 12, 15, ... (Sloane's A003401), while it is not constructible for $m = 7$, 9, 11, 13, 14, 18, 19, 21, 22, 23, ... (Sloane's A004169).

4.5.2. It is known that *a Mersenne number greater than 1 can not be a square number*:

$$M_n \neq k^2, \ k \in \mathbb{N}\backslash\{1\}.$$

Moreover, *any Mersenne number greater than 1 can not be a nontrivial integer power*:

$$M_n \neq k^s \quad \text{for an integer } k > 1 \quad \text{and a positive integer } s > 1.$$

In fact, if $2^n - 1 = k^s$, then k is odd. If s is even, then $k^s \equiv 1 (mod\, 8)$, and $2^n = k^s + 1 = 2(4t + 1)$, a contradiction. If s is odd, then $2^n = k^s + 1 = (k+1)T$, where T is odd, and, hence, is equal to 1. Then $2^n = k + 1$, i.e., $s = 1$, a contradiction.

Therefore, there are no **square Mersenne numbers**, **cubic Mersenne numbers**, and **biquadratic Mersenne numbers**, except the trivial 1. In general, there are no **k-dimensional hypercube Mersenne numbers**, greater than 1, for any dimension k, $k \geq 2$.

4.5.3. On the other hand, it is easy to check that there are **triangular Mersenne numbers**: $M_1 = S_3(1)$, $M_2 = S_3(2)$, $M_4 = S_3(4)$, and $M_{12} = S_3(90)$. However, it is proven that *the numbers* $1, 3, 15$ *and* 4095 *are the only triangular Mersenne numbers*, i.e.,

$$M_v = \frac{u(u+1)}{2} \text{ only for } v = 1, 2, 4, 12.$$

The problem of finding all triangular Mercenne numbers is equivalent to the problem of finding all positive integer solutions of the *Ramanujan-Nagell equation* $2^n - 7 = x^2$. The solutions of this equation in positive integers n and x exist just when n = 3, 4, 5, 7 and 15 (Sloane's A060728) with corresponding x = 1, 3, 5, 11, and 181 (Sloane's A038198), as was conjectured by Ramanujan ([Rama00]) and proved by Nagell ([Nage61]). In this case, if the number $M_v = 2^v - 1$ is triangular, then the values of v are just those of $n - 3$, so that the triangular Mersenne numbers are 1, 3, 15, 4095 (see Sloane's A076046) and no more.

The numbers 0, 1, 3, 15, 4095, which are the only non-negative integers which have simultaneously the form $\frac{u(u+1)}{2}$, $u = 0, 1, 2, \ldots$, and the form $2^v - 1$, $v = 0, 1, 2, \ldots$, are called **Ramanujan-Nagell numbers**.

4.5.4. Similarly, *a Fermat number can not be a square number*, i.e.,

$$F_n \neq k^2, \quad k \in \mathbb{N}.$$

In fact, if $2^m + 1 = k^2$, then k is odd, and $2^m = (k-1)(k+1)$. Hence, $2^t = k - 1$, and $2^{m-t} = k + 1$, where $t \in \mathbb{N}$, $t < m - t$. Then $2^{m-t} - 2^t = 2^t(2^{m-2t} - 1) = (k+1) - (k-1) = 2$, i.e., $t = 1$, $k = 3$, and $m = 3$. Therefore, the only square number of the form $2^m + 1$ is 9, which is not a Fermat number.

The same result can be obtained using modular arguments. In fact, any integer k has reminder $0, 1$ or 2 modulo 3, i.e., $k^2 \equiv 0(mod\,3)$, or $k^2 \equiv 1(mod\,3)$ for any integer k; but for any $n \geq 1$ the number $F_n \equiv 2^{2m} + 1 \equiv (-1)^{2m} + 1 \equiv 2(mod\,3)$, therefore, $F_n \neq k^2$. For $n = 0$ the number $F_0 = 3$ also is not a perfect square.

If we consider modulo 7, we can prove similarly, that *a Fermat number can not be a cubic number*, i.e.,

$$F_n \neq k^3, \ k \in \mathbb{N}.$$

Therefore, there are no **square Fermat numbers**, and **cubic Fermat numbers**.

4.5.5. Moreover, one can prove that *a Fermat number can not be a non-trivial integer power*:

$$F_n \neq k^s, \ k, s \in \mathbb{N}, \ s > 1.$$

In order to prove it, one should use the previous fact. Since $2^m + 1, m \neq 3$, is not a square, then in the equation $2^m+1 = k^s$ the number s should be odd. Then $2^m = k^s - 1 = (k-1)(k^{s-1}+k^{s-2}+\cdots+k+1)$, that impossible, as the second factor, as a sum of odd number of odd numbers, should be odd.

Therefore, there are no **biquadratic Fermat numbers**, and, in general, there are no k**-dimensional hypercube Fermat numbers** for any dimension k, $k \geq 2$.

4.5.6. It is easy to see that *the number* $F_0 = 3$ *is the only* **triangular Fermat number**, i.e.,

$$F_n \neq \frac{k(k+1)}{2}, \ n, k \in \mathbb{N}.$$

In fact, for $n \geq 2$, it holds $F_n = 2^{2^n} + 1 \equiv 4^{2^{n-1}} + 1 \equiv (-1)^{2^{n-1}} + 1 \equiv 1 + 1 \equiv 2(mod\,5)$. On the other hand, the number $\frac{k(k+1)}{2} \neq 2(mod\,5)$: for $k \equiv 0, 4(mod\,5)$, we get $\frac{k(k+1)}{2} \equiv 0(mod\,5)$, for $k \equiv 1, 3(mod\,5)$, we get $\frac{k(k+1)}{2} \equiv 1(mod\,5)$, and, for $k \equiv 2(mod\,5)$, we get $\frac{k(k+1)}{2} \equiv 3(mod\,5)$. For $n = 1$, the number $F_1 = 5$ also is not a triangular number (see [Sier64]).

4.5.7. Moreover, it is known (see [Luca01]), that *the number* $F_0 = 3$ *is the only* k*-dimensional hypertetrahedron number in any dimension* k, $k \geq 2$:

$$F_n \neq S_3^k(m) \quad \text{for any } k, m \in \mathbb{N} \quad \text{except} \quad k = 2, m = 1.$$

This result can be interpreted by saying that *the Fermat numbers sit in the Pascal's triangle only in the trivial way*:

$$\text{if } F_m = \binom{n}{k} \quad \text{for some } n \geq 2k \geq 2, \quad \text{then } k = 1.$$

On the other hand, there exist non-trivial solutions of $F_m(a) = \binom{n}{k}$, where $F_m(a) = a^{2^m} + 1$, $a > 1$, is a **generalized Fermat number**. This fact can be illustrated by the example

$$F_1(3) = \binom{5}{3}.$$

4.5.8. As was shown in Chapter 2, the Mersenne and Fermat number appear unexpectedly in the theory of *polite numbers.*

In fact, we have shown, that the only polite numbers, that may be non-trapezoidal, are the triangular numbers with only one non-trivial odd divisor. Thus, polite non-trapezoidal numbers must have the form of a power of two multiplied by a prime number.

There are exactly two types of triangular numbers of such form:

- the even perfect numbers $2^{k-1}(2^k - 1)$ formed by the product of a Mersenne prime $2^k - 1$ with half the nearest power of two;
- the products $2^{k-1}(2^k + 1)$ of a Fermat prime $2^k + 1 = 2^{2^n} + 1$ with half the nearest power of two.

Indeed, if we have equality $\frac{n(n+1)}{2} = 2^{k-1} \cdot p$ with odd prime number p, then $n(n+1) = 2^k \cdot p$, and, hence, $p|n(n+1)$. So, we get that either $p|n$, or $p|(n+1)$.

If $p|n$, then, by elementary arguments, $p = n$ and $2^k = n + 1$. Therefore, it holds $p = 2^k - 1$, i.e., p is a Mersenne prime, and the corresponding triangular number $2^{k-1}(2^k - 1)$ is an even perfect number.

If $p|(n + 1)$, then, by elementary arguments, $p = n + 1$ and $2^{t+1} = n$. Hence, it holds $p = 2^{t+1} + 1$, i.e., $p = 2^{2^n} + 1$ is a Fermat prime, and the corresponding triangular number $2^{k-1}(2^k+1)$ has the form $2^{2^n-1}(2^{2^n}+1)$.

4.5.9. The Mersenne and Fermat numbers have interesting connections with Pascal's triangle.

It is obviously, that *the sum of all entries of the first n rows of Pascal's triangle is equal to n-th Mersenne number M_n*. In other words, *n-th Mersenne number M_n can be represented as the sum of all k-dimensional hypertetrahedron numbers $S_3^k(i)$, $0 \le i \le n-1, 0 \le k \le i$.*

In fact, it was proven that the sum of all elements of the i-th row of Pascal's triangle is 2^i, $i = 0, 1, 2, 3, \ldots$. Then the sum of all elements of the first n rows of Pascal's triangle is equal to $2^0+2^1+\cdots+2^{n-1} = 2^n - 1$.

The Mersenne numbers arise also in the Sierpiński sieve, given by the representation of Pascal's triangle modulo 2. As it was shown before, the number of zeros in each central zero-triangle is equal to $2^{k-1}(2^k - 1) = 2^{k-1}M_k$ (find, on the picture below, first two such central triangles, containing $1 = 2^0 \cdot M_1$ and $6 = 2^1 \cdot M_2$ zeros, respectively).

On the other hand, the number of sides of constructible polygons with an odd number of sides are given by the first 32 rows in the above representation of the Pascal's triangle, interpreted as binary numbers, giving 1, 3, 5, 15, 17, 51, 85, 255, 257, 771, ... (Sloane's A004729).

$$\begin{array}{ccccccccccccccccl}
&&&&&&&1&&&&&&&&=&1\\
&&&&&&1&&1&&&&&&&=&3\\
&&&&&1&&0&&1&&&&&&=&5\\
&&&&1&&1&&1&&1&&&&&=&15\\
&&&1&&0&&0&&0&&1&&&&=&17\\
&&1&&1&&0&&0&&1&&1&&&=&51\\
&1&&0&&1&&0&&1&&0&&1&&=&85\\
1&&1&&1&&1&&1&&1&&1&&1&=&255\\
&&&&&&&\cdots&&&&&&&&&
\end{array}$$

In other words, *every n-th row, $n \le 32$, of Sierpiński sieve, represented by Pascal's triangle modulo 2, is product of Fermat primes, with terms given by binary counting* (see [Gard77], [CoGu96]).

4.6 Fibonacci and Lucas numbers

4.6.1. The **Fibonacci numbers** are defined as members of the following well-known recurrent sequence:

$$u_{n+2} = u_{n+1} + u_n, \; u_1 = u_2 = 1.$$

The first few Fibonacci numbers are 1, 1, 2, 3, 5, 8, 13, 21, 34, 55, ... (Sloane's A000045).

The **Lucas numbers** are defined by the same recurrent equation with different initial conditions:

$$L_{n+2} = L_{n+1} + L_n, \; L_1 = 1, L_2 = 3.$$

The first few Lucas numbers are 1, 3, 4, 7, 11, 18, 29, 47, 76, 123, ... (Sloane's A000032).

The *Binet's Fibonacci number formula* and its analogue for Lucas numbers have the form

$$u_n = \frac{\alpha^n - \beta^n}{\sqrt{5}}, \; L_n = \alpha^n + \beta^n, \quad \text{where } \alpha = \frac{1+\sqrt{5}}{2}, \; \beta = \frac{1-\sqrt{5}}{2}.$$

Often these numbers are considered for all integer values of argument. It is easy to see that $u_0 = 0$, and $L_0 = 2$, while, for negative values of indices, we get

$$u_{-n} = (-1)^{n-1} u_n, \; L_{-n} = (-1)^{n-1} L_n, \; n \in \mathbb{N}.$$

The Fibonacci and Lucas numbers have many interesting properties (see [Hogg69], [Voro61], [Weis11]). We list below some of these properties, needed for further consideration.

- $2u_{m+n} = u_m L_n + u_n L_m$; $2L_{m+n} = 5u_m u_n + L_m L_n$;
- $u_m^2 = u_{m-1}u_{m+1} + (-1)^{m+1}$; $L_m^2 = L_{2m} + 2(-1)^m$;
- $gcd(u_n, L_n) = 2$ if $3|n$; $gcd(u_n, L_n) = 1$ if $gcd(3, n) = 1$;
- $2|u_m \Leftrightarrow 3|m$; $2|L_m \Leftrightarrow 3|m$; $3|u_m \Leftrightarrow 4|m$; $3|L_m \Leftrightarrow m \equiv 2(mod 4)$;
- $L_n|L_m \Leftrightarrow n$ divides into m an odd number of times;
- $L_n|u_m \Leftrightarrow n$ divides into m an odd number of times;
- $L_{m+2k} \equiv -L_m(mod L_k)$; $u_{m+2k} \equiv -u_m(mod L_k)$;
- $L_{m+12} \equiv L_m(mod 8)$; $L_k \equiv 3(mod 4)$ if $2|k$ and $gcd(3, k) = 1$;

4.6.2. It is known ([Cohn64]), that *the numbers 1 and 4 are the only* **square Lucas numbers**, *i.e., square numbers which belong to the Lucas sequence.* Moreover, *the numbers 2 and 18 are the only doubled square Lucas numbers.*

Similarly, *the numbers 0, 1, and 144 are the only* **square Fibonacci numbers**, *i.e., square numbers which belong to the Fibonacci sequence,* while *the numbers 0, 2 and 8 are the only doubled square Fibonacci numbers.*

I. Let us show that $L_n = x^2$ only for $n = 1$ and 3. In fact, let $L_n = x^2$ for some integers n and x.

If n is even, the property $L_{2m} = L_m^2 + (-1)^{m-1} \cdot 2$ gives a representation $L_n = y^2 \pm 2$. Obviously, in this case $L_n \neq x^2$.

If $n \equiv 1(mod 4)$, then $L_1 = 1$, whereas if $n \neq 1$ we can write $n = 1+2\cdot 3^r \cdot k$, where k is an even integer not divisible by 3. In this case we can obtain, using the property $L_{m+2k} \equiv -L_m(mod\ L_k)$, that $L_n \equiv -L_1(mod\, L_{3^r \cdot k})$. Due to the form of k and the divisibility property of Lucas numbers it holds $L_k | L_{3^r \cdot k}$ and, hence, $L_n \equiv -L_1(mod\, L_k)$. So, $L_n \equiv -1(mod\ L_k)$. Due to the form of k it holds that $L_k \equiv 3(mod\, 4)$, and, hence, -1 is a non-residue modulo L_k. It implies that $L_n \neq x^2$.

Finally, if $n \equiv 3(mod\, 4)$, then $n = 3$ gives $L_3 = 2^2$, whereas if $n \neq 3$, we write as before $n = 3 + 2 \cdot 3^r \cdot k$ and obtain $L_n \equiv -L_3(mod\, L_k)$, or $L_n \equiv -4(mod\, L_k)$. Since -1 is a non-residue modulo L_k, it holds that -4 is also a non-residue modulo L_k. It implies that $L_n \neq x^2$.

II. Let us show that $L_n = 2x^2$ only for $n = 0$ and ± 6. In fact, let $L_n = 2 \cdot x^2$ for some integers n and x.

If n is odd and L_n is even, then $n \equiv \pm 3(mod\, 12)$, since $2 | L_n \Leftrightarrow 3 | n$. As $L_{m+12} \equiv L_m(mod\, 8)$, and $L_{-n} = (-1)^{n-1} L_n$, it holds $L_n \equiv 4(mod\, 8)$, and so, $L_n \neq 2 \cdot x^2$.

If $n \equiv 0(mod\, 4)$, then $n = 0$ gives $L_0 = 2$, whereas, if $n \neq 0$, we can write $n = 0 + 2 \cdot 3^r \cdot k$, and so, $2L_n \equiv -2L_0 = -4(mod\, L_k)$, whence $2L_n \neq y^2$, i.e., $L_n \neq 2 \cdot x^2$.

If $n \equiv 6(mod\, 8)$, then $n = 6$ gives $L_6 = 2 \cdot 3^2$, whereas, if $n \neq 6$, we can write $n = 6 + 2 \cdot 3^r \cdot k$, where now 4 divides into k, and 3 does not divide into k, and so, $2L_n \equiv -2L_6 = -36(mod\, L_k)$; hence, again, -36 is a non-residue modulo L_k. Thus, as before, $L_n \neq 2 \cdot x^2$.

Finally, if $n \equiv 2(mod\, 8)$, then $L_{-n} = L_n$, where now $-n \equiv 6(mod\, 8)$, and so, the only admissible value is $-n = 6$, i.e., $n = -6$.

III. Let us show that $u_n = x^2$ only for $n = 0, \pm 1, 2$, and 12. In fact, let $u_n = x^2$ for some integers n and x.

If $n \equiv 1 (mod\, 4)$, then $n = 1$ gives $u_1 = 1$, whereas if $n \neq 1$, we can write $n = 1 + 2 \cdot 3^r \cdot k$ with some even k not divisible by 3. Then the property $u_{m+2k} \equiv -u_m (mod\, L_k)$ and arguments, using in the above consideration, imply that $u_n \equiv -u_1 (mod\, L_k)$, i.e., $u_n \equiv -1 (mod\, L_k)$, whence $u_n \neq x^2$.

If $n \equiv 3 (mod\, 4)$, then $u_{-n} = u_n$, and $-n \equiv 1 (mod\, 4)$. As before, in this case we get only $n = -1$.

If n is even, then the property $2u_{m+n} = u_m L_n + u_n L_m$, taken for $m = n$, implies the identity $u_n = u_{\frac{1}{2}n} L_{\frac{1}{2}n}$.

If n is divisible by 3, then conditions $u_n = x^2$, and $gcd(u_{3m}, L_{3m}) = 2$ imply that $u_{\frac{1}{2}n} = 2 \cdot y^2$, and $L_{\frac{1}{2}n} = 2 \cdot z^2$. The latter is possible only for $n = 0, 6$, or -6. The first two values also satisfy the former, while the last must be rejected because it does not.

If 3 does not divide n, then conditions $u_n = x^2$, and $gcd(u_n, L_n) = 1$ imply that $u_{\frac{1}{2}n} = y^2$, $L_{\frac{1}{2}n} = z^2$. The latter is possible only for $n = 1$ or 3, and again the second value must be rejected.

IY. Finally, let us show that $u_n = 2x^2$ only for $n = 0, \pm 1, 2$, and 12. In fact, let $u_n = 2 \cdot x^2$ for some integers n and x.

If $n \equiv 3 (mod\, 4)$, then $n = 3$ gives $u_3 = 2$, whereas if $n \neq 3$, we can write $n = 3 + 2 \cdot 3^r \cdot k$. Therefore, it holds $2u_n \equiv -2u_3 (mod\, L_k)$, i.e., $2u_n \equiv -4 (mod\, L_k)$, and, hence, $u_n \neq 2 \cdot x^2$.

If $n \equiv 1 (mod 4)$, then, as before, $u_{-n} = u_n$, and we get only $n = -3$.

If n is even, then conditions $u_n = u_{\frac{1}{2}n} L_{\frac{1}{2}n}$, and $u_{\frac{1}{2}n} = 2 \cdot x^2$ imply that it holds either $u_{\frac{1}{2}n} = y^2$ and $L_{\frac{1}{2}n} = 2 \cdot z^2$, or $u_{\frac{1}{2}n} = 2 \cdot y^2$ and $L_{\frac{1}{2}n} = z^2$. In the first case, we see that the only value which satisfies both of these is $n = 0$. In the second case, the condition $L_{\frac{1}{2}n} = z^2$ is satisfied only for $n = 1$ or 3. But the former of these does not satisfy the first equation.

4.6.3. It was proven by Ming in 1989 ([Ming89]), that *the numbers 1, 3, 21, and 55 are the only* **triangular Fibonacci numbers**, i.e., triangular numbers with belong to the Fibonacci sequence (see Sloane's A039595).

In 1991 Ming ([Ming91]) also proved that *the numbers 1, 3, and 5778 form the complete set of* **triangular Lucas numbers**, i.e., triangular numbers with belong to the Lucas sequence.

4.6.4. McDaniel, 1998 ([McDa98], [McDa98a]) proved that *the only* **pronic Fibonacci numbers** *are* $u_0 = 1$ *and* $u_3 = 2$, *while the only* **pronic Lucas number** *is* $L_0 = 2$, rediscovering a result by Ming, 1995 ([Ming95]).

It was proven also ([Ming96]), that *the numbers 1 and 5 are the only* **pentagonal Fibonacci numbers**, *the number 1 is the only* **pentagonal Lucas number**, and *the only generalized pentagonal numbers in the Lucas sequence are* $L_0 = 2, L_1 = 1$ *and* $L_4 = 7$.

It was established by Rao ([Rao02], [Rao03]) that *the numbers 0, 1, 13, 34 and 55 are the only generalized* **heptagonal Fibonacci numbers**, while *the numbers 1, 4, 7, and 18 are the only generalized* **heptagonal Lucas numbers**.

4.6.5. It is well-known, that sums of the rising diagonals of the Pascal's triangle

$$\begin{array}{ccccc} 1 & & & & \\ 1 & 1 & & & \\ 1 & 2 & 1 & & \\ 1 & 3 & 3 & 1 & \\ 1 & 4 & 6 & 4 & 1 \\ \vdots & \vdots & \vdots & \vdots & \vdots \end{array}$$

give consecutive Fibonacci numbers: $1 = u_1,\ 1 = u_2,\ 1+1 = u_3,\ 1+2 = u_4,\ 1+3+1 = u_5,\ \ldots$. So, we can say, that *any Fibonacci number is a sum of figurate numbers*:

$$u_n = S_3^0(n) + S_3^1(n-2) + S_3^2(n-4) + S_3^3(n-6) + \cdots$$

In fact, due to the property $\binom{n+1}{k+1} = \binom{n}{k+1} + \binom{n}{k}$ of the binomial coefficients, it holds

$$S_3^0(n) + S_3^1(n-2) + S_3^2(n-4) + S_3^3(n-6) + \cdots = \binom{n-1}{0}$$

$$+\binom{n-2}{1} + \binom{n-3}{2} + \cdots = \binom{n-2}{0} + \left(\binom{n-3}{0}\right.$$

$$+\binom{n-3}{1}\Bigg)+\left(\binom{n-4}{1}+\binom{n-4}{2}\right)+\left(\binom{n-5}{2}\right.$$
$$\left.+\binom{n-5}{3}\right)+\cdots=\left(\binom{n-2}{0}+\binom{n-3}{1}+\binom{n-4}{2}+\cdots\right)$$
$$+\left(\binom{n-3}{0}+\binom{n-4}{1}+\binom{n-5}{2}+\cdots\right).$$

In other words, the sum $X_n = \binom{n-1}{0} + \binom{n-2}{1} + \binom{n-3}{2} + \cdots$ satisfies the same recurrent equation $X_n = X_{n-1} + X_{n-2}$ and the same initial conditions $X_1 = X_2 = 1$, as Fibonacci numbers. Therefore, $X_n = u_n$ for any $n = 1, 2, 3, \ldots$.

4.6.6. It is known ([Hons85]), that either $5S_4(x) + 4 = S_4(y)$, or $5S_4(x) - 4 = S_4(y)$ has a solution in positive integers if and only if, for some n, it holds $(x, y) = (u_n, L_n)$.

4.6.7. There exist many sequences, giving some interesting connections between figurate numbers and some relatives of Fibonacci numbers. For example, the Sloane's A086213 give the first **Tribonacci numbers** $u_{n+3} = u_{n+2} + u_{n+1} + u_n$ that start with first three cubes: 1, 8, 27, 36, 71, 134, 241, 446, 821, 1508, ...

4.7 Palindromic numbers

4.7.1. A **palindromic number** (or **Scheherazade number**[1]) in some base $g > 1$ is a number that is the same when written forwards or backwards, i.e., of the form

$$c_n \cdot g^n + c_{n-1}g^{n-1} + \cdots + c_{n-1}g + c_n, \ 0 \le c_i \le g-1, \ c_n \neq 0.$$

The term *palindromic* is derived from *palindrome*, which refers to a word that remains unchanged under reversal of its letters. The first palindromic numbers in decimal are 0, 1, 2, 3, 4, 5, 6, 7, 8, 9, 11, 22, 33, 44, 55, 66, 77, 88, 99, 101, ... (Sloane's A002113).

[1]Buckminster Fuller referred to palindromic numbers as Scheherazade numbers in his book *Synergetics*, because Scheherazade was the name of the story-telling wife in "1001 Nights"

Palindromic numbers receive most attention in the realm of recreational mathematics. A typical problem asks for numbers that possess a certain property and are palindromic (see [Weis11], [Gees11]). For instance, the **palindromic prime numbers** are 2, 3, 5, 7, 11, 101, 131, 151, 181, 191, ... (Sloane's A002385).

4.7.2. The **palindromic triangular numbers** are 1, 3, 6, 55, 66, 171, 595, 666, 3003, 5995, ... (Sloane's A003098). They correspond to the indices 1, 2, 3, 10, 11, 18, 34, 36, 77, 109, ... (Sloane's A008509). Choosing from these indices palindromic numbers, we get the sequence 1, 2, 3, 11, 77, 363, 1111, 2662, 111111, 246642,.... (Sloane's A008510) of the numbers n such that both n and n-th triangular number are palindromic. The number of palindromic triangular numbers with n digits is 3, 2, 3, 3, 2, 2, 6, 2, 1, 4, ... (Sloane's A054263) for $n = 1, 2, 3, 4, 5, 6, 7, 8, 9, 10, \ldots$, respectively.

The largest known palindromic triangular number corresponds to a 24-digital index and has 46 digits:

$$S_3(128184152897963669861552)$$
$$= 8215588527084263951180440811593624807258855128.$$

The largest known palindromic triangular number with a palindromic index is

$$S_3(3654345456545434563)$$
$$= 6677120357887130286820317887530217766.$$

4.7.3. The **palindromic pronic numbers** are 2, 6, 272, 6006, 289982, 2629262, 6039306, 27999972, 28233282, 2704884072, ... (Sloane's A028337). They correspond to the indices 1, 2, 16, 77, 538, 1621, 2457, 5291, 5313, 52008, ... (Sloane's A028336). The numbers n such that there are no palindromic pronic numbers of length n, are 2, 5, 9, 12, 18, 20, 30, 34 (Sloane's A034307).

The largest known palindromic pronic number corresponds to the 30-digital palindromic index 255455445544554455445544554552 and has 59 digits:

$$P(255455445544554455445544554552)$$
$$= 65257484658366826781688069846664896088618762866385648475256.$$

4.7.4. The **palindromic square numbers** are 1, 4, 9, 121, 484, 676, 10201, 12321, 14641, 40804, ... (Sloane's A002779). They correspond to the indices 1, 2, 3, 11, 22, 26, 101, 111, 121, 202, ... (Sloane's A002778). Choosing from these indices palindromic numbers, one gets the sequence 1, 2, 3, 11, 22, 101, 111, 121, 202, 212, ... (Sloane's A057135) of palindromes whose squares are palindromes.

There are no palindromic square n-digital numbers for $n = 2, 4$, 8, 10, 14, 18, 20, 24, 30, 38, 40 (Sloane's A034822).

Unlike palindromic triangular and palindromic pronic numbers, where it is impossible to predict a next higher one, whether its index is palindromic or not, with the palindromic squares we have an opposite situation. For example, finding large palindromic squares is easy if we start with 11 and square it, obtaining a palindromic square 121. Then we add a zero between the two unities and square the number 101, obtaining a palindromic square 10201. Adding next zero between unities gives the number 1001, such that 1001^2 is a palindromic square 1002001, etc.

n	$S_4(n)$
11	121
101	10201
1001	1002001
10001	100020001

In this way, we obtain an infinite sequence $10^{2n} + 2 \cdot 10^n + 1$ of palindromic square numbers, corresponding to the palindromic indices of the form $10^n + 1$. There are other indices with the similar property, for example, the sequence 111, 10101, 1001001, ... of the form $10^{2n} + 10^n + 1$, $n = 1, 2, 3, \ldots$, giving the pronic squares $10^{4n} + 2 \cdot 10^{3n} + 3 \cdot 10^{2n} + 2 \cdot 10^n + 1$.

The **repunit numbers** 1, 11, 111, 1111, 11111, 111111, 1111111, 11111111, 111111111, 1111111111, ... (Sloane's A002275), consisting of copies of the single digit 1, looks also a good candidate for finding palindromic squares but the expansion does not go on forever. Nine unities still produce a palindrome, but not ten unities: $1111111111^2 =$ 1234567900987654321 is not palindromic in decimal.

n	$S_4(n)$
1	1
11	121
111	12321
1111	1234321
11111	123454321
111111	12345654321
1111111	1234567654321
11111111	123456787654321
111111111	12345678987654321

However, the squares of the first nine repuints 1, 11, 111, ..., 111111111 correspond to the first nine **Demlo numbers**, which are defined as the initially palindromic numbers 1, 121, 12321, 1234321, 123454321, 12345654321, 1234567654321, 123456787654321, 12345678987654321 (see Sloane's A002477).

It is shown in the above table and follows immediately from schoolbook multiplication, but can be checked also algebraically:

$$\begin{aligned} D_n &= 1 + 2\cdot 10 + 3\cdot 10^2 + \cdots + 2\cdot 10^{n-1} + n\cdot 10^{n-1} + (n-1) \\ &\quad \cdot 10^n + \cdots + 2\cdot 10^{2n-3} + 1\cdot 10^{2n-2} \\ &= 1 + (1+1)\cdot 10 + (1+1+1)\cdot 10^2 + \cdots \\ &\quad + (1+1)\cdot 10^{2n-2} + 1\cdot 10^{2n-1} \\ &= (1 + 10 + 10^2 + \cdots + 10^{n-1}) + 10(1 + 10 + \cdots + 10^{n-1}) \\ &\quad + \cdots + 10^{n-1}(1 + 10 + 10^2 + \cdots + 10^{n-1}) \\ &= \sum_{k=0}^{n-1} 10^k \cdot R_n = R_n \sum_{k=0}^{n-1} = R_n \cdot R_n = R_n^2, \end{aligned}$$

where D_n is n-th Demlo number, and R_n is n-th repuint, $n \leq 9$.[2]

It is easy to see that the sums of digits of the Demlo numbers for $n \leq 9$ are equal to n^2:

$$\sum_{k=1}^{n} k + (k-1) = \sum_{k=1}^{n} 2k - 1 = n^2.$$

4.7.5. The **palindromic cubic numbers** are 1, 8, 343, 1331, 1030301, 1367631, 1003003001, 10662526601, 1000300030001, 1030607060301, ...(Sloane's A002781). They correspond to the indices 1, 2, 7, 11, 101, 111, 1001, 2201, 10001, 10101, ...(Sloane's A002780). Choosing from these indices palindromic numbers, we get the sequence 1, 2, 7, 11, 101, 111, 1001, 10001, 10101, ...(Sloane's A069748) of the numbers n such that n and n^3 are both palindromic.

[2]So, for $n \geq 10$ it is convenient to define D_n as R_n^2, to obtain 1234567900987 654321, 123456790120987654321, 12345679012320987654321, ...(see Sloane's A002477).

Kamenetsky (see [Sloa11]) conjectured that this sequence is the same as A002780, except for 2201.

Like palindromic squares, for a given palindromic cube finding a next higher palindromic cube number is very easy. For example, starting with index 11 and repeatedly adding a zero between the two unities, we get an infinite sequence $10^n + 1$ of palindromic numbers such that their cubes $10^{3n} + 3 \cdot 10^{2n} + 3 \cdot 10^n + 1$ are palindromic, $n = 1, 2, 3, \ldots$:

$$11^3 = 1331, \ 101^3 = 1030301, \ 1001^3 = 1003003001,$$
$$10001^3 = 1000300030001, \ldots .$$

The palindromic indices $111, 10101, 1001001, \ldots$ of the form $10^{2n} + 10^n + 1$, $n = 1, 2, 3, \ldots$, give an other infinite sequence of palindromic cubes, which has the form

$$10^{6n} + 3 \cdot 10^{5n} + 6 \cdot 10^{4n} + 7 \cdot 10^{3n} + 6 \cdot 10^{2n} + 3 \cdot 10^n + 1,$$

starting with 1367631, 1030607060301, 1003006007006003001, ...

The formula $1331 \cdot 10^{3n} + 3 \cdot 1331 \cdot 10^{2n} + 3 \cdot 1331 \cdot 10^n + 1331$, $n = 3, 4, 5, \ldots$, also gives infinite many palindromic cubes, starting with 1334996994331, 1331399339931331, 1331039930399301331, 1331003993003993001331, ...

Finally, the formula $10^{6n+3} + 33 \cdot 10^{5n+2} + 393 \cdot 10^{4n+1} + 1991 \cdot 10^{3n} + 393 \cdot 10^{2n} + 33 \cdot 10^{n+1}$ gives infinite many palindromic cubes for $n = 2, 3, 4, \ldots$: 1033394994933301, 1003303931991393033001, 1000330039301991039300330001, ...

4.7.6 The first five powers of 11 are palindromic numbers:

$$\begin{aligned} 11^0 &= 1, \\ 11^1 &= 11, \\ 11^2 &= 121, \\ 11^3 &= 1331, \\ 11^4 &= 14641. \end{aligned}$$

Moreover, these numbers correspond to the first five rows of the Pascal's Triangle, i.e., are obtained as concatenations of corresponding figurate numbers. However, the pattern does not continue after this.

In base 18, some powers of seven are palindromic:

$$7^0 = 1_{18};\quad 7^3 = 111_{18};\quad 7^4 = 777_{18};$$
$$7^6 = 12321_{18};\quad 7^9 = 1367631_{18}.$$

Hence, we get a palindromic representation in base 18 for 7-th k-dimensional hypercube number $C^k(7)$ for any dimension k, $k \in \{0, 3, 4, 6, 9\}$.

In base 24, the first eight powers of five are palindromic. More exactly, it holds

$$5^0 = 1_{24};\quad 5^1 = 5_{24};\quad 5^2 = 11_{24};\quad 5^3 = 55_{24};\quad 5^4 = 121_{24};$$
$$5^5 = 5A5_{24};\quad 5^6 = 1331_{24};\quad 5^7 = 5FF5_{24};\quad 5^8 = 14641_{24};$$
$$5^{10} = 15AA51_{24};\quad 5^{12} = 16FLF61_{24}.$$

So, we get a palindromic representation in the base 24 for 5-th hypercube number $C^k(5)$ for any dimension k, $0 \le k \le 8$.

4.7.7. The sequence of **palindromic k-dimensional hypertetrahedron numbers**, $k \ge 2$, starts from the elements 3, 6, 55, 66, 171, 252, 595, 666, 969, 1001, 1771, 2002, 3003, 3003, 3003, 5005, 5995, 8008, 8778, 15051, ... (Sloane's A051641).

4.8 Other special numbers

4.8.1. The **Catalan numbers** are the members of the following recurrent sequence:

$$C_n = C_1C_{n-1} + C_2C_{n-2} + \cdots + C_{n-1}C_1,\ C_1 = 1,\ C_2 = 1.$$

The first few Catalan numbers are 1, 1, 2, 5, 14, 42, 132, 429, 1430, 4862, ... (Sloane's A000108).

These numbers give the solution to the *Euler's polygon division problem: in how many ways can a regular n-gon be divided into triangles if different orientations are counted separately?*.

They can be obtained by the formula

$$C_{n+1} = \frac{1}{n+1}\binom{2n}{n},$$

i.e., are closely connected with the centered binomial coefficients $\binom{2n}{n}$, $n = 0, 1, 2, \ldots$. In other words, it holds

$$(n+1)C_{n+1} = S_3^n(n+1),$$

where $S_3^n(n+1)$ is $(n+1)$-th n-dimensional hypertetrahedron number.

On the other hand, it is easy to check the formula

$$C_{n+1} = \binom{2n}{n} - \binom{2n}{n-1},$$

which shows that the Catalan sequence can be obtained by substraction from any element of the main diagonal of the Pascal's triangle its closest left neighbor in the same row of the triangle. In other words, it holds

$$C_{n+1} = S_3^n(n+1) - S_3^{n-1}(n+2),$$

i.e., *any Catalan number is a difference between n-dimensional and* $(n-1)$*-dimensional hypertetrahedron numbers* (see [Gard88], [Weis11]).

4.8.2. A **Stirling number of the second kind** $S(n, m)$ is defined as the number of ways of partitioning a set of n elements into m non-empty sets, while a **Stirling number of the first kind** give the number of permutations of n elements which contain exactly m permutation cycles (see [GKP94], [CoGu96]). The first few Stirling numbers $S(n, m)$, $n \geq 1$, $1 \leq m \leq n$, are 1; 1, 1; 1, 3, 1; 1, 7, 6, 1; ... (Sloane's A008277).

The Stirling numbers of the second kind can be computed by the recurrent equation

$$S(n, m) = S(n-1, m-1) + mS(n-1, k),$$
$$S(n, 1) = S(n, n) = 1,$$

or from the sum

$$S(n, m) = \frac{1}{m!} \sum_{i=0}^{m} (-1)^i \binom{m}{i} (m-i)^n.$$

The special cases include $S(n, 2) = 2^n - 1$, $S(n, 3) = \frac{1}{6}(3^n - 3 \cdot 2^n + 3)$, and $S(n, n-1) = \binom{n}{2}$.

So, *any triangular number is a Stirling number of the second kind:*

$$S_3(n) = S(n+1, n).$$

Moreover, *every second k-dimensional hypercube number, $k \geq 1$, is exceeded a Stirling number of the second kind by unity:*

$$C^k(2) = S(k, 2) + 1.$$

Also (see [Sloa11]), *any four-dimensional* 5*-gonal pyramidal number* $S_5^4(n) = \frac{n(n+1)(n+2)(3n+1)}{24} = \frac{3n+1}{4} \cdot \binom{n+2}{3}$ *is a Stirling number of the second kind:*

$$S_5^4(n) = S(n+2, n).$$

It is known that $\sum_{m=0}^{k} S(k, m)x^{\underline{m}} = x^k$, where $x^{\underline{m}} = x \cdot (x-1) \cdot \ldots \cdot (x-m+1)$ is *falling factorial* . So, we get the following decomposition of n-th k-dimensional hypercube number $C^k(n)$:

$$C^k(n) = \sum_{m=0}^{k} S(k, m)x^{\underline{m}}.$$

4.8.3. The **Bell number** $B(n)$, $n \geq 0$, is defined as the number of all partitions of a set with n members, i.e., is a sum of corresponding Stirling numbers of the second kind: $B(n) = S(n, 0) + S(n, 1) + \cdots + S(n, n)$ (see [GKP94], [CoGu96], [Weis11]). Starting with $B(0) = B(1) = 1$, the first few Bell numbers are 1, 1, 2, 5, 15, 52, 203, 877, 4140, 21147, ... (Sloane's A000110).

The Bell numbers satisfy the recurrent equation

$$B(n+1) = \sum_{k=0}^{n} \binom{n}{k} B(k), \ B(0) = 1.$$

In other words, *$(n+1)$-th Bell number can be represented as the sum of the products of each previous Bell number on an k-dimensional hypertetrahedron number, $0 \leq k \leq n$:*

$$B_{n+1} = S_3^0(n+1)B(0) + S_3^1(n)B(1) + \cdots + S_3^k(n-k+1)B(k) + \cdots + S_3^n(1)B(n).$$

Moreover, the Bell numbers satisfy the well-known *Dobinski's formula:* $B(n) = \frac{1}{e}\sum_{i=0}^{\infty} \frac{i^n}{i!}$. Therefore, we get a decomposition of n-th Bell number in the infinite sum, including the entire sequence

$0^n, 1^n, 2^n, 3^n, 4^n, \dots$ of n-dimensional hypercube numbers $C^n(i)$, and the entire sequence $1, 1, 2, 6, 24, \dots$ of factorial numbers $i!$:

$$B(n) = \frac{1}{e}\left(\frac{C^n(0)}{0!} + \frac{C^n(1)}{1!} + \frac{C^n(2)}{2!} + \cdots + \frac{C^n(i)}{i!} + \cdots\right).$$

4.8.4. As k-dimensional hypercube numbers form the sequence of k-degrees of positive integers, then their partial sums give the sums of k-degrees: $C^k(1) + \cdots + C^k(n) = 1^k + \cdots + n^k$. In turn, these sums are closely connected with **Bernoulli numbers** $1, -\frac{1}{2}, 0, \frac{1}{6}, 0, -\frac{1}{30}, 0\ \frac{1}{42}, 0, -\frac{1}{30}, \dots$ (see Sloane's A000367 and A002445), which can be obtained from the following recurrent equation:

$$B_n = -\frac{1}{n+1}\sum_{k=1}^{n}\binom{n+1}{k+1}B_{n-k},\ B_0 = 0.$$

In fact, the sequence $B_0, B_1, B_2, \dots, B_n, \dots$ of Bernoulli numbers can be defined as the sequence of coefficients at n in the representation of the sum $1^k + 2^k + \cdots + (n-1)^k$, $k = 0, 1, 2, 3, \dots$, as a polynomial in n (see [CoGu96], [GKP94]). Thus,

$$\begin{aligned}
1 + 2 + \cdots + (n-1) &= \frac{n(n-1)}{2} \\
&= \frac{1}{2}n^2 - \frac{1}{2}n, \quad \text{giving } B_1 = -\frac{1}{2}; \\
1^2 + 2^2 + \cdots + (n-1)^2 &= \frac{(n-1)n(2n-1)}{6} \\
&= \frac{1}{3}n^3 - \frac{1}{2}n^2 + \frac{1}{6}n, \quad \text{giving } B_2 = \frac{1}{6}, \\
1^3 + 2^3 + \cdots + (n-1)n^3 &= \frac{n^2(n-1)^2}{4} = \frac{1}{4}n^4 \\
&\quad -\frac{1}{2}n^3 + \frac{1}{4}n^2, \quad \text{giving } B_3 = 0, \quad \text{etc.}
\end{aligned}$$

Moreover, all coefficients of such decomposition are connected with Bernoulli numbers:

$$1^k + 2^k + \cdots + (n-1)^k = \frac{1}{k+1}\sum_{i=0}^{k}\binom{k+1}{i}B_i n^{k+1-i}.$$

Hence, we obtain the following representation of the sum of the first $n-1$ k-dimensional hypercube numbers, including Bernoulli numbers, and i-dimensional hypertetrahedron numbers, $0 \le i \le k$:

$$C^k(1) + C^k(2) + \cdots + C^k(n-1)$$
$$= \frac{1}{k+1} \sum_{i=0}^{k} S_3^i(k-i+2) B_i C^{k+1-i}(n).$$

4.8.5. It is obvious that any **binomial number**, i.e., the number of the form $a^k \pm b^k$, where $a, b \in \mathbb{Z}$, and $n \in \mathbb{N}\backslash\{1\}$ (see [Weis11]), is a sum or a difference of two k-dimensional hypercube numbers:

$$a^k \pm b^k = C^k(a) \pm C^k(b).$$

4.8.6. Moreover, every k-dimensional hypercube number is between two **Cunningham numbers** $C^-(b,k) = b^k - 1$ and $C^+(b,k) = b^k + 1$ (see [Weis11]):

$$C^-(b,k) + 1 = C^k(b) = C^+(b,k) - 1.$$

4.8.7. Also, the number $C^k(k)$ plus 1 is a **Sierpiński number of the first kind**, i.e., a number of the form $k^k + 1$ (see [Weis11]). The first few such numbers are 2, 5, 28, 257, 3126, 46657, 823544, 16777217, 16777217, 387420490, ... (Sloane's A014566).

4.8.8. The number of *domino tilings* of n-th order Aztec diamond is $2^{S_3(n)}$, where $S_3(n)$ is n-th triangular number ([EKLP92]).

4.9 Prime numbers

4.9.1. It is easy to check that *a prime number p can not be a polygonal number, except of the second p-gonal number $S_p(2) = p$.*

In fact, n-th m-gonal number $S_m(n)$ can be written as $\frac{1}{2}(2+(n-1)(m-2))n$. If n is even, then $\frac{n}{2}$ is a positive integer, i.e., $S_m(n)$ is a product of two positive integers $(2+(n-1)(m-2))$ and $\frac{n}{2}$. If n is odd, then $n-1$ is even, and $\frac{2+(n-1)(m-2)}{2}$ is a positive integer, i.e., $S_m(n)$ is a product of two positive integers $\frac{2+(n-1)(m-2)}{2}$ and n. In the case $n=1$ this decomposition is trivial: $S_m(1) = 1 \cdot 1$. In the case $n=2$ this decomposition has the form $S_m(2) = m \cdot 1$. Hence, the number $S_m(2)$ is a prime only for prime m. For $n > 2$, both elements of the decomposition are greater than 1, and the number $S_m(n)$ is composite.

4.9.2. However, primes occur often in the sequences of centered polygonal numbers.

- A **centered triangular prime number** is a centered triangular number that is prime. In other words, it is a prime of the form $\frac{3n^2-3n+2}{2}$. The first such numbers are 19, 31, 109, 199, 409, ... (Sloane's A125602), corresponding to $n = 4, 5, 9, 12, 17, \ldots$.
- A **centered square prime number** is a centered square number that is prime. In other words, it is a prime of the form $n^2 + (n-1)^2$. The first such numbers are 5, 13, 41, 61, 113, ... (Sloane's A027862), corresponding to $n = 2, 3, 5, 6, 8, \ldots$.
- A **centered pentagonal prime number** is a centered pentagonal number that is prime. In other words, it is a prime of the form $\frac{5n^2-5n+2}{2}$. The first such numbers are 31, 181, 331, 391, 601, ... (Sloane's A145838), corresponding to $n = 4, 9, 12, 13, 16, \ldots$.
- A **centered hexagonal prime number** (or **hex prime number**) is a centered hexagonal number that is prime. The prime hex numbers are sometimes called **cuban prime numbers**, because they can be expressed as differences between successive cubic numbers, i.e., have the form $n^3 - (n-1)^3$. The first cuban primes are 7, 19, 37, 61, 127, ... (Sloane's A002407). They correspond to indices $n = 2, 3, 4, 5, 7, \ldots$ (Sloane's A002504). The numbers of cuban primes less than 1, 10, 100, 1000, 10000, ... are 0, 1, 4, 11, 28, ... (Sloane's A113478), which is well approximated by $\pi_c(x) = \ln x - 0.8$.
- A **centered heptagonal prime number** is a prime of the form $\frac{7n^2-7n+2}{2}$. The first such numbers are 43, 71, 197, 463, 547, ... (Sloane's A144974), corresponding to $n = 4, 5, 8, 12, 13, \ldots$.
- There are no **centered octagonal prime numbers**, since any centered octagonal number is a perfect square: $CS_8(n) = (2n-1)^2$.
- There are no **centered nonagonal prime numbers**, since it holds $CS_9(n) = \frac{(3n-2)(3n-1)}{2}$, i.e., any centered nonagonal number except 1 is a product of two positive integers greater than 1.

- The first **centered decagonal prime numbers**, i.e., primes of the form $5(n^2 - n) + 1$, are 11, 31, 61, 101, 151, ... (Sloane's A090562), corresponding to $n = 2, 3, 4, 5, 6, \ldots$.
- The first **centered hendecagonal prime numbers**, i.e., primes of the form $\frac{11n^2-11n+2}{2}$, are 67, 397, 727, 859, 1321, ..., corresponding to $n = 4, 9, 12, 13, 16, \ldots$.
- A **star prime number** is a star (i.e., centered 12-gonal) number that is prime. In other words, it is a prime of the form $6n(n - 1) + 1$. The first star primes are 13, 37, 73, 181, 337, ... (Sloane's A083577), corresponding to $n = 2, 3, 4, 6, 8, \ldots$.

4.9.3. There exist many geometrical constructions, related to figurate numbers and using prime numbers. (see [PrPe11]).

Thus, the **congruent prime numbers** correspond to "shapes" of identical digits nested about the center when drawn in the form of symmetrical polygons. The original prime can be constructed by concatenating each row left-to-right, top-to-bottom, as oriented in the drawing.

For example, 111181111 is a **square-congruent prime number**, since it can be represented in the form of a "symmetric" 3×3 square, as shown on the picture below. In fact, it is the smallest square-congruent prime number.

```
1 1 1
1 8 1
1 1 1
```

On the next picture two square-congruent primes of order 5 are given.

```
1 1 1 1 1        1 1 1 1 1
1 6 6 6 1        1 8 8 8 1
1 6 1 6 1        1 8 8 8 1
1 6 6 6 1        1 8 8 8 1
1 1 1 1 1        1 1 1 1 1
```

The number 1111118881188811888111111 on the right is the only square-congruent prime of order 5, which is **strobogrammatic number** (i.e., a positive integer, giving the same number when turned upside down), and **tetradic number** (i.e., a positive integer, giving the same number in four ways, whether viewed from right to left, left to right, top to bottom, or upside down).

The only square-congruent prime of prime order 7, that contains two prime digits, has the following form:

$$\begin{matrix}
7 & 7 & 7 & 7 & 7 & 7 & 7 \\
7 & 3 & 3 & 3 & 3 & 3 & 7 \\
7 & 3 & 7 & 7 & 7 & 3 & 7 \\
7 & 3 & 7 & 3 & 7 & 3 & 7 \\
7 & 3 & 7 & 7 & 7 & 3 & 7 \\
7 & 3 & 3 & 3 & 3 & 3 & 7 \\
7 & 7 & 7 & 7 & 7 & 7 & 7
\end{matrix}$$

The next picture gives the smallest near-repdigit square-congruent prime of order 9, and the smallest square-congruent prime of order 9, containing all four prime digits 2, 3, 5, 7.

$$\begin{matrix}
1 & 1 & 1 & 1 & 1 & 1 & 1 & 1 & 1 \\
1 & 1 & 1 & 1 & 1 & 1 & 1 & 1 & 1 \\
1 & 1 & 1 & 1 & 1 & 1 & 1 & 1 & 1 \\
1 & 1 & 1 & 1 & 1 & 1 & 1 & 1 & 1 \\
1 & 1 & 1 & 1 & 6 & 1 & 1 & 1 & 1 \\
1 & 1 & 1 & 1 & 1 & 1 & 1 & 1 & 1 \\
1 & 1 & 1 & 1 & 1 & 1 & 1 & 1 & 1 \\
1 & 1 & 1 & 1 & 1 & 1 & 1 & 1 & 1 \\
1 & 1 & 1 & 1 & 1 & 1 & 1 & 1 & 1
\end{matrix}
\qquad
\begin{matrix}
3 & 3 & 3 & 3 & 3 & 3 & 3 & 3 & 3 \\
3 & 2 & 2 & 2 & 2 & 2 & 2 & 2 & 3 \\
3 & 2 & 5 & 5 & 5 & 5 & 5 & 2 & 3 \\
3 & 2 & 5 & 7 & 7 & 7 & 5 & 2 & 3 \\
3 & 2 & 5 & 7 & 7 & 7 & 5 & 2 & 3 \\
3 & 2 & 5 & 7 & 7 & 7 & 5 & 2 & 3 \\
3 & 2 & 5 & 5 & 5 & 5 & 5 & 2 & 3 \\
3 & 2 & 2 & 2 & 2 & 2 & 2 & 2 & 3 \\
3 & 3 & 3 & 3 & 3 & 3 & 3 & 3 & 3
\end{matrix}$$

A square-congruent prime of order 13, which is a **triadic number** (i.e., a positive integer that is invariant upon reflection only along the line it is written on), is given below.

$$\begin{matrix}
1 & 1 & 1 & 1 & 1 & 1 & 1 & 1 & 1 & 1 & 1 & 1 & 1 \\
1 & 1 & 1 & 1 & 1 & 1 & 1 & 1 & 1 & 1 & 1 & 1 & 1 \\
1 & 1 & 8 & 8 & 8 & 8 & 8 & 8 & 8 & 8 & 8 & 1 & 1 \\
1 & 1 & 8 & 3 & 3 & 3 & 3 & 3 & 3 & 3 & 8 & 1 & 1 \\
1 & 1 & 8 & 3 & 8 & 8 & 8 & 8 & 8 & 3 & 8 & 1 & 1 \\
1 & 1 & 8 & 3 & 8 & 1 & 1 & 1 & 8 & 3 & 8 & 1 & 1 \\
1 & 1 & 8 & 3 & 8 & 1 & 1 & 1 & 8 & 3 & 8 & 1 & 1 \\
1 & 1 & 8 & 3 & 8 & 1 & 1 & 1 & 8 & 3 & 8 & 1 & 1 \\
1 & 1 & 8 & 3 & 8 & 8 & 8 & 8 & 8 & 3 & 8 & 1 & 1 \\
1 & 1 & 8 & 3 & 3 & 3 & 3 & 3 & 3 & 3 & 8 & 1 & 1 \\
1 & 1 & 8 & 8 & 8 & 8 & 8 & 8 & 8 & 8 & 8 & 1 & 1 \\
1 & 1 & 1 & 1 & 1 & 1 & 1 & 1 & 1 & 1 & 1 & 1 & 1 \\
1 & 1 & 1 & 1 & 1 & 1 & 1 & 1 & 1 & 1 & 1 & 1 & 1
\end{matrix}$$

The smallest *titanic*, i.e., with 1000 or more decimal digits, square-congruent prime is

```
1 1 1 1 1 1 1 1 1 1 1 1 1 1 1 1 1 1 1 1 1 1 1 1 1 1 1 1 1 1 1 1 1
1 0 0 0 0 0 0 0 0 0 0 0 0 0 0 0 0 0 0 0 0 0 0 0 0 0 0 0 0 0 0 0 1
1 0 0 0 0 0 0 0 0 0 0 0 0 0 0 0 0 0 0 0 0 0 0 0 0 0 0 0 0 0 0 0 1
1 0 0 0 0 0 0 0 0 0 0 0 0 0 0 0 0 0 0 0 0 0 0 0 0 0 0 0 0 0 0 0 1
1 0 0 0 0 0 0 0 0 0 0 0 0 0 0 0 0 0 0 0 0 0 0 0 0 0 0 0 0 0 0 0 1
1 0 0 0 0 0 0 0 0 0 0 0 0 0 0 0 0 0 0 0 0 0 0 0 0 0 0 0 0 0 0 0 1
1 0 0 0 0 0 0 0 0 0 0 0 0 0 0 0 0 0 0 0 0 0 0 0 0 0 0 0 0 0 0 0 1
1 0 0 0 0 0 0 0 0 0 0 0 0 0 0 0 0 0 0 0 0 0 0 0 0 0 0 0 0 0 0 0 1
1 0 0 0 0 0 0 0 0 0 0 0 0 0 0 0 0 0 0 0 0 0 0 0 0 0 0 0 0 0 0 0 1
1 0 0 0 0 0 0 0 0 0 0 0 0 0 0 0 0 0 0 0 0 0 0 0 0 0 0 0 0 0 0 0 1
1 0 0 0 0 0 0 0 0 0 0 0 0 0 0 0 0 0 0 0 0 0 0 0 0 0 0 0 0 0 0 0 1
1 0 0 0 0 0 0 0 0 0 0 0 0 0 0 0 0 0 0 0 0 0 0 0 0 0 0 0 0 0 0 0 1
1 0 0 0 0 0 0 0 0 0 0 0 0 0 0 0 0 0 0 0 0 0 0 0 0 0 0 0 0 0 0 0 1
1 0 0 0 0 0 0 0 0 0 0 0 0 0 0 0 0 0 0 0 0 0 0 0 0 0 0 0 0 0 0 0 1
1 0 0 0 0 0 0 0 0 0 0 0 0 0 7 7 7 7 7 0 0 0 0 0 0 0 0 0 0 0 0 0 1
1 0 0 0 0 0 0 0 0 0 0 0 0 0 7 1 1 1 7 0 0 0 0 0 0 0 0 0 0 0 0 0 1
1 0 0 0 0 0 0 0 0 0 0 0 0 0 7 1 5 1 7 0 0 0 0 0 0 0 0 0 0 0 0 0 1
1 0 0 0 0 0 0 0 0 0 0 0 0 0 7 1 1 1 7 0 0 0 0 0 0 0 0 0 0 0 0 0 1
1 0 0 0 0 0 0 0 0 0 0 0 0 0 7 7 7 7 7 0 0 0 0 0 0 0 0 0 0 0 0 0 1
1 0 0 0 0 0 0 0 0 0 0 0 0 0 0 0 0 0 0 0 0 0 0 0 0 0 0 0 0 0 0 0 1
1 0 0 0 0 0 0 0 0 0 0 0 0 0 0 0 0 0 0 0 0 0 0 0 0 0 0 0 0 0 0 0 1
1 0 0 0 0 0 0 0 0 0 0 0 0 0 0 0 0 0 0 0 0 0 0 0 0 0 0 0 0 0 0 0 1
1 0 0 0 0 0 0 0 0 0 0 0 0 0 0 0 0 0 0 0 0 0 0 0 0 0 0 0 0 0 0 0 1
1 0 0 0 0 0 0 0 0 0 0 0 0 0 0 0 0 0 0 0 0 0 0 0 0 0 0 0 0 0 0 0 1
1 0 0 0 0 0 0 0 0 0 0 0 0 0 0 0 0 0 0 0 0 0 0 0 0 0 0 0 0 0 0 0 1
1 0 0 0 0 0 0 0 0 0 0 0 0 0 0 0 0 0 0 0 0 0 0 0 0 0 0 0 0 0 0 0 1
1 0 0 0 0 0 0 0 0 0 0 0 0 0 0 0 0 0 0 0 0 0 0 0 0 0 0 0 0 0 0 0 1
1 0 0 0 0 0 0 0 0 0 0 0 0 0 0 0 0 0 0 0 0 0 0 0 0 0 0 0 0 0 0 0 1
1 0 0 0 0 0 0 0 0 0 0 0 0 0 0 0 0 0 0 0 0 0 0 0 0 0 0 0 0 0 0 0 1
1 0 0 0 0 0 0 0 0 0 0 0 0 0 0 0 0 0 0 0 0 0 0 0 0 0 0 0 0 0 0 0 1
1 0 0 0 0 0 0 0 0 0 0 0 0 0 0 0 0 0 0 0 0 0 0 0 0 0 0 0 0 0 0 0 1
1 0 0 0 0 0 0 0 0 0 0 0 0 0 0 0 0 0 0 0 0 0 0 0 0 0 0 0 0 0 0 0 1
1 1 1 1 1 1 1 1 1 1 1 1 1 1 1 1 1 1 1 1 1 1 1 1 1 1 1 1 1 1 1 1 1
```

The smallest **triangle-congruent prime number** is 1111211111, while the smallest triangle-congruent prime with only prime digits is 3333323333. Moreover, 9999959999 is the only triangle-congruent prime containing one odd prime digit in the center. Their triangle-representations are given below.

```
      1            3   3   3   3        9   9   9   9
    1   1            3   2   3            9   5   9
  1   2   1            3   3                9   9
1   1   1   1            3                    9
```

Next picture gives a triangle-congruent prime consisting of only prime digits.

```
7   7   7   7   7   7   7   7   7   7
  7   3   3   3   3   3   3   3   7
    7   3   5   5   5   5   3   7
      7   3   5   2   5   3   7
        7   3   5   5   3   7
          7   3   5   3   7
            7   3   3   7
              7   3   7
                7   7
                  7
```

The smallest **centered hexagonal-congruent prime number** of order 3 has the following form:

```
    1   1   1
  1   0   0   1
1   0   1   0   1
  1   0   0   1
    1   1   1
```

The smallest centered hexagonal-congruent prime that contains all the prime digits is given below.

```
        3   3   3   3   3
      3   3   3   3   3   3
    3   3   2   2   2   3   3
  3   3   2   5   5   2   3   3
3   3   2   5   7   5   2   3   3
  3   3   2   5   5   2   3   3
    3   3   2   2   2   3   3
      3   3   3   3   3   3
        3   3   3   3   3
```

4.9.4. Many hidden connections with primes numbers there exist in Pascal's triangle.

For example, the smallest prime formed by concatenating of several consecutive rows of Pascal's triangle is 111121, while the number

12940636542375111875547502015607804292145100150052003001034597290518
95935678639157755876077558760678639155189593534597
290200300101001500542921451560780475020118755237513654406291

is a prime formed by concatenating all the numbers of the 30-th row of Pascal's triangle, i.e., by concatenating a sequence of k-dimensional hypertetrahedron numbers.

Furthermore, the number 18285670562881 is the only known **emirp**, i.e., a prime that gives a different prime when its digits are reversed, to be formed by concatenating a row of Pascal's triangle.

The number 13311464115101051 is the smallest emirp formed by concatenating several consecutive rows of Pascal's triangle.

4.9.5. There exist interesting number sequences, supplying different, often amusing, connections between prime and figurate numbers.

Thus, triangular numbers with prime indices, $S_3(p), p \in P$, are represented in the sequence 3, 10, 21, 55, 78, 136, 171, 253, 406, ... (see Sloane's A008837).

Squares of primes, i.e., the numbers $S_4(p), p \in P$, form the sequence 4, 9, 25, 49, 121, 169, 289, 361, 529, 841, ... (Sloane's A001248).

Cubes of primes, i.e., the numbers $C(p), p \in P$, form the sequence 8, 27, 125, 343, 1331, 2197, 4913, 6859, 12167, 24389, ... (Sloane's A030078).

Prime powers of prime numbers, i.e., the numbers $C^q(p)$, $p, q \in P$, form the sequence 4, 8, 9, 25, 27, 32, 49, 121, 125, 128, ... (Sloane's A053810), while distinct-digital prime powers of prime numbers are 4, 8, 9, 25, 27, 32, 49, 125, 128, 169, 243, ... (Sloane's A076702).

Prime numbers, for which the sum of the digits is a triangular number, form the sequence 3, 19, 37, 73, 109, 127, 163, 181, 271, 307, ... (Sloane's A117674).

Prime numbers, for which the product of the digits is a triangular number, are 3, 11, 13, 23, 31, 37, 47, 53, 59, 61, ... (Sloane's A117682).

The sequence of prime numbers with every digit a perfect square, i.e., consisting of only digits 0, 1, 4, 9, starts with the elements 11, 19, 41, 101, 109, 149, 191, 199, 401, 409, ... (Sloane's A061246).

4.10 Magic constructions

4.10.1. A *magic square* (see, for example, [Andr60]) is a square array of numbers consisting of the distinct positive integers $1, 2, \ldots, S_4(n) = n^2$, arranged so that the sum of the numbers in any horizontal, vertical, or main diagonal line is always the same number, known as the *magic constant* $M_2(n)$.

16	3	2	13
5	10	11	8
9	6	7	12
4	15	14	1

For instance, above magic square of 4-th order has the magic constant $M_2(4) = 34$.

For $n = 0, 1, 2, 3, 4, 5, 6, 7, 8, 9, \ldots$ the values of magic constants are given by 0, 1, 5, 15, 34, 65, 111, 175, 260, 369, ... (Sloane's A006003).

By definition, we get $M_2(n) = \frac{1}{n}\sum_{k=1}^{n^2} k = \frac{n(n^2+1)}{2}$. In other words, *$n$ copies of the magic constant $M_2(n)$ give a triangular number with square index*:

$$nM_2(n) = S_3(S_4(n)).$$

Also, *any magic constant is the sum of integers between two triangular numbers*:

$$M_2(n) = (S_3(n-1)+1) + (S_3(n-1)+2)\ldots + S_3(n).$$

Indeed, it holds $(S_3(n-1)+1) + \ldots + S_3(n) = \left(\frac{(n-1)n}{2}+1\right) + \left(\frac{(n-1)n}{2}+2\right) + \ldots + \left(\frac{(n-1)n}{2}+n\right) = \frac{(n-1)n^2}{2} + \frac{n(n+1)}{2} = \frac{n(n^2+1)}{2} = M_2(n)$.

Moreover, *one can obtain the magic constant $M_2(n)$ by adding the first n centered triangular numbers*:

$$M_2(n) = CS_3(1) + CS_3(2) + \cdots + CS_3(n).$$

In fact, we have $\sum_{k=1}^{n} \frac{3k^2-3k+2}{2} = \frac{1}{2}(3\sum_{k=1}^{n} k^2 - 3\sum_{k=1}^{n} k + 2\sum_{k=1}^{n} 1) = \frac{n(n+1)(2n+1)}{4} - \frac{3n(n+1)}{4} + n = \frac{1}{2}(n(n+1)(n-1)+2n) = \frac{1}{2}(n^3+n)$.

4.10.2. The above magic square is the *Dürer's magic square*, used in an engraving *Melancholia* by Dürer (The British Museum). Dürer's magic square is located in the upper right-hand corner of the engraving. The numbers 15 and 14 appear in the middle of the bottom row, indicating the date of the engraving, 1514.

Dürer's magic square has additional property that the sums in any of the four quadrants, as well as the sum of the middle four numbers, are all 34. It is thus a *gnomon magic square* (i.e., a magic square of 4-th order, in which the elements in each corner have the same sum). In addition, any pair of numbers symmetrically placed about the center of the square sums to 17, a property making the square even more "magical".

4.10.3. A *magic cube* (see, for example, [Andr60]) is a version of a magic square in which n^2 rows, n^2 columns, n^2 pillars, and four space diagonals each sum to a single number $M_3(n)$ called the *cube's magic constant.*

If, in addition, the diagonals of each $n \times n$ orthogonal slice sum to $M_3(n)$, then the magic cube is called *perfect.* Magic cubes are most commonly assumed to be *normal*, i.e., to have elements that are the consecutive integers 1, 2, ..., $C(n) = n^3$.

If it exists, a normal magic cube has magic constant $M_3(n) = \frac{1}{n^2}\sum_{k=1}^{n^3} k = \frac{n(n^3+1)}{2}$. For $n = 0, 1, 2, 3, 4, 5, 6, 7, 8, 9, \dots$, the cube's magic constants are given by 0, 1, 9, 42, 130, 315, 651, 1204, 2052, 3285, ... (Sloane's A027441).

So, one can see, that n^2 *copies of the cube's magic constant* $M_3(n)$ *give a triangular number with cubic index*:

$$n^2 M_3(n) = S_3(n^3), \quad \text{or} \quad S_4(n)M_3(n) = S_3(C(n)).$$

There is a trivial perfect magic cube of 1-st order, but no perfect cubes exist for orders 2, 3, and 4. While normal perfect magic cubes of orders 7 and 9 have been known since the late 1800s, it was long not known if perfect magic cubes of orders 5 or 6 could exist. An $5 \times 5 \times 5$ perfect magic cube was subsequently discovered by Boyer and Trump in 2003 (see [Weis11]).

4.10.4. Centered hexagonal numbers appear in the construction of the *magic hexagon*, i.e., a hexagon, constructed from the hexagonal cells, containing the numbers $1, 2, \ldots, CS_6(n) = 3n(n-1)+1$, such that the sums of numbers along every straight line coincide. The only such hexagon, except of trivial case of 1-point hexagon, is given below.

15
14 13
9 8 10
6 4
11 5 12
1 2
18 7 16
17 19
3

In the above magic hexagon of 3-rd order, each line (of lengths 3, 4, and 5) adds up to 38.

This problem and its solution have a long history. Above magic hexagon was discovered independently by Haselberg in 1887, Radcliffe in 1895, Kuhl in 1940, etc. Adams worked on the problem from 1910 to 1957 by trial and error and after many years arrived at the solution which he transmitted to Gardner. Gardner sent Adams' magic hexagon to Trigg, who by mathematical analysis found that it was unique disregarding rotations and reflections. Later Trigg ([Trig64]) did further research and summarized known results as well as the history of the problem.

In order to prove, that no magic hexagons exist except those of order 1 and 3, one should note that the the sum of all entries of a magic hexagon of order n is a triangular number: the numbers in the hexagon are consecutive, and run from 1 to $3n^2 - 3n + 1$; hence, their sum is equal to $S_3(3n^2 - 3n + 1)$. Furthermore, there are $2n - 1$ rows in a magic hexagon of order n, running along any given direction. Each of these rows sum up to the same number $M(n)$. So, it holds

$$M(n) = \frac{S_3(3n^2 - 3n + 1)}{2n - 1} = \frac{(3n^2 - 3n + 1)(3n^2 - 3n)}{2(2n - 1)}$$
$$= \frac{9n^4 - 18n^3 + 18n^2 - 9n + 2}{2(2n - 1)}.$$

Rewriting this as $32M(n) = 72n^3 - 108n^2 + 90n - 27 + \frac{5}{2n-1}$ shows that $\frac{5}{2n-1}$ must be an integer. The only $n \geq 1$, that meet this condition, are $n = 1$ and $n = 3$ (see [Weis11], [Wiki11]).

Although there are no normal magical hexagons of order greater than 3, certain abnormal ones exist. In this case, abnormal means starting the sequence of numbers other than with 1. Hexagons of such kind are known for any order $n, 4 \leq n \leq 7$. An order 4 hexagon starts with 3 and ends with 39, its rows summing to 111. An order 5 hexagon starts with 6 and ends with 66, its row summing to 244. An order 6 hexagon starts with 21, ends with 111, and its sum is 546. An order 7 hexagon starts with 2, ends with 128 and its sum is 635. However, a slightly larger, order 8 magic hexagon can be obtained, if we start with -84 and ends with 84; the sum of such hexagon is equal to 0 (see [Wiki11]).

4.10.5. If a magic hexagon is constructed with triangles instead of hexagons, as in classical case, it is called *T-hexagon*. An example of T-hexagon of order 2 is given on the picture below.

```
          5       11
      7       20      22
      19      9       8
  21      2       6       10
  13      4       1       24
      14      3       16
      15      23      18
          12      7
```

Such magic constructions have much more properties than the hexagons made by hexagons. Thus, an T-hexagon of order n has $6n^2$ triangles. Therefore, the sum of all its entries $1, 2, \ldots, 6n^2$ is a triangular number, namely, $S_3(6n^2)$. If we try to construct a magic T-hexagon of order n, we have to choose n to be even, because there are $2n$ rows; so, the sum in each row must be $\frac{3n^2(6n^2+1)}{2n}$. To date, magic T-hexagons of order 2, 4, 6 and 8 have been discovered (see [Wiki11]).

4.10.6. An (n, k)-*talisman hexagon* is an arrangement of nested hexagons containing the integers 1, 2, ..., $CS_6(n) = 3n(n-1) + 1$, such that the difference between all adjacent hexagons is at least as large as a number k.

```
         16
     2        11
17       7         6
    12        15
3        1        10
     8         5
18       19       14
     4         9
         13
```

The hexagon illustrated above is an $(3, 4)$-talisman hexagon (see [Mada79]).

4.10.7. The **tetractys** (or *tetraktys*) is an equilateral triangular figure consisting of ten points arranged in four rows: one, two, three, and four points in each row. It represents geometrically the 4-th triangular number: $10 = 1 + 2 + 3 + 4$. This simple figure was the main metaphysical symbol in Pythagoreanism, where it embraced within itself in seedlike quantized form the principles/harmony of the natural world, the cosmos, and the divine. In this secret religion, initiates were required to swear a secret oath by the tetractys.

The four rows of the tetractys were seen as the point, line, triangle and tetrahedron and were loaded with somehow related general notions. The tetractus was used then in Judaic Kabbalah and later, in Christian Cabbala and Hermetic Qabalah. Neoplatonist's interpretation of the tetractus was used in tarot's reading as follows. *Monad*, the point in the 1-st row, represents the the Premise of the reading. *Dyad* of the 2-nd row represents the Light and the Darkness. *Triad* of the 3-nd row represents the spiritual forces: Destroyer, Sustainer, and Creator. *Tetrad* of the 4-th row represents the material world: Earth, Water, Air and Fire (see, for example, [Wiki11]).

The tetrahedron with basic length 4 (summing up to 20) can be seen as a three-dimensional analogue of the tetractys.

4.10.8. The **beast number** 666 is triangular. In fact, $666 = S_3(36)$. More exactly, 666 is a triangular number with the index 36, which is a square with the index 6, while 6 is a pronic number whose index 2 is also a pronic number (see [Dick05]).

From other point of view, the number 666 is formed from the triangular number 6 repeated 3 (the only prime triangular number) times.

Moreover, this is the largest **repdigit number**, i.e., a positive integer composed of repetition of a single digit, which is a triangular number ([BaWe76]). The others are 1, 3, 6, 55 and 66 (see Sloane's A045914).

Furthermore, it is easy to check, that 666 is the smallest triangular number of the form n^2+m^2 such that n, m and $n+m$ are triangular numbers:

$$666 = 15^2 + 21^2,\ 15 = S_3(5),\ 21 = S_3(6),\ 15 + 21 = 36 = S_3(8).$$

This equation can be written also with all the single-digital primes:

$$666 = (3 \cdot 5)^2 + (3 \cdot 7)^2.$$

Well-known are two factorial numbers, related to 666: they are **Leviathan number** $(10^{666})!$, having $\approx 6,656 \times 10^{668}$ decimal digits, and the **baby Leviathan number** 666!, having 1594 decimal digits.

The **Legion's numbers** of the first and second kind are 666^{666} and $666!^{666!}$. The first number has 1181 decimal digits; the second has $\approx 1.61 \times 10^{1596}$ digits and ends with $165 \times 666!$ trailing zeros (see [PrPe11]).

4.11 Unrestricted partitions

The generalized pentagonal numbers 0, 1, 2, 5, 7, 12, 15, 22, 26, ... (Sloane's A001318), obtained as values of the formula $S_5(n) = \frac{n(3n-1)}{2}$ for $n = 0, \pm 1, \pm 2, \pm 3, \pm 4, \ldots$, arise unexpectedly in the theory of partitions.

4.11.1. Let $p(n)$ be *the number of unrestricted partitions of* n, i.e., the number of ways of writing the positive integer n as a sum of positive integers, where the order of summands is not considered significant. For example, $p(4) = 5$, since 4 can be written as

$$4 = 4,\ 4 = 3 + 1,\ 4 = 2 + 2,\ 4 = 2 + 1 + 1,\ 4 = 1 + 1 + 1 + 1.$$

The values of $p(n)$ for $n = 0, 1, 2, 3, 4, 5, 6, 7, 8, 9, \ldots$ are 1, 1, 2, 3, 5, 7, 11, 15, 22, 30, ... (Sloane's A000041). The values of $p(10^n)$

for $n = 0, 1, 2, 3, \ldots$ are given by

$$1, 42, 190569292, 2406146786403262247369214 9$$
$$727991, \ldots \quad \text{(Sloane's A070177)}.$$

4.11.2. It is known (see, for example, [Andr98]), that the generating function for $p(n)$ has the form $\prod_{n=1}^{\infty}(1-x^n)^{-1}$, i.e., it holds

$$\sum_{n=0}^{\infty} p(n)x^n = \prod_{n=1}^{\infty}(1-x^n)^{-1}, \quad \text{with} \quad p(0) = 1.$$

A formal derivation of this identity, ignoring questions of convergence, is very simple. Thus, each factor $(1-x^n)^{-1}$, viewing as a sum of infinite geometric progression, can be represent in the form $\frac{1}{1-x^n} = 1 + x^n + x^{2n} + x^{4n} + \ldots$, and, hence,

$$\prod_{n=1}^{\infty}(1-x^n)^{-1} = (1 + x + x^2 + \ldots) \cdot (1 + x^2 + x^4 + \ldots)$$
$$\cdot (1 + x^3 + x^6 + \ldots) \cdot \ldots .$$

Multiplication of the above series gives a power series of the form $1 + \sum_{k=1}^{\infty} a(k)x^k$, where x^k is obtained as a product of an element x^{k_1}, $k_1 \geq 0$, from the first series, an element x^{2k_2}, $k_2 \geq 0$, from the second series, x^{3k_3}, $k_3 \geq 0$, from the third series, $\ldots$, x^{mk_m}, $k_m \geq 0$, from the m-th series, and so on. In other words, $x^k = x^{k_1} \cdot x^{2k_2} \cdot x^{3k_3} \cdot \ldots \cdot x^{mk_m} \cdot \ldots$, or

$$k = 1 \cdot k_1 + 2 \cdot k_2 + 3 \cdot k_3 + \cdots + m \cdot k_m + \ldots,$$
$$k_i \geq 0, \ i = 1, 2, 3, \ldots .$$

This decomposition corresponds to an unrestricted partition of the number k, which contains k_1 unities, k_2 twos, k_3 threes, etc. So, each term x^k comes from an unrestricted partition of k, and, conversely, any unrestricted partition of k produces one such term x^k. Hence, the coefficient $a(k)$ of x^k is equal to the number $p(k)$ of unrestricted partitions of k. Therefore,

$$\frac{1}{(1-x) \cdot (1-x^2) \cdot (1-x^3) \cdot \ldots \cdot (1-x^m) \cdot \ldots}$$
$$= p(0) + p(1)x + p(2)x^2 + p(3)x^3 + \cdots + p(n)x^n + \ldots .$$

4.11.3. On the other hand, Euler's *pentagonal number theorem* (1783, [Eule83]) states that

$$\prod_{n=1}^{\infty}(1-x^n)=\sum_{k=-\infty}^{+\infty}(-1)^k x^{\frac{k(3k-1)}{2}},$$

or, in other words,

$$(1-x)\cdot(1-x^2)\cdot(1-x^3)\cdot\cdots\cdot(1-x^m)\cdot\cdots$$
$$=1-x-x^2+x^5+x^7-x^{12}-x^{15}+x^{22}+x^{26}-\cdots,$$

where the exponents $\frac{k(3k-1)}{2}$, $k=0,\pm1,\pm2,\pm3,\pm4,\ldots$ in the right-hand part are generalized pentagonal numbers 0, 1, 2, 5, 7, 12, 15, 22, 26, ...

We are going to consider a combinatorial proof of this fact, given by Franklin in 1881 ([Fran81]).

At first, we have that

$$\prod_{n=1}^{\infty}(1-x^n)=(1-x)\cdot(1-x^2)\cdot(1-x^3)\cdot\ldots\cdot(1-x^n)\cdot\ldots$$
$$=1+\sum_{k=1}^{\infty}a(k)x^k,$$

where each term x^k on the right is obtained as a product of distinct factors, chosen from the set $1,x,x^2,x^3,\ldots$, i.e., the exponent k is represented as a sum of distinct positive integers. It corresponds to *a partition of* k *into unequal parts.* Conversely, any partition of k into unequal parts produces a term x^k with coefficient ±1: the coefficient is $+1$, if x^k is a product of an even number of terms, and is -1, otherwise. Therefore, $a(k)=p_e(k)-p_o(k)$, where $p_e(k)$ is the number of *partitions of* k *into an even number of unequal parts*, and $p_o(k)$ is the number of *partitions of* k *into an odd number of unequal parts.*

Now we will use graphical representations of partitions in order to show that there is a bijection between partitions of k into an odd and an even number of unequal parts, so that $p_o(k)=p_e(k)$, except when k is a generalized pentagonal number.

Consider a graphical representation of any partition k into unequal parts, such that the parts are arranged in decreasing order, as shown in the picture below, corresponding to the partition $24 = 7+6+5+4+2$.

Call the line segment connecting points in the last (i.e., shortest) row the *base*, and denote the number of points in the base by n. Call the 45^0 line segment joining the last point in the first row with all possible points in the graph the *slope*, and denote the number of points in the slope by s. On our picture the base with $n = 2$ is written by $\star$, while the slope with $s = 4$ is written by $\bullet$.

Now we define two operations on the above graph. The first operation moves the points on the base so they lie parallel to the slope; the second one moves the points on the slope so that they lie on the line parallel to the base, as shown in the picture below.

The operation is called *admissible*, if it gives the graphical representation of a new partition of k into unequal parts. For the first operation the number of parts in this new partition is one less than before; for the second operation the number of parts in the new partition is one greater than before. Hence, if for every partition of k into unequal parts exactly one from these two operations is admissible, then there exists a bijection between partitions of k into odd and even unequal parts, so $p_e(k) = p_o(k)$ for such k.

Consider three possible cases: $n < s$, $n = s$, and $n > s$.

If $n < s$, then $n \leq s - 1$ and the first operation is admissible, while the second is not (see the above picture).

If $n = s$, then the second operation is not admissible, while the first one is admissible except the case when the base and the slope intersect (see the picture below).

If $n > s$, then the first operation is not admissible, while the second one is admissible except the case when $n = s + 1$ and the base and the slope intersect (see the picture below).

$$\begin{array}{lllllll} * & * & * & * & * & * & \bullet \\ * & * & * & * & * & \bullet & \\ * & * & * & * & \bullet & & \\ \bullet & \bullet & \bullet & \bullet & & & \end{array} \qquad n = s. \qquad \begin{array}{llllllll} * & * & * & * & * & * & * & \bullet \\ * & * & * & * & * & * & \bullet & \\ * & * & * & * & * & \bullet & & \\ \bullet & \bullet & \bullet & \bullet & \bullet & & & \end{array} \qquad n = s + 1.$$

Consider two obtained exceptional cases, supposing that there are l rows in the graph. For the first case $n = s$, but $s = l$, therefore, $n = l$, and the number k is given by

$$k = l + (l + 1) + \cdots + (2l - 1) = \frac{3l^2 - l}{2} = S_5(l).$$

For such partition of k we have an extra partition into even parts if l is even, and an extra partition into odd parts if l is odd, so, $p_e(k) - p_o(k) = (-1)^l$.

For the second case $n = s + 1$ there is an additional point in each row, so,

$$k = \frac{3l^2 - l}{2} + k = \frac{3k^2 + k}{2} = S_5(-l),$$

and again, $p_e(k) - p_o(k) = (-1)^l$. This concluded the proof.

4.11.4. Two above results imply that

$$\sum_{n=0}^{\infty} p(n)x^n = \frac{1}{\sum_{k=-\infty}^{+\infty}(-1)^k x^{\frac{k(3k-1)}{2}}}.$$

In other words, we get

$$(1 + p(1)x + p(2)x^2 + p(3)x^3 + \ldots) \cdot (x^{S_5(0)} - x^{S_5(1)} - x^{S_5(-1)} \\ + x^{S_5(2)} + x^{S_5(-2)} - x^{S_5(3)} - x^{S_5(-3)} + \ldots) = 1.$$

Since the values of $p(n)$ for $n = 1, 2, 3, 4, 5, 6, 7, 8, 9, 10, \ldots,$ are $1, 2, 3, 5, 7, 11, 15, 22, 30, 42, \ldots,$ and the values of generalized pentagonal numbers for $k = 0, \pm 1, \pm 2, \pm 3, \pm 4, \ldots$ are 0, 1, 2, 5, 7,

12, 15, 22, 26, ..., one obtains the following beautiful identity:

$$(1 + x + 2x^2 + 3x^3 + 5x^4 + 7x^5 + \dots)$$
$$\cdot(1 - x - x^2 + x^5 + x^7 - x^{12} + x^{15} + \dots) = 1.$$

The right multiplication gives now the corresponding recurrent formula for $p(n)$, involving generalized pentagonal numbers:

$$p(n) = \sum_{k=1}^{n} (-1)^{k+1} \left(p\left(n - \frac{k(3k-1)}{2}\right) + p\left(n - \frac{k(3k+1)}{2}\right) \right).$$

In other words, we have

$$p(n) = p(n-1) + p(n-2) - p(n-5) - p(n-7) + \cdots$$
$$+ (-1)^{n+1} p(n - S_5(n)) + (-1)^{n+1} p(n - S_5(-n)).$$

4.12 Waring's problem

4.12.1. In the terms of square, cubic, biquadratic etc. numbers one can easy formulate the classical results about representation of a given positive integer as a sum of some powers of integer numbers, in particular, the well-known results in the *Waring's problem* (see [Dick05], [Weis11], [Wiki11], [Sloa11], [Kara83], [Sier64], [CoGu96]).

The *Waring's problem*, proposed in 1770 by Waring in his *Meditationes algebraicae* ([Wari70]), asks whether for every natural number n there exists an associated positive integer $g(n)$ such that every positive integer N is a sum of at most $g(n)$ n-th powers of positive integers:

$$N = x_1^n + x_2^n + \cdots + x_{g(n)}^n.$$

The affirmative answer, known as the *Hilbert-Waring's theorem*, was provided by Hilbert in 1909 (see [RaTo57]).

Hardy and Littlewood (see [HaLi25]) introduced more fundamental function $G(n)$, which is defined to be the least positive integer, such that every sufficiently large integer (i.e., every integer greater than some constant) can be represented as a sum of at most $G(n)$ n-th powers of positive integers:

$$N = x_1^n + x_2^n + \cdots + x_{G(n)}^n \quad \text{for any } N \geq N_0.$$

Since $k^n = C^n(k)$, one can formulate general results in the Waring's problem in terms of n-dimensional hypercube numbers. So, *every positive integer number N can be written as the sum of at most $g(n)$ n-dimensional hypercube numbers*:

$$N = C^n(x_1) + C^n(x_2) + \cdots + C^n(x_{g(n)}).$$

Moreover, *every sufficiently large positive integer can be written as the sum of at most $G(n)$ n-dimensional hypercube numbers*:

$$N = C^n(x_1) + C^n(x_2) + \cdots + C^n(x_{G(n)}) \quad \text{for any } N \geq N_0.$$

4.12.2. In the simplest case $n = 2$, the *Lagrange's four-square theorem* (1770, [Lagr70]) states that *any positive integer can be written as the sum of at most 4 perfect squares.* Since three squares are not enough, this theorem establishes $g(2) = 4$.

In the other words, *every positive integer N can be represented as a sum of at most four square numbers* (exactly four square numbers, if we add to the list $S_4(0) = 0$):

$$N = S_4(m) + S_4(n) + S_4(k) + S_4(l).$$

It implies, that the Lagrange's four-square theorem is a special case of the *Fermat's polygonal number theorem* (see Chapter 5).

Numbers expressible as a sum of three squares are those not of the form $4^k(8m+7)$ for $k, m \geq 0$. So, any positive integer $N \neq 4^k(8m+7)$, $k, m \geq 0$, can be represented as a sum of three square numbers (1798, [Lege30]).

As for sums of two squares, it is known, that a a positive integer can be represented as a sum of two squares precisely if its prime factorization contains no odd powers of primes of the form $4k + 3$ (see, for example, [Sier64]). This is equivalent the requirement that all odd factors of the square-free part n' of n are equal to 1 modulo 4.

Since there are infinitely many positive integer numbers $N = 4^k(8m+7)$, $k, m \geq 0$, which require four squares to represent them, the least integer $G(2)$ such that every sufficiently large positive integer requires $G(2)$ squares is given by $G(2) = 4$.

The following table gives the first few numbers which require $t = 1, 2, 3$, and 4 squares to represent them as a sum.

t		Sloane
1	1, 4, 9, 16, 25, 36, 49, 64, 81, 100, ...	A000290
2	2, 5, 8, 10, 13, 17, 18, 20, 26, 29, ...	A000415
3	3, 6, 11, 12, 14, 19, 21, 22, 24, 27, ...	A000419
4	7, 15, 23, 28, 31, 39, 47, 55, 60, 63, ...	A004215

As a part of the study of Waring's problem, it is known also that *every integer is a sum of at most 3 signed squares*: $eg(2) = 3$ (see [Weis11]).

Furthermore, *all numbers greater than* 188 *can be expressed as the sum of at most five distinct squares*. In fact, there are only 31 numbers that can not be expressed as a sum of distinct squares: 2, 3, 6, 7, 8, 11, 12, 15, 18, 19, 22, 23, 24, 27, 28, 31, 32, 33, 43, 44, 47, 48, 60, 67, 72, 76, 92, 96, 108, 112, 128 (Sloane's A001422). Only two numbers require six distinct squares: $124 = 1+4+9+25+36+49$, and $188 = 1+4+9+25+49+100$. Moreover, all known numbers less than 10^5, which can not be represented using fewer than five distinct squares, are 55, 88, 103, 132, 172, 176, 192, 240, 268, 288, 304, 368, 384, 432, 448, 496, 512, and 752, together with all numbers obtained by multiplying these numbers by a power of 4 (see [Weis11]).

The following table gives the numbers that can be represented in exactly w different ways as a sum of s squares. For example, $50 = 1^2 + 7^2 = 5^2 + 5^2$ can be represented in two ways ($w = 2$) by two squares ($s = 2$).

s	w		Sloane
1	0	2, 3, 5, 6, 7, 8, 10, 11, 12, 13, ...	A000037
1	1	1, 4, 9, 16, 25, 36, 49, 64, 81, 100, ...	A000290
2	1	2, 5, 8, 10, 13, 17, 18, 20, 25, 26, ...	A025284
2	2	50, 65, 85, 125, 130, 145, 170, 185, 200, 205, ...	A025285
2	3	325, 425, 650, 725, 845, 850, 925, 1025, 1250, 1300, ...	A025286
2	4	1105, 1625, 1885, 2125, 2210, 2405, 2465, 2665, 3145, 3250, ...	A025287
2	5	8125, 8450, 10625, 14450, 16250, 18125, 21250, 23125, 25625, 32500, ...	A025287
3	1	3, 6, 9, 11, 12, 14, 17, 18, 19, 21, ...	A025321
3	2	27, 33, 38, 41, 51, 57, 59, 62, 69, 74, ...	A025322
3	3	54, 66, 81, 86, 89, 99, 101, 110, 114, 126, ...	A025323
3	4	129, 134, 146, 153, 161, 171, 189, 198, 201, 234, ...	A025324
4	1	4, 7, 10, 12, 13, 15, 16, 18, 19, 20, ...	A025357
4	2	31, 34, 36, 37, 39, 43, 45, 47, 49, 50, ...	A025358
4	3	28, 42, 55, 60, 66, 67, 73, 75, 78, 85, ...	A025359
4	4	52, 58, 63, 70, 76, 84, 87, 91, 93, 97, ...	A025360

4.12.3. In the case $n = 3$, Pollock (1850, see [Dick05]) conjectured that every positive integer number is the sum of at most nine perfect cubes. It was proved by Wieferich ([Wief09]) and Kempner ([Kemp12]) in the early XX-th century, giving the solution $g(3) = 9$ of the Waring's problem.

In 1939, Dickson proved that the only integers requiring nine positive cubes are

$$23 = 2^3 + 2^3 + 1^3 + 1^3 + 1^3 + 1^3 + 1^3 + 1^3 + 1^3, \quad \text{and } 239.$$

Wieferich proved that only 15 integers require eight cubes: 15, 22, 50, 114, 167, 175, 186, 212, 231, 238, 303, 364, 420, 428, and 454 (Sloane's A018889). Hence, any positive integer greater than 454 can be represented as a sum of at most 7 cubes, i.e., every sufficiently large positive integer is a sum of no more than 7 positive cubes. The quantity $G(3)$ in Waring's problem therefore satisfies the inequality $G(3) \leq 7$, and the largest number known requiring seven cubes is 8042. However, it is not known if 7 can be reduced (see [Weis11], [Well86]).

The following table gives the first few numbers which require at least $t = 1, 2, 3, \ldots, 9$ positive cubes to represent them as a sum.

t		Sloane
1	1, 8, 27, 64, 125, 216, 343, 512, 729, 1000, ...	A000578
2	2, 9, 16, 28, 35, 54, 65, 72, 91, 126, 128, ...	A003325
3	3, 10, 17, 24, 29, 36, 43, 55, 62, 66, ...	A003072
4	4, 11, 18, 25, 30, 32, 37, 44, 51, 56, ...	A003327
5	5, 12, 19, 26, 31, 33, 38, 40, 45, 52, ...	A003328
6	6, 13, 20, 34, 39, 41, 46, 48, 53, 58, ...	A003329
7	7, 14, 21, 42, 47, 49, 61, 77, 85, 87, ...	A018890
8	15, 22, 50, 114, 167, 175, 186, 212, 231, 238, ...	A018889
9	23, 239	A018888

It is known that *every integer is a sum of at most 5 signed cubes*, i.e., $eg(3) \leq 5$ in the Waring's problem. It is believed that 5 can be reduced to 4, so that $N = a^3 + b^3 + c^3 + d^3$ for any integer N, although this has not been proved for numbers of the form $9n \pm 4$. In fact, all numbers $N < 1000$ and not of the form $9n \pm 4$ are known to be expressible as the sum $N = a^3 + b^3 + c^3$ of three (positive or negative) cubes with the exception of $N = 33$, 42, 74, 114, 156, 165, 318, 366, 390, 420, 501, 530, 534, 564, 579, 588, 600, 606, 609, 618, 627, 633, 732, 735, 758, 767, 786, 789, 795, 830, 834, 861, 894, 903, 906, 912, 921, 933, 948, 964, 969, and 975 (Sloane's A046041), while it is known that equation $N = a^3 + b^3 + c^3$ has no solutions for N of the form $9n \pm 4$ (see [Weis11]).

There are exactly 65 numbers which can not be expressed as the sum of distinct positive cubes: 2, 3, 4, 5, 6, 7, 10, 11, 12, 13, 14, 15, 16, 17, 18, 19, 20, 21, 22, 23, 24, 25, 26, 29, 30, 31, 32, 33, 34, 37, 38, 39, 40, 41, 42, 43, 44, 45, 46, 47, 48, 49, 50, 51, 52, 53, 54, 55, 56, 57, 58, 59, 60, 61, 62, 63, 66, 67, 68, 69, 70, 71, 74, 75, 76 (Sloane's A001476).

The following table gives the numbers which can be represented in exactly w different ways as a sum of s positive cubes. For example, $157 = 4^3 + 4^3 + 3^3 + 1^3 + 1^3 = 5^3 + 2^3 + 2^3 + 2^3 + 2^3$ can be represented in $w = 2$ ways by $s = 5$ cubes.

s	w		Sloane
1	0	2, 3, 4, 5, 6, 7, 9, 10, 11, 12, ...	A007412
1	1	1, 8, 27, 64, 125, 216, 343, 512, 729, 1000, ...	A000578
2	0	1, 3, 4, 5, 6, 7, 8, 10, 11, 12, ...	A057903
2	1	2, 9, 16, 28, 35, 54, 65, 72, 91, 126, ...	cf. A003325, A001235
2	2	1729, 4104, 13832, 20683, 32832, 39312, 40033, 46683, 64232, 65728, ...	cf. A001235, A018787
2	3	87539319, 119824488, 143604279, ...	cf. A018787, A051167
2	4	6963472309248, 12625136269928, ...	
3	0	1, 2, 4, 5, 6, 7, 8, 9, 11, 12,	A057904
3	1	3, 10, 17, 24, 29, 36, 43, 55, 62, 66, ...	A025395
3	2	251, 1009, 1366, 1457, 1459, 1520, 1730, 1737, 1756, 1763, ...	A025396
3	3	5104, 9729, 12104, 12221, 12384, 14175, 17604, 17928, 19034, 20691, ...	A025397
4	0	1, 2, 3, 5, 6, 7, 8, 9, 10, 12, 13, 14, ...	A057905
4	1	4, 11, 18, 25, 30, 32, 37, 44, 51, 56, ...	A025403
4	2	219, 252, 259, 278, 315, 376, 467, 522, 594, 702, ...	A025404
4	3	1225, 1521, 1582, 1584, 1738, 1764, 2009, 2249, 2366, 2415, ...	A025405
5	0	1, 2, 3, 4, 6, 7, 8, 9, 10, 11, ...	A057906
5	1	5, 12, 19, 26, 31, 33, 38, 40, 45, 52, ...	A048926
5	2	157, 220, 227, 246, 253, 260, 267, 279, 283, 286, ...	A048927
6	0	1, 2, 3, 4, 5, 7, 8, 9, 10, 11, ...	A057907
6	1	6, 13, 20, 27, 32, 34, 39, 41, 46, 48, ...	A048929
6	2	158, 165, 184, 221, 228, 235, 247, 256, 261, 268, ...	A048930
6	3	221, 254, 369, 411, 443, 469, 495, 502, 576, 595, ...	A048931

The smallest number representable in $w = 2$ ways as a sum of $s = 2$ cubes,

$$1729 = 1^3 + 12^3 = 9^3 + 10^3,$$

is called **Hardy-Ramanujan number** and has special significance in the history of Mathematics after a famous anecdote of Hardy regarding a hospital visit to Ramanujan. Once, in the taxi from London, Hardy noticed its number, 1729. Entering the room of Ramanujan, Hardy presented it as "rather a dull number". "No, Hardy", said Ramanujan, "it is a very interesting number. It is the smallest number expressible as the sum of two cubes in two different ways" ([Hard99]). However, this number was also found in one of

Ramanujan's notebooks dated years before the incident.

In general, n-th **taxicab number**, denoted by $Ta(n)$, is defined as the smallest number which can be expressed as a sum of two positive cubes in n distinct ways, up to order of summands. Hardy and Wright proved in 1954 (see [HaWr79]) that such numbers exist for all positive integers n; however, their proof does not help in constructing them, and so far, only six taxicab numbers are known: 2, 1729, 87539319, 6963472309248, 48988659276962496 24153319581254312065344 (Sloane's A011541). In fact, one has

$$\begin{aligned}
Ta(1) &= 2 = 1^3 + 1^3,\ Ta(2) = 1729 = 1^3 + 12^3 = 9^3 + 10^3, \\
Ta(3) &= 87539319 = 167^3 + 436^3 = 228^3 + 423^3 = 255^3 + 414^3, \\
Ta(4) &= 6963472309248 = 2421^3 + 19083^3 = 5436^3 + 18948^3 \\
&= 10200^3 + 18072^3 = 13322^3 + 16630^3, \\
Ta(5) &= 48988659276962496 = 38787^3 + 365757^3 \\
&= 107839^3 + 362753^3 = 205292^3 + 342952^3 \\
&= 221424^3 + 336588^3 = 231518^3 + 331954^3. \\
Ta(6) &= 24153319581254312065344 = 582162^3 + 28906206^3 \\
&= 3064173^3 + 28894803^3 = 8519281^3 + 28657487^3 \\
&= 16218068^3 + 27093208^3 = 17492496^3 \\
&\quad + 26590452^3 = 18289922^3 + 26224366^3.
\end{aligned}$$

The number $Ta(2)$, i.e., the Hardy-Ramanujan number, was first published by Bessy in 1657. The subsequent taxicab numbers were found with the help of computers: Leech obtained $Ta(3)$ in 1957, Rosenstiel, Dardis and Rosenstiel found $Ta(4)$ in 1991, and Wilson found $Ta(5)$ in 1997. In 2008 it was proved by Hollerbach, that $T(6) = 24153319581254312065344$ (see [Weis11]).

The n-th **cabtaxi number**, denoted by $CT(n)$, is defined as the smallest positive integer that can be written as the sum of two integer (positive, negative or 0) cubes in n ways. Such numbers exist for all n, since taxicab numbers exist for all n. However, only nine are known: 1, 91, 728, 2741256, 6017193, 1412774811, 11302198488, 137513849003496, 424910390480793000,

933528127886302221000 (Sloane's A047696). In fact,

$$\begin{aligned}
CT(1) &= 1 = 1^3 \pm 0^3,\ CT(2) = 91 = 3^3 + 4^3 = 6^3 - 5^3,\\
CT(3) &= 728 = 6^3 + 8^3 = 9^3 - 1^3 = 12^3 - 10^3,\\
CT(4) &= 2741256 = 108^3 + 114^3 = 140^3 - 14^3\\
&= 168^3 - 126^3 = 207^3 - 183^3,\\
CT(5) &= 6017193 = 166^3 + 113^3 = 180^3 + 57^3\\
&= 185^3 - 68^3 = 209^3 - 146^3 = 246^3 - 207^3,\\
CT(6) &= 1412774811 = 963^3 + 804^3 = 1134^3 - 357^3 = 1155^3\\
&\quad - 504^3 = 1246^3 - 805^3 = 2115^3 - 2004^3 = 4746^3 - 4725^3,\\
CT(7) &= 11302198488 = 1926^3 + 1608^3 = 1939^3 + 1589^3\\
&= 2268^3 - 714^3 = 2310^3 - 1008^3 = 2492^3 - 1610^3\\
&= 4230^3 - 4008^3 = 9492^3 - 9450^3,\\
CT(8) &= 137513849003496 = 22944^3 + 50058^3 = 36547^3 + 44597^3\\
&= 36984^3 + 44298^3 = 52164^3 - 16422^3 = 53130^3 - 23184^3\\
&= 57316^3 - 37030^3 = 97290^3 - 92184^3 = 218316^3 - 217350^3,\\
CT(9) &= 424910390480793000 = 645210^3 + 538680^3 = 649565^3\\
&\quad + 532315^3 = 752409^3 - 101409^3 = 759780^3 - 239190^3\\
&= 773850^3 - 337680^3 = 834820^3 - 539350^3 = 1417050^3\\
&\quad - 1342680^3 = 3179820^3 3165750^3 = 5960010^3 - 5956020^3,\\
CT(10) &= 933528127886302221000 = 7002840^3 + 8387730^3\\
&= 6920095^3 + 8444345^3 = 77480130^3 - 77428260^3\\
&= 41337660^3 - 41154750^3 = 18421650^3 - 17454840^3\\
&= 10852660^3 - 7011550^3 = 10060050^3 - 4389840^3 = 9877140^3\\
&\quad - 3109470^3 = 9781317^3 - 1318317^3 = 9773330^3 - 84560^3.
\end{aligned}$$

The numbers $CT(5)$, $CT(6)$ and $CT(7)$ were found by Rathbun; $CT(8)$ was found by Bernstein; $CT(9)$ was found by Moore, 2005; $CT(1)$ was found by Boyer, 2006, and was independently verified by Hollerbach, 2008 (see [Wiki11], [Weis11]).

The **generalized taxicab number** $Ta(k, j, n)$ is the smallest number which can be expressed as the sum of j k-th powers in n

different ways. For $k = 3$ and $j = 2$, they coincide with taxicab numbers. It has been shown by Euler that

$$Ta(4, 2, 2) = 635318657 = 59^4 + 158^4 = 133^4 + 134^4.$$

However, $Ta(4, 3, n)$ is not known for any $n \geq 2$, and neither is $Ta(5, 2, n)$; in fact, no positive integer is known at all which can be written as the sum of three fourth powers or two fifth powers in more than one way (see [Wiki11]).

4.12.4. In the case $n = 4$, it is known, that every positive integer can be represented as a sum of at most 19 biquadratic numbers, giving the solution $g(4) = 19$ of the Waring's problem. This result was established in 1986 by Balasubramanian, Dress, and Deshouillers ([BDD86]).

Moreover, Davenport, 1939 ([Dave39]) proved, that all sufficiently large integers require only 16 biquadratic numbers, i.e., $G(4) = 16$ in the Waring's problem.

The following table gives the first few numbers which require $t = 1, 2, 3, \ldots, 19$ biquadratic numbers to represent them as a sum, with the sequences for 17, 18, and 19 being finite (49, 24, and 7 entries, respectively).

t		Sloane
1	$1, 16, 81, 256, 625, 1296, 2401, 4096, \ldots$	*A*000290
2	$2, 17, 32, 82, 97, 162, 257, 272, \ldots$	*A*003336
3	$3, 18, 33, 48, 83, 98, 113, 163, \ldots$	*A*003337
4	$4, 19, 34, 49, 64, 84, 99, 114, 129, \ldots$	*A*003338
5	$5, 20, 35, 50, 65, 80, 85, 100, 115, \ldots$	*A*003339
6	$6, 21, 36, 51, 66, 86, 96, 101, 116, \ldots$	*A*003340
7	$7, 22, 37, 52, 67, 87, 102, 112, 117, \ldots$	*A*003341
8	$8, 23, 38, 53, 68, 88, 103, 118, 128, \ldots$	*A*003342
9	$9, 24, 39, 54, 69, 89, 104, 119, 134, \ldots$	*A*003343
10	$10, 25, 40, 55, 70, 90, 105, 120, 135, \ldots$	*A*003344
11	$11, 26, 41, 56, 71, 91, 106, 121, 136, \ldots$	*A*003345
12	$12, 27, 42, 57, 72, 92, 107, 122, 137, \ldots$	*A*003346
13	$13, 28, 43, 58, 73, 93, 108, 123, 138, \ldots$	*A*046044
14	$14, 29, 44, 59, 74, 94, 109, 124, 139, \ldots$	*A*046045
15	$15, 30, 45, 60, 75, 95, 110, 125, 140, \ldots$	*A*046046
16	$31, 46, 61, 76, 111, 126, 141, 156, \ldots$	*A*046047
17	$47, 62, 77, 127, 142, 157, 207, 222, \ldots$	*A*046048
18	$63, 78, 143, 158, 223, 238, 303, 318, \ldots$	*A*04649
19	$79, 159, 239, 319, 399, 479, 559$	*A*046050

It is known, that *every integer is a sum of at most ten signed bi-squares*, i.e., $eg(4) \leq 10$ in the Waring's problem. It is not known if ten can be reduced to nine. The numbers 2, 3, 4, 5, 6, 7, 8, 9, 10, 11, 12, 13, 14, 15, 18, 19, 20, 21, ... (Sloane's A046039) can not be represented using distinct bi-squares (see [Weis11]).

The following table gives the numbers which can be represented in at least w different ways as a sum of s bi-squares.

s	w		Sloane
1	1	$1, 16, 81, 256, 625, 1296, 2401, 4096, 6561, 10000, \dots$	A000583
2	2	$635318657, 3262811042, 8657437697, 10165098512, 51460811217, \dots$	A018786

4.12.5. In the case $n \geq 5$, the result $g(5) = 37$ was established in 1964 by Chen ([Chen64]), and the result $g(6) = 73$ in 1940 by Pillai ([Pill40]).

Apart from a certain ambiguity, all the other values of $g(n), n > 6$, are now also known, as a result of works (1936 - 44) by Dickson, Pillai, Rubugunday, and Niven (see [Weis11], [Wiki11]). In fact, the formula

$$g(n) = \left\lfloor \left(\frac{3}{2}\right)^n \right\rfloor + 2^n - 2$$

is verified for $n \leq 471600000$ ([KuWu90]), giving the values 1, 4, 9, 19, 37, 73, 143, 279, 548, 1079, ... (Sloane's A002804).

So, *every positive integer can be written as the sum of at most 4 square numbers, at most* 9 *cubic numbers, at most 19 biquadratic numbers, at most 37 five-dimensional hypercube numbers, at most 73 six-dimensional hypercube numbers*, etc.

The exact value of $G(n)$ is unknown for any $n \neq 2, 4$, but Vinogradov (1947, see [Kara83]) has shown, using his improved Hardy-Littlewood method, that $G(n) \leq n(3 \log n + 11)$. Moreover, the following upper bounds for $G(n)$ are known: if $n = 3$, 5, 6, 7, 8, 9, 10, 11, 12, 13, 14, 15, 16, 17, 18, 19, 20, then $G(n) \leq 7$, 17, 24, 33, 42, 51, 59, 67, 76, 84, 92, 100, 109, 117, 125, 134, 142, respectively.

So, *every sufficiently large positive integer can be written as a sum of at most 4 square numbers, at most 7 cubic numbers, at most 16 biquadratic numbers, at most 17 five-dimensional hypercube numbers, at most 21 six-dimensional hypercube numbers*, and so on.

Chapter 5

Fermat's polygonal number theorem

In this chapter, we give full and detailed proof of the Fermat's polygonal theorem. Until now, it was scattered in many, often difficult to find publications and never united in some book. For example, the proofs by Cauchy and Pepin were only in extinct french mathematical journals, accessible only in several libraries.

5.1 History of the problem

5.1.1. In 1636, Fermat made the statement that *every number is a sum of at most three triangular numbers, four squares, five pentagonal numbers, and so on.*

He wrote in a letter to Mersenne (September 1636): *I was the first to discover the very beautiful and entirely general theorem that every number is either triangular or the sum of 2 or 3 triangular numbers; every number is either a square or the sum of 2, 3 or 4 squares; either pentagonal or the sum of 2, 3, 4 or 5 pentagonal numbers; and so on ad infinitum, whether it is a question of hexagonal, heptagonal or any polygonal numbers. I can not give the proof here, which depends upon numerous and abstruse mysteries of numbers; for I intend to devote an entire book to this subject and to effect in this part of arithmetic astonishing advances over the previously known limits.*

In a letter to Pascal in 1654 he described it as being so far his most important result, but Fermat's proof has never been found.

5.1.2. The square case was proved by Lagrange in 1770 ([Lagr70]). It is known as *Lagrange's four-square theorem.*

Gauss proved the triangular case of Fermat's polygonal number theorem (see [Gaus01], [Well86]) and noted the event in his diary on July 10, 1796, with the notation

$$**E\Upsilon RHKA \quad num = \Delta + \Delta + \Delta.$$

Thus, Gauss has found a proof that every positive integer N may be represented by the particular quadratic polynomial in three variables with non-negative integers k, m and n:

$$N = \frac{k(k+1)}{2} + \frac{m(m+1)}{2} + \frac{n(n+1)}{2}.$$

This fact is equivalent to the statement that every number of the form $8N + 3$ is a sum of three odd squares:

$$8N + 3 = (2k+1)^2 + (2m+1)^2 + (2n+1)^2.$$

Last decomposition is a special case of the more general theorem that a number is a sum of three squares precisely when it is not of the form $4^k(8m+7)$ for $k, m \geq 0$. This theorem was proved by Legendre in 1798 ([Lege79]). However, the proof of Legendre was incomplete, leaving a gap which was later filled by Gauss in his *Disquisitiones* in 1801 ([Gaus01]).

Gauss proved the result by means of the theory of ternary quadratic forms. The number of ways a number M can be decomposed into a sum of three triangular numbers depends in a definite manner on the prime factors of $N = 8M + 3$ and the number of classes of binary quadratic forms of determinant $-N$ (see [Dick05]).

In 1813, Cauchy ([Cauc13]) gave the first proof of Fermat's assertion in total by deriving it in an elementary but involved way from the three-triangular-number theorem of Gauss. Cauchy's theorem is sharp in the sense that there are numbers which can not be represented by fewer than m m-gonal numbers, for example, $N = 2m - 1$.

In the simplified proof by Legendre ([Lege30]), the case $m = 3$ is not presupposed, as it was done by Cauchy. Moreover, Legendre proved in effect that every sufficiently large integer is a sum of four $(m+2)$-gonal numbers if $m+2$ is odd, while, for $m+2$ even, every sufficiently large integer is a sum of five $(m+2)$-gonal numbers one of which is either 0, or 1.

5.1.3. Later Dirichlet gave a beautiful formula for the number of ways in which N can be expressed as the sum of three triangular numbers. In the special case when $8N+3$ is a prime, it says that this number is the excess of the number of *quadratic residues* (i.e., squares modulo $8N+3$) over non-residues modulo $8N+3$ in the segment from 1 to $4N+1$ (see [Weis11]).[1]

In 1834, Jacobi found an exact formula for the total number of ways a given positive integer N can be represented as the sum of four squares. This number is eight times the sum of the divisors of N if N is odd, and 24 times the sum of the odd divisors of N if N is even (see [Wiki11]).

5.1.4. Some interesting and difficult problems about representing integers as sums of polygonal numbers are still open.

Thus, Legendre, in the third edition of his *Théorie des Nombres* of 1830, proved by elementary means that *every number larger than 1791 is a sum of four hexagonal numbers* ([Lege30]). The question arose whether or not three hexagonal numbers eventually suffice. Duke and Schulze-Pillot ([DuSc90]) in 1990 proved, that every sufficiently large number is a sum of three hexagonal numbers. Since n-th hexagonal number is $(2n-1)$-th triangular number, this result strengthens the three-triangular-number theorem of Gauss. In fact, there are only two positive integers that can not be represented using five hexagonal numbers: $11 = 1+1+1+1+1+6$ and $26 = 1+1+6+6+6+6$. Moreover, there are exactly 13 positive integers that can not be represented using four hexagonal numbers ([Guy94]): 5, 10, 11, 20, 25, 26, 38, 39, 54, 65, 70, 114, and 130 (Sloane's A007527).

[1]For instance, when $N=2$ (so, $8N+3=19$), the quadratic residues modulo 19 up to $9=4N+1$ are $1,4,5,6,7,9$ and the non-residues are $2,3,8$; so, the excess is 3, which is the number of ways of expressing 2 as a sum of three triangular numbers: $2=1+1+0=1+0+1=0+1+1$.

The *Pollock's conjecture* (1850, see [Dick05]) states that every number is the sum of at most five tetrahedral numbers. In fact, it was proved by Watson (1952), that any positive integer is a sum of at most eight tetrahedral numbers. Moreover, all $N \leq 10^7$ can be represented as a sum of at most five tetrahedral numbers. The numbers that are not the sum of at most four tetrahedral numbers are given by the sequence 17, 27, 33, 52, 73, 82, 83, 103, 107, 137, ... (Sloane's A000797), containing 241 terms, with 343867 being almost certainly the last such number (see [Weis11]).

Pollock octahedral numbers conjecture (1850, see [Dick05]) states that every number is the sum of at most seven octahedral numbers. Lerner in 2005 has conjectured that there is an upper bound on numbers requiring 5 terms or more. Computer results show that 309 seems to be the last number requiring seven terms, 11579 seems to be the last number requiring six terms, and 65285683 seems to be the last number requiring five terms. If the conjecture is true, then it can be easily proven that there must be arbitrary high numbers requiring four terms (see [Wiki11]).

5.2 Lagrange's four-square theorem

In this section we prove Lagrange's result on four square numbers: *for every positive integer N there exist non-negative integers a,b,c,d such that*

$$N = a^2 + b^2 + c^2 + d^2.$$

5.2.1. It holds, that *the product of two numbers, which are representable as sums of four squares, is also representable as a sum of four squares.*

This statement is based on the famous *Euler's four-square identity* (1749): *for any integers a, b, c, d, w, x, y, z one has*

$$\begin{aligned}(a^2+b^2+c^2+d^2)&(w^2+x^2+y^2+z^2)\\ &= (aw+bx+cy+dz)^2 + (ax-bw-cz+dy)^2\\ &\quad +(ay+bz-cw-dx)^2 + (az-by+cx-dw)^2.\end{aligned}$$

The reader can check it by direct computation. Due to the *fundamental theorem of Arithmetics*, it is sufficient now to check the Lagrange's four-square theorem only for prime numbers.

5.2.2. In order to prove the four-square theorem for odd prime numbers, we need the following simple proposition: *if an even number $2m$ is a sum of two squares, then so is m.*

In fact, if $2m = x^2 + y^2$, then x and y have the same parity, and in the identity $m = \left(\frac{x-y}{2}\right)^2 + \left(\frac{x+y}{2}\right)^2$ both fractions are integers.

5.2.3. The last preliminary result states that *for any odd prime p there exist a positive integer k such that $kp = a^2 + b^2 + 1,\ a, b \in \mathbb{Z}$.*

In order to prove this fact, let us consider two sets of integers: $A = \{a^2 | a = 0, 1, \dots, \frac{p-1}{2}\}$, and $B = \{-b^2 - 1 | b = 0, 1, \dots, \frac{p-1}{2}\}$. For any $x, y \in \{0, 1, \dots, \frac{p-1}{2}\}$ it holds

$$x^2 \equiv y^2 (mod\ p) \Leftrightarrow (x-y)(x+y) \equiv 0 (mod\ p) \Leftrightarrow$$
$$x - y \equiv 0 (mod\ p), \quad \text{or } x + y \equiv 0 (mod\ p) \Leftrightarrow x = y.$$

Therefore, no two elements of the set A are congruent modulo p. Similarly, no two elements of the set B are congruent modulo p. Since each set has $\frac{p+1}{2}$ elements, but there are only p residue classes modulo p, there exists an element from the first set which is congruent to some element from the second set:

$$a^2 \equiv -b^2 - 1 (mod\ p), \quad \text{i.e., } a^2 + b^2 + 1 = kp, \quad k \in \mathbb{N}.$$

5.2.4. Now we are ready to show that *any prime number can be represented as a sum of four squares.*

In fact, for $p = 2$ one has $2 = 1^2 + 1^2 + 0^2 + 0^2$. For any odd prime p, there exists a positive integer k such that $kp = a^2 + b^2 + 1^2 + 0^2$, i.e., kp is represented as a sum of four squares:

$$kp = a^2 + b^2 + c^2 + d^2.$$

If $k = 1$, then p is a sum of four squares.

If $k > 1$, then it is easy to show that there exists a positive integer n, such that $n < k$, and np is a sum of four squares.

In fact, if k is even, then none, two or four of a, b, c, d in the above decomposition of kp are even. In any of those cases, kp is a sum of two even numbers, both represented as sums of two squares. Hence, by the above preliminary proposition, the number $\frac{k}{2}p$ is represented as a sum of four squares, i.e., one can take $n = \frac{k}{2}$.

If k is odd, let us find four integer numbers w, x, y, and z, such that

$$a \equiv w(mod\ k),\ b \equiv x(mod\ k),\ c \equiv y(mod\ k),\ d \equiv z(mod\ k),$$

and $w, x, y, z \in \left(-\frac{k}{2}, \frac{k}{2}\right)$. In other words, let w, x, y, z be the smallest absolute residues modulo k of the numbers a, b, c, d, respectively. The choice of w, x, y, z implies

$$w^2 + x^2 + y^2 + z^2 \equiv a^2 + b^2 + c^2 + d^2(mod\ k),\ \text{with}$$
$$w^2 + x^2 + y^2 + z^2 < 4 \cdot \frac{k^2}{4} = k^2.$$

Since $a^2 + b^2 + c^2 + d^2 \equiv 0(mod\ k)$, it holds

$$w^2 + x^2 + y^2 + z^2 \equiv 0(mod\ k), \quad \text{with } w^2 + x^2 + y^2 + z^2 < k^2.$$

In other words, we get that

$$w^2 + x^2 + y^2 + z^2 = nk, \quad \text{with } 1 \leq n < k.$$

The representations $kp = a^2 + b^2 + c^2 + d^2$ and $nk = w^2 + x^2 + y^2 + +z^2$ allow to obtain the equality

$$k^2np = (a^2 + b^2 + c^2 + d^2)(w^2 + x^2 + y^2 + +z^2).$$

By the above Euler's four-square identity, it holds

$$k^2np = (aw + bx + cy + dz)^2 + (ax - bw - cz + dy)^2$$
$$+(ay + bz - cw - dx)^2 + (az - by + cx - dw)^2.$$

Since $ax \equiv bw(mod\ k)$, and $dy \equiv cz(mod\ k)$, we get $ax - bw - cz + dy \equiv 0(mod\ k)$. Similarly, $ay + bz - cw - dx \equiv 0(mod\ k)$, and $az - by + cx - dw \equiv 0(mod\ k)$. Therefore, the sum $(ax - bw - cz + dy)^2 + (ay + bz - cw - dx)^2 + (az - by + cx - dw)^2$ is divisible by k^2. It implies that $(aw + bx + cy + dz)^2$ is divisible by k^2 as a difference of two integers, which are divisible by k^2.

So, it is possible to divide both sides of above representation for k^2np by k^2. The result gives an expression for np, $n < k$, as a sum of four squares.

Therefore, starting from a representation $kp = a^2 + b^2 + c^2 + d^2$, with $k > 1$, we can obtain a similar representation for np with some positive integer n, such that $n < k$. It will imply, after a finite number of steps, representation for p as a sum of four squares.

5.2.5. Now we can finish the proof of the Lagrange's four-square theorem: *any positive integer N is a sum of four squares.*

In fact, for $N = 1$ one has the trivial decomposition $1 = 1^2 + 0^2 + 0^2 + 0^2$.

Furthermore, any $N > 1$ is decomposable into a product of primes. Since any prime is representable as a sum of four squares, and a product of numbers, which are representable as sums of four squares, has a similar representation, the number N can be represented as a sum of four squares. This completes the proof.

5.3 Gauss's three-triangular-number theorem; elementary considerations

In this section we are going to consider some preliminary reasons and elementary results (see, for example, [Sier64]) connected with Gauss's theorem on three triangular numbers: *for every positive integer N there exist non-negative integers m, n and k such that*

$$N = \frac{k(k+1)}{2} + \frac{m(m+1)}{2} + \frac{n(n+1)}{2}.$$

5.3.1. First of all, in order to prove Gauss's three-triangular theorem, it is sufficient to show that *any positive integer of the form $8N + 3$ can be represented as a sum of three squares.*

In fact, every such representation of the number $8N + 3$ (if it exists) is a sum of three odd squares, since a sum of three even squares, as well as a sum of one even square and two odd squares, are even, while a sum of two even squares and one odd square has the form $4t + 1$. So, in this case we obtain

$$8N + 3 = (2n+1)^2 + (2m+1)^2 + (2k+1)^2,$$

where m, n, k are non-negative integers. It implies

$$8N + 3 = 4(n^2 + n) + 4(m^2 + m) + 4(k^2 + k) + 3, \quad \text{and}$$
$$N = \tfrac{n(n+1)}{2} + \tfrac{m(m+1)}{2} + \tfrac{k(k+1)}{2}.$$

5.3.2. The existence of a decomposition of $8N+3$ into a sum of three squares is a particular case of more general theorem, stating that *any positive integer* $N \neq 4^k(8m+7), k, m \geq 0$, *can be represented as a sum of three squares.*

Without loss of generality, one can prove this theorem only for the case $N \neq 8m+7$, since the general case easily reduces to it. In fact, if $N = x^2+y^2+z^2$, and $4^k|N$, then $2^k|x$, $2^k|y$, $2^k|z$, i.e., it holds $\frac{N}{4^k} = \left(\frac{x}{2^k}\right)^2 + \left(\frac{y}{2^k}\right)^2 + \left(\frac{z}{2^k}\right)^2$.

5.3.3. It is easy to show, that any positive integer $N = 8m+7$ can not be represented as a sum of three squares. In fact, it holds

$$0^2 \equiv 0(mod\ 8), \quad (\pm 1)^2 \equiv 1(mod\ 8), \quad (\pm 2)^2 \equiv 4(mod\ 8),$$
$$(\pm 3)^2 \equiv 1(mod\ 8), \quad 4^2 \equiv 0(mod\ 8).$$

Hence, the residues r_{x^2} modulo 8 of perfect squares x^2 belong to the set $\{0,1,4\}$. Therefore, one has $r_{x^2+y^2} \in \{0,1,2,4,5\}$, and $r_{x^2+y^2+z^2} \in \{0,1,2,3,4,5,6\}$. In other words,

$$8m+7 \neq x^2+y^2+z^2,$$

i.e., *any positive integer* $N = 8m+7$, *as well as any positive integer* $N = 4^k(8m+7)$, *can not be represented as a sum of three squares.*

On the other hand, the proof of the existence of such representation for any positive integer $N \neq 4^k(8m+7)$ is, in general, very complicate. It was given by Gauss in 1801 ([Gaus01]), using the general theory of quadratic forms.

However, for our consideration it is necessary to prove only one particular case of above theorem, finding a decomposition into three squares for any positive integer of the form $8m+3$. We will consider in the next section the general proof of this fact, using the theory of ternary quadratic forms. However, several particular cases can be obtained by elementary reasons.

5.3.4. A simple consideration, based on properties of the *Legendre symbol*, allows to prove that *any prime number* $p = 8m+3$ *can be represented as a sum of a square and a doubled square.*

In fact, we will prove a more general proposition: *any prime number equal to 1 or 3 modulo 8 is represented as a sum of a square and a doubled square, while any prime number equal to 5 or 7 modulo 8 has no such representation:*

$$p = x^2 + 2y^2 \quad \text{if } p \equiv 1, 3(mod\ 8), \quad \text{and}$$
$$p \neq x^2 + 2y^2 \quad \text{if } p \equiv 5, 7(mod\ 8).$$

In order to prove this theorem, let us consider an odd prime number p, which can be represented as a sum of a square and a doubled square: $p = a^2 + 2b^2$ with some positive integers a and b.

Clearly, the numbers p and b, as well as p and a, are relatively prime. Hence, there is a solution c for the congruence $bz \equiv a(mod\ p)$: $bc \equiv a(mod\ p)$, $c \in \mathbb{Z}$. Since $a^2 + 2b^2 = p$, the number $a^2 + 2b^2$ is equal to 0 modulo p. So, we get the following chain of congruences:

$$a^2 + 2b^2 \equiv 0(mod\ p), \quad (bc)^2 + 2b^2 \equiv 0(mod\ p),$$
$$c^2 + 2 \equiv 0(mod\ p), \quad c^2 \equiv -2(mod\ p).$$

Therefore, the congruence $z^2 \equiv -2(mod\ p)$ has a solution $c \in \mathbb{Z}$. It implies that -2 is a square residue modulo p. In particular, it holds $\left(\frac{-2}{p}\right) = 1$, where $\left(\frac{a}{p}\right)$ is the *Legendre symbol*. Since

$$\left(\frac{-2}{p}\right) = \left(\frac{-1}{p}\right)\left(\frac{2}{p}\right) = (-1)^{\frac{p-1}{2}}(-1)^{\frac{p^2-1}{8}}$$

(see, for example, [Buch09]), we get

$$\left(\frac{-2}{p}\right) = \begin{cases} 1, & \text{if } p \equiv 1, 3(mod\ 8) \\ -1, & \text{if } p \equiv 5, 7(mod\ 8). \end{cases}$$

So, it is proven, that prime numbers equal to 5 or 7 modulo 8 can not be represented as a sum of a square and a doubled square.

For a prime p equal to 1 or 3 modulo 8, let us consider the numbers $cx + y$, where $x, y = 0, 1, ..., \lfloor\sqrt{p}\rfloor$. Since $(\lfloor\sqrt{p}\rfloor + 1)^2 > (\sqrt{p})^2 = p$, we have more than p such numbers. It implies that there are two numbers $cx_1 + y_1$ and $cx_2 + y_2$, coinciding modulo p:

$$cx_1 + y_1 \equiv cx_2 + y_2(mod\ p), \quad \text{i.e., } c(x_1 - x_2) \equiv y_2 - y_1(mod\ p).$$

It is easy to show that in above congruence $x_1 \neq x_2$, and $y_1 \neq y_2$. In fact, if $y_1 = y_2$, then $c(x_1 - x_2) \equiv 0(mod\ p)$, and $x_1 - x_2 \equiv 0(mod\ p)$. Since $|x_1 - x_2| < \sqrt{p}$, we obtain $x_1 = x_2$. Similarly, if $x_1 = x_2$, then $y_2 - y_1 \equiv 0(mod\ p)$. Since $|y_1 - y_2| < \sqrt{p}$, we obtain $y_1 = y_2$.

Since $x_1 \neq x_2$, and $y_1 \neq y_2$, we get that the numbers $k = |x_1 - x_2|$ and $t = |y_1 - y_2|$ are positive integers. Therefore, there are positive integers k and t, such that

$$ck \equiv t(mod\ 8), \quad \text{or } ck \equiv -t(mod\ 8).$$

Since above congruence $c^2 + 2 \equiv 0(mod\ p)$ implies $k^2(c^2+2) \equiv 0(mod\ p)$, we have

$$k^2(c^2+2) \equiv (kc)^2 + 2k^2 \equiv t^2 + 2k^2(mod\ p).$$

So, it holds $t^2 + 2k^2 = mp$, where m is a positive integer. Since $k < \sqrt{p}$ and $t < \sqrt{p}$, we obtain $k^2 < p$, and $t^2 < p$. Hence, it holds $t^2 + 2k^2 < 3p$, i.e., $m = 1$, or $m = 2$.

If $m = 1$, we have the representation $p = t^2 + 2k^2$ of p as a sum of a square and a doubled square.

If $m = 2$, then $2p = t^2 + 2k^2$. In this case $2|t$, i.e., $t = 2u$. Dividing by 2 gives the decomposition $p = k^2 + 2u^2$ of p into a sum of a square and a doubled square.

5.3.5. Moreover, it holds the following proposition: *the product of two positive integers, which are represented as sums of a square and a doubled square, is also representable as a sum of a square and a doubled square.*

This property is a simple corollary of the identity $(a^2+2b^2)(c^2+2d^2) = (ac \mp 2bd)^2 + 2(ad \pm bc)^2$.

It is easy to see also, that *any perfect square is represented as a sum of a square and a doubled square:* $n^2 = n^2 + 2 \cdot 0^2$.

5.3.6. Above propositions allow to formulate the general result of our elementary consideration: *any positive integer of the form* $8m+3$ *is represented as a sum of a square and a doubled square if its canonical decomposition has no odd powers of prime numbers equal to 5 or 7 modulo 8.*

In fact, any positive integer N of the form $8m+3$ with canonical decomposition having no odd powers of prime numbers equal to 5 or 7 modulo 8, can be represented as a product of prime numbers equal to 1 or 3 modulo 8, and a perfect square, containing all prime divisors of N equal to 5 and 7 modulo 8. Since each factor of this decomposition is represented as a sum of a square and a doubled square, then their product N is also representable as a sum of a square and a doubled square.

So, we proved the existence of decomposition into a sum of three squares for all numbers $8m+3$, except of those of the form $N = n^2 n^{'}$, where the *square-free part* $n^{'}$ of N, which is a product of distinct primes, contains prime factors equal to 5 and 7 modulo 8.

5.4 Proof of the Gauss's three-triangular-number theorem

In this section we will prove the Gauss three-triangular number theorem in all its entirety. The proof is based on the theory of quadratic forms and the Dirichlet's theorem about primes in an arithmetic progression (see [Land09]).

The first part of our consideration contains the main properties of binary quadratic forms, leading to the *two-square theorem: an integer can be represented as a sum of two squares if and only if it has the form* $4m+1$ *and its canonical decomposition has no odd degrees of the primes equal to 3 modulo 4.*

In the second part, the base of the theory of ternary quadratic forms is constructed. It allows to obtain a decomposition of any positive integer $N \neq 4^k(8m+7)$ into a sum of three squares.

5.4.1. For a given 2×2 symmetric matrix $A = \begin{pmatrix} a & b \\ b & c \end{pmatrix}$ with integer entries and two integer variables x and y, the *binary quadratic form* $F(x, y)$, associated with the matrix A, is defined as

$$F(x, y) = ax^2 + 2bxy + cy^2.$$

The *determinant* D *of the binary quadratic form* $F(x, y)$ is defined as the determinant of the corresponding matrix A:

$$D = |A| = \begin{vmatrix} a & b \\ b & c \end{vmatrix} = ac - b^2.$$

For a given 2×2 matrix $S = \begin{pmatrix} \alpha & \beta \\ \gamma & \delta \end{pmatrix}$ with integer entries and non-zero determinant $|S| = \alpha\delta - \beta\gamma$, let u, t be new integer variables such that

$$x = \alpha t + \beta u, \ y = \gamma t + \delta u, \quad \text{for some } \alpha, \beta, \gamma, \delta \in \mathbb{Z}.$$

Then one gets

$$\begin{aligned} ax^2 + 2bxy + by^2 &= a(\alpha t + \beta u)^2 + 2b(\alpha t + \beta u)(\gamma t + \delta u) + b(\gamma t + \delta u)^2 \\ &= (a\alpha^2 + 2b\alpha\gamma + b\gamma^2)t^2 + 2(a\alpha\beta + b(\alpha\delta + \beta\gamma) \\ &\quad + c\gamma\delta)tu + (a\beta^2 + 2b\beta\delta + c\delta^2)u^2 \\ &= a' t^2 + 2b' tu + c' u^2, \end{aligned}$$

where $a' = a\alpha^2 + 2b\alpha\gamma + b\gamma^2$, $b' = a\alpha\beta + b(\alpha\delta + \beta\gamma) + c\gamma\delta$, $c' = a\beta^2 + 2b\beta\delta + c\delta^2$. Therefore, using the matrix S, we obtain from a given binary quadratic form $F(x, y) = ax^2 + 2bxy + c^2$, associated with the matrix A, a new binary quadratic form $G(t, u) = a' t^2 + 2b' tu + c' u^2$, associated with the matrix $A' = \begin{pmatrix} a' & b' \\ b' & c' \end{pmatrix}$.

It is easy to check that $\begin{pmatrix} a' & b' \\ b' & c' \end{pmatrix} = \begin{pmatrix} \alpha & \gamma \\ \beta & \delta \end{pmatrix} \cdot \begin{pmatrix} a & b \\ b & c \end{pmatrix} \cdot \begin{pmatrix} \alpha & \beta \\ \gamma & \delta \end{pmatrix}$. In other words, it holds $A' = S^T \cdot A \cdot S$. Then

$$|A'| = \begin{vmatrix} a' & b' \\ b' & c' \end{vmatrix} = \begin{vmatrix} a & b \\ b & c \end{vmatrix} \cdot \begin{vmatrix} \alpha & \beta \\ \gamma & \delta \end{vmatrix}^2 = |A| \cdot |S|^2,$$

and one gets, that the determinant D' of the form $G(t, u)$, equal to the determinant of A', is obtained from the determinant D of the form $F(x, y)$ by the equation $D' = D \cdot |S|^2$. If the matrix S has the determinant equal to 1, then $D = D'$.

Two binary quadratic forms $F(x, y)$ and $G(t, u)$ are called *equivalent*, $F(x, y) \sim G(t, u)$, if there exists an 2×2 matrix S with determinant 1, which takes $F(x, y)$ into $G(t, u)$.

It is easy to see that this relation is an equivalence relation: $F(x, y) \sim F(x, y)$; if $F(x, y) \sim G(t, u)$ then $G(t, u) \sim F(x, y)$; if $F(x, y) \sim G(t, u)$ and $G(t, u) \sim H(z, w)$, then $F(x, y) \sim H(z, w)$.

In fact, the matrix $S = \begin{pmatrix} 1 & 0 \\ 0 & 1 \end{pmatrix}$, having the determinant 1, takes $F(x, y)$ into $F(x, y)$. If the matrix $S = \begin{pmatrix} \alpha & \beta \\ \gamma & \delta \end{pmatrix}$ with determinant 1 takes $F(x, y)$ into $G(t, u)$, then the matrix $S^{-1} = \begin{pmatrix} \delta & -\beta \\ -\gamma & \alpha \end{pmatrix}$ which also has determinant 1, takes $G(t, u)$ into $F(x, y)$. Finally, if the matrices $S = \begin{pmatrix} \alpha & \beta \\ \gamma & \delta \end{pmatrix}$ and $S_1 = \begin{pmatrix} \alpha_1 & \beta_1 \\ \gamma_1 & \delta_1 \end{pmatrix}$ with determinants 1 take $F(x, y)$ into $G(t, u)$ and $G(t, u)$ into $H(z, w)$, then the matrix $SS_1 = \begin{pmatrix} \alpha\alpha_1 + \beta\gamma_1 & \alpha\beta_1 + \beta\delta_1 \\ \gamma\alpha_1 + \delta\gamma_1 & \gamma\beta_1 + \delta\delta_1 \end{pmatrix}$, which determinant also equal to 1, takes $F(x, y)$ into $H(z, w)$.

5.4.2. We say, that a binary quadratic from $F(x, y) = ax^2 + 2bxy + cx^2$ *represent a given integer* N, if there exist integers x_0, y_0, such that $N = F(x_0, y_0) = ax_0^2 + 2bx_0y_0 + cy_0^2$.

Obviously, the equivalent forms $F(x, y)$ and $G(t, u)$ represent the same integers:

$$N = F(x_0, y_0) = ax_0^2 + 2bx_0y_0 + cy_0^2 \Leftrightarrow N = a^{'}t_0^2 + 2b^{'}t_0u_0 + c^{'}u_0^2.$$

A binary quadratic form $F(x, y) = ax^2 + 2bxy + cx^2$ is called *positive definite*, if it represents only non-negative integers, and represents 0 only in the case $x = y = 0$.

In this case, a is non-negative since it is represented by $F(x, y)$: $a = F(1, 0, 0)$. Moreover, $a \neq 0$, since in the case $a = 0$ we get that $0 = F(1, 0, 0)$, a contradiction. Therefore, for a positive-definite binary quadratic form $F(x, y) = ax^2 + 2bxy + cy^2$ we have that $a > 0$. Then

$$aF(x, y) = a(ax^2 + 2bxy + cy^2) = (ax + by)^2 + Dy^2 \geq 0$$

for all values x and y. It implies that $D > 0$.

In fact, in the case of $D < 0$ the value $aF(b, -a) = (ab - ba)^2 + Da^2 < 0$, a contradiction. In the case of $D = 0$, it holds $F(b, -a) = ab^2 - 2ab^2 + ca^2 = a(ca - b^2) = aD = 0$, i.e., $F(b, -a)$ represents 0, a contradiction.

So, *for any positive-definite binary quadratic form* $F(x,y) = ax^2 + 2bxy + cy^2$ *it holds*

$$a > 0, \quad \text{and } D = ac - b^2 > 0.$$

5.4.3. A positive-definite binary quadratic form $F(x,y) = ax^2 + 2bxy + cy^2$ is called *reduced* if $2|b| \leq a \leq c$.

Now we are going to show that *any positive-definite binary quadratic form* $F(x,y) = ax^2 + 2bxy + cy^2$ *is equivalent to a reduced binary quadratic form, i.e., to the form* $G(t,u) = a^{'}t^2 + 2b^{'}tu + c^{'}u^2$, *such that* $2|b^{'}| \leq a^{'} \leq c^{'}$.

In fact, let $a^{''}$ be the smallest positive integer, represented by the form $F(x,y)$. Then there exist integers α, and γ, such that $a^{''} = F(\alpha,\gamma) = a\alpha^2 + 2b\alpha\gamma + c\gamma^2$. It is easy to show that the numbers α and γ are relatively prime, i.e., $gcd(\alpha,\gamma) = 1$. In fact, if $gcd(\alpha,\gamma) = d$, $d > 1$, then $d|\alpha$, $d|\beta$, so, $d^2|a^{''}$, and we obtain the decomposition $\frac{a^{''}}{d^2} = a\left(\frac{\alpha}{d}\right)^2 + 2b\frac{\alpha}{d}\cdot\frac{\beta}{d} + c\left(\frac{\gamma}{d}\right)^2$. It means that the positive integer $\frac{a^{''}}{d^2} < a^{''}$ is represented by $F(x,y)$, a contradiction with the minimality of $a^{''}$.

As $gcd(\alpha,\beta) = 1$, there exist integers β and δ, such that $\alpha\delta - \beta\gamma = 1$. Hence, the matrix $S = \begin{pmatrix} \alpha & \beta \\ \gamma & \delta \end{pmatrix}$ has the determinant 1 and takes the form $F(x,y) = ax^2 + 2bxy + cy^2$ into the form $G(t,u) = a^{''}t^2 + 2b^{''}tu + c^{''}u^2$.

Furthermore, for a given integer $\beta^{''}$, the matrix $\begin{pmatrix} 1 & \beta^{''} \\ 0 & 1 \end{pmatrix}$ has the determinant 1 and takes the form $G(t,u) = a^{''}x^2 + 2b^{''}tu + c^{''}u^2$ into the form $H(z,w) = a^{'}z^2 + 2b^{'}zw + c^{'}w^2$, where $a^{'} = a^{''}$, and $b^{'} = a^{'}\beta^{''} + b^{''}$.

Since $\beta^{''}$ is an arbitrary integer, one can choose it so that $2|b^{'}| \leq a^{'}$: it is sufficient to consider the decomposition $b^{''} = a^{'}q + r$, where r is the smallest residue modulo $a^{'}$, i.e., $|r| \leq \frac{a^{''}}{2}$, and to take $\beta^{''} = -q$; then $b^{'} = -a^{'}q + b^{''} = r$, i.e., $|b^{'}| \leq \frac{a^{'}}{2}$, and, hence, $2|b^{'}| < a^{'}$.

Since $H(0,1) = a^{'}\cdot 0^2 + 2b^{'}\cdot 0\cdot 1 + c^{'}\cdot 1^2 = c^{'}$, it holds that $c^{'}$ is represented by the form $H(z,w)$. On the other hand, the equivalent forms $F(x,y)$ and $H(z,w)$ represent the same integers, and, therefore, the number $a^{'} = a^{''}$ is the smallest positive integer represented by the form $H(z,w)$. Hence, we get that $a^{'} \leq c^{'}$.

So, we have proved, that *any class of positive-definite binary quadratic forms contains a reduced binary quadratic form.*

It is easy to show that *for any reduced binary quadratic form* $F(x,y) = ax^2 + 2bxy + cy^2$ *it holds*

$$a \leq \frac{2}{\sqrt{3}}\sqrt{D}.$$

In fact, for a reduced binary quadratic form $F(x,y) = ax^2+2bxy+cy^2$ it holds $2|b| \leq a \leq c$. Therefore, one has $a \leq c$, and $b^2 \leq \frac{a^2}{4}$. Since $D = ac - b^2$, we get that

$$a^2 \leq ac = b^2 + D \leq \frac{a^2}{4} + D, \quad \text{i.e.,} \ \frac{3}{4}a^2 \leq D, \quad \text{and } a \leq \frac{2}{\sqrt{3}}\sqrt{D}.$$

Now we can check that *the only reduced binary quadratic form of the determinant 1 is the form* $x^2 + y^2$.

In fact, for a reduced binary quadratic form $F(x,y) = ax^2+2bxy+cy^2$ with determinant 1 it holds: $0 < a \leq \frac{2}{\sqrt{3}}$, i.e., $a = 1$; $|b| \leq \frac{a}{2} = \frac{1}{2}$, i.e., $b = 0$; at last, $c = \frac{b^2+D}{a} = \frac{1+1}{1} = 1$. So, $F(x,y) = x^2 + y^2$.

The above results allow to state that *any positive definite form* $F(x,y)$ *of determinant 1 is equivalent to the form* x^2+y^2*, and, hence, represent the positive integers, which can be written as a sum of two squares.*

5.4.4. We can prove now the *two-square theorem: a positive integer can be represented as a sum of two squares if and only if its canonical decomposition has no odd powers of primes equal to 3 modulo 4.*

Let n be some positive integer. Let $n = a^2 \cdot n'$, where n' is a *square-free number*, i.e., a product, possibly empty, of distinct prime numbers.

Obviously, if n' is represented as a sum of two squares, then also n is:

$$n' = x^2 + y^2 \Leftrightarrow a^2n' = (ax)^2 + (ay)^2.$$

So, it is sufficient to prove the theorem for a square-free number n'. Since $1 = 1^2 + 0^2$, then without loss of generality we can suppose that $n' > 1$, i.e., $n = p_1 \cdot \ldots \cdot p_k$, $p_i \in P$, $p_i \neq p_j$ for $i \neq j$.

If a square-free number n' has a prime divisor equal to 3 modulo 4, it can not be represented as a sum of two squares. In fact, if $n' = x^2+y^2$, and $p|n'$, then $x^2+y^2 \equiv 0(mod\, p)$, i.e., $x^2 \equiv -y^2(mod\, p)$. If $p \in P\backslash\{2\}$ and $gcd(x,p) = gcd(y,p) = 1$, we get that $\left(\frac{-1}{p}\right) = 1$, i.e., $p \equiv 1(mod\, 4)$. If $p \in P\backslash\{2\}$ and $gcd(x,p) = gcd(y,p) = p$, then $p^2|x^2+y^2$, i.e., $p^2|n'$, a contradiction.

On the other hand, any square-free number n', which has no prime divisors equal to 3 modulo 4, is represented as a sum of two squares.

In fact, it is easy to show that for a given such number $n' > 1$ the congruence $x^2 \equiv -1(mod\, n')$ has an integer solition, i.e., there exists $x_0 \in \mathbb{Z}$ such that $x_0^2 \equiv -1(mod\, n')$.

Indeed, any odd prime divisor p_i of n' is equal to 1 modulo 4 and, hence, $\left(\frac{-1}{p_i}\right) = 1$, i.e., the congruence $x^2 \equiv -1(mod\, p_i)$ has integer solutions. If $2|n'$, then the congruence $x^2 \equiv -1(mod\, 2)$ also has integer solutions. So, there exists an integer solution x_0 modulo $p_1 \cdot \ldots \cdot p_k = n'$ of the system

$$\begin{cases} x^2 \equiv -1 \ (mod\, p_1) \\ \ldots \\ x^2 \equiv -1 \ (mod\, p_k). \end{cases}$$

Then $n' \cdot m - x_0^2 = 1$ for some positive integer m. It means that n' is represented by the positive definite quadratic form $F(x,y) = n'x^2 + 2x_0xy + my^2$ of determinant 1: $n' = F(1,0)$, and $D = n'm - x_0^2 = 1$.

So, n' can be represented by the form x^2+y^2, i.e., $n' = a^2+b^2$ for some integers a and b.

5.4.5. For a given 3×3 symmetric matrix $A = \begin{pmatrix} a_{11} & a_{12} & a_{13} \\ a_{21} & a_{22} & a_{23} \\ a_{31} & a_{32} & a_{33} \end{pmatrix}$ with integer entries and three integer variables x_1, x_2 and x_3, the *ternary quadratic form* $F(x_1,x_2,x_3)$, associated with the matrix A, is defined as

$$\begin{aligned} F(x_1,x_2,x_3) &= \sum_{1\le i,j\le 3} a_{ij}x_ix_j \\ &= a_{11}x_1^2 + 2a_{12}x_1x_2 + a_{22}x_2^2 + 2a_{13}x_1x_3 \\ &\quad +2a_{23}x_2x_3 + a_{33}x_3^2. \end{aligned}$$

The *determinant D of the ternary quadratic form $F(x_1, x_2, x_3)$* is defined as the determinant of the corresponding matrix A:

$$D = |A| = \begin{vmatrix} a_{11} & a_{12} & a_{13} \\ a_{21} & a_{22} & a_{23} \\ a_{31} & a_{32} & a_{33} \end{vmatrix}.$$

For a given 3×3 matrix $S = \begin{pmatrix} \alpha_{11} & \alpha_{12} & \alpha_{13} \\ \alpha_{21} & \alpha_{22} & \alpha_{23} \\ \alpha_{31} & \alpha_{32} & \alpha_{33} \end{pmatrix}$ with integer entries and non-zero determinant, let u_1, u_2, u_3 be new integer variables such that

$$x_1 = \alpha_{11}t_1 + \alpha_{12}t_2 + \alpha_{13}t_3, \quad x_2 = \alpha_{21}t_1 + \alpha_{22}t_2 + \alpha_{23}t_3,$$
$$x_3 = \alpha_{31}t_1 + \alpha_{32}t_2 + \alpha_{33}t_3.$$

Then one gets

$$\sum_{1\le i,j\le 3} a_{ij}x_ix_j = \sum_{1\le i,j\le 3}\left(\sum_{1\le k\le 3}\alpha_{ik}t_k\right)\left(\sum_{1\le m\le 3}\alpha_{jm}t_m\right)$$
$$= \sum_{1\le k,m\le 3} a'_{km}t_kt_m, \; a'_{km} = \sum_{1\le i,j\le 3}\alpha_{ik}a_{ij}\alpha_{jm}.$$

Therefore, using the matrix S, we obtain from a given ternary quadratic form $F(x_1, x_2, x_3) = \sum_{1\le i,j\le 3} a_{ij}x_ix_j$, associated with the matrix A, a new ternary quadratic form $G(u_1, u_2, u_3) = \sum_{1\le k,m\le 3} a'_{km}$ t_kt_m, associated with the matrix $A' = \begin{pmatrix} a'_{11} & a'_{12} & a'_{13} \\ a'_{21} & a'_{22} & a'_{23} \\ a'_{31} & a'_{32} & a'_{33} \end{pmatrix}$.

It is easy to check that

$$\begin{pmatrix} a'_{11} & a'_{12} & a'_{13} \\ a'_{21} & a'_{22} & a'_{23} \\ a'_{31} & a'_{32} & a'_{33} \end{pmatrix} = \begin{pmatrix} \alpha_{11} & \alpha_{21} & \alpha_{31} \\ \alpha_{12} & \alpha_{22} & \alpha_{32} \\ \alpha_{13} & \alpha_{23} & \alpha_{33} \end{pmatrix} \cdot \begin{pmatrix} a_{11} & a_{12} & a_{13} \\ a_{21} & a_{22} & a_{23} \\ a_{31} & a_{32} & a_{33} \end{pmatrix} \cdot \begin{pmatrix} \alpha_{11} & \alpha_{12} & \alpha_{13} \\ \alpha_{21} & \alpha_{22} & \alpha_{23} \\ \alpha_{31} & \alpha_{32} & \alpha_{33} \end{pmatrix}.$$

In other words, it holds $A' = S^T \cdot A \cdot S$. Then $|A'| = |A| \cdot |S|^2$, and one gets, that the determinant D' of the form $G(u_1, u_2, u_3)$, equal to the determinant of A', is obtained from the determinant D of the form

$F(x_1, x_2, x_3)$ by the equation $D' = D \cdot |S|^2$. If the determinant $|S|$ of the matrix S is equal to 1, then $D = D'$.

Two ternary quadratic forms $F(x_1, x_2, x_3)$ and $G(u_1, u_2, u_3)$ are called *equivalent*, $F(x_1, x_2, x_3) \sim G(u_1, u_2, u_3)$, if there exists an 3×3 matrix S with determinant 1, which takes $F(x_1, x_2, x_3)$ into $G(u_1, u_2, u_3)$.

It is easy to see that this relation is an equivalence relation: $F(x_1, x_2, x_3) \sim F(x_1, x_2, x_3)$; if $F(x_1, x_2, x_3) \sim G(u_1, u_2, u_3)$ then $G(u_1, u_2, u_3) \sim F(x_1, x_2, x_3)$; if $F(x_1, x_2, x_3) \sim G(u_1, u_2, u_3)$ and $G(u_1, u_2, u_3) \sim H(z_1, z_2, z_3)$, then $F(x_1, x_2, x_3) \sim H(z_1, z_2, z_3)$.

5.4.6. We say, that a ternary quadratic from $F(x_1, x_2, x_3) = \sum_{1 \leq i,j \leq 3} a_{ij} x_i x_j$ *represent a given integer* N, if there exist integers x_1^0, x_2^0, x_3^0, such that $N = F(x_1^0, x_2^0, x_3^0) = \sum_{1 \leq i,j \leq 3} a_{ij} x_i^0 x_j^0$.

Obviously, the equivalent forms $F(x_1, x_2, x_3)$ and $G(u_1, u_2, u_3)$ represent the same integers:

$$N = F(x_1^0, x_2^0, x_3^0) = \sum_{1 \leq i,j \leq 3} a_{ij} x_i^0 x_j^0 \Leftrightarrow N = G(u_1^0, u_2^0, u_3^0)$$
$$= \sum_{0 \leq i,j \leq 3} a''_{ij} u_i^0 u_j^0.$$

A ternary quadratic form $F(x_1, x_2, x_3) = \sum_{1 \leq i,j \leq 3} a_{ij} x_i x_j$ is called *positive definite* if it represents only non-negative integers, and represents 0 only in the case $x_1 = x_2 = x_3$.

Consider a form $F(x_1, x_2, x_3) = \sum_{1 \leq i,j \leq 3} a_{ij} x_i x_j$ with determinant D, which represents only non-negative integers, and represents 0 only in the case $x_1 = x_2 = x_3 = 0$. Then a_{11} should be non-negative, as it is represented by F: $a_{11} = F(1, 0, 0)$. Moreover, $a_{11} > 0$, since in the case $a_{11} = 0$ we get, that $F(1, 0, 0) = 0$, i.e., 0 is represented by $F(x_1, x_2, x_3)$ with non-zero values of variables, a contradiction.

Therefore, $a_{11} > 0$, and $a_{11} F(x_1, x_2, x_3) \geq 0$ for all $x_1, x_2, x_3 \in \mathbb{Z}$. But we have that

$$a_{11} F(x_1, x_2, x_3) = (a_{11} x_1 + a_{12} x_2 + a_{13} x_3)^2 + (a_{11} a_{22} - a_{12}^2) x_2^2$$
$$+ 2(a_{11} a_{23} - a_{12} a_{13}) x_2 x_3 + (a_{11} a_{33} - a_{13}^2) x_3^2.$$

So, the binary quadratic form

$$\begin{aligned}M(x_2,x_3) = (a_{11}a_{22} - a_{12}^2)x_2^2 + 2(a_{11}a_{23} - a_{12}a_{13})x_2x_3\\ +(a_{11}a_{33} - a_{13}^2)x_3^2\end{aligned}$$

should be positive definite. In fact, if it takes a negative value for some x_2^0 and x_3^0, then

$$\begin{aligned}&a_{11}F(-a_{12}x_2^0 - a_{13}x_3^0, a_{11}x_2^0, a_{11}x_3^0)\\ &\quad= (-a_{11}a_{12}x_2^0 - a_{11}a_{13}x_3^0 + a_{12}a_{11}x_2^0 + a_{13}a_{11}x_3^0)^2\\ &\qquad+M(x_2^0, x_3^0) = M(x_2^0, x_3^0) < 0,\end{aligned}$$

a contradiction. Similarly, if it has an non-trivial representation of 0 with some x_2^0 and x_3^0, then

$$\begin{aligned}&F(-a_{12}x_2^0 - a_{13}x_3^0, a_{11}x_2^0, a_{11}x_3^0)\\ &\quad= (-a_{12}x_2^0 - a_{13}x_3^0 + a_{12}x_2^0 + a_{13}x_3^0)^2 = 0\end{aligned}$$

is also a non-trivial representation of 0, a contradiction.

So, the form $M(x_2, x_3)$ is positive definite. It means that its first coefficient is positive, and its determinant is positive:

$$\begin{aligned}&a_{11}a_{22} - a_{12}^2 > 0, \quad \text{and}\\ &\qquad(a_{11}a_{22} - a_{12}^2)(a_{11}a_{33} - a_{13}^2) - (a_{11}a_{23} - a_{12}a_{13})^2 > 0.\end{aligned}$$

Since

$$\begin{aligned}&(a_{11}a_{22} - a_{12}^2)(a_{11}a_{33} - a_{13}^2) - (a_{11}a_{23} - a_{12}a_{13})^2\\ &\quad= a_{11}^2a_{22}a_{33} - a_{11}a_{12}^2a_{33} - a_{11}a_{22}a_{13}^2 + a_{12}^2a_{13}^2 - a_{11}^2a_{23}^2\\ &\qquad+2a_{11}a_{12}a_{13}a_{23} - a_{12}^2a_{13}^2\\ &\quad= a_{11}(a_{11}a_{22}a_{33} - a_{11}a_{23}^2 + 2a_{12}a_{13}a_{23} - a_{12}^2a_{33} - a_{13}^2a_{22}) = a_{11}D,\end{aligned}$$

and $a_{11} > 0$, we get that $D > 0$.

So, *for any positive definite ternary quadratic form* $F(x_1, x_2, x_3) = \sum_{1\le i,j\le 3} a_{ij}x_ix_j$ *it holds*

$$a_{11} > 0, \quad \begin{vmatrix} a_{11} & a_{12} \\ a_{21} & a_{22} \end{vmatrix} > 0, \quad \text{and} \quad \begin{vmatrix} a_{11} & a_{12} & a_{13} \\ a_{21} & a_{22} & a_{23} \\ a_{31} & a_{32} & a_{33} \end{vmatrix} > 0.$$

5.4.7. Now we are going to show that *any positive definite ternary quadratic form* $F(x_1, x_2, x_3) = \sum_{1\le i,j\le 3} a_{ij}x_ix_j$ *is equivalent to the ternary quadratic form* $G(u_1, u_2, u_3) = \sum_{0\le i,j\le 3} a'_{ij}u_iu_j$, *such that*

$$a'_{11} \le \frac{4}{3}\sqrt[3]{D'},\ 2|a'_{12}| \le a'_{11},\ 2|a'_{13}| \le a'_{11},$$

and $a'_{11}G(u_1, u_2, u_3) - (a'_{11}u_1 + a'_{12}u_2 + a'_{13}u_3)^2$ *is a reduced binary quadratic form.*

In fact, let a''_{11} is the smallest positive integer which is represented by $F(x_1, x_2, x_3)$. Then $a''_{11} = F(\alpha_{11}, \alpha_{21}, \alpha_{31})$ for some integers α_{11}, $\alpha_{21},$, α_{31}.

It is easy to see that the numbers α_{11}, α_{21}, α_{31} are relatively primes, i.e., $gcd(\alpha_{11}, \alpha_{21}, \alpha_{31}) = 1$.

Indeed, if $gcd(\alpha_{11}, \alpha_{12}, \alpha_{13}) = d > 1$, then $d^2|a''_{11}$, and it holds $\frac{a''_{11}}{d^2} = F\left(\frac{\alpha_{11}}{d}, \frac{\alpha_{12}}{d}, \frac{\alpha_{13}}{d}\right)$, i.e., the positive integer $\frac{a''_{11}}{d^2} < a''_{11}$ is represented by $F(x_1, x_2, x_3)$, a contradiction with minimality of a''_{11}.

Now we will try to find six other numbers α_{i2}, α_{i3}, $i = 1, 2, 3$, such that $\begin{vmatrix} \alpha_{11} & \alpha_{12} & \alpha_{13} \\ \alpha_{21} & \alpha_{22} & \alpha_{23} \\ \alpha_{31} & \alpha_{32} & \alpha_{33} \end{vmatrix} = 1$. Let $gcd(\alpha_{11}, \alpha_{21}) = g$. Then there exist integers α_{12} and α_{22} such that $\alpha_{11}\alpha_{22} - a_{12}\alpha_{21} = g$. Since $gcd(g, \alpha_{31}) = 1$, there exist integers k and q such that $gk - \alpha_{31}q = 1$. Then, taking $\alpha_{13} = \frac{\alpha_{11}}{g}q$, $\alpha_{23} = \frac{\alpha_{21}}{g}q$, $\alpha_{32} = 0$, and $\alpha_{33} = k$, we get

$$\begin{vmatrix} \alpha_{11} & \alpha_{12} & \frac{\alpha_{11}}{g}q \\ \alpha_{21} & \alpha_{22} & \frac{\alpha_{21}}{g}q \\ \alpha_{31} & 0 & k \end{vmatrix} = \alpha_{31}\left(\frac{\alpha_{12}\alpha_{21}}{g}q - \frac{\alpha_{11}\alpha_{22}}{g}q\right) + k\left(\alpha_{11}\alpha_{22} - \alpha_{12}\alpha_{21}\right)$$

$$= kg - q\alpha_{31} = 1.$$

So, the matrix $\begin{pmatrix} \alpha_{11} & \alpha_{12} & \alpha_{13} \\ \alpha_{21} & \alpha_{22} & \alpha_{23} \\ \alpha_{31} & \alpha_{32} & \alpha_{33} \end{pmatrix} = \begin{pmatrix} \alpha_{11} & \alpha_{12} & \frac{\alpha_{11}}{g}q \\ \alpha_{21} & \alpha_{22} & \frac{\alpha_{21}}{g}q \\ \alpha_{31} & 0 & k \end{pmatrix}$ with determinant 1 takes the form $F(x_1, x_2, x_3) = \sum_{1\le i,j\le 3} a_{ij}x_ix_j$ into the form $H(t_1, t_2, t_3) = \sum_{1\le i,j\le 3} a''_{ij}t_it_j$, such that $a''_{11} = \sum_{1\le i,j\le 3} \alpha_{i1}a_{ij}\alpha_{j1} = F(\alpha_{11}, \alpha_{21}, \alpha_{31})$ is the smallest positive integer represented by $F(x_1, x_2, x_3)$ and, therefore, represented by $H(t_1, t_2, t_3)$.

Furthermore, for a given integers α, β, γ and δ such that $\alpha\delta - \beta\gamma = 1$, the matrix $\begin{pmatrix} 1 & r & s \\ 0 & \alpha & \beta \\ 0 & \gamma & \delta \end{pmatrix}$ has determinant 1 and takes the form $H(t_1, t_2, t_3)$ into the form $G(u_1, u_2, u_3) = \sum_{1\le i,j\le 3} a'_{ij}u_iu_j$, where

$$a'_{11} = a''_{11},\ a'_{12} = ra''_{11} + \alpha a''_{12} + \gamma a''_{13},\ a'_{13} = sa''_{11} + \beta a''_{12} + \delta a''_{13},$$

and, moreover,

$$a'_{11}u_1 + a'_{12}u_2 + a'_{13}u_3 = a''_{11}t_1 + a''_{12}t_2 + a''_{13}t_3.$$

It holds that

$$a''_{11}H(t_1, t_2, t_3) = (a''_{11}t_1 + a''_{12}t_2 + a''_{13}t_3)^2 + L(t_2, t_3),$$

where $L(t_2, t_3) = (a''_{11}a''_{22} - (a''_{12})^2)t_2^2 + 2(a''_{11}a''_{23} - a''_{12}a''_{13})t_2t_3 + (a''_{11}a''_{33} - (a''_{13})^2)t_3^2$ is a positive definite binary quadratic form in t_2, t_3. Similarly,

$$a'_{11}G(u_1, u_2, u_3) = (a'_{11}u_1 + a'_{12}u_2 + a'_{13}u_3)^2 + M(u_2, u_3),$$

where $M(u_2, u_3) = (a'_{11}a'_{22} - (a'_{12})^2)u_2^2 + 2(a'_{11}a'_{23} - a'_{12}a'_{13})u_2u_3 + (a'_{11}a'_{33} - (a'_{13})^2)u_3^2$ is a positive definite binary quadratic form in u_2, u_3.

The conditions $a'_{11} = a''_{11}$ and $a'_{11}u_1 + a'_{12}u_2 + a'_{13}u_3 = a''_{11}t_1 + a''_{12}t_2 + a''_{13}t_3$ imply that the matrix $\begin{pmatrix} \alpha & \beta \\ \gamma & \delta \end{pmatrix}$ takes $L(t_2, t_3)$ into $M(u_2, u_3)$.

As was shown before, we can choose α, β, γ and δ so that the form

$$M(u_2, u_3) = (a'_{11}a'_{22} - (a'_{12})^2)u_2^2 + 2(a'_{11}a'_{23} - a'_{12}a'_{13})u_2u_3 + (a'_{11}a'_{33} - (a'_{13})^2)u_3^2$$

be reduced. It means that

$$2|a'_{11}a'_{23} - a'_{12}a'_{13}| \le a'_{11}a'_{22} - (a'_{12})^2 \le a'_{11}a'_{33} - (a'_{13})^2, \quad \text{and}$$
$$a'_{11}a'_{22} - (a'_{12})^2 \le \frac{2}{\sqrt{3}}\sqrt{a'_{11}D'},$$

since the determinant of $M(u_2, u_3)$ has the form $a'_{11}D'$, where $D' = D'' = D$ are the determinants of the form $G(u_1u_2, u_3)$, $H(t_1, t_2, t_3)$, and $F(x_1, x_2, x_3)$, respectively.

Since r is an arbitrary integer, one can choose it so that

$$|a_{12}^{'}| = |ra_{11}^{''} + \alpha a_{12}^{''} + \gamma a_{13}^{''}| \le \frac{a_{11}^{'}}{2} = \frac{a_{11}^{''}}{2}.$$

In fact, it is sufficient to consider the decomposition $\alpha a_{12}^{''} + \gamma a_{13}^{''} = a_{11}^{'}Q + R$, where R is the smallest residue modulo $a_{11}^{'}$, i.e., $|R| \le \frac{a_{11}^{'}}{2}$, and take $r = -Q$. Then $ra_{11}^{''} + \alpha a_{12}^{''} + \gamma a_{13}^{''} = R$, i.e., $|ra_{11}^{''} + \alpha a_{12}^{''} + \gamma a_{13}^{''}| \le \frac{a_{11}^{'}}{2}$.

Similarly, since s is an arbitrary integer, one can choose it so that

$$|a_{13}^{'}| = |sa_{11}^{''} + \beta a_{12}^{''} + \delta a_{13}^{''}| \le \frac{a_{11}^{'}}{2} = \frac{a_{11}^{''}}{2}.$$

In fact, it is sufficient to consider the decomposition $\beta a_{12}^{''} + \delta a_{13}^{''} = a_{11}^{'}P + W$, where W is the smallest residue modulo $a_{11}^{'}$, i.e., $|W| \le \frac{a_{11}^{'}}{2}$, and take $s = -P$. Then $sa_{11}^{''} + \beta a_{12}^{''} + \delta a_{13}^{''} = W$, i.e., $|sa_{11}^{''} + \beta a_{12}^{''} + \delta a_{13}^{''}| \le \frac{a_{11}^{'}}{2}$.

Since $a_{22}^{'} = G(0,1,0)$, it holds that $a_{22}^{'}$ is represented by $G(u_1, u_2, u_3)$. On the other hand, the equivalent forms $F(x_1, x_2, x_3)$ and $G(u_1, u_2, u_3)$ represent the same integers, and, therefore, the number $a_{11}^{'} = a_{11}^{''}$ is the smallest positive integer represented by the form $G(u_1, u_2, u_3)$. Hence, we get that $a_{11}^{'} \le a_{22}^{'}$. Therefore, it holds

$$(a_{11}^{'})^2 \le a_{11}^{'}a_{22}^{'} = a_{11}^{'}a_{22}^{'} - (a_{12}^{'})^2 + (a_{12}^{'})^2 \le \frac{2}{\sqrt{3}}\sqrt{a_{11}^{'}D'} + \frac{(a_{11}^{'})^2}{4},$$

$$\frac{3(a_{11}^{'})^2}{4} \le \frac{2}{\sqrt{3}}\sqrt{a_{11}^{'}D'}, \ (a_{11}^{'})^{\frac{3}{2}} \le \frac{8}{3\sqrt{3}}\sqrt{D'}, \quad \text{and } a_{11}^{'} \le \frac{4}{3}\sqrt[3]{D'}.$$

5.4.8. Now we can show that *every positive definite ternary quadratic form $F(x_1, x_2, x_3)$ of determinant 1 is equivalent to the form $u_1^2 + u_2^2 + u_3^3$, i.e., any positive integer, represented by such form, can be written as a sum of three squares.*

Indeed, in this case the form $G(u_1, u_2, u_3)$ also has the determinant 1 and we get that $0 < a_{11}^{''} = a_{11}^{'} \le \frac{4}{3}$, i.e., $a_{11}^{'} = 1$. Then

$|a'_{12}| \leq \frac{1}{2}$ and $|a'_{13}| \leq \frac{1}{2}$, i.e., $a'_{12} = 0$, and $a'_{13} = 0$. Therefore, it holds $G(u_1, u_2, u_3) = u_1^2 + M(u_2, u_3)$, where

$$\begin{aligned} M(u_2, u_3) &= (a'_{11}a'_{22} - (a'_{12})^2)u_2^2 + 2(a'_{11}a'_{23} - a'_{12}a'_{13})u_2u_3 \\ &\quad + (a'_{11}a'_{33} - (a'_{13})^2)u_3^2 \\ &= a'_{22}u_2^2 + 2a'_{23}u_2u_3 + a'_{33}u_3^2. \end{aligned}$$

Since $M(u_2, u_3)$ is reduced binary quadratic form with determinant $a'_{11}D' = 1 \cdot 1 = 1$, we get that $a'_{22} = 1$, $a'_{23} = 0$ and $a'_{33} = 1$, i.e., $M(u_2, u_3) = u_2^2 + u_3^2$, and

$$G(u_1, u_2, u_3) = u_1^2 + u_2^2 + u_3^2.$$

5.4.9. Now we can show that *any positive integer number N of the form $8m + 3$ can be represented as a sum of three squares.*

In order to prove this fact, it is sufficient to find a ternary positive definite quadratic form of determinant 1, representing N. If such form exists, then, by the previous result, it is equivalent to the form $u_1^2 + u_2^2 + u_3^2$ and, hence, the number N is represented as a sum of three squares.

So, for a given positive integer $N = 8m+3$ we should find an 3×3 symmetric matrix $((a_{ij}))$ such that the following conditions hold:

$$N = a_{11}x_1^2 + 2a_{12}x_1x_2 + a_{22}x_2^2 + 2a_{13}x_1x_3 + 2a_{23}x_2x_3 + a_{33}x_3^2,$$

$$a_{11} > 0, \ a_{11}a_{22} - a_{12}^2 > 0, \quad \text{and} \begin{vmatrix} a_{11} & a_{12} & a_{13} \\ a_{21} & a_{22} & a_{23} \\ a_{31} & a_{32} & a_{33} \end{vmatrix} = 1.$$

In order to simplify the situation, let $a_{13} = 1$, $a_{23} = 0$, $x_1 = 0$, $x_2 = 0$ and $x_3 = 1$. Then we can rewrite our conditions in the following form:

$$N = a_{33}, \ a_{11} > 0, \ a_{11}a_{22} - a_{12}^2 > 0, \quad \text{and} \begin{vmatrix} a_{11} & a_{12} & 1 \\ a_{21} & a_{22} & 0 \\ 1 & 0 & a_{33} \end{vmatrix}$$

$$= (a_{11}a_{22} - a_{12}^2)N - a_{22} = 1.$$

In other words, it holds

$$a_{11} > 0, \ \Delta = a_{11}a_{22} - a_{12}^2 > 0, \quad \text{and } a_{22} = \Delta N - 1.$$

Since $N > 1$, the first condition $a_{11} > 0$ follows from others ones: if $\Delta > 0$ and $N > 1$, then $a_{22} = \Delta N - 1 > 0$; if $a_{22} > 0$ and $\Delta = a_{11}a_{22} - a_{12}^2 > 0$, then $a_{11}a_{22} > 0$; if $a_{22} > 0$ and $a_{11}a_{22} > 0$, then $a_{11} > 0$. So, we can rewrite our conditions as follows:

$$\Delta = a_{11}a_{22} - a_{12}^2 > 0, \quad \text{and } a_{22} = \Delta N - 1.$$

Since $a_{12}^2 \equiv -\Delta(mod\, a_{22})$, and $a_{22} = \Delta N - 1$, it holds that the congruence $x^2 \equiv -\Delta(mod\ \Delta N - 1)$ has integer solutions; so, we are looking for a positive integer Δ, such that $-\Delta$ is a quadratic residue modulo $\Delta N - 1$.

We are going to find a number $2p$, where p is an odd prime, such that $2p = \Delta N - 1$ and the *Legendre symbol* $\left(\frac{-\Delta}{p}\right) = 1$. Since $N \equiv 3(mod\, 4)$ and $2p \equiv 2(mod\, 4)$, it holds that $\Delta \equiv 1(mod\, 4)$. Then, considering $\left(\frac{-\Delta}{p}\right)$ as *Jacobi symbol*, we get

$$\begin{aligned}\left(\frac{-\Delta}{p}\right) &= (-1)^{\frac{p-1}{2}}\left(\frac{\Delta}{p}\right) = (-1)^{\frac{p-1}{2}+\frac{p-1}{2}\cdot\frac{\Delta-1}{2}}\left(\frac{p}{\Delta}\right)\\ &= (-1)^{\frac{p-1}{2}+\frac{p-1}{2}\cdot\frac{\Delta-1}{2}+\frac{\Delta^2-1}{8}}\left(\frac{2p}{\Delta}\right)\\ &= (-1)^{\frac{p-1}{2}+\frac{p-1}{2}\cdot\frac{\Delta-1}{2}+\frac{\Delta^2-1}{8}}\left(\frac{\Delta N-1}{\Delta}\right)\\ &= (-1)^{\frac{p-1}{2}+\frac{p-1}{2}\cdot\frac{\Delta-1}{2}+\frac{\Delta^2-1}{8}}\left(\frac{-1}{\Delta}\right)\\ &= (-1)^{\frac{p-1}{2}+\frac{p-1}{2}\cdot\frac{\Delta-1}{2}+\frac{\Delta^2-1}{8}+\frac{\Delta-1}{2}}.\end{aligned}$$

So, if $p \equiv 1(mod\, 4)$, we get, that $\Delta \equiv 1\,(mod\, 8)$ and, hence, $\left(\frac{-\Delta}{p}\right) = 1$. It means that there exists integer x such that $x^2 \equiv -\Delta(mod\, p)$. Obviously, in this case one can choose x odd: for x even one can take instead of x the number $x + p$. Therefore, it holds $x^2 \equiv -\Delta(mod\, 2)$, and then $x^2 \equiv -\Delta(mod\, 2p)$, i.e., $-\Delta$ is a quadratic residue modulo $2p = \Delta N - 1$.

So, we should find a prime p equal to 1 modulo 4, such that for some Δ equal to 1 modulo 8 it holds $2p = \Delta N - 1$. In other words, one should find a prime in the arithmetic progression $\frac{\Delta N-1}{2} = \frac{(8k+1)N-1}{2} = (4N)\cdot k + \frac{N-1}{2}$ with the first term $\frac{N-1}{2}$ and the difference $4N$.

It is easy to see that for $N \equiv 3(mod\,8)$ it holds $8N = (N-1) \cdot 8 + 8$, $N - 1 = 8t + 2$, and $8 = 4 \cdot 2 + 0$. Therefore, due to the *Euclidean algorithm* we get that $gcd(8N, N-1) = 2$. It implies that $gcd(4N, \frac{N-1}{2}) = 1$, i.e., the difference $4N$ and the first term $\frac{N-1}{2}$ of the above arithmetic progression are relatively primes. By the *Dirichlet's theorem about primes in arithmetic progression* (see, for example, [Kara83]), one can find infinitely many such primes p.

5.5 Sums of squares and Minkowski's convex body theorem

The *Minkowski's convex body theorem* is often regarded as the fundamental theorem upon which an entire field, the geometry of numbers, is based. It was first proved in an (1896) paper of Minkowksi, and was treated at further length in Minkowski's (1910) text *Geometrie der Zahlen* ([Mink10]). Another proof was given in his (1927) *Diophantische Approximationen* ([Mink27]).

In this section we consider a short and elementary proof of the *three-square theorem*, based on the Minkowski convex body theorem (see [OLD00], [Clar11], [Land09], [Anke57]).

In the first part of our consideration we prove the Minkowski convex body theorem. In the next three parts we apply this result to obtain the *two-square*, *four-square*, and *three-square theorems.*

5.5.1. A set $\Omega \subset \mathbb{R}^n$ is called *convex* if for all pairs of points $x, y \in \Omega$ it contains also the entire *line segment* $[x, y] = \{(1-t)x + ty | 0 \leq t \leq 1\}$.

A set $\Omega \subset \mathbb{R}^n$ is called *centrally symmetric* if, whenever it contains a point $x \in \mathbb{R}^n$, it also contains $-x \in \mathbb{R}^n$, the reflection of x through the origin.

A set $\Omega \subset \mathbb{R}^n$ is called *bounded* if there exists some positive integer r such that the centered at the origin ball of the radius r contains Ω: $\Omega \subset \{x \in \mathbb{R}^n | x_1^2 + x_2^2 + \cdots + x_n^2 \leq r^2\}$.

A non-empty, bounded, centrally symmetric convex set $\Omega \subset \mathbb{R}^n$ is called *convex body.*

It is known that *a bounded convex set* $\Omega \subset \mathbb{R}^n$ *is Jordan measurable: that is, the function* $1_\Omega : \mathbb{R}^n \to \mathbb{R}^n$ *defined by* $1_\Omega(x) = \begin{cases} 1, x \in \Omega, \\ 0, x \notin \Omega \end{cases}$ *is Riemann integrable.* Therefore, one can define the *volume* $Vol(\Omega)$ of Ω as $Vol(\Omega) = \int_{\mathbb{R}^n} 1_\Omega dx_1 \dots dx_n$.

Let $M = ((m_{ij}))$ be an $n \times n$ matrix with real entries and non-zero determinant. Then the map $M : \mathbb{R}^n \to \mathbb{R}^n$, defined by $M(x) = (M \cdot x^T)^T$ for any $x = (x_1, \dots, x_n) \in \mathbb{R}^n$, is an *invertible linear map.*

It holds that *if* $\Omega \subset \mathbb{R}^n$ *is a convex body and* $M : \mathbb{R}^n \to \mathbb{R}^n$ *is an invertible linear map, then* $M(\Omega) = \{M(x)|x \in \Omega\}$ *is a convex body, and* $Vol(M(\Omega)) = |det\, M| Vol(\Omega)$.

In fact, the image $M(\Omega)$ of any bounded set $\Omega \subset \mathbb{R}^n$ is bounded. It holds with M replaced by any *homeomorphism* of $\mathbb{R}^n$, i.e., a continuous map from $\mathbb{R}^n$ to itself with continuous inverse, because a subset of $\mathbb{R}^n$ is bounded if and only if it is contained in a compact subset, and the image of a compact subset under a continuous function is bounded.

Moreover, the image $M(\Omega)$ of any convex set $\Omega \subset \mathbb{R}^n$ is convex. It is true because the image of a line segment under a linear map is a line segment.

Next, the image $M(\Omega)$ of any centrally symmetric set $\Omega \subset \mathbb{R}^n$ is centrally symmetric. It follows from the property $M(-x) = -M(x)$ of linear maps.

The preservation of Jordan measurability follows from the fact that an image of a set of measure zero under a linear map has measure zero.

Finally, the statement about volumes is precisely what one gets by applying the change of variables $(x_1, \dots, x_n) \to (y_1, \dots, y_n) = M(x_1, \dots, x_n)$ in the integral $\int_{\mathbb{R}^n} 1_\Omega dx_1 \dots dx_n$.

Given a set $\Omega \subset \mathbb{R}^n$ and a positive real number α, the *dilate* $\alpha\Omega$ of Ω is defined as the set

$$\alpha\Omega = \{\alpha \cdot x = (\alpha x_1, \dots, \alpha x_n)|x = (x_1, \dots, x_n) \in \Omega\}.$$

Obviously, the dilation by $\alpha > 0$, that is, the map $x \to \alpha x$, is an invertible linear map $M : \mathbb{R}^n \to \mathbb{R}^n$, defined by the matrix

$$M = \begin{pmatrix} \alpha & 0\ 0 & \dots & 0\ 0 \\ 0 & \alpha\ 0 & \dots & 0\ 0 \\ \dots & & & \\ 0 & 0\ 0 & \dots & 0\ \alpha \end{pmatrix} \text{ with determinant } \alpha^n.$$

An immediate consequence is that *if $\Omega \subset \mathbb{R}^n$ is a convex body of volume V, then for any positive real number α the set $\alpha\Omega$ is a convex body of volume $\alpha^n V$.*

By definition, for any convex body $\Omega \subset \mathbb{R}^n$ there exists a point $x \in \Omega$. Then $-x \in \Omega$ and, therefore, $0 \in \Omega$ as $0 \in [-x, x]$: $0 = \frac{1}{2}x + \frac{1}{2}(-x)$. So, any convex body contains the origin. In other words, it contains at least one point in the *lattice* $\mathbb{Z}^n = \{x = (x_1, \dots, x_n) \in \mathbb{R}^n | x_1, \dots, x_n \in \mathbb{Z}\}$.

Obviously, there exist convex bodies meet no non-trivial lattice points. Thus, the *open cube* $(-1, 1)^n$ is a symmetric convex subset of volume 2^n which meets no non-trivial lattice point, whereas for any $0 < V < 2^n$ the convex body $[-\frac{V^{\frac{1}{n}}}{2}, \frac{V^{\frac{1}{n}}}{2}]^n$ meets no non-trivial lattice point and has volume V.

On the other hand, the *Minkowski's convex body theorem* states, that *if $\Omega \subset \mathbb{R}^n$ is a convex body with $Vol(\Omega) > 2^n$, then Ω has at least one integer point, different from the origin.*

We will consider the proof of this result for $n = 2$ and $n = 3$ (see [OLD00]).

I. Let $n = 2$, and let $\Omega \subset \mathbb{R}^2$ be an convex body in the plane with area $Vol(\Omega)$ such that $Vol(\Omega) > 4$.

Using dilation with $\alpha = \frac{1}{2}$, we get the convex body $\Omega' = \frac{1}{2}\Omega$ with area $Vol(\Omega') = \frac{1}{4}Vol(\Omega)$, so that $Vol(\Omega') > 1$.

Given a lattice point $p \in \mathbb{Z}^2$, the *translate* $p + \Omega'$ of Ω' is defined as the set $\{p + x | x \in \Omega'\}$. In other words, we translate the set Ω' from the origin to the point $p = (p_1, p_2) \in \mathbb{Z}^2$: thus, if $x = (x_1, x_2)$ is a point of Ω', then $x + p = (x_1 + p_1, x_2 + p_2)$ is the corresponding point of the translate of Ω' centered at p.

Consider the set $\{p + \Omega' | p \in \mathbb{Z}\}$ of all translates of Ω'. We will show that these translates overlap.

We begin by considering the square with vertices $(0,0)$, $(k,0)$, $(0,k)$ and (k,k) fore some positive integer k. This square has $(k+1)^2$ lattice points in his interior and on its boundary. The sum of the areas of the $(k+1)^2$ translates centered of these lattice points is $(k+1)^2 Vol(\Omega^{'})$.

Let d be the maximum distance of any point of $\Omega^{'}$ from its center $0 = (0,0)$. Then all $(k+1)^2$ above convex bodies are contained in the square with the vertices $(-d,-d)$, $(k+d,-d)$, $(k+d,k+d)$, $(-d,k+d)$ of the side $k+2d$ and of the area $(k+2d)^2$. Let as show that the sum of the areas of the $(k+1)^2$ translates of $\Omega^{'}$ exceeds the area of the square containing them: $(k+1)^2 Vol(\Omega^{'}) > (k+2d)^2$. In fact,

$$(k+1)^2 Vol(\Omega^{'}) > (k+2d)^2 \Leftrightarrow (k+1)^2 Vol(\Omega^{'}) - (k+2d)^2 > 0$$
$$\Leftrightarrow (Vol(\Omega^{'}) - 1)k^2 + 2(Vol(\Omega^{'}) - 2d)k$$
$$+(Vol(\Omega^{'}) - 4d^2) > 0.$$

Since $Vol(\Omega^{'}) > 0$, the leading coefficient of the quadratic expression in k is positive, and the entire expression $(Vol(\Omega^{'})-1)k^2 + 2(Vol(\Omega^{'} - 2d))k + (Vol(\Omega^{'}) - 4d^2)$ is, therefore, positive for sufficiently large k. For such large k we have that $(k+1)^2 Vol(\Omega^{'}) > (k+2d)^2$, i.e., the translates overlap in the corresponding square.

Consider two overlapping translates $p + \Omega^{'}$ and $q + \Omega^{'}$ of $\Omega^{'}$ centered at $p = (p_1, p_2)$ and $q = (q_1, q_2)$. A third translate $a + \Omega^{'}$ centered at any lattice point $a = (a_1, a_2)$ must then have points in common with a fourth translate $a+q-p+\Omega^{'}$ centered at $a+p-q = (a_1 + q_1 - p_1, a_2 + q_2 - p_2)$. In particular, the set $\Omega^{'}$ itself, centered at $(0,0)$, has points in common with the translate $q - p + \Omega^{'} = \Omega^{''}$, centered at $q-p = (q_1 - p_1, q_2 - p_2)$. Remind that, by definition, any point $x^{''} = (x_1^{''}, x_2^{''}) \in \Omega^{''}$ can be written as $x^{''} = x^{'} + q - p$, where $x^{'} \in \Omega^{'}$ is the corresponding point of $\Omega^{'}$, i.e., $x_1^{''} = x_1^{'} + q_1 - p_1$, and $x_2^{''} = x_2^{'} + q_2 - p_2$.

Now we consider any point $c = (c_1, c_2) \in \Omega^{'} \cap \Omega^{''}$. Since $c = (c_1, c_2) \in \Omega^{''}$, there exists a point $b = (b_1, b_2) \in \Omega^{'}$ such that $c = b + (q-p)$, i.e., $c_1 = b_1 + (q_1 - p_1)$, $c_2 = b_2 + (q_2 - p_2)$. Because $\Omega^{'}$

is symmetric, it also contains the point $-b = (-b_1, -b_2)$. And because Ω' is convex, it contains the midpoint of the segment connecting any two of its points. The midpoint, connecting points $c = (c_1, c_2) \in \Omega'$ and $-b = (-b_1, -b_2) \in \Omega'$, is $\frac{1}{2}c + \frac{1}{2}(-b) = \frac{q-p}{2}$:

$$\left(\frac{c_1 - b_1}{2}, \frac{c_2 - b_2}{2}\right) = \left(\frac{q_1 - p_1}{2}, \frac{q_2 - p_2}{2}\right).$$

So, the set Ω' contains the point $\frac{q-p}{2} = \left(\frac{q_1-p_1}{2}, \frac{q_2-p_2}{2}\right)$, where $p, q \in \mathbb{Z}^2$. In other words, the set Ω' contains a half-integral point $\frac{q-p}{2}$, and, therefore, the set Ω contains integral point $q - p = (q_1 - p_1, q_2 - p_2)$, different from the origin.

II. The proof for $n = 3$ is similar. Let $n = 3$, and let $\Omega \subset \mathbb{R}^3$ be an convex body with volume $Vol(\Omega)$, such that $Vol(\Omega) > 2^3 = 8$. Using dilation with $\alpha = \frac{1}{2}$, we get the convex body $\Omega' = \frac{1}{2}\Omega$ with area $Vol(\Omega') = \frac{1}{8}Vol(\Omega)$, so that $Vol(\Omega') > 1$.

Now we consider the set of translates $p + \Omega'$ of Ω' on every lattice point $p \in \mathbb{Z}^3$. In other words, we translate the set Ω' from the origin to every point $p = (p_1, p_2, p_3) \in \mathbb{Z}^3$: thus, if $x = (x_1, x_2, x_3)$ is a point of Ω', then $x + p = (x_1 + p_1, x_2 + p_2, x_3 + p_3)$ is the corresponding point of the translate of Ω' centered at p.

We will show that these translates overlap. We begin by considering the cube with vertices $(0,0,0)$, $(k,0,0)$, $(0,k,0)$, $(k,k,0)$, $(0,0,k)$, $(k,0,k)$, $(0,k,k)$, (k,k,k) fore some positive integer k. This cube has $(k+1)^3$ lattice points in his interior and on its boundary. The sum of the volumes of the $(k+1)^3$ translates centered on these lattice points is $(k+1)^3 Vol(\Omega')$.

Let d be the maximum distance of any point of Ω' from the its center $(0,0,0)$. Then all $(k+1)^3$ above convex bodies are contained in the cube with the vertices $(-d,-d,-d)$, $(k+d,-d,-d)$, $(-d,k+d,-d)$, $(k+d,k+d,-d)$, $(-d,-d,k+d)$, $(k+d,-d,k+d)$, $(-d,k+d,k+d)$, $(k+d,k+d,k+d)$ of the side $k+2d$ and of the volume $(k+2d)^3$.

Let as show that the sum of the volumes of the $(k+1)^3$ translates of Ω' exceeds the volume of the cube containing them: $(k+1)^3Vol(\Omega^{'}) > (k+2d)^3$. In fact,

$$(k+1)^3Vol(\Omega^{'}) > (k+2d)^3 \Leftrightarrow (k+1)^3Vol(\Omega^{'}) - (k+2d)^3 > 0$$
$$\Leftrightarrow (Vol(\Omega^{'})-1)k^3 + 3(Vol(\Omega^{'})-2d)k^2$$
$$+3(Vol(\Omega^{'})-4d^2) + (Vol(\Omega^{'})-8d^3) > 0.$$

Since $Vol(\Omega^{'}) > 1$, the leading coefficient of the cubic expression in k is positive, and the entire expression is therefore positive for sufficiently large k. For such large k we have that $(k+1)^3Vol(\Omega^{'}) > (k+2d)^3$, i.e., the translates overlap in the corresponding cube.

So, there exist $p, q \in \mathbb{Z}^3$ such that $(p+\Omega^{'}) \cap (q+\Omega^{'}) \neq \emptyset$. In other words, there exist $c, b \in \Omega^{'}$ such that $p+c = q+b$ or $c-b = q-p \in \mathbb{Z}^3$. Since $\Omega^{'}$ is symmetric, it holds also $-b \in \Omega^{'}$ and, since $\Omega^{'}$ is convex and $c, -b \in \Omega^{'}$, their midpoint $\frac{1}{2}c + \frac{1}{2}(-b) = \frac{c-b}{2} \in \Omega^{'}$. Since $\frac{c-b}{2} = \frac{q-p}{2}$, we get that $\Omega^{'}$ contains a half-integral point $\frac{q-p}{2}$ and, hence, the set Ω contains the integral point $q-p \in \mathbb{Z}^3$, different from $(0,0,0)$.

III. The proof for general n can be obtained in similar way. However, one can use also the *Blichfeldt's Theorem*, which stated that *any bounded Jordan measurable subset* $\Omega^{'}$ *of* $\mathbb{R}^n$ *with* $Vol(\Omega^{'}) > 1$ *is not packable, i.e., there exist translates* $p+\Omega^{'}$ *and* $q+\Omega^{'}$, $p, q \in \mathbb{Z}^n$, *with non-empty intersection* (see [Clar11]).

Now it is easy to prove the following more general version of Minkowski's convex body theorem: *if* $M : \mathbb{R}^n \to \mathbb{R}^n$ *is an invertible linear map,* $\Lambda_M = M(\mathbb{Z}^n)$, *and* $\Omega \subset \mathbb{R}^n$ *is a convex body with* $Vol(\Omega) > 2^n|det\, M|$, *then there exists* $x \in \Omega \cap (\Lambda_M \backslash \{(0, \ldots, 0)\})$.

In fact, for any $n \times n$ matrix M with real entries and non-zero determinant put $\Lambda_M = M(\mathbb{Z}^n) = \{M(x) | x = (x_1, \ldots, x_n) \in \mathbb{Z}^n\}$. The map $\mathbb{Z}^n \to M(\mathbb{Z}^n)$ is an isomorphism of groups; so, $M(\mathbb{Z}^n)$ is, abstractly, just another copy of $\mathbb{Z}^n$. However, it is embedded inside $\mathbb{R}^n$ differently.

A simple geometric way to look at it is that $\mathbb{Z}^n$ is the vertex set of a tiling of $\mathbb{R}^n$ by unit hypercubes, whereas Λ_M is the vertex set of a tiling of $\mathbb{R}^n$ by hyperparallelepipeds. A single parallelpiped is called a *fundamental domain* for Λ_M, and the volume of a fundamental domain is given by $|det\, M|$. We refer to the volume of the fundamental domain as the volume of Λ_M, and write $Vol(\Lambda_M) = |det\, M|$.

It holds the following fundamental fact: *let* $\Omega \subset \mathbb{R}^n$ *and let* $M : \mathbb{R}^n \to \mathbb{R}^n$ *be an invertible linear map. Then* M *induces a bijection between* $M(\mathbb{Z}^n) \cap \Omega$ *and* $\mathbb{Z}^n \cap M^{-1}(\Omega)$. Once the statement is understood, the proof is immediate.

Applying this, gives the following: if we have a lattice $\Lambda_M = M(\mathbb{Z}^n)$ and a convex body Ω, the number of points of $\Lambda_M \cap \Omega$ is the same as the number of points of $\mathbb{Z}^n \cap M^{-1}(\Omega)$. Since

$$Vol(M^{-1}(\Omega)) = |det\, M^{-1}|Vol(\Omega) = \frac{Vol(\Omega)}{|det\, M|} = \frac{Vol(\Omega)}{Vol(\Lambda_M)},$$

we immediately deduce the proof of the above more general version of Minkowski's theorem. (see [Clar11]).

5.5.2. Now we consider the application of the Minkowski's convex body theorem to the decomposition into two squares (see [Clar11]).

Let $p \equiv 1(mod\, 4)$ be a prime number. Then there exists $u \in \mathbb{Z}$ such that $u^2 \equiv -1(mod\, p)$. Consider the 2×2 matrix

$$M = \begin{pmatrix} 1 & 0 \\ u & p \end{pmatrix}.$$

Its determinant $det\, M = p$, so, $\Lambda_M = M(\mathbb{Z}^2)$ defines a lattice in $\mathbb{R}^2$ with $Vol(\Lambda_M) = |det\, M|Vol(\mathbb{Z}^2) = p$. If $x = (x_1, x_2), t = (t_1, t_2) \in \mathbb{Z}^2$, and $x = M(t)$, then

$$x_1^2 + x_2^2 = t_1^2 + (ut_1 + pt_2)^2 \equiv (1 + u^2)t_1^2 \equiv 0(mod\, p).$$

Consider, as a convex body Ω, the open ball of radius $\sqrt{2p}$, centered at the origin:

$$\Omega = \{x \in \mathbb{R}^n | x_1^2 + x_2^2 < 2p\}.$$

Then $Vol(\Omega) = 2\pi p > 2^2 p = 2^2 Vol(\Lambda_M)$. So, by Minkowski's theorem, there exists $(x_1, x_2) \in \Lambda_M$ with $0 < x_1^2 + x_2^2 < 2p$. Since $p|(x_1^2 + x_2^2)$, the only possible conclusion is $x_1^2 + x_2^2 = p$.

Now we can easily obtain the general version of the two-square theorem, if we note that $a^2 = a^2 + 0^2$, $2 = 1^2 + 1^2$, and that if numbers n and m are represented as a sum of two squares, so is nm: $(x^2 + y^2)(a^2 + b^2) = (ax + by)^2 + (ay - bx)^2$.

5.5.3. Now we consider the application of the Minkowski's theorem to the four-square theorem (see [Clar11]).

At first we will show that *for any prime number p and any integer a there exist $r, s \in \mathbb{Z}$ such that $r^2 + s^2 \equiv a (mod\, p)$.*

We may assume that $p > 2$. Then there are precisely $\frac{p-1}{2}$ quadratic residues modulo p, and hence precisely $\frac{p-1}{2} + 1 = \frac{p+1}{2}$ squares modulo p, including 0. Therefore, as r, s range over all residues $0, 1, \ldots, p-1$ modulo p, both the left and right hand sides of the congruence $r^2 \equiv a - s^2 (mod\, p)$ take $\frac{p+1}{2}$ distinct values. Since $\frac{p+1}{2} + \frac{p+1}{2} > p$, these subsets can not be disjoint, and any common value gives a solution to the above congruence.

For $a = -1$ we get, that there exist $r, s \in \mathbb{Z}$ such that $r^2 + s^2 + 1 \equiv 0 (mod\, p)$.

Consider the 4×4 matrix

$$M = \begin{pmatrix} p & 0 & r & s \\ 0 & p & s & -r \\ 0 & 0 & 1 & 0 \\ 0 & 0 & 0 & 1 \end{pmatrix}.$$

Its determinant $det\, M = p^2$, therefore $\Lambda_M = M(\mathbb{Z}^4)$ defines a lattice in $\mathbb{R}^4$ with $Vol(\Lambda_M) = |det\, M| Vol(\mathbb{Z}^4) = p^2$. If $x = (x_1, \ldots, x_4), t = (t_1, \ldots, t_4) \in \mathbb{Z}^4$, and $x = M(t)$, then

$$\begin{aligned} x_1^2 + x_2^2 + x_3^2 + x_4^2 &= (pt_1 + rt_3 + st_4)^2 + (pt_2 + st_3 - rt_4)^2 + t_3^2 + t_4^2 \\ &\equiv t_3^2(r^2 + s^2 + 1) + t_4^2(r^2 + s^2 + 1) \equiv 0 (mod\, p). \end{aligned}$$

Consider, as a convex body $\Omega \subset \mathbb{R}^4$, the open ball of of radius $\sqrt{2p}$ centered at the origin:

$$\Omega = \{x \in \mathbb{R}^4 | x_1^2 + x_2^2 + x_3^2 + x_4^2 < 2p\}.$$

It is known that the four-dimensional volume of a ball of radius r in $\mathbb{R}^4$ is equal to $\frac{\pi^2}{2} r^4$. Therefore, $Vol(\Omega) = 2\pi^2 p^2 > 2^4 Vol(\Lambda_M)$, and, by Minkowski's Theorem, there exists $(x_1, \ldots, x_4) \in \Lambda_M$ with $0 < x_1^2 + x_2^2 + x_3^2 + x_4^2 < 2p$. Since $p | (x_1^2 + x_2^2 + x_3^2 + x_4^2)$, the only possible conclusion is $x_1^2 + x_2^2 + x_3^2 + x_4^2 = p$.

Now we can easy obtain the general version of the four-square theorem, if we note that $2 = 1^2 + 1^2 + 0^2 + 0^2$, and, due to the *Euler's four-square identity*, if numbers n and m are represented as a sum of two squares, so is nm.

5.5.4. Now we consider the application of the Minkowski's theorem to the case of $8m+3$ of the three-square theorem: *any positive integer* $N = 8m+3$ *can be written as a sum of three squares* (see [Anke57]).

Without loss of generality we will prove the theorem only when positive integer N equal to 3 modulo 8 is square-free.

Then the canonical decomposition of n has the form $N = p_1 \cdot \ldots \cdot p_k$, where $p_1, \ldots, p_k$ are distinct odd primes. Denote by q a prime number equal to 1 modulo 4 which satisfies the condition

$$\left(\frac{-2q}{p_i}\right) = 1 \quad \text{for any } i = 1, 2, \ldots, k,$$

where $(\frac{a}{b})$ is the *Legendre symbol.*

It is easy to see that such a prime exists, by Dirichlet's theorem regarding primes in an arithmetic progression, as the above conditions merely necessitate that q lie within certain relatively prime residue classes modulo $4N$.

In fact, since $(\frac{-2q}{p_i}) = 1$ for any $i = 1, \ldots, k$, then all congruences $x^2 \equiv -2q (mod\, p_i)$, $i = 1, \ldots, k$, have integer solutions. Since $gcd(p_i, p_j) = 1$ for $i \neq j$, it holds that the congruence $x^2 \equiv -2q(mod\, p_1 \cdot \ldots \cdot p_k)$ has an integer solution. Therefore, there exist $x_0 \in \mathbb{Z}$, such that $x_0^2 \equiv -2q(mod\, N)$. Moreover, the number x_0 always can be chosen odd: if x_0 is even, one should take instead of x_0 the odd number $x_0 + N$.

Hence, one gets that $-2q = x_0^2 + Na$ for some integer a. Since $q \equiv 1(mod\, 4)$, it holds that $-2q \equiv -2(mod\, 8)$. Since x_0 is odd, it holds that $x_0^2 \equiv 1(mod\, 8)$. Therefore, using the condition $N \equiv 3(mod\, 8)$, we get that $-2 \equiv 1 + 3a(mod\, 8)$, and, hence, $a \equiv -1(mod\, 8)$.

Therefore, it holds $q = -\frac{x_0^2+Na}{2} = -\frac{x_0^2+N(8k-1)}{2} = (4N)\cdot s + \frac{N-x_0^2}{2}$. It means that q belongs to the arithmetic progression with the first term $\frac{N-x_0^2}{2}$ and the difference $4N$. Since $8N = (N-1)\cdot 8 + 8$, $N-1 = 8k+2$, and $8 = 2\cdot 4$, we get due to the *Euclidean algorithm* that $gcd(N - x_0^2, 8N) = gcd(N-1, 8N) = 2$. So, $gcd(\frac{N-x_0^2}{2}, 4N) = 1$.

Considering $(\frac{a}{b})$ as *Jacobi symbol*, and using that $N \equiv 3(mod\, 8)$, while $q \equiv 1(mod\, 4)$, we get

$$\begin{aligned} 1 &= \left(\frac{-2q}{p_1}\right) \cdot \ldots \cdot \left(\frac{-2q}{p_k}\right) \\ &= \left(\left(\frac{-2}{p_1}\right) \cdot \ldots \cdot \left(\frac{-2}{p_k}\right)\right) \cdot \left(\left(\frac{q}{p_1}\right) \cdot \ldots \cdot \left(\frac{q}{p_k}\right)\right) \\ &= \left(\frac{-2}{N}\right) \cdot \left(\left(\frac{p_1}{q}\right) \cdot \ldots \cdot \left(\frac{p_k}{q}\right)\right) = \left(\frac{N}{q}\right) = \left(\frac{-N}{q}\right). \end{aligned}$$

Therefore, $\left(\frac{-N}{q}\right) = 1$, i.e., the number $-N$ is a quadratic residue modulo q. Hence, there exists an odd integer b such that $b^2 \equiv -N(mod\, q)$. In other words, $b^2 - qh_1 = -N$ for some integer h_1.

Considering last equation modulo 4 yields that $1 - h_1 \equiv 1(mod\, 4)$, or $h_1 \equiv 0(mod\, 4)$. So, $h_1 = 4h$, where h is an integer, and the above condition can be rewritten as

$$b^2 - 4qh = -N, \ h \in \mathbb{Z}.$$

Utilizing the condition $\left(\frac{-2q}{p_i}\right) = 1$, $i = 1, 2, \ldots, k$, we can find integers u_i such that $u_i^2 \equiv -2q(mod\, p_i)$, $i = 1, 2, \ldots, k$, and, hence, an integer u such that $u^2 \equiv -2q(mod\, N)$. Since $gcd(2q, N) = 1$, there exists an *inverse element* for $2q$: an integer g, such that $(2q) \cdot g \equiv 1(mod\, m)$. Let us define it by $\frac{1}{2q}$. In this case we have $(gu)^2 \equiv -((2q)g) \cdot g \equiv -g(mod\, N)$, i.e., there exists an integer t, such that

$$t^2 = -\frac{1}{2q}(mod\, N).$$

Now we consider the figure $R^2 + S^2 + T^2 < 2N$, where

$$R = 2tqx + tby + Nz, \ S = \sqrt{2q}x + \frac{b}{\sqrt{2q}}y, \ T = \frac{\sqrt{N}}{\sqrt{2q}}y.$$

In the (R, S, T)-space the condition $R^2 + S^2 + T^2 < 2N$ defines an open ball of radius $\sqrt{2N}$, centered at the origin, i.e., a convex body of volume $\frac{4}{3}\pi(2N)^{\frac{3}{2}}$. The determinant of the transformation (x, y, z)

into (R, S, T) is

$$\begin{vmatrix} 2tq & tb & N \\ \sqrt{2q} & \frac{b}{\sqrt{2q}} & 0 \\ 0 & \frac{\sqrt{N}}{\sqrt{2q}} & 0 \end{vmatrix} = (N)^{\frac{3}{2}}.$$

Hence, in the (x, y, z)-space the condition $R^2 + S^2 + T^2 < 2N$ represents a convex symmetric body of volume $\frac{4}{3}\pi(2N)^{\frac{3}{2}} \cdot N^{-\frac{3}{2}} = \frac{2^{7/2}}{3}\pi$, and certainly $\frac{2^{7/2}}{3}\pi > 8$.

So, by Minkowski's convex body theorem, there exist integer values of x, y, z not all zero which satisfy $R^2+S^2+T^2 < 2N$. Let x_1, y_1, z_1 be the integers which satisfy the above conditions and R_1, S_1, and T_1 be the corresponding values of R, S, T. So, we have

$$\begin{aligned} R_1^2 + S_1^2 + T_1^2 &= (2tqx_1 + tby_1 + Nz_1)^2 \\ &\quad +(\sqrt{2q}x_1 + \frac{b}{\sqrt{2q}}y_1)^2 + (\frac{\sqrt{N}}{\sqrt{2q}}y_1)^2 \\ &\equiv t^2(2qx_1 + by_1)^2 + \frac{1}{2q}(2qx_1 + by_1)^2 \equiv 0(mod\, N) \end{aligned}$$

by the selection of t. Furthermore,

$$\begin{aligned} R_1^2 + S_1^2 + T_1^2 &= R_1^2 + \left(\sqrt{2q}x_1 + \frac{b}{\sqrt{2q}}y_1\right)^2 + \left(\frac{\sqrt{m}}{\sqrt{2q}}y_1\right)^2 \\ &= R_1^2 + \frac{1}{2q}(2qx_1 + by_1)^2 + \frac{m}{2q}y_1^2 \\ &= R_1^2 + 2(qx_1^2 + bx_1y_1 + hy_1^2). \end{aligned}$$

Let v be the positive integer defined by $v = qx_1^2 + bx_1y_1 + hy_1^2$. So, we have that R_1 and v are integers, such that $N|R_1^2 + 2v$ and $0 \leq R_1^2 + 2v < 2N$. Furthermore, $R_1^2 + 2v \neq 0$ by the non-degenerateness of the transformation $(x, y, z) \to (R, S, T)$ and the fact that not all x_1, y_1, z_1 are equal to zero. Hence, it holds

$$R_1^2 + 2v = N.$$

Suppose that there exists an odd prime p which exactly divides v to an odd power, i.e., $p^{2n+1}|v$, but p^{2n+2} is not divides v. Then $N \equiv R_1^2(mod\, p)$.

If p does not divide N, then $\left(\frac{N}{p}\right)=1$. By the equation $v = qx_1^2 + bx_1y_1 + hy_1^2$ we get that

$$4qv = (2qx_1 + by_1)^2 + Ny_1^2.$$

If $p|q$ it implies $-Ny_1^2 \equiv (2qx_1 + by_1)^2 (mod\, p)$ and, therefore, $\left(\frac{-N}{p}\right) = 1$.

If p is not divided q, then by the chose of p we get that that $p^{2n+1}|(2qx_1 + by_1)^2 + Ny_1^2$, but p^{2n+2} is not divided $(2qx_1 + by_1)^2 + Ny_1^2$, and therefore $\left(\frac{-N}{p}\right) = 1$. Thus, in either case, $\left(\frac{-N}{p}\right) = 1$, which combined with $\left(\frac{N}{p}\right) = 1$ implies $\left(\frac{-1}{p}\right) = 1$ or $p \equiv 1(mod 4)$.

If $p|v$, $p|N$, then by the formula $N = R_1^2 + 2v$ we get that $p|R_1$, and by the formula $4qv = (2qx_1 + by_1)^2 + Ny_1^2$ we get that $p|2qx_1 + by_1$. Using the formula

$$N = R_1^2 + 2v = R_1^2 + \frac{1}{2q}4qv = R_1^2 + \frac{1}{2q}((2qx_1 + by_1)^2 + my_1^2)$$

and the fact that N is square-free, we get by dividing both sides above equation by p that

$$\frac{1}{2p} \cdot \frac{N}{p} y_1^2 \equiv \frac{N}{p} (mod\, p)$$

or $y_1^2 \equiv 2q(mod\, p)$, i.e., $\left(\frac{2q}{p}\right) = 1$, which combined with the condition $\left(\frac{-2q}{p}\right) = 1$ gives $\left(\frac{-1}{p}\right) = 1$, i.e., $p = 1(mod\, 4)$.

Thus, all odd primes which exactly divide v to an odd power are equal to 1 modulo 4. Due to the two-square theorem it implies that $2v$ can be represented as a sum of two squares.

Hence, $N = R^2 + 2v$ can be represented as a sum of of three squares, which proves the theorem for $N \equiv 3(mod\, 8)$.[2]

[2]If $N \equiv 1, 2, 5, 6(mod\, 8)$, we alter the proof in the following ways. Let $q \equiv 1(mod\, 4)$ be a prime, such that $\left(\frac{-q}{p_i}\right) = 1$ for all odd prime divisors of N. Moreover, if $N = 2N_1$, then $\left(\frac{-2}{q}\right) = (-1)^{\frac{N_1-1}{2}}$. Let $t^2 = -\frac{1}{q}(mod\, p_i)$, t is odd, $b^2 - qh = -N$, and $R = tqx + tby + mz$, $S = \sqrt{q}x + \frac{b}{\sqrt{q}}y$, $T = \frac{\sqrt{m}}{\sqrt{q}}y$. The proof will proceed exactly as in the previous case, which will complete the proof for any admissible square-free N.

5.6 Cauchy's proof of the polygonal number theorem

In this section we consider the general case of Fermat's polygonal number theorem, which was proved by Cauchy ([Cauc13]): *for* $m \geq 3$, *every positive integer* N *can be represented as a sum of* $m+2$ $(m+2)$-*gonal numbers, at most four of which are different from* 0 *or* 1.

5.6.1. Following the original Cauchy's proof, we will show that, for a given $m \geq 3$, any positive integer N can be represented as

$$N = m\frac{k-s}{2} + s + r,$$

where $r \in \{0, \ldots, m-2\}$,while k and s are some positive integers, for which the system

$$\begin{cases} k = t^2 + u^2 + v^2 + w^2 \\ s = t + u + v + w \end{cases}$$

has a non-negative integer solution (t, u, v, w). In this case we get

$$N = \left(m\frac{t^2-t}{2} + t\right) + \left(m\frac{u^2-u}{2} + u\right) + \left(m\frac{v^2-v}{2} + v\right) + \left(m\frac{w^2-w}{2} + w\right) + r.$$

It is a representation N as a sum of $m+2$ $(m+2)$-gonal numbers, at most four of which are different from 0 or 1. In fact, $m\frac{n^2-n}{2} + n = S_{m+2}(n)$, and any $r \in \{0, \ldots, m-2\}$ can be considered as the sum of $m-2$ $(m+2)$-gonal numbers, r of which are equal to $S_{m+2}(1) = 1$, and others are equal to $S_{m+2}(0) = 0$.

5.6.2. The idea of the proof is the following.

At first, we will show that, for any odd numbers k and s, such that s belong to the segment $[\sqrt{3k-2} - 1, \sqrt{4k}]$, there exists a non-negative integer solution (t, u, v, w) of the system

$$\begin{cases} k = t^2 + u^2 + v^2 + w^2 \\ s = t + u + v + w. \end{cases}$$

This result is usually called *Cauchy's lemma.*

In fact, it is easy to show that for any odd positive integer k the segment $[\sqrt{3k-2}-1, \sqrt{4k}]$ contains at least one odd integer s, and for such k and s the above system has a non-negative integer solution (t, u, v, w). Therefore, all consecutive positive integers

$$m\frac{k-s}{2}+s, m\frac{k-s}{2}+s+1, \dots, m\frac{k-s}{2}+s+m-2$$
$$= m\left(\frac{k-s}{2}+1\right)+s-2$$

are represented as a sum of $m+2$ $(m+2)$-gonal numbers.

Moreover, usually the segment $[\sqrt{3k-2}-1, \sqrt{4k}]$ contains at least two odd integers, say, s and $s+2$. In particular, it is true for all odd $k \geq 123$.

In this case we get two finite sequences of consecutive positive integers, represented as a sum of $m+2$ $(m+2)$-gonal numbers. The first sequence, corresponding to s, is given before. The second sequence, corresponding to $s+2$, has the form

$$m\frac{k-(s+2)}{2}+(s+2), m\frac{k-(s+2)}{2}+(s+2)+1, \dots,$$
$$m\frac{k-(s+2)}{2}+(s+2)+m-2.$$

Since $m\frac{k-(s+2)}{2} = m\left(\frac{k-s}{2}-1\right)$, it can be rewrite as follows:

$$m\left(\frac{k-s}{2}-1\right)+s+2, m\left(\frac{k-s}{2}-1\right)+s+3, \dots,$$
$$m\left(\frac{k-s}{2}-1\right)+s+m = m\frac{k-s}{2}+s.$$

The union of these two overlapping sequences gives the following finite sequence of positive integers, represented as a sum of $m+2$ $(m+2)$-gonal numbers:

$$m\left(\frac{k-s}{2}-1\right)+s+2, m\left(\frac{k-s}{2}-1\right)+s+3, \dots,$$
$$m\left(\frac{k-s}{2}+1\right)+s-2.$$

Starting from $k=1$, we can construct, for any odd k, corresponding finite sequence of positive integers, representable as a sum of $m+2$

$(m+2)$-gonal numbers. It will be shown, that these sequences cover, with several exceptions for small k, the set $\mathbb{N}$ of all positive integers.

Moreover, it will be proven that for a given k the only possible exception has the form

$$m\frac{k-s}{2}+s+m-1.$$

In this case, the pairs $(k+1, s+1)$ or $(k+1, s-1)$ of even positive integers should be used. In the first case, the number $m\frac{k-s}{2}+s+m-1$ can be written as $m\frac{(k+1)-(s+1)}{2}+(s+1)+m-2$. In the second case, $m\frac{k-s}{2}+s+m-1$ can be written as $m\frac{(k+1)-(s-1)}{2}+(s-1)+0$.

So, the number $m\frac{k-s}{2}+s+m-1$ can be represented as a sum of $m+2$ $(m+2)$-gonal numbers, if the above system has positive integer solutions either for the pair $(k+1, s+1)$, or for the pair $(k+1, s-1)$. More exactly, we obtain such representation for $m\frac{k-s}{2}+s+m-1$, if one of the systems

$$\begin{cases} k+1 = t^2+u^2+v^2+w^2 \\ s+1 = t+u+v+w \end{cases} \quad \text{or} \quad \begin{cases} k+1 = t^2+u^2+v^2+w^2 \\ s-1 = t+u+v+w \end{cases}$$

has non-negative integer solutions. As the number of exceptions is finite, we will obtain below exact solution for any such pair.

5.6.3. Let us prove the *Cauchy's lemma: for any odd numbers k and s, such that s belong to the segment $[\sqrt{3k-2}-1, \sqrt{4k}]$, there exists a non-negative integer solution (t, u, v, w) of the system*

$$\begin{cases} k = t^2+u^2+v^2+w^2 \\ s = t+u+v+w. \end{cases}$$

In fact, for odd k and s, the number $4k-s^2$ has the form $8l+3$:

$$\begin{aligned} 4k-s^2 &= 4(2b+1)-(2a+1)^2 = 8b+4-4a^2-4a-1 \\ &= 8b-4a(a+1)+3 = 8l+3. \end{aligned}$$

Since $s \leq \sqrt{4k}$, one has $4k \geq s^2$. Therefore, the number $4k-s^2$ is a positive integer of the form $8m+3$. Hence, due to the Gauss's three-triangular-number theorem, there exist positive odd numbers

$x \geq y \geq z$, such that

$$4k - s^2 = x^2 + y^2 + z^2.$$

Clearly, for non-negative integers x, y, z one has

$$(x+y+z)^2 = x^2 + y^2 + z^2 + 2x + 2y + 2z \geq x^2 + y^2 + z^2, \quad \text{and}$$

$$\begin{aligned}(x+y+z)^2 &\leq (x+y+z)^2 + (x-y)^2 + (x-z)^2 + (y-z)^2 \\ &= 3(x^2+y^2+z^2).\end{aligned}$$

In particular, the equality $x^2 + y^2 + z^2 = 4k - s^2$ implies $x + y + z \leq \sqrt{3(4k - s^2)}$. Since $s \geq \sqrt{3k-2} - 1$, one has $(s+1)^2 \geq 3k - 2$, and $3k \leq s^2 + 2s + 3$. It implies

$$\begin{aligned}x + y + z &\leq \sqrt{12k - 3s^2} \leq \sqrt{s^2 + 8s + 12} < \sqrt{s^2 + 8s + 16} \\ &= s + 4.\end{aligned}$$

In other words, it holds $\frac{s-x-y-z}{4} > -1$. So, the least integer number, obtained by the formula $\frac{s \pm x \pm y \pm z}{4}$, should be non-negative.

Since the numbers x, y and z are odd, the number $x + y + z$ is also odd, i.e., $x + y + z \equiv \pm 1 (mod\ 4)$. Therefore, for odd s it holds

$$s - x - y - z \equiv 0 (mod\ 4), \quad \text{or } s + x + y + z \equiv 0 (mod\ 4).$$

In other words, either $\frac{s-x-y-z}{4}$ or $\frac{s+x+y+z}{4}$ is an integer.

In the first case, let us define integers t, u, v and w by

$$t = \frac{s-x-y-z}{4}, \quad u = t + \frac{y+z}{2},$$

$$v = t + \frac{x+z}{2}, \quad w = t + \frac{x+y}{2},$$

or, in other words, by

$$t = \frac{s-x-y-z}{4}, \quad u = \frac{s-x+y+z}{4},$$

$$v = \frac{s+x-y+z}{4}, \quad t = \frac{s+x+y-z}{4}.$$

In the second case, let us define integers t, u, v and w by

$$t = \frac{s+x+y+z}{4}, \quad u = t - \frac{y+z}{2},$$

$$v = t - \frac{x+z}{2}, \quad w = t - \frac{x+y}{2},$$

or, in other words, by

$$t = \frac{s+x+y+z}{4}, \quad u = \frac{s+x-y-z}{4},$$
$$v = \frac{s-x+y-z}{4}, \quad w = \frac{s-x-y+z}{4}.$$

A direct checking shows, that in both cases one obtains a non-negative integer solution (t, u, v, w) of the above system.[3]

5.6.4. Let us consider the general case $k \geq 123$. In this case, the difference $\sqrt{4k} - (\sqrt{3k-2} - 1)$ is greater than 4:

$$\begin{aligned}
\sqrt{4k} - (\sqrt{3k-2} - 1) > 4 &\Leftrightarrow \sqrt{4k} - \sqrt{3k-2} > 3 \\
&\Leftrightarrow 4k + 3k - 2 - 2\sqrt{4k(3k-2)} > 9 \\
&\Leftrightarrow 7k - 11 > 4\sqrt{k(3k-2)} \\
&\Leftrightarrow 49k^2 + 121 - 154k > 16k(3k-2) \\
&\Leftrightarrow k^2 - 122k + 121 > 0 \Leftrightarrow k > 121 \quad \text{or } k < 1.
\end{aligned}$$

Hence, the segment $[\sqrt{3k-2} - 1, \sqrt{4k}]$ has at least two odd integers. Taking two smallest such numbers, s and $s+2$, one obtains, for a given k, the following finite sequence of consecutive positive integers, represented as a sum of $m+2$ $(m+2)$-gonal numbers:

$$m\left(\frac{k-s}{2} - 1\right) + s + 2, m\left(\frac{k-s}{2} - 1\right) + s + 3, \ldots,$$
$$m\left(\frac{k-s}{2} + 1\right) + s - 2.$$

Going from k to $k' = k+2$, one can see that the difference $(\sqrt{3k'-2} - 1) - (\sqrt{3k-2} - 1)$ is less than 2:

$$\begin{aligned}
\sqrt{3k+4} - \sqrt{3k-2} < 2 &\Leftrightarrow 3k + 4 + 3k - 2 - 2\sqrt{(3k+4)(3k-2)} < 4 \\
&\Leftrightarrow 6k - 2 < 2\sqrt{(3k+4)(3k-2)} \\
&\Leftrightarrow 36k^2 + 4 - 24k < 4(3k+4)(3k-2) \\
&\Leftrightarrow 48k > 36 \Leftrightarrow k > \frac{1}{8}.
\end{aligned}$$

---

[3] In [Cauc13] it is shown, that for even k and s, such that $s \in [\sqrt{3k-2} - 1, \sqrt{4k}]$, the system above also has non-negative integer solutions, with the only exception for the case $4k - s^2 = 4^{\alpha}(8l + 7)$.

Hence, there exists at most one odd number between these bounds. In other words, the smallest number s', corresponding to k', is either s, or $s+2$.

In the first case, it holds $\frac{k'-s'}{2} = \frac{k-s}{2} + 1$, and the finite sequence (of positive integers, represented as a sum of $m+2$ $(m+2)$-gonal numbers), corresponding to k', has the following form:

$$m\frac{k-s}{2} + s + 2, m\frac{k-s}{2} + 3, \ldots, m\left(\frac{k-s}{2} + 2\right) + s - 2.$$

In the second case, it holds $\frac{k'-s'}{2} = \frac{k-s}{2}$, and the finite sequence (of positive integers, represented as a sum of $m+2$ $(m+2)$-gonal numbers), corresponding to k', has the following form:

$$m\left(\frac{k-s}{2} - 1\right) + s + 4, m\left(\frac{k-s}{2} - 1\right) + s + 5, \ldots,$$
$$m\left(\frac{k-s}{2} + 1\right) + s.$$

For any $m \geq 3$, it is easy to check that

$$\max\left\{m\frac{k-s}{2} + s + 2, m\left(\frac{k-s}{2} - 1\right) + s + 4\right\}$$
$$\leq m\left(\frac{k-s}{2} + 1\right) + s - 1.$$

So, two finite sequences, constructed for k and k', glued together, always cover a segment of consecutive positive integers without gaps. In the first case we obtain the segment $[m(\frac{k-s}{2} - 1) + s + 2, m(\frac{k-s}{2} + 2) + s - 2]$. In the second case it is the segment $[m(\frac{k-s}{2} - 1) + s + 2, m(\frac{k-s}{2} + 1) + s]$.

On the other hand, we have the following obvious inequality:

$$\min\left\{m\left(\frac{k-s}{2} + 1\right) + s, m\left(\frac{k-s}{2} + 2\right) + s - 2\right\}$$
$$> m\left(\frac{k-s}{2} + 1\right) + s - 2.$$

It shows that the length of the segment, constructed above, increases without limit when k goes to infinity.

Finally, for $k = 123$ one gets $s = 19$, and $m\left(\frac{k-s}{2} - 1\right) + s + 2 = m\left(\frac{123-19}{2} - 1\right) + 19 + 2 = 51m + 21$. Hence, it is proven the following fact: *for a given $m \geq 3$ all positive integers greater than or equal to $51m + 21$ can be represented as a sum of $m + 2$ $(m + 2)$-gonal numbers.*

5.6.5. Let us consider now the case of small odd k: $1 \leq k \leq 121$. The table below gives the values of $\sqrt{3k - 2} - 1$, $\sqrt{4k}$, the smallest numbers s and $s + 2$ (if $s + 2$ exists), belonging to the segment $[\sqrt{3k - 2} - 1, \sqrt{4k}]$, as well as the corresponding finite sequences (of positive integers, represented as a sum of $m + 2$ $(m + 2)$-gonal numbers) with possible gaps.

k	$\sqrt{3k-2}-1$	$\sqrt{4k}$	s	$s+2$	finite sequence	exceptions
1	0	2	1	–	$0, \ldots, m-1$	no, since $(m-1)+1 \geq 3$
3	$1,64\ldots$	$3,46\ldots$	3	–	$3, \ldots, m+1$	$m+2$
5	$2,60\ldots$	$4,47\ldots$	3	–	$m+3, \ldots, 2m+1$	no, since $2m+2 \geq m+5$
7	$3,35\ldots$	$5,29\ldots$	5	–	$m+5, \ldots, 2m+3$	$2m+4$
9	4	6	5	–	$2m+5, \ldots, 3m+3$	$3m+4$
11	$4,56\ldots$	$6,63\ldots$	5	–	$3m+5, \ldots, 4m+3$	no, since $4m+4 \geq 3m+7$
13	$5,08\ldots$	$7,21\ldots$	7	–	$3m+7, \ldots, 4m+5$	$4m+6$
15	$5,55\ldots$	$7,74\ldots$	7	–	$4m+7, \ldots, 5m+5$	$5m+6$
17	6	$8,24\ldots$	7	–	$5m+7, \ldots, 6m+5$	$6m+6$
19	$6,41\ldots$	$8,71\ldots$	7	–	$6m+7, \ldots, 7m+5$	no, since $7m+6 \geq 6m+9$
21	$6,81\ldots$	$9,16\ldots$	7	9	$6m+9, \ldots, 8m+5$	no, since $8m+6 \geq 7m+9$
23	$7,18\ldots$	$9,59\ldots$	9	–	$7m+9, \ldots, 8m+7$	$8m+8$
25	$7,54\ldots$	10	9	–	$8m+9, \ldots, 9m+7$	$9m+8$
27	$7,88\ldots$	$10,39\ldots$	9	–	$9m+9, \ldots, 10m+7$	$10m+8$
29	$8,21\ldots$	$10,77\ldots$	9	–	$10m+9, \ldots, 11m+7$	no, since $11m+8 \geq 10m+11$
31	$8,53\ldots$	$11,13\ldots$	9	11	$10m+11, \ldots, 12m+7$	no, since $12m+8 \geq 11m+11$
33	$8,84\ldots$	$11,48\ldots$	11	–	$11m+11, \ldots, 13m+7$	no, since $13m+8 \geq 12m+11$
35	$9,14\ldots$	$11,83\ldots$	11	–	$12m+11, \ldots, 13m+9$	$13m+10$
37	$9,44\ldots$	$12,16\ldots$	11	–	$13m+11, \ldots, 14m+9$	$14m+10$
39	$9,72\ldots$	$12,48\ldots$	11	–	$14m+11, \ldots, 15m+9$	$15m+10$
41	10	$12,80\ldots$	11	–	$15m+11, \ldots, 16m+9$	no, since $16m+10 \geq 15m+13$
43	$10,26\ldots$	$13,11\ldots$	11	13	$15m+13, \ldots 17m+9$	no, since $17m+10 \geq 16m+13$
45	$10,53\ldots$	$13,41\ldots$	11	13	$16m+13, \ldots, 18m+9$	no, since $18m+10 \geq 17m+13$
47	$10,78\ldots$	$13,71\ldots$	11	13	$17m+13, \ldots, 19m+9$	no, since $19m+10 \geq 18m+13$
49	$11,04\ldots$	14	13	–	$18m+13, \ldots, 19m+11$	$19m+12$
51	$11,28\ldots$	$14,28\ldots$	13	–	$19m+13, \ldots, 20m+11$	$20m+12$
53	$11,52\ldots$	$14,56\ldots$	13	–	$20m+13, \ldots, 21m+11$	$21m+12$
55	$11,76\ldots$	$14,83\ldots$	13	–	$21m+13, \ldots, 22m+11$	no, since $22m+12 \geq 21m+15$
57	12	$15,09\ldots$	13	15	$21m+15, \ldots, 23m+11$	no, since $23m+12 \geq 22m+15$
59	$12,22\ldots$	$15,36\ldots$	13	15	$22m+15, \ldots, 24m+11$	no, since $24m+12 \geq 23m+15$
61	$12,45\ldots$	$15,62\ldots$	13	15	$23m+15, \ldots, 25m+11$	no, since $25m+12 \geq 24m+15$
63	$12,67\ldots$	$15,87\ldots$	13	15	$24m+15, \ldots, 26m+11$	no, since $26m+12 \geq 25m+15$
65	$12,89\ldots$	$16,12\ldots$	13	15	$25m+15, \ldots, 27m+11$	no, since $27m+12 \geq 26m+15$
67	$13,10\ldots$	$16,37\ldots$	15	–	$26m+15, \ldots, 27m+13$	$27m+14$
69	$13,31\ldots$	$16,61\ldots$	15	–	$27m+15, \ldots, 28m+13$	$28m+14$
71	$13,52\ldots$	$16,85\ldots$	15	–	$28m+15, \ldots, 29m+13$	no, since $29m+14 \geq 28m+17$
73	$13,73\ldots$	$17,08\ldots$	15	17	$28m+17, \ldots, 30m+13$	no, since $30m+14 \geq 29m+17$
75	$13,93\ldots$	$17,32\ldots$	15	17	$29m+17, \ldots, 31m+13$	no, since $31m+14 \geq 30m+17$
77	$14,13\ldots$	$17,54\ldots$	15	17	$30m+17, \ldots, 32m+13$	no, since $32m+14 \geq 31m+17$
79	$14,32\ldots$	$17,77\ldots$	15	17	$31m+17, \ldots, 33m+13$	no, since $33m+14 \geq 32m+17$
81	$14,52\ldots$	18	15	17	$32m+17, \ldots, 34m+13$	no, since $34m+14 \geq 33m+17$
83	$14,71\ldots$	$18,22\ldots$	15	17	$33m+17, \ldots, 35m+13$	no, since $35m+14 \geq 34m+17$
85	$14,90\ldots$	$18,43\ldots$	15	17	$34m+17, \ldots, 36m+13$	no, since $36m+14 \geq 35m+17$
87	$15,09\ldots$	$18,65\ldots$	17	–	$35m+17, \ldots, 36m+15$	$36m+16$
89	$15,27\ldots$	$18,86\ldots$	17	–	$36m+17, \ldots, 37m+15$	no, since $37m+16 \geq 36m+19$
91	$15,46\ldots$	$19,07\ldots$	17	19	$36m+19, \ldots, 38m+15$	no, since $38m+16 \geq 37m+19$
93	$15,64\ldots$	$19,28\ldots$	17	19	$37m+19, \ldots, 39m+15$	no, since $39m+16 \geq 38m+17$
95	$15,82\ldots$	$19,49\ldots$	17	19	$38m+19, \ldots, 40m+15$	no, since $40m+16 \geq 39m+17$
97	16	$19,69\ldots$	17	19	$39m+19, \ldots, 41m+15$	no, since $41m+16 \geq 40m+17$
99	$16,17\ldots$	$19,89\ldots$	17	19	$40m+19, \ldots, 42m+15$	no, since $42m+16 \geq 41m+17$
101	$16,34\ldots$	$20,09\ldots$	17	19	$41m+19, \ldots, 43m+15$	no, since $43m+16 \geq 42m+17$
103	$16,52\ldots$	$20,29\ldots$	17	19	$42m+19, \ldots, 44m+15$	no, since $44m+16 \geq 43m+17$
105	$16,69\ldots$	$20,49\ldots$	17	19	$43m+19, \ldots, 45m+15$	no, since $45m+16 \geq 44m+17$

(Continued)

(Continued)

k	$\sqrt{3k-2}-1$	$\sqrt{4k}$	s	$s+2$	finite sequence	exceptions
107	$16,87\ldots$	$20,68\ldots$	17	19	$44m+19,\ldots,46m+15$	no, since $46m+16\geq 45m+19$
109	$17,02\ldots$	$20,88\ldots$	19	–	$45m+19,\ldots,46m+17$	no, since $46m+18\geq 45m+21$
111	$17,19\ldots$	$21,07\ldots$	19	21	$45m+21,\ldots,47m+17$	no, since $47m+18\geq 46m+21$
113	$17,35\ldots$	$21,26\ldots$	19	21	$46m+21,\ldots,48m+17$	no, since $48m+18\geq 47m+21$
115	$17,52\ldots$	$21,44\ldots$	19	21	$47m+21,\ldots,49m+17$	no, since $49m+18\geq 48m+21$
117	$17,68\ldots$	$21,63\ldots$	19	21	$48m+21,\ldots,50m+17$	no, since $50m+18\geq 49m+21$
119	$17,84\ldots$	$21,81\ldots$	19	21	$49m+21,\ldots,51m+17$	no, since $51m+18\geq 50m+21$
121	18	22	19	21	$50m+21,\ldots,52m+17$	no, since $52m+18\geq 51m+21$

So, one has exactly 18 possible exceptions, which will be considered separately.

5.6.6. For $k=3$ (and $s=3$), the only possible exception is the number $m+2$. Considering $\overline{k}=k+1=4$, one finds the only one even number $\overline{s}=4=s+1$ between $\sqrt{3\overline{k}-2}-1=2,16\ldots$ and $\sqrt{4\overline{k}}=4$. Since $4=1^2+1^2+1^2+1^2$, and $4=1+1+1+1$, one gets representation of $m+2=m\frac{\overline{k}-\overline{s}}{2}+\overline{s}+m-2$ as a sum of $m+2$ $(m+2)$-gonal numbers.

For $k=7$ (and $s=5$), the only possible exception is the number $2m+4$. Considering $\overline{k}=k+1=8$, one finds the only one even number $\overline{s}=4=s-1$ between $\sqrt{3\overline{k}-2}-1=3,69\ldots$ and $\sqrt{4\overline{k}}=5,65\ldots$. Since $8=2^2+2^2+0^2+0^2$, and $4=2+2+0+0$, one gets representation of $2m+4=m\frac{\overline{k}-\overline{s}}{2}+\overline{s}+0$ as a sum of $m+2$ $(m+2)$-gonal numbers.

For $k=9$ (and $s=5$), the only possible exception is the number $3m+4$. Considering $\overline{k}=k+1=10$, one finds the only one even number $\overline{s}=6=s+1$ between $\sqrt{3\overline{k}-2}-1=4,29\ldots$ and $\sqrt{4\overline{k}}=6,32\ldots$. Since $10=2^2+2^2+1^2+1^2$, and $6=2+2+1+1$, one gets representation of $3m+4=m\frac{\overline{k}-\overline{s}}{2}+\overline{s}+(m-2)$ as a sum of $m+2$ $(m+2)$-gonal numbers.

For $k=13$ (and $s=7$), the only possible exception is the number $4m+6$. Considering $\overline{k}=k+1=14$, one finds the only one even number $\overline{s}=6=s-1$ between $\sqrt{3\overline{k}-2}-1=5,32\ldots$ and $\sqrt{4\overline{k}}=7,48\ldots$. Since $14=3^2+2^2+1^2+0^2$, and $6=3+2+1+0$, one gets representation of $4m+6=m\frac{\overline{k}-\overline{s}}{2}+\overline{s}+0$ as a sum of $m+2$ $(m+2)$-gonal numbers.

For $k=15$ (and $s=7$), the only possible exception is the number $5m+6$. Considering $\overline{k}=k+1=16$, one finds the only two even numbers 6 and 8 between $\sqrt{3\overline{k}-2}-1=5,78\ldots$ and $\sqrt{4\overline{k}}=8$. Taking, for example, $\overline{s}=8=s+1$, and noting that $16=2^2+2^2+2^2+2^2$, and $8=2+2+2+2$, one gets representation of

$5m+6 = m\frac{\overline{k}-\overline{s}}{2} + \overline{s} + (m-2)$ as a sum of $m+2$ $(m+2)$-gonal numbers.

For $k = 17$ (and $s = 7$), the only possible exception is the number $6m+6$. Considering $\overline{k} = k+1 = 18$, one finds the only one even number $\overline{s} = 8 = s+1$ between $\sqrt{3\overline{k}-2} - 1 = 6,21\ldots$ and $\sqrt{4\overline{k}} = 8,48\ldots$. Since $18 = 3^2+2^2+2^2+1^2$, and $8 = 3+2+2+1$, one gets representation of $6m+6 = m\frac{\overline{k}-\overline{s}}{2} + \overline{s} + (m-2)$ as a sum of $m+2$ $(m+2)$-gonal numbers.

For $k = 23$ (and $s = 9$), the only possible exception is the number $8m+8$. Considering $\overline{k} = k+1 = 24$, one finds the only one even number $\overline{s} = 8 = s-1$ between $\sqrt{3\overline{k}-2} - 1 = 7,36\ldots$ and $\sqrt{4\overline{k}} = 9,79\ldots$. Since $24 = 4^2+2^2+2^2+0^2$, and $8 = 4+2+2+0$, one gets representation of $8m+8 = m\frac{\overline{k}-\overline{s}}{2} + \overline{s} + 0$ as a sum of $m+2$ $(m+2)$-gonal numbers.

For $k = 25$ (and $s = 9$), the only possible exception is the number $9m+8$. Considering $\overline{k} = k+1 = 26$, one finds two even numbers 8 and 10 between $\sqrt{3\overline{k}-2} - 1 = 7,71\ldots$ and $\sqrt{4\overline{k}} = 10,19\ldots$. Taking, for example, $\overline{s} = 8 = s-1$, and noting that $26 = 4^2+3^2+1^2+0^2$, and $8 = 4+3+1+0$, one gets representation of $9m+8 = m\frac{\overline{k}-\overline{s}}{2} + \overline{s} + 0$ as a sum of $m+2$ $(m+2)$-gonal numbers.

For $k = 27$ (and $s = 9$), the only possible exception is the number $10m+8$. Considering $\overline{k} = k+1 = 28$, one finds the only one even number $\overline{s} = 10 = s+1$ between $\sqrt{3\overline{k}-2} - 1 = 8,05\ldots$ and $\sqrt{4\overline{k}} = 10,58\ldots$. Noting that $28 = 4^2+2^2+2^2+2^2$, and $10 = 4+2+2+2$, one gets representation of $10m+8 = m\frac{\overline{k}-\overline{s}}{2} + \overline{s} + (m-2)$ as a sum of $m+2$ $(m+2)$-gonal numbers.

For $k = 35$ (and $s = 11$), the only possible exception is the number $13m+10$. Considering $\overline{k} = k+1 = 36$, one finds two even numbers 10 and 12 between $\sqrt{3\overline{k}-2} - 1 = 9,29\ldots$ and $\sqrt{4\overline{k}} = 12$. Taking, for example, $\overline{s} = 10 = s-1$, and noting that $36 = 4^2+4^2+2^2+0^2$, and $10 = 4+4+2+0$, one gets representation of $13m+10 = m\frac{\overline{k}-\overline{s}}{2} + \overline{s} + 0$ as a sum of $m+2$ $(m+2)$-gonal numbers.

For $k = 37$ (and $s = 9$), the only possible exception is the number $14m+10$. Considering $\overline{k} = k+1 = 38$, one finds two even numbers 10 and 12 between $\sqrt{3\overline{k}-2} - 1 = 9,58\ldots$ and $\sqrt{4\overline{k}} = 12,32\ldots$. Taking, for example, $\overline{s} = 10 = s-1$, and noting that $38 = 5^2+3^2+2^2+0^2$, and $10 = 5+3+2+0$, one gets representation of $14m+10 = m\frac{\overline{k}-\overline{s}}{2}+\overline{s}+0$ as a sum of $m+2$ $(m+2)$-gonal numbers.

For $k = 39$ (and $s = 9$), the only possible exception is the number $15m+10$. Considering $\overline{k} = k+1 = 40$, one finds two even numbers 10 and 12 between $\sqrt{3\overline{k}-2}-1 = 9,86\ldots$ and $\sqrt{4\overline{k}} = 12,64\ldots$. Taking, for example, $\overline{s} = 12 = s+1$, and noting that $40 = 4^2+4^2+2^2+2^2$, and $12 = 4+4+2+2$, one gets representation of $15m+10 = m\frac{\overline{k}-\overline{s}}{2}+\overline{s}+(m-2)$ as a sum of $m+2$ $(m+2)$-gonal numbers.

For $k = 49$ (and $s = 13$), the only possible exception is the number $19m+12$. Considering $\overline{k} = k+1 = 50$, one finds two even numbers 12 and 14 between $\sqrt{3\overline{k}-2}-1 = 11,16\ldots$ and $\sqrt{4\overline{k}} = 14,14\ldots$. Taking, for example, $\overline{s} = 12 = s-1$, and noting that $50 = 5^2+4^2+3^2+0^2$, and $12 = 5+4+3+0$, one gets representation of $19m+11 = m\frac{\overline{k}-\overline{s}}{2}+\overline{s}+0$ as a sum of $m+2$ $(m+2)$-gonal numbers.

For $k = 51$ (and $s = 13$), the only possible exception is the number $20m+12$. Considering $\overline{k} = k+1 = 52$, one finds two even numbers 12 and 14 between $\sqrt{3\overline{k}-2}-1 = 11,40\ldots$ and $\sqrt{4\overline{k}} = 14,42\ldots$. Taking, for example, $\overline{s} = 12 = s-1$, and noting that $52 = 5^2+5^2+1^2+1^2$, and $12 = 5+5+1+1$, one gets representation of $20m+12 = m\frac{\overline{k}-\overline{s}}{2}+\overline{s}+0$ as a sum of $m+2$ $(m+2)$-gonal numbers.

For $k = 53$ (and $s = 9$), the only possible exception is the number $21m+12$. Considering $\overline{k} = k+1 = 54$, one finds two even numbers 12 and 14 between $\sqrt{3\overline{k}-2}-1 = 11,64\ldots$ and $\sqrt{4\overline{k}} = 14,69\ldots$. Taking, for example, $\overline{s} = 12 = s-1$, and noting that $54 = 5^2+5^2+2^2+0^2$, and $12 = 5+5+2+0$, one gets representation of $21m+12 = m\frac{\overline{k}-\overline{s}}{2}+\overline{s}+0$ as a sum of $m+2$ $(m+2)$-gonal numbers.

For $k = 67$ (and $s = 15$), the only possible exception is the number $27m+14$. Considering $\overline{k} = k+1 = 68$, one finds two even numbers 14 and 16 between $\sqrt{3\overline{k}-2}-1 = 13,21\ldots$ and $\sqrt{4\overline{k}} = 16,49\ldots$. Taking, for example, $\overline{s} = 14 = s-1$, and noting that $68 = 6^2+4^2+4^2+0^2$, and $14 = 6+4+4+0$, one gets representation of $27m+14 = m\frac{\overline{k}-\overline{s}}{2}+\overline{s}+0$ as a sum of $m+2$ $(m+2)$-gonal numbers.

For $k = 69$ (and $s = 15$), the only possible exception is the number $28m + 14$. Considering $\overline{k} = k + 1 = 70$, one finds two even numbers 14 and 16 between $\sqrt{3\overline{k} - 2} - 1 = 13,42\ldots$ and $\sqrt{4\overline{k}} = 16,73\ldots$. Taking, for example, $\overline{s} = 16 = s + 1$, and noting that $70 = 5^2 + 5^2 + 4^2 + 2^2$, and $16 = 5 + 5 + 4 + 2$, one gets representation of $28m + 14 = m\frac{\overline{k} - \overline{s}}{2} + \overline{s} + (m - 2)$ as a sum of $m + 2$ $(m + 2)$-gonal numbers.

For $k = 87$ (and $s = 17$), the only possible exception is the number $36m + 16$. Considering $\overline{k} = k + 1 = 88$, one finds two even numbers 16 and 18 between $\sqrt{3\overline{k} - 2} - 1 = 15,18\ldots$ and $\sqrt{4\overline{k}} = 18,76\ldots$. Taking, for example, $\overline{s} = 16 = s - 1$, and noting that $88 = 6^2 + 6^2 + 4^2 + 0^2$, and $16 = 6 + 6 + 4 + 0$, one gets representation of $36m + 16 = m\frac{\overline{k} - \overline{s}}{2} + \overline{s} + 0$ as a sum of $m + 2$ $(m + 2)$-gonal numbers.

Therefore, for any possible exception we have obtained a representation of the corresponding positive integer as a sum of $m + 2$ $(m + 2)$-gonal numbers. This completes the Cauchy's proof of the polygonal number theorem.

5.6.7. If one want to use the Cauchy's ideas for a decomposition of a given positive integer N into a sum of $m + 2$ $(m + 2)$-gonal numbers, $m \geq 3$, one should start with finding the corresponding number k. Supposing that $N \approx m\frac{k-s}{2} + m - 2$, and $s \approx \sqrt{3k - 2}$, one obtains

$$N \approx \frac{m}{2}k + m - 2 - \frac{m-2}{2}\sqrt{3k - 2},$$

and can find a approximate value of k from the equation

$$\left(\frac{m}{2}k - (N - m + 2)\right)^2 = \frac{(m-2)^2}{4}(3k - 2).$$

This approximative value of k allows, after several steps, to obtain the exact value of k, and, solving above system for given k and s, to obtain the expected decomposition.

For example, let us decompose 114 into a sum of six hexagonal numbers. One has $m = 4$, $N = 114$, and the equation above takes the form $4(k - 56)^2 = 3k - 2$, or $4k^2 - 451k + 12542 = 0$, giving $k \approx 49$, or $k \approx 63$.

The finite sequence (of numbers, represented as a sum of six hexagonal numbers), corresponding to 49, is (see above table)

$85 = 18\cdot 4+13, 86, 87, 88, 89 = 20\cdot 4+9$, and so, 49 can not be used for a decomposition of the number 114.

The finite sequence (of numbers, represented as a sum of 6 hexagonal numbers), corresponding to 63, is $111 = 24\cdot 4 + 15, 112, 113, 114, 115 = 26\cdot 4+11 = 115$. So, taking $s = 13$, one has

$$114 = 4\cdot\frac{63-13}{2}+13+1.$$

Since $63 = 6^2+5^2+1^2+1^2$, and $13 = 6+5+1+1$, one obtains

$$114 = \left(4\cdot\frac{6^2-6}{2}+6\right)+\left(4\cdot\frac{5^2-5}{2}+5\right)+\left(4\cdot\frac{1^2-1}{2}+1\right)$$
$$+\left(4\cdot\frac{1^2-1}{2}+1\right)+1.$$

In other words, it holds

$$\begin{aligned}114 &= 66+45+1+1+1+0\\ &= S_6(6)+S_6(5)+S_6(1)+S_6(1)+S_6(1)+S_6(0).\end{aligned}$$

5.7 Pepin's proof of the polygonal number theorem

The simplified proof of the polygonal number theorem was given by Pepin in 1892 ([Pepi92]). The main result was that, for $m \geq 3$ and $N \leq 120m$, Pepin published tables of explicit representations of a given positive integer N as a sum of $m+2$ $(m+2)$-gonal numbers, at most four of which are different from 0 or 1.

In this section we consider these tables, as well as the main ideas of the Pepin's proof of the polygonal number theorem.

5.7.1. In our consideration we will use for n-th $(m+2)$-gonal number the formula

$$S_{m+2}(n) = \frac{m}{2}(n^2-n)+n.$$

The first sixteen $(m+2)$-gonal numbers are given in the table below.

$S_{m+2}(1) = 1$	$S_{m+2}(2) = m+2$	$S_{m+2}(3) = 3m+3$	$S_{m+2}(4) = 6m+4$
$S_{m+2}(5) = 10m+5$	$S_{m+2}(6) = 15m+6$	$S_{m+2}(7) = 21m+7$	$S_{m+2}(8) = 28m+8$
$S_{m+2}(9) = 36m+9$	$S_{m+2}(10) = 45m+10$	$S_{m+2}(11) = 55m+11$	$S_{m+2}(12) = 66m+12$
$S_{m+2}(13) = 78m+13$	$S_{m+2}(14) = 91m+14$	$S_{m+2}(15) = 105m+15$	$S_{m+2}(16) = 120m+16$

Using these representations, it is easy to obtain the table of all numbers less than or equal to $30m + 9$, which are represented in the form

$$S_{m+2}(n) + S_{m+2}(k) + S_{m+2}(l) + rS_{m+2}(1), \ r \le m - 2,$$

i.e., as a sum of $m + 1$ $(m + 2)$-gonal numbers from which at most three are different from 0 or 1.

This table, containing all such numbers with corresponding representations, as well as possible exceptions, is given below.

representation	boundary for r	smallest number	biggest number	exceptions
$rS_{m+2}(1)$	$r \le m+1$	0	$m+1$	
$S_{m+2}(2) + rS_{m+2}(1)$	$r \le m$	$m+2$	$2m+2$	$2m+3$
$S_{m+2}(2) + S_{m+2}(2) + rS_{m+2}(1)$	$r \le m-1$	$2m+4$	$3m+3$	
$S_{m+2}(3) + rS_{m+2}(1)$	$r \le m$	$3m+3$	$4m+3$	
$S_{m+2}(2) + S_{m+2}(2) + S_{m+2}(2) + rS_{m+2}(1)$	$r \le m-2$	$3m+6$	$4m+4$	
$S_{m+2}(3) + S_{m+2}(2) + rS_{m+2}(1)$	$r \le m-1$	$4m+5$	$5m+4$	$5m+5$, $5m+6$
$S_{m+2}(3) + S_{m+2}(2) + S_{m+2}(2) + rS_{m+2}(1)$	$r \le m-2$	$5m+7$	$6m+5$	
$S_{m+2}(4) + rS_{m+2}(1)$	$r \le m$	$6m+4$	$7m+4$	
$S_{m+2}(3) + S_{m+2}(3) + rS_{m+2}(1)$	$r \le m-1$	$6m+6$	$7m+5$	
$S_{m+2}(4) + S_{m+2}(2) + rS_{m+2}(1)$	$r \le m-1$	$7m+6$	$8m+5$	
$S_{m+2}(3) + S_{m+2}(3) + S_{m+2}(2) + rS_{m+2}(1)$	$r \le m-2$	$7m+8$	$8m+6$	$8m+7$
$S_{m+2}(4) + S_{m+2}(2) + S_{m+2}(2) + rS_{m+2}(1)$	$r \le m-2$	$8m+8$	$9m+6$	
$S_{m+2}(4) + S_{m+2}(3) + rS_{m+2}(1)$	$r \le m-1$	$9m+7$	$10m+6$	
$S_{m+2}(5) + rS_{m+2}(1)$	$r \le m$	$10m+5$	$11m+5$	
$S_{m+2}(4) + S_{m+2}(3) + S_{m+2}(2) + rS_{m+2}(1)$	$r \le m-2$	$10m+9$	$11m+7$	
$S_{m+2}(5) + S_{m+2}(2) + rS_{m+2}(1)$	$r \le m-1$	$11m+7$	$12m+6$	$12m+7$
$S_{m+2}(4) + S_{m+2}(4) + rS_{m+2}(1)$	$r \le m-1$	$12m+8$	$13m+7$	
$S_{m+2}(5) + S_{m+2}(3) + rS_{m+2}(1)$	$r \le m-1$	$13m+8$	$14m+7$	
$S_{m+2}(4) + S_{m+2}(4) + S_{m+2}(2) + rS_{m+2}(1)$	$r \le m-2$	$13m+10$	$14m+8$	$14m+9$
$S_{m+2}(5) + S_{m+2}(3) + S_{m+2}(2) + rS_{m+2}(1)$	$r \le m-2$	$14m+10$	$15m+8$	
$S_{m+2}(6) + rS_{m+2}(1)$	$r \le m$	$15m+6$	$16m+6$	
$S_{m+2}(4) + S_{m+2}(4) + S_{m+2}(3) + rS_{m+2}(1)$	$r \le m-2$	$15m+11$	$16m+9$	
$S_{m+2}(5) + S_{m+2}(4) + rS_{m+2}(1)$	$r \le m-1$	$16m+9$	$17m+8$	
$S_{m+2}(5) + S_{m+2}(3) + S_{m+2}(3) + rS_{m+2}(1)$	$r \le m-2$	$16m+11$	$17m+9$	
$S_{m+2}(6) + S_{m+2}(2) + S_{m+2}(2) + rS_{m+2}(1)$	$r \le m-2$	$17m+10$	$18m+8$	
$S_{m+2}(6) + S_{m+2}(3) + rS_{m+2}(1)$	$r \le m-1$	$18m+9$	$19m+8$	
$S_{m+2}(4) + S_{m+2}(4) + S_{m+2}(4) + rS_{m+2}(1)$	$r \le m-2$	$18m+12$	$19m+10$	
$S_{m+2}(6) + S_{m+2}(3) + S_{m+2}(2) + rS_{m+2}(1)$	$r \le m-2$	$19m+11$	$20m+9$	
$S_{m+2}(5) + S_{m+2}(5) + rS_{m+2}(1)$	$r \le m-1$	$20m+10$	$21m+9$	
$S_{m+2}(6) + S_{m+2}(4) + rS_{m+2}(1)$	$r \le m-1$	$21m+10$	$22m+9$	
$S_{m+2}(7) + S_{m+2}(2) + rS_{m+2}(1)$	$r \le m-1$	$22m+9$	$23m+8$	
$S_{m+2}(6) + S_{m+2}(4) + S_{m+2}(2) + rS_{m+2}(1)$	$r \le m-2$	$22m+12$	$23m+10$	
$S_{m+2}(7) + S_{m+2}(2) + S_{m+2}(2) + rS_{m+2}(1)$	$r \le m-2$	$23m+11$	$24m+9$	
$S_{m+2}(7) + S_{m+2}(3) + rS_{m+2}(1)$	$r \le m-1$	$24m+10$	$25m+9$	
$S_{m+2}(6) + S_{m+2}(4) + S_{m+2}(3) + rS_{m+2}(1)$	$r \le m-2$	$24m+13$	$25m+11$	
$S_{m+2}(6) + S_{m+2}(5) + rS_{m+2}(1)$	$r \le m-1$	$25m+11$	$26m+10$	$26m+11$, $26m+12$
$S_{m+2}(6) + S_{m+2}(5) + S_{m+2}(2) + rS_{m+2}(1)$	$r \le m-2$	$26m+13$	$27m+11$	
$S_{m+2}(7) + S_{m+2}(4) + rS_{m+2}(1)$	$r \le m-1$	$27m+11$	$28m+10$	
$S_{m+2}(8) + rS_{m+2}(1)$	$r \le m$	$28m+8$	$29m+8$	$29m+9$
$S_{m+2}(8) + S_{m+2}(2) + rS_{m+2}(1)$	$r \le m-1$	$29m+10$	$30m+9$	

It implies that all numbers from zero to $30m + 9$ are represented as a sum of $m + 1$ $(m + 2)$-gonal numbers, $m \ge 3$, at most three of which are different from 0 and 1, with the only exceptions $2m + 3$, $5m + 5$, $5m + 6$, $8m + 7$, $12m + 7$, $14m + 9$, $26m + 11$, $26m + 12$, and $29m + 9$.

For the last number $29m+9$ we have no exception, as this number can be represented in above way, but this representation has different form for $m > 3$ and for $m = 3$. In the first case, one has

$$29m + 9 = S_{m+2}(7) + S_{m+2}(4) + S_{m+2}(2) + (m-4)S_{m+2}(1),$$

while in the second case we get

$$29m + 9 = S_{m+2}(6) + S_{m+2}(4) + S_{m+2}(4) + S_{m+2}(1)$$

$$\text{i.e., } 29 \cdot 3 + 9 = 51 + 22 + 22 + 1.$$

However, for the above exceptional numbers the general theorem holds, i.e., they are represented as a sum of $(m+2)$ $(m+2)$-gonal numbers, at most four of which are different from 0 and 1:

$$\begin{aligned}
2m + 3 &= S_{m+2}(2) + (m+1)S_{m+2}(1),\\
5m + 5 &= S_{m+2}(3) + S_{m+2}(2) + mS_{m+2}(1),\\
5m + 6 &= 4S_{m+2}(2) + (m-2)S_{m+2}(1),\\
8m + 7 &= 2S_{m+2}(3) + S_{m+2}(2) + (m-1)S_{m+2}(1),\\
12m + 7 &= S_{m+2}(5) + S_{m+2}(2) + mS_{m+2}(1),\\
14m + 9 &= 2S_{m+2}(4) + S_{m+2}(2) + (m-1)S_{m+2}(1),\\
26m + 11 &= S_{m+2}(6) + S_{m+2}(5) + mS_{m+2}(1),\\
26m + 12 &= S_{m+2}(6) + S_{m+2}(4) + S_{m+2}(3)\\
&\quad + S_{m+2}(2) + (m-3)S_{m+2}(1).
\end{aligned}$$

So, the polygonal number theorem is proven for all numbers, less than or equal to $30m + 9$.

5.7.2. Now we continue this verification for all numbers, which are at most $S_{m+2}(16) = 120m + 16$.

In fact, let us consider a positive integer N between $S_{m+2}(8) = 28m + 8$ and $S_{m+2}(16) = 120m + 16$:

$$S_{m+2}(8) < N < S_{m+2}(16).$$

Let x be a positive integer such that $S_{m+2}(x) < N < S_{m+2}(x+1)$. Then it holds $8 \leq x < 16$. Let

$$N = S_{m+2}(x) + R = S_{m+2}(x-1) + R',$$

where $R' = R + (S_{m+2}(x) - S_{m+2}(x-1)) = R + m(x-1) + 1$, and $R < S_{m+2}(x+1) - S_{m+2}(x) = mx + 1$.

If one of the numbers R or $R^{'}$ is represented as a sum of $m+1$ $(m+2)$-gonal numbers, at most three of which are different from 0 or 1, then, adding $S_{m+2}(x)$ or $S_{m+2}(x-1)$, respectively, one obtains that N is represented by $m+2$ $(m+2)$-gonal numbers, at most four of which are different from 0 or 1.

Suppose now, that the number R can not be represented as a sum of $m+1$ $(m+2)$-gonal numbers, at most three of which are different from 0 or 1. Since $x \leq 15$, we have $R < 15m+1$, i.e., R should be of the form

$$2x+3,\ 5m+5,\ 5m+6,\ 8m+7,\ 12m+7,\ 14m+9.$$

For such R the number $R^{'} = R + m(x-1)+1$ should be of the following forms:

$$m(x+1)+4,\ m(x+4)+6,\ m(x+4)+7,\ m(x+7)+8,$$

$$m(x+11)+8,\ m(x+13)+10.$$

Using the condition $8 \leq x \leq 15$, we can see that for all such cases

$$9m+4 = m(8+1)+4 \leq R^{'} \leq m(15+13)+10 = 28m+10.$$

If $R^{'}$ has a representation as a sum of $m+1$ $(m+2)$-gonal numbers, at most three of which are different from 0 or 1, we get for N the corresponding decomposition. Since $5m+5 < 9m+4$ for all $m \geq 3$, and $8m+7 < 9m+4$ for $m > 3$, but $8m+7 = 9m+4$ for $m=3$, we should consider separately only the following exceptions for $R^{'}$:

$$8m+7 = 9m+4 \quad \text{for } m=3,\ 12m+7,\ 14m+9,$$

$$26m+11,\ 26m+12.$$

I. $R^{'} = m(x+13)+10$. Therefore, $21m+10 \leq R^{'} \leq 26m+11$, and, since $21m+10 > 14m+9$ for all $m \geq 3$, we should consider only two exceptions: $26m+11$, and $26m+12$.

In the first case, we have $m(x+13)+10 = 26m+11 \Leftrightarrow 10 \equiv 11 (mod\, m) \Leftrightarrow 1 \equiv 0 (mod\, m)$, that is impossible for $m \geq 3$.

In the second case, we have $m(x+13)+10 = 26m+12 \Leftrightarrow 10 \equiv 12 (mod\, m) \Leftrightarrow 2 \equiv 0 (mod\, m)$, that is also impossible for $m \geq 3$.

II. $R^{'} = m(x+7)+8$. So, $15m+8 \leq R^{'} \leq 26m+8$, and, since $15m+8 > 14m+9$ for all $m \geq 3$, we have no exceptions into these limits.

III. $R' = m(x+4)+7$. So, $12m+7 \le R' \le 19m+7$, and we should consider only two exceptions: $12m+7$, and $14m+9$.

In the first case, we have that $R' = m(8+4)+7$, i.e., $x = 8$, and, so, $R = R' - 7m - 1 = 12m + 7 - 7m - 1 = 5m + 6$. Hence, $N = S_{m+2}(8) + R = S_{m+2}(7) + R' = 33m + 14$, and we can write that

$$N = S_{m+2}(2) + S_{m+2}(4) + S_{m+2}(5) + S_{m+2}(6) + (m-3)S_{m+2}(1).$$

Hence, the theorem holds for such N.

In the second case, we have that $R' = m(x+4)+7 = 14m+9$, i.e., $m(x+4)+7 = 14m+9$, or $m(x-10) = 2$, but this equation has no positive integer solutions for $m \ge 3$.

IV. $R' = m(x+4)+6$. So, $12m+6 \le R' \le 19m+6$, and we should consider only two exceptions: $12m+7$, and $14m+9$.

In the first case, we have that $R' = m(x+4)+6 = 12m+7$, i.e., $m(x+4)+6 = 12m+7$, or $m(x-8) = 1$, but this equation has no positive integer solutions for $m \ge 3$.

In the second case, we have that $R' = m(x+4)+6 = 14m+9$, i.e., $m(x+4)+6 = 14m+9$, or $m(x-10) = 3$. So, we get $m = 3$, and $x - 10 = 1$, i.e., $m = 3$, and $x = 11$. In this case $R' = 14 \cdot 3 + 9 = 51 = S_5(6)$, and

$$N = S_5(x-1) + R' = S_5(10) + S_5(6), \quad \text{i.e., } 70 = 55 + 15.$$

Therefore, the theorem holds for $N = 70$ in the case of pentagonal numbers.

V. $R' = m(x+1)+4$. So, $9m+4 \le R' \le 16m+4$, and we should consider only three exceptions: $8m+7 = 9m+4$ for $m = 3$, $12m+7$, and $14m+9$.

In the first case, we have that $R' = m(x+1)+4 = 8m+7$, i.e., $m(x+1)+4 = 8m+7$, or $m(x-7) = 3$. So, we get $m = 3$, and $x - 7 = 1$, i.e., $m = 3$, and $x = 8$. In this case $R' = 8 \cdot 3 + 7 = 31$, and $N = S_5(x-1) + R' = S_5(6) + 31 = 101$. So, one can write

$$N = S_5(4) + S_5(4) + S_5(4) + S_5(5), \quad \text{i.e.,}$$

$$101 = 22 + 22 + 22 + 35.$$

Hence, the theorem holds for $N = 105$ in the case of pentagonal numbers.

In the second case, we have that $R^{'} = m(x+1)+4 = 12m+7$, i.e., $m(x+1)+4 = 12m+7$, or $m(x-11) = 3$. So, we get $m = 3$, and $x-11 = 1$, i.e., $m = 3$, and $x = 12$. In this case $R^{'} = 12 \cdot 3 + 7 = 43$, and $N = S_5(x-1) + R^{'} = S_5(10) + R^{'} = 176 + 43 = 219$. So, we have $N = S_5(11) + (12m+7) = S_5(10) + (10m+1) + (12m+7) = S_5(10) + 22m + 8 = S_5(10) + (15m+6) + (6m+4) + (m-2) = S_5(10) + S_5(6) + S_5(4) + (m-2)S_5(1)$. Hence, one can write

$$N = S_5(10)+S_5(6)+S_5(4)+(m-2)S_5(1), \quad \text{i.e., } 219 = 145+51+22+1.$$

Therefore, the theorem holds for $N = 219$ in the case of pentagonal numbers.

In the third case, we have that $R^{'} = m(x+1)+4 = 14m+9$, i.e., $m(x+1)+4 = 14m+9$, or $m(x-13) = 5$. So, we get $m = 5$, and $x-13 = 1$, i.e., $m = 5$, and $x = 14$. In this case $R^{'} = 14 \cdot 5 + 9 = 79$, and $N = S_7(x-1) + R^{'} = S_7(13) + R^{'} = 403 + 79 = 482$. So, we have $N = S_7(13) + (14m+9) = S_7(12) + (12m+1) + (14m+9) = S_7(12) + 26m + 10 = S_7(12) + (15m+6) + (10m+5) + (m-1) = S_7(12) + S_7(6) + S_7(5) + (m-1)S_7(1)$. Hence, one can write

$$N = S_7(12) + S_7(6) + S_7(5) + (m-1)S_7(1), \quad \text{i.e.,}$$

$$482 = 342 + 81 + 55 + 1 + 1 + 1 + 1.$$

Therefore, the theorem holds for $N = 482$ in the case of heptagonal numbers.

So, we have proved the polygonal number theorem for all positive integers less than or equal to $120m + 16$.

5.7.3. In order to finish the proof of the polygonal number theorem, we should show that any positive integer $N > 120m + 16$, $m \geq 3$, can be represented in the form

$$N = \frac{m}{2}(t^2 - t + u^2 - u + v^2 - v + w^2 - w) + (t + u + v + w) + r,$$

where t, u, v, w and r are non-negative integers, and $r \leq m - 2$. Considering the system

$$\begin{cases} k = t^2 + u^2 + v^2 + w^2 \\ s = \quad t + u + v + w, \end{cases}$$

we obtain that $N = \frac{m}{2}(k-s)+s+r$, $0 \le r \le m-2$. For a given N, it is easy to obtain the decomposition

$$N = mB + c, \ 0 < c < m+1,$$

in which $c = rest(N,m)$ if $rest(N,m) \neq 0$, and $c = m$, if $rest(N,m) = 0$. Then it holds

$$mB + c = \frac{m}{2}(k-s) + s + r, \quad \text{i.e., } mB - \frac{m}{2}(k-s) = s + r - c.$$

So, $m|(s+r-c)$. Then, for some integer l, we have

$$s = (c-r) + ml, \quad \text{and } k = 2B - 2l + s.$$

Denoting $c - r = \mu$, we get

$$s = ml + \mu, \quad \text{and } k = 2B + (m-2)l + \mu,$$

while the conditions $0 \le r \le m-2$ and $0 < c < m+1$ imply the following boundaries for μ: $-m+2 < \mu \le m$. For a given N, the numbers B and c are completely determined.

For $m \ge 3$, the number r can obtained at least two values 0 and 1, and we will choose r so that s (and, therefore, k) is odd.

Now let as see, in which limits should be the number s, if we want to have an solution for the system

$$\begin{cases} k = t^2 + u^2 + v^2 + w^2 \\ s = \quad t + u + v + w. \end{cases}$$

For it, we get from the above system that

$$4k - s^2 = (t+u-v-w)^2 + (t+v-u-w)^2 + (t+w-u-v)^2.$$

It is easy to see that for odd s and k the number $4k - s^2$ has the form $8t+3$. So, by the three-square theorem there exist odd positive integers x, y, and z such that

$$4k - s^2 = x^2 + y^2 + z^2, \quad x \ge y \ge z > 0.$$

Now we can determine numbers t, u, v, w from the conditions

$$t+u-v-w = x, \quad t+v-u-w = y, \quad t+w-u-v = \pm z.$$

Combining these conditions with the condition $t+u+v+w = s$, we get

$$t = \frac{s+x+y\pm z}{4}, \quad u = \frac{s+x}{2} - t,$$

$$v = \frac{s+y}{2} - t, \quad w = \frac{s\pm z}{2} - t.$$

All these numbers are integers if t is. But, since all numbers s, x, y, z are odd, one always can choose the sign of z so that $s + x + y \pm z \equiv 0 (mod\, 4)$.

The above numbers t, u, v, w are non-negative, if the smallest one, w, is non-negative in the worst case, when $w = \frac{s-z}{2} - t = \frac{s-x-y-z}{4}$. So, it is sufficient to get the condition

$$\frac{s - x - y - z}{4} > -1, \quad \text{i.e., } s + 4 > x + y + z.$$

It is known, that the biggest value of $x+y+z$ in the case of constant $x^2 + y^2 + z^2$ is obtained for $x = y = z$. Therefore, the equality $x^2 + y^2 + z^2 = 4k - s^2$ implies that $x + y + z \leq 3x \leq \sqrt{12k - 3s^2}$, and it is sufficient to check the condition

$$s + 4 > \sqrt{12k - 3s^2}.$$

Since $k = 2B - 2l - s$, we obtain the following chain of the inequalities:

$$s^2 + 2s + 4 - 3k > 0; \quad (s + 1)^2 + 3 - 3k > 0, \quad (s + 1)^2 > 3k - 3,$$
$$3 + (s + 1)^2 > 3k = 6B - 6l + 3s, \quad s^2 - s + 4 > 6B - 6l.$$

For non-negative values of l it is sufficient to check the condition $s^2 - s + 4 > 6B$, which can be rewritten as $s^2 - s - (6B - 4) > 0$, or, finally, as

$$s > \frac{1}{2} + \sqrt{6B - 3}.$$

The second natural condition, requiring that $4k - s^2$ is a positive integer, has the form

$$s^2 < 4k.$$

Since $k = 2B + s - 2l$, and $lm = s + \mu$, i.e., $8l = \frac{8s}{m} + \frac{8\mu}{m}$, we obtain the following chain of inequalities:

$$s^2 - 4k < 0, \; s^2 - 4(2B + s - 2l) < 0, \quad s^2 - 4s - 8B + 8l < 0,$$
$$s^2 - 4s - 8B + \tfrac{8s}{m} + \tfrac{8\mu}{m} < 0, \quad s^2 - 2(2 - \tfrac{4}{m})s - (8B - \tfrac{8\mu}{m}) < 0.$$

In order to find such positive integers s, we should check that the discriminant $(2 - \frac{4}{m})^2 + (8B - \frac{8\mu}{m})$ of the quadratic function $s^2 - 2(2 - \frac{4}{m})s - (8B - \frac{8\mu}{m})$ on s is non-negative, and then find positive integers s such that $s \in \left(0, (2-\frac{4}{m}) + \sqrt{(2-\frac{4}{m})^2 + (8B - \frac{8\mu}{m})}\right)$. Taking smallest $m = 3$ in the first term $2 - \frac{4}{m}$, taking biggest μ in the term $\frac{8\mu}{m}$, and dropping the term $(2-\frac{4}{m})^2$, we get for s the following boundaries:

$$0 < s < \frac{2}{3} + \sqrt{8B - 8}.$$

Combining two obtained boundaries for s, we can state, that *in order to get a non-negative integer solution* (t, u, v, w) *of the system*

$$\begin{cases} k = t^2 + u^2 + v^2 + w^2 \\ s = t + u + v + w \end{cases}$$

it is sufficient take an odd positive integer s *such that*

$$\frac{1}{2} + \sqrt{6B - 3} < s < \frac{2}{3} + \sqrt{8B - 8}.$$

If such s is found, one should only choose r, $0 \le r \le m - 2$, such that

$$s + r \equiv c (mod\, m).$$

But the last congruence is always satisfied if there are two odd numbers s and $s + 2$ in above limits. In this case we can use $m + 2$ values $s, s+1, s+2, \ldots, s+m-1$, $(s+2)+(m-2) = s+m$, $(s+2)+(m-1) = s+m+1$, from which two are always congruent modulo m.

Therefore, let us find B for which we have

$$\frac{2}{3} + \sqrt{8B - 8} - \frac{1}{2} - \sqrt{6B - 3} > 4.$$

One can rewrite this condition as $(8B - 8)(6B - 3) - 2\sqrt{48B^2 - 72B + 24} > (4 - \frac{1}{6})^2$. Then, putting 4 instead of $4 - \frac{1}{6}$, we get $(7B - 13)^2 > 48B^2 - 72B + 24$, or $B^2 - 110B + 145 > 0$. The last condition obviously holds if $B \ge 110$.

So, we have shown that for $N \ge 120m + 16$, i.e., for $B \ge 120$, the difference between two above limits is greater than 4. Then there are two odd numbers s and $s + 2$ between them, and we can represent N as a sum of $m + 2$ $(m + 2)$-gonal numbers, at most four of which are different from 0 or 1. So, the polygonal number theorem is proven for $N \geq 120m + 16$.

5.7.4. If B is sufficiently large, so that the difference of the above limits is greater or equal to $2m$, the number $N = Bm + c$ can be represented as a sum of at most four $(m + 2)$-gonal numbers, if m is odd, or if m is even and N is odd.

Indeed, in this case we can find among $2m$ positive integers between the above limits two numbers k and $k + m$, giving solutions of the congruence $x \equiv c(mod\, m)$, and one of this numbers is odd for odd m. So, the congruence $s + r \equiv c(mod\, m)$ holds for some odd s and $r = 0$. If m is even and N is odd, then c is odd, and the congruence $s + r \equiv c(mod\, m)$ also holds for some odd s and $r = 0$. In all these cases $N = S_{m+2}(t) + S_{m+2}(u) + S_{m+2}(v) + S_{m+2}(w)$.

If N and m are both even, the congruence $s + r \equiv c(mod\, m)$ holds for some odd s and $r = 1$. In this case we can represent N as a sum of five $(m + 1)$-gonal numbers, one of which is equal to 1: $N = S_{m+2}(t) + S_{m+2}(u) + S_{m+2}(v) + S_{m+2}(w) + 1$.

Since the difference of the limits is at least $2m$, if $B \geq 28m^2 - 2$, and, hence, if $N > 28m^3 - 2m$, we prove the following result: *if $m \geq 3$ is odd, then every positive integer $N > 28m^3 - 2m$ is the sum of four $(m + 2)$-gonal numbers; if $m \geq 3$ is even, then every positive integer $N > 28m^3 - 2m$ is the sum of five $(m + 2)$-gonal numbers, one of which is either 0 or 1.*

5.7.5. If we want to use the Pepin's method in order to represent a given positive integer N as a sum of $(m+2)$-gonal numbers, $m \geq 3$, we should start with the representation N in the form $N = Bm + c$, $0 < c < m + 1$, and check if it is smaller or bigger $120m + 16$.

If $N \leq 120m + 16$, then we check if $N \leq 30m + 9$. If yes, we find this number in our table or in our exceptions to the table. If $30m + 9 \leq N < 120m + 16$, then we find x such that $S_{m+2}(x) < N < S_{m+2}(x + 1)$, and represent N as $N = S_{m+2}(x) + R = S_{m+2}(x - 1) + R'$. If one of the numbers R or R' is contained in our table, we represent it as a sum of $m + 1$ $(m + 2)$-gonal numbers and then get the corresponding representation of N. If both R and R' are not in the table, we find that N is one of five above exceptions and use the corresponding representations.

Finally, if $N > 120m + 16$, we use general method. For given B and c, we find the smallest odd number l such that $l > \frac{1}{2} + \sqrt{6B-3}$. If the congruence $l + r \equiv c (mod\, m)$ holds for some r, $0 \le r \le m-2$, then we take $s = k$. If no, we take $s = l + 2$, and find r from the congruence $l + r \equiv c - 2 (mod\, m)$.

For given N, m, s and r, we find k from the equation $N = \frac{m}{2}(k-s) + s + r$, and then represent $4k - s^2$ as a sum of three squares $x^2 + y^2 + z^2$, $x \ge y \ge z > 0$. Finally, we find corresponding t, u, v, and w. It gives desired decomposition of N:

$$N = S_{m+2}(t) + S_{m+2}(u) + S_{m+2}(v) + S_{m+2}(w) + rS_{m+2}(1).$$

For example, consider $N = 114$, $m = 4$. Then $114 = 28m + 2 = 26m + 10$. So, we find in the table, that $N = S_6(6) + S_6(5) + 3S_6(1)$.

If $N = 210$, $m = 5$, then $N = 42m + 0 = 41m + m$. So, $30m + 9 < N < 120m + 16$. It is easy to see that $S_{m+2}(9) = 36m + 9 < N < S_{m+2}(10) = 45m + 10$. Then $N = S_{m+2}(9) + R = S_{m+2}(8) + R'$. Here $R = N - S_{m+2}(9) = 42m - 36m - 9 = 6m - 9 = 5m - 4 = 4m + 1$. So, $R = (3m + 3) + (m - 2) = S_{m+2}(3) + (m - 2)S_{m+2}(1)$, and $N = S_7(9) + S_7(3) + 5S_7(1)$.

If $N = 603$, $m = 5$, then $N = 121m + 4$, and $N > 120m + 16$. So, $B = 121$, $\sqrt{6B-3} = \sqrt{726} = 26,8\ldots$, $\frac{1}{2} + \sqrt{6B-3} = 27,3\ldots$, and $l = 29$. The congruence $29 + r \equiv 4 (mod\, 5)$ is true for $r = 0$. So, $s = 29$, and $r = 0$. Therefore, $603 = \frac{5}{2}(k - 29) + 29 + 0$, and $k = 261$. Furthermore, it holds $4k - s^2 = 4 \cdot 261 - 29^2 = 1044 - 841 = 203 = 11^2 + 9^2 + 1^2$, and $x = 11, y = 9, z = \pm 1$. So, $t = \frac{29+11+9\pm 1}{4} = \frac{29+11+9-1}{4} = 12$, $u = \frac{29+11}{2} - 12 = 8$, $v = \frac{29+9}{2} - 12 = 7$, $w = \frac{29-1}{2} - 12 = 2$. Therefore, $N = S_7(12) + S_7(8) + S_7(7) + S_7(2)$.

5.8 Other results related to the problem

In this section we consider several results of Dickson and Natanson, improving and generalizing the Fermat's polygonal number theorem (see [Dick27], [Dick28], [Dick28a], [Nata87]).

5.8.1. In 1927, Dickson ([Dick27]) obtained a new proof of the polygonal number theorem, as well as many other interesting results, related to the problem. In particular, he listed all positive integers up to $200m + 34$, $m \geq 3$, which are representable as a sum of $m + 2$ $(m + 2)$-gonal numbers.

More exactly, the Dickson's original list, given below, contains all non-negative integers up to $200m$, which are sums of four $(m + 2)$-gonal numbers, $m \geq 3$.

$0 - 4$, $m + 2 - 5$, $2m + 4 - 6$, $3m + 3 - 7$, $4m + 5 - 8$, $5m + 7 - 8$, $6m + 4 - 9$, $7m + 6 - 9$, $8m + 8 - 10$, $9m + 7 - 10$, $10m + 5 - 11$, $11m + 7 - 9$, 11, $12m + 8 - 12$, $13m + 8 - 12$, $14m + 10 - 12$, $15m + 6 - 9$, $11 - 13$, $16m + 8 - 13$, $17m + 10 - 13$, $18m + 9 - 14$, $19m + 11 - 14$, $20m + 10 - 14$, $21m + 7 - 13$, 15, $22m + 9 - 15$, $23m + 11 - 15$, $24m + 10 - 16$, $25m + 11 - 13$, 15, 16, $26m + 13 - 16$, $27m + 11 - 16$, $28m + 8 - 11$, $13 - 17$, $29m + 10 - 12$, $15 - 17$, $30m + 12 - 17$, $31m + 11 - 17$, $32m + 13 - 18$, $33m + 15 - 18$, $34m + 12 - 18$, $35m + 14 - 17$, $36m + 9 - 19$, $37m + 11 - 13$, $15 - 19$, $38m + 13 - 15$, $17 - 19$, $39m + 12 - 17$, 19, $40m + 14 - 20$, $41m + 16 - 20$, $42m + 13 - 20$, $43m + 14 - 17$, 19, 20, $44m + 16 - 20$, $45m + 10 - 13$, $16 - 21$, $46m + 12 - 21$, $47m + 14 - 17$, $19 - 21$, $48m + 13 - 21$, $49m + 15 - 21$, $50m + 17 - 22$, $51m + 14 - 17$, $19 - 22$, $52m + 16 - 22$, $53m + 18 - 21$, $54m + 17 - 19$, 21, 22, $55m + 11 - 17$, $19 - 23$, $56m + 13 - 23$, $57m + 15 - 21$, 23, $58m + 14 - 23$, $59m + 16$, 17, $19 - 23$, $60m + 16 - 24$, $61m + 15 - 21$, 23, 24, $62m + 17 - 24$, $63m + 19 - 24$, $64m + 17 - 24$, $65m + 16 - 21$, 23, 24, $66m + 12 - 15$, $17 - 25$, $67m + 14 - 16$, $19 - 25$, $68m + 16$, 17, $19 - 25$, $69m + 15 - 18$, 20, 21, $23 - 25$, $70m + 17 - 19$, $21 - 25$, $71m + 19 - 25$, $72m + 16 - 26$, $73m + 18 - 21$, $23 - 26$, $74m + 20 - 23$, 25, 26, $75m + 19 - 25$, $76m + 17 - 26$, $77m + 19 - 21$, $23 - 26$, $78m + 13 - 16$, $20 - 27$, $79m + 15 - 17$, $20 - 25$, 27, $80m + 17$, 18, $22 - 27$, $81m + 16 - 21$, $23 - 27$, $82m + 18 - 27$, $83m + 19 - 25$, 27, $84m + 17 - 28$, $85m + 19 - 21$, $23 - 28$, $86m + 21 - 28$, $87m + 19 - 25$, 27, 28, $88m + 18 - 22$, $24 - 28$, $89m + 20$, 21, $23 - 28$, $90m + 20 - 23$, $25 - 28$, $91m + 14 - 17$, $20 - 25$, $27 - 29$, $92m + 16 - 18$, $22 - 29$, $93m + 18 - 21$, $23 - 29$, $94m + 17 - 29$, $95m + 19$, 20, $22 - 25$, $27 - 29$, $96m + 21 - 29$, $97m + 18 - 21$, $23 - 29$, $98m + 20 - 27$, 29, 30, $99m + 20 - 25$, $27 - 30$, $100m + 21 - 30$, $101m + 19 - 21$, $23 - 29$, $102m + 21 - 30$, $103m + 22 - 25$, $27 - 30$, $104m + 22 - 30$, $105m + 15 - 18$, $24 - 29$, 31, $106m + 17 - 23$, $25 - 31$, $107m + 19$, 20, $22 - 25$, $27 - 31$, $108m + 18 - 21$, $24 - 31$, $109m + 20$, 21, $23 - 29$, 31, $110m + 22 - 31$, $111m + 19 - 25$, $27 - 31$, $112m + 21 - 32$, $113m + 23 - 29$, 31, 32, $114m + 22 - 27$, $29 - 32$, $115m + 20 - 22$, 24, 25, $27 - 32$, $116m + 22$, 23, $25 - 32$, $117m + 23 - 29$, 31, 32, $118m + 23 - 32$, $119m + 22 - 25$, $27 - 32$, $120m + 16 - 19$, $21 - 33$, $121m + 18 - 20$, $23 - 29$, $31 - 33$, $122m + 20$, 21, $25 - 33$, $123m + 19 - 25$, $27 - 33$, $124m + 21$, 22, $25 - 33$, $125m + 23$, $25 - 29$, $31 - 33$, $126m + 20 - 31$, 33, $127m + 22 - 25$, $27 - 33$, $128m + 24 - 26$, $28 - 34$, $129m + 23 - 29$, $31 - 34$, $130m + 21 - 23$, $25 - 27$, $29 - 34$, $131m + 23$, 24, $28 - 33$, $132m + 24 - 34$, $133m + 23 - 29$, $31 - 34$, $134m + 25 - 34$, $135m + 22 - 24$, $27 - 33$, $136m + 17 - 20$, $24 - 35$, $137m + 19 - 21$, $26 - 29$, $31 - 35$, $138m + 21$, 22, 25, 26, $28 - 35$, $139m + 20 - 23$, $27 - 33$, 35, $140m + 22$, 23, 26, 27, $29 - 35$, $141m + 23 - 29$, $31 - 35$, $142m + 21 - 31$, $33 - 35$, $143m + 23 - 25$, $27 - 33$, 35, $144m + 25 - 36$, $145m + 24 - 29$, $31 - 36$, $146m + 22 - 27$, $29 - 36$, $147m + 24$, 25, $27 - 33$, 35, 36, $148m + 24 - 27$, $29 - 36$, $149m + 25 - 29$, $31 - 36$, $150m + 25 - 36$, $151m + 23 - 25$, $27 - 33$, 35, 36, $152m + 25 - 36$, $153m + 18 - 21$, $27 - 29$, $31 - 37$, $154m + 20 - 22$, $26 - 37$, $155m + 22$, 23, $28 - 33$, $35 - 37$, $156m + 21 - 37$, $157m + 23 - 29$, $31 - 37$, $158m + 25 - 30$, $33 - 35$, 37, $159m + 22 - 25$, $28 - 33$, $35 - 37$, $160m + 24 - 37$, $161m + 26$, 28, 29, $31 - 37$, $162m + 25 - 27$, $29 - 38$, $163m + 23 - 25$, $27 - 33$, $35 - 38$, $164m + 25 - 27$, $30 - 38$, $165m + 26 - 28$, $31 - 37$, $166m + 26 - 38$, $167m + 28 - 33$, $35 - 38$, $168m + 24 - 26$, $29 - 31$, $33 - 38$, $169m + 26 - 29$, $31 - 37$, $170m + 28 - 38$, $171m + 19 - 22$, $27 - 33$, $35 - 39$, $172m + 21 - 23$, $26 - 39$, $173m + 23$, 24, 28, 29, $31 - 37$, 39, $174m + 22 - 31$, $33 - 35$, $37 - 39$, $175m + 24$, 25, $27 - 33$, $35 - 39$, $176m + 26$, $29 - 39$, $177m + 23 - 26$, 28, 29, $31 - 37$, 39, $178m + 25 - 27$, $29 - 39$, $179m + 27$, $31 - 33$, $35 - 39$, $180m + 26 - 40$, $181m + 24 - 29$, $31 - 37$, 39, 40, $182m + 26 - 40$, $183m + 27 - 33$, $35 - 40$, $184m + 27 - 40$, $185m + 29$, $31 - 37$, 39, 40, $186m + 25 - 40$, $187m + 27 - 33$, $35 - 40$, $188m + 29$, 30, $32 - 40$, $189m + 27 - 29$, $31 - 37$, 39, 40, $190m + 20 - 23$, $29 - 31$, $33 - 35$, $37 - 41$, $191m + 22 - 24$, $28 - 33$, $35 - 41$, $192m + 24 - 41$, $193m + 23 - 26$, 28, 29, $31 - 37$, $39 - 41$, $194m + 25$, 26, $30 - 32$, $34 - 39$, 41, $195m + 27$, $29 - 33$, $35 - 41$, $196m + 24 - 27$, $29 - 41$, $197m + 26 - 28$, $31 - 37$, $39 - 41$, $198m + 28 - 41$, $199m + 27$, 28, $30 - 32$, 35, 36.

Using this list, one can see that any positive integer up to $200m + 34$ can be represented as a sum of $m + 2$ $(m + 2)$-gonal numbers, at most four of which are different from 0 and 1.

We give below the full description of given in Dickson's list numbers up to $13m+12$, with exact representations of all such numbers as sums of at most four $(m+2)$ gonal numbers, as well as list of all numbers up to $14m+10$, which can be represented as a such sum plus some r, $r \leq m-2$.

sum of ≤ 4 $(m+2)$-gons	bondary for p	smallest number	biggest number	biggest number + $(m-2)$
$pS_{m+2}(1)$	$p \leq 4$	0	4	$m+2$
$S_{m+2}(2) + pS_{m+2}(1)$	$p \leq 3$	$m+2$	$m+5$	$2m+3$
$S_{m+2}(2) + S_{m+2}(2) + pS_{m+2}(1)$	$p \leq 2$	$2m+4$	$2m+6$	$3m+4$
$S_{m+2}(3) + pS_{m+2}(1)$	$p \leq 3$	$3m+3$	$3m+6$	$4m+4$
$S_{m+2}(2) + S_{m+2}(2) + S_{m+2}(2) + pS_{m+2}(1)$	$p \leq 1$	$3m+6$	$3m+7$	$4m+5$
$S_{m+2}(3) + S_{m+2}(2) + pS_{m+2}(1)$	$p \leq 2$	$4m+5$	$4m+7$	$5m+5$
$S_{m+2}(2) + S_{m+2}(2) + S_{m+2}(2) + S_{m+2}(2)$		$4m+8$	$4m+8$	$5m+6$
$S_{m+2}(3) + S_{m+2}(2) + S_{m+2}(2) + pS_{m+2}(1)$	$p \leq 1$	$5m+7$	$5m+8$	$6m+6$
$S_{m+2}(4) + pS_{m+2}(1)$	$p \leq 3$	$6m+4$	$6m+7$	$7m+5$
$S_{m+2}(3) + S_{m+2}(3) + pS_{m+2}(1)$	$p \leq 2$	$6m+6$	$6m+8$	$7m+6$
$S_{m+2}(3) + S_{m+2}(2) + S_{m+2}(2) + S_{m+2}(2)$		$6m+9$	$6m+9$	$7m+7$
$S_{m+2}(4) + S_{m+2}(2) + pS_{m+2}(1)$	$p \leq 2$	$7m+6$	$7m+8$	$8m+6$
$S_{m+2}(3) + S_{m+2}(3) + S_{m+2}(2) + pS_{m+2}(1)$	$p \leq 1$	$7m+8$	$7m+9$	$8m+7$
$S_{m+2}(4) + S_{m+2}(2) + S_{m+2}(2) + pS_{m+2}(1)$	$p \leq 1$	$8m+8$	$8m+9$	$9m+7$
$S_{m+2}(3) + S_{m+2}(3) + S_{m+2}(2) + S_{m+2}(2)$		$8m+10$	$8m+10$	$9m+8$
$S_{m+2}(4) + S_{m+2}(3) + pS_{m+2}(1)$	$p \leq 2$	$9m+7$	$9m+9$	$10m+7$
$S_{m+2}(4) + S_{m+2}(2) + S_{m+2}(2) + S_{m+2}(2)$		$9m+10$	$9m+10$	$10m+8$
$S_{m+2}(5) + pS_{m+2}(1)$	$p \leq 3$	$10m+5$	$10m+8$	$11m+6$
$S_{m+2}(4) + S_{m+2}(3) + S_{m+2}(2) + pS_{m+2}(1)$	$p \leq 1$	$10m+9$	$10m+10$	$11m+8$
$S_{m+2}(3) + S_{m+2}(3) + S_{m+2}(3) + S_{m+2}(2)$		$10m+11$	$10m+11$	$11m+9$
$S_{m+2}(5) + S_{m+2}(2) + pS_{m+2}(1)$	$p \leq 2$	$11m+7$	$11m+9$	$12m+7$
$S_{m+2}(4) + S_{m+2}(3) + S_{m+2}(2) + S_{m+2}(2)$		$11m+11$	$11m+11$	$12m+9$
$S_{m+2}(4) + S_{m+2}(4) + pS_{m+2}(1)$	$p \leq 2$	$12m+8$	$12m+10$	$13m+8$
$S_{m+2}(4) + S_{m+2}(3) + S_{m+2}(3) + pS_{m+2}(1)$	$p \leq 1$	$12m+10$	$12m+11$	$13m+9$
$S_{m+2}(3) + S_{m+2}(3) + S_{m+2}(3) + S_{m+2}(3)$		$12m+12$	$12m+12$	$13m+10$
$S_{m+2}(5) + S_{m+2}(3) + pS_{m+2}(1)$	$p \leq 2$	$13m+8$	$13m+10$	$14m+8$
$S_{m+2}(5) + S_{m+2}(3) + S_{m+2}(3) + S_{m+2}(2)$		$13m+11$	$13m+11$	$14m+9$
$S_{m+2}(4) + S_{m+2}(3) + S_{m+2}(3) + S_{m+2}(2)$		$13m+12$	$13m+12$	$14m+10$

For example, the number $10m+11 = S_{m+2}(3)+S_{m+2}(3)+S_{m+2}(3)+S_{m+2}(2)$ if represented as a sum of four $(m+2)$-gonal numbers. Then $10m+12 = S_{m+2}(3)+S_{m+2}(3)+S_{m+2}(3)+S_{m+2}(2)+S_{m+2}(1)$ is represented as a sum of five $(m+2)$-gonal numbers, and, in general, any number $10m+11+r$, $r \leq m-2$, is represented as a sum of $4+r$ $(m+2)$-gonal numbers, where $4+r \leq m+2$.

This table shows that any number up to $14m+10$ is represented as a sum of at most $m+2$ $(m+2)$-gonal numbers. The only possible exception, $11m+10$, can be written as $11m+10 = S_{m+2}(5)+S_{m+2}(2)+S_{m+2}(1)+S_{m+2}(1)+S_{m+2}(1)$, i.e., is a sum of five $(m+2)$-gonal numbers; such representation is possible for any $m \geq 3$.

The next table shows, how we can obtain from the Dickson's list a representation of such kind for any number $N \leq 120m+30$.

$\leq 4\,(m+2)$-gons		exc.
$0-4$	$m+2$	
$m+2-5$	$2m+3$	
$2m+4-6$	$3m+4$	
$3m+3-7$	$4m+5$	
$4m+5-8$	$5m+6$	
$5m+7-8$	$6m+6$	
$6m+4-9$	$7m+7$	
$7m+6-9$	$8m+7$	
$8m+8-10$	$9m+8$	
$9m+7-10$	$10m+8$	
$10m+5-11$	$11m+9$	
$11m+7-9, 11$	$12m+9$	$11m+10$
$12m+8-12$	$13m+10$	
$13m+8-12$	$14m+10$	
$14m+10-12$	$15m+10$	
$15m+6-9, 11-13$	$16m+11$	no
$16m+8-13$	$17m+11$	
$17m+10-13$	$18m+11$	
$18m+9-14$	$19m+12$	
$19m+11-14$	$20m+12$	
$20m+10-14$	$21m+12$	
$21m+7-13, 15$	$22m+13$	$21m+14$
$22m+9-15$	$23m+13$	
$23m+11-15$	$24m+13$	
$24m+10-16$	$25m+14$	
$25m+11-13, 15, 16$	$26m+14$	no
$26m+13-16$	$27m+14$	
$27m+11-16$	$28m+14$	
$28m+8-11, 13-17$	$29m+15$	no
$29m+10-12, 15-17$	$30m+15$	no
$60m+16-24$	$61m+22$	
$61m+15-21, 23, 24$	$62m+22$	no
$62m+17-24$	$63m+22$	
$63m+19-24$	$64m+22$	
$64m+17-24$	$65m+22$	
$65m+16-21, 23, 24$	$6m+22$	no
$66m+12-15, 17-25$	$67m+23$	no
$67m+14-16, 19-25$	$68m+23$	no
$68m+16, 17, 19-25$	$69m+23$	no
$69m+15-18, 20, 21, 23-25$	$70m+23$	no
$70m+17-19, 21-25$	$71m+23$	no
$71m+19-25$	$72m+23$	
$72m+16-26$	$73m+24$	
$73m+18-21, 23-26$	$74m+24$	no
$74m+20-23, 25, 26$	$75m+24$	no
$75m+19-25$	$76m+23$	
$76m+17-26$	$77m+24$	
$77m+19-21, 23-26$	$78m+24$	no
$78m+13-16, 20-27$	$79m+25$	no
$79m+15-17, 20-25, 27$	$80m+25$	$79m+26$
$80m+17, 18, 22-27$	$81m+25$	no
$81m+16-21, 23-27$	$82m+25$	no
$82m+18-27$	$83m+25$	
$83m+19-25, 27$	$84m+25$	$83m+26$
$84m+17-28$	$85m+26$	
$85m+19-21, 23-28$	$86m+26$	no
$86m+21-28$	$87m+26$	
$87m+19-25, 27, 28$	$88m+26$	no
$88m+18-22, 24-28$	$89m+26$	no
$89m+20, 21, 23-28$	$90m+26$	no

$\leq 4\,(m+2)$-gons		exc.
$30m+12-17$	$31m+15$	
$31m+11-17$	$32m+15$	
$32m+13-18$	$33m+16$	
$33m+15-18$	$34m+16$	
$34m+12-18$	$36m+16$	
$35m+14-17$	$36m+15$	
$36m+9-19$	$37m+17$	
$37m+11-13, 15-19$	$38m+17$	no
$38m+13-15, 17-19$	$39m+17$	no
$39m+12-17, 19$	$40m+17$	$39m+18$
$40m+14-20$	$41m+18$	
$41m+16-20$	$42m+18$	
$42m+13-20$	$43m+18$	
$43m+14-17, 19, 20$	$44m+18$	no
$44m+16-20$	$45m+18$	
$45m+10-13, 16-21$	$46m+19$	no
$46m+12-21$	$47m+19$	
$47m+14-17, 19-21$	$48m+19$	no
$48m+13-21$	$49m+19$	
$49m+15-21$	$50m+19$	
$50m+17-22$	$51m+20$	
$51m+14-17, 19-22$	$52m+20$	no
$52m+16-22$	$53m+20$	
$53m+18-21$	$54m+19$	$54m+20$
$54m+17-19, 21, 22$	$55m+20$	no
$55m+11-17, 19-23$	$56m+21$	no
$56m+13-23$	$57m+21$	$57m+22$
$57m+15-21, 23$	$58m+19$	no
$58m+14-23$	$59m+21$	
$59m+16, 17, 19-23$	$60m+21$	no
$90m+20-23, 25-28$	$91m+26$	no
$91m+14-17, 20-25, 27-29$	$92m+27$	no
$92m+16-18, 22-29$	$93m+27$	no
$93m+18-21, 23-29$	$94m+27$	no
$94m+17-29$	$95m+27$	
$95m+19, 20, 22-25, 27-29$	$96m+27$	no
$96m+21-29$	$97m+27$	
$97m+18-21, 23-29$	$98m+27$	no
$98m+20-27, 29, 30$	$99m+28$	$98m+28$
$99m+20-25, 27-30$	$100m+28$	no
$100m+21-30$	$101m+28$	
$101m+19-21, 23-29$	$102m+27$	no
$102m+21-30$	$103m+28$	
$103m+22-25, 27-30$	$104m+28$	no
$104m+22-30$	$105m+28$	$105m+30$
$105m+15-18, 24-29, 31$	$106m+29$	no
$106m+17-23, 25-31$	$107m+29$	no
$107m+19, 20, 22-25, 27-31$	$108m+29$	no
$108m+18-21, 24-31$	$109m+29$	no
$109m+20, 21, 23-29, 31$	$110m+29$	$109m+30$
$110m+22-31$	$111m+29$	
$111m+19-25, 27-31$	$112m+29$	no
$112m+21-32$	$113m+30$	
$113m+23-29, 31, 32$	$114m+30$	no
$114m+22-27, 29-32$	$115m+30$	no
$115m+20-22, 24, 25, 27-32$	$116m+30$	no
$116m+22, 23, 25-32$	$117m+30$	no
$117m+23-29, 31, 32$	$118m+30$	no
$118m+23-32$	$119m+30$	
$119m+22-25, 27-32$	$120m+30$	no

In the first column of the table there are numbers, which can be represented as a sum of at most four $(m+2)$-gonal numbers. The second column contains the numbers $N+(m-2)$, where N is the biggest number from the corresponding first column's place, i.e., all numbers $N, N+1, \ldots, N+(m-2)$ are represented as a sum of at most $m+2$ $(m+2)$-gonal numbers. The third column contains possible exceptions, not covered by previous construction.

For example, the numbers from $36m + 9$ to $36m + 19$ are represented as a sum of at most four $(m + 2)$-gonal numbers. So, the numbers $36m+20, \ldots, 37m+17$ are representable as a sum of at most $m+2$ $(m+2)$-gonal numbers. Similarly, the numbers from $37m+11$ to $36m + 13$, as well as the numbers from $37m + 15$ to $37m + 19$, are represented as a sum of at most four $(m+2)$-gonal numbers. So, the numbers $37m + 20, \ldots, 38m + 17$ are representable as a sum of at most $m + 2$ $(m + 2)$-gonal numbers. The possible gap $37m + 14$ is included to the previous interval $36m + 20, \ldots, 37m + 17$. On the other hand, the number $39m+18$ is not covered by this construction, and is considered as possible exception. However, all possible exceptions $11m+10$, $21m+14$, $39m+18$, $54m+20$, $57m+22$, $79m+26$, $83m + 26$, $98m + 28$, $105m + 30$, and $109m + 30$ can be represented as a sum of five $(m+2)$-gonal numbers, since for any such exception, written in the form $N + 1$, the number N can be represented as a sum of four $(m + 2)$-gonal numbers.

Hence, it is shown, that the above table covers all positive integers $N \leq 120m + 30$, and the polygonal number theorem is proven for all positive integers up to $120m + 30$. Moreover, applying similar considerations to the full version of the Dickson's list, we can state that the theorem is proven for all positive integers $N \leq 200m+34$. So, in order to obtain now the entire proof of the theorem, it is sufficient consider only the general case, which was proven before by Cauchy's or Pepin's methods.

5.8.2. So, Dickson ([Dick27]) gave a new proof of the classical theorem. Moreover, in the series of papers (1927–1929) he gave several more general results, related to the problem ([Dick27], [Dick28], [Dick28a]).

In particular, he proved, that *the functions derived from* $\frac{1}{2}m(x^2 - x)+x$ *by replacing* x *by* $x-a$ *or* $a-x$ *are the only quadratic functions of* x *which give non-negative integers for every integer* $x \geq 0$, *and every positive integer* N *is a sum of* b *of these values, where* b *depends on* $f(x)$, *but not on* N.

In the simplest case $a = 0$ it gives the classical $(m + 2)$-gonal numbers $\frac{1}{2}m(x^2 - x) + x$, and the generalized $(m+2)$-gonal numbers $\frac{1}{2}m(x^2 + x) - x$ with negative indices, respectively.

Dickson studied all possible cases of such functions and obtained full description of corresponding decompositions of positive integers. In particular, he had shown that *any positive integer can be represented as a sum of* $m+2$ *generalized* $(m+2)$*-gonal numbers* $\frac{m}{2}(x^2-x)+x$ *with negative indices, at most four of which are different from 0 and 1.*

The Dickson's list, containing all positive integers up to $92m+25$, represented as a sum of at most four generalized $(m+2)$-gonal numbers $\frac{m}{2}(x^2-x)+x$, is given below.

0, $m-1$, $2m-2$, $3m-2-3$, $4m-3-4$, $5m-4$, $6m-3-5$, $7m-4-5$, $8m-5-6$, $9m-5-6$, $10m-4, 6, 7$, $11m-5, 7$, $12m-6-8$, $13m-6-8$, $14m-7, 8$, $15m-5, 8, 9$, $16m-6-9$, $17m-7-9$, $18m-7-10$, $19m-8-10$, $20m-8-10$, $21m-6, 8, 9, 11$, $22m-7, 9-11$, $23m-8, 10, 11$, $24m-8-12$, $25m-9, 11, 12$, $26m-10-12$, $27m-9-12$, $28m-7, 10-13$, $29m-8, 11-13$, $30m-9-13$, $31m-9-13$, $32m-10-14$, $33m-11-14$, $34m-10-14$, $35m-11-13$, $36m-8, 11-15$, $37m-9, 12-15$, $38m-10, 11, 13-15$, $39m-10-13, 15$, $40m-11, 13-16$, $41m-12-16$, $42m-11, 12, 14-16$, $43m-12, 13, 15, 16$, $44m-13-16$, $45m-9, 13-17$, $46m-10, 12, 14-17$, $47m-11, 13, 15-17$, $48m-11, 12, 14-17$, $49m-12-17$, $50m-13-18$, $51m-12, 13, 15-18$, $52m-13-18$, $53m-14-17$, $54m\ 14, 15, 17, 18$, $55m-10, 13, 15-19$, $56m-11, 14, 16-19$, $57m-12, 14-17, 19$, $58m-12, 13, 15-19$, $59m-13, 16-19$, $60m-14, 16-20$, $61m-13-17, 19, 20$, $62m-14-20$, $63m-15-20$, $64m-15-20$, $65m-14, 16, 17, 19, 20$, $66m-11, 15, 17-21$, $67m-12, 16-21$, $68m-13, 16-21$, $69m-13, 14, 17, 19-21$, $70m-14, 15, 18-21$, $71m-15-21$, $72m-14-22$, $73m-15-17, 19-22$, $74m-16-19, 21, 22$, $75m-16-22$, $76m-15-22$, $77m-16, 17, 19-22$, $78m-12, 17-23$, $79m-13, 17-21, 23$, $80m-14, 18-23$, $81m-14-17, 19-23$, $82m-15, 17-23$, $83m-16-21, 23$, $84m-15, 16, 18-24$, $85m-16, 17, 19-24$, $86m-17, 19-24$, $87m-17-21, 23, 24$, $88m-16, 18, 20-24$, $89m-17, 19-24$, $90m-18, 19, 21-25$, $91m-13, 18-21, 23-25$, $92m-14, 19-25$.

Using this list, one can check that any positive integer up to $92m+23$ is representable as a sum of $m+2$ generalized $(m+2)$-gonal numbers, and then prove the general case of the corresponding theorem, following the standard consideration.

5.8.3. Because of Pepin's and Dickson's tables, in order to prove the polygonal number theorem it suffices to consider only $N \geq 120m$. This case was studied by Natanson ([Nata87]), who obtained a short and easy proof of Cauchy's lemma and some improvement of the classical results in this general case.

In particular, he had proven, that *if* $m \geq 3$ *and* $N \geq 120m$, *then* N *is the sum of* $m+1$ $(m+2)$*-gonal numbers, at most four of which are different from* 0 *or* 1. The proof of this fact is given below.

At first, let us show that *for odd positive integers* k *and* s, *such that* $s^2 < 4k$ *and* $3k < s^2+2s+4$, *there exist non-negative integers* t, u, v, w *such that* $k = t^2+u^2+v^2+w^2$ *and* $s = t+u+v+w$ (*Cauchy's lemma*).

In fact, since k and s are odd, it follows that $4k - s^2 \equiv 3(mod\ 8)$, and so, there exist odd integers $x \geq y \geq z > 0$ such that $4k - s^2 = x^2 + y^2 + z^2$. Let the sign of $\pm z$ be chosen, so that $s + x + y \pm z \equiv 0 (mod\ 4)$. Let integers t, u, v, w be defined by

$$t = \frac{s + x + y \pm z}{4}, \quad u = \frac{s + x}{2} - t = \frac{s + x - y \mp z}{4},$$
$$v = \frac{s + y}{2} - t = \frac{s - x + y \mp z}{4}, \quad w = \frac{s \pm z}{2} - t = \frac{s - x - y \pm z}{4}.$$

It implies that $k = t^2 + u^2 + v^2 + w^2$, and $s = t + u + v + w$. Moreover, it holds $t \geq u \geq v \geq w$. In order to show that these integers are non-negative, it suffices to prove that $w \geq 0$, or $w > -1$. This is true if $s - x - y - z > -4$ or, equivalently, if $x + y + z < s + 4$. The maximum value of $x + y + z$ subject to the constraint $4k - s^2 = x^2 + y^2 + z^2$ is $\sqrt{12k - 3s^2}$, and the inequality $3k < s^2 + 2s + 4$ implies that $x + y + z \leq \sqrt{12k - 3s^2} < s + 4$. This completes the proof.

Now let s_1 and s_2 be consecutive odd integers. The set of numbers of the form $s + r$, where $s \in \{s_1, s_2\}$ and $r \in \{0, 1, \ldots, m - 3\}$, contains a complete set of residue classes modulo m. So, it holds $N \equiv s + r (mod\ m)$ for some $s \in \{s_1, s_2\}$, and $r \in \{0, 1, \ldots m - 3\}$. Let

$$k = 2\left(\frac{N - s - r}{m}\right) + s = \left(1 - \frac{2}{m}\right) s + 2\left(\frac{N - r}{m}\right).$$

Then k is an odd integer, and

$$N = \frac{m}{2}(k - s) + s + r.$$

If $0 < s < \frac{2}{3} + \sqrt{8\frac{N}{m} - 8}$, then $s^2 - 4k = s^2 - 4\left(1 - \frac{2}{m}\right) s - 8\left(\frac{N-r}{m}\right) < 0$, and so, $s^2 < 4k$. Similarly, if $s > \frac{1}{2} + \sqrt{6\frac{N}{m} - 3}$, then $3k < s^2 + 2s + 4$. Since the length of the interval

$$I = \left(\frac{1}{2} + \sqrt{6\frac{N}{m} - 3}, \frac{2}{3} + \sqrt{8\frac{N}{m} - 8}\right)$$

is greater than 4, it follows that I contains two consecutive odd positive integers s_1, and s_2.

Thus, there exist odd positive integers k and s that satisfy the condition $N = \frac{m}{2}(k - s) + s + r$ and the inequalities $s^2 < 4k$,

$3k < s^2 + 2s + 4$. Cauchy's lemma implies that there exist t, u, v, w such that $k = t^2 + u^2 + v^2 + w^2$, and $s = t + u + v + w$. So,

$$\begin{aligned} N &= \frac{m}{2}(k - s) + s + r \\ &= (\frac{m}{2}(t^2 - t) + t) + \ \dots \ + \frac{m}{2}(w^2 - w) + w + r. \end{aligned}$$

5.8.4. Moreover, Natanson gave a short proof of the classical Legendre's result ([Lege79]), stating that *every sufficiently large integer is the sum of five* $(m+2)$*-gonal numbers, one of which is either* 0 *or* 1. More exactly, the following proposition holds: *if* $m \geq 3$ *is odd, then every sufficiently large integer is the sum of four* $(m + 2)$*-gonal numbers; if* m *is even, then every sufficiently large integer is the sum of five* $(m + 2)$*-gonal numbers, one of which is either* 0 *or* 1.

In order to prove this result, note that there is a constant C such that if $N > Cm^3$, then the length of the interval I defined before is greater than $2m$, and so, I contains at least m consecutive odd numbers. If m is odd, these numbers form a complete set of residues modulo m, and so, $N \equiv s(mod\ m)$ for some odd number $s \in I$. Let $r = 0$. Let $k = 2(\frac{N-s}{m}) + s = (1 - \frac{2}{m})s + 2(\frac{N}{m})$. If m is even, then $N \equiv s + r(mod\ m)$ for some odd number $s \in I$, and $r \in \{0, 1\}$. Let $k = 2(\frac{N-s-r}{m}) + s = (1 - \frac{2}{m})s + 2(\frac{N-r}{m})$.

In both cases, the theorem follows immediately from Cauchy's lemma.

Chapter 6

Zoo of figurate-related numbers

In this chapter we collected some remarkable individual figurate-related numbers (see [PrPe11], [Weis11], [Wiki11], [LeLi83]).

- **3**: the only prime p such that a sum of divisors of p^4 is a perfect square: $1 + 3 + 3^2 + 3^3 + 3^4 = S_4(11)$; the only prime of the form $S_4(n) - 1$; the only Fibonacci prime that is also a triangular number.
- **5**: the only prime of the form $4BC(n) + BC(m)$; the only prime, digit in which a perfect square can end; the only prime whose the square is composed of only prime digits.
- **6**: the only mean $\frac{5+7}{2}$ between a pair of twin primes which is triangular; the only known even number n such that both numbers $C^n(n) - (n+1)$ and $C^n(n) + (n+1)$ are primes; the largest known number n such that there are n integers for which all pairwise sums are perfect squares.
- **7**: the only prime p such that $p+1$ is a cubic number; the smallest **cuban number**, i.e., a number of the form $(n+1)^3 - n^3$; the biggest known solution of the *Brocard's problem*: to find integers n such that $n! + 1$ is a perfect square; the largest known prime that is not the sum of a triangular number, a square, and a cube, all of them positive.
- **9**: the only known composite number n such that both $C^n(2) + S_4(n)$ and $C^n(2) - S_4(n)$ are primes.

- **11**: the largest prime that can not be written as a sum of four hexagonal numbers; the only prime that can not be represented using five hexagonal numbers.
- **13**: a concatenation of the first two triangular numbers; the only prime sum $(2^2 + 3^2)$ of squares of two consecutive primes; the only Fibonacci prime sum of squares of two consecutive Fibonacci primes; the prime whose square is equal to the sum of squares of all digits in which primes end: $13^2 = 1^2 + 2^2 + 3^2 + 5^2 + 7^2 + 9^2$; the only prime index of a triangular number that is also a square pyramidal number: $S_3(13) = S_4^3(6)$; the smallest prime having exactly one representation as a sum of squares greater than one (31, its reversal, is the largest one).
- **16**: the only integer of the form $C^m(n) = C^n(m)$ for distinct integers n and m: $16 = 2^4 = 4^2$.
- **17**: the only prime of the form $C^q(p) + C^p(q)$, where p and q are primes: $17 = 2^3 + 3^2$; the smallest prime whose sum of the digits is a cubic number; the only known prime that is equal to the sum of digits of its cube: $17^3 = 4913$, and $4 + 9 + 1 + 3 = 17$; the smallest prime that is the **quartan**, i.e., the sum of two biquadratic numbers: $17 = 1^4 + 2^4$; $S_4(17)$ can be expressed as the sum of $1, 2, 3, 4, 5, 6, 7, 8$ distinct squares.
- **23**: the largest integer n such that no factor of a binomial coefficient $\binom{n}{k}$ is a perfect square; the biggest prime which is uniquely expressible as a sum of at most four squares.
- **24**: the only integer $n > 1$ such that $\sum_{i=1}^{n} i^2$ is a perfect square: $\sum_{i=1}^{24} i^2 = S_4(70)$.
- **25**: the only perfect square of the form $C(n) - 2$.
- **28**: the only perfect number of the form $C^k(n)+C^k(m)$ with $k > 1$: $28 = C(3) + C(1)$.
- **29**: the only two-digital prime whose square is the sum of squares of two consecutive two-digital numbers: $S_4(29) = S_4(20) + S_4(21)$; the smallest prime equal to the sum of three consecutive squares: $29 = S_4(2) + S_4(3) + S_4(4)$; the smallest multi-digital prime which on adding its reverse gives perfect square: $29 + 92 = S_4(11)$.
- **31**: the smallest prime that can be represented as a sum of two triangular numbers in two different ways: $31 = 21+10 = 28+3$; the smallest prime that can be represented as a sum of two triangular

numbers with prime indices; there are only 31 numbers that can not be expressed as a sum of distinct squares; $3+5+7+11+\cdots+89 = S_4(31)$, and a sum of the first 31 odd primes is a square of prime.

- **33**: the largest integer which is not a sum of distinct triangular numbers.
- **36**: the smallest triangular number whose sum of divisors, as well as sum of its proper divisors, is also a triangular number; the smallest perfect square expressible as a sum of four consecutive primes which are also two couples of prime twins: $36 = 5 + 7 + 11 + 13$.
- **41**: the only known prime number that is not a sum of two triangular numbers and a non-negative cube; the smallest prime whose cube can be written as a sum of three cubes in two ways: $C(41) = C(40) + C(17) + C(2) = C(33) + C(32) + C(6)$.
- **53**: the largest known integer that can be expressed as a sum of three non-negative triangular numbers in exactly one way.
- **65**: the only number which gives a square of prime on adding as well as subtracting its reverse from it: $65+56 = S_4(11)$, $65-56 = S_4(3)$; the only number which is the difference $(BC(3) - BC(2))$ of two biquadratic numbers with prime indices.
- **67**: the smallest multi-digital prime whose square 4489 and cube 300763 consist of different digits.
- **81**: the only known square n such that $n \cdot C^n(2) - 1$ is a prime.
- **83**: the largest known prime that can be expressed as a sum of three positive triangular numbers in exactly one way; the smallest prime whose square 6889 is a strobogrammatic number; the only prime equal to a sum of squares of odd primes: $83 = S_4(3) + S_4(5) + S_4(7)$; the only prime of the form $BC(p) + 2$, where p is a prime.
- **89**: the largest integer that can not be expressed as a sum of four pentagonal numbers (its rightmost digit is the smallest positive integer with this property); the smallest positive integer whose square 7921 and cube 704969 are likewise primes upon reversal.
- **100**: the smallest perfect square whose summation of the differences between itself and each of its digits, where each difference is raised to the power of the corresponding digit, is equal to a prime: $101 = (100 - 1)^1 + (100 - 0)^0 + (100 - 0)^0$ is a prime.

- **108**: the smallest positive integer which can be written as a sum of a cubic number and a perfect square in two ways.
- **113**: the smallest prime which is a sum of three biquadratic numbers with prime indices: $113 = BC(2) + BC(2) + BC(3)$.
- **121**: the only perfect square of the form $1 + p + S_4(p) + C(p) + BC(p)$, $p \in P$: $121 = 1 + 3 + S_4(3) + C(3) + BC(3)$; the only, besides 4, perfect square of the form $C(n) - 4$.
- **128**: the largest integer, which is not a sum of distinct perfect squares (but $129 = S_4(8) + S_4(7) + S_4(4)$).
- **144**: the largest perfect square, which is a Fibonacci number.
- **149**: the only known prime in the concatenate square sequence.
- **169**: the largest **Pell number**, i.e., a member of the recurrent sequence $P_{n+2} = 2P_{n+1} + P_n$, $P_0 = 0, P_1 = 1$, which is a perfect square.
- **173**: the largest known prime whose square 29929 and cube 5177717 consist of different digits.
- **191**: a palindromic prime whose square 36481 is a distinct-digital number whose first two digits, central digit, and last two digits, are perfect squares.
- **196**: the smallest candidate **Lychrel number**, i.e., a natural number which can not form a palindrome through the iterative process of repeatedly reversing its base 10 digits and adding the resulting numbers; in fact, 196 does not yield a palindrome after 700 000 000 iterations.
- **211**: the largest known prime that can not be written as a sum of a prime and a positive triangular number.
- **216**: the smallest cubic number, which is a sum of three cubic numbers: $216 = C(6) = C(5) + C(4) + C(3)$.
- **239**: the largest integer, which is not a sum of less than 9 cubic numbers: $239 = 2 \cdot C(4) + 4 \cdot C(3) + 3 \cdot C(1)$.
- **289**: the square of the sum of the first four primes: $289 = S_4(2 + 3 + 5 + 7)$.
- **343**: the only, besides 1, cubic number such that the sum of its divisors is a perfect square: $1 + 7 + 7^2 + 7^3 = S_4(20)$.
- **367**: the largest number whose square 134689 has strictly increasing digits.

- **400**: the only known square of the form $1+k+S_4(k)+C(k)$, where $k \in \mathbb{N}\backslash\{1\}$.
- **407**: the largest integer, which is the sum of cubes of their decimal digits.
- **463**: the smallest multi-digital prime such that both a sum of digits and product of digits of its square remain squares.
- **496**: the third perfect number; the smallest triangular number such that the sum of the cubes of its digits is prime: $C(4) + C(9) + C(6) = 1009$.
- **576**: the only known perfect square represented as a difference between a squared sum of consecutive primes and the sum of their squares: $576 = S_4(24) = S_4(2+3+5+7+11) - (S_4(2) + S_4(3) + S_4(5) + S_4(7) + S_4(11))$; it is the only such case for all primes up to 2×10^9.
- **613**: a prime which gives a mathematical enigma in the story *Number of the End* by Jason Earls: *bring the first digit back to get* 136, *it is triangular, now bring the first digit of that back to get* 361, *it is a square*; the square 375769 of 613 is the largest known perfect square that divides a number of the form $n! + 1$, which happens when $n = 229$, another prime.
- **631**: a prime which is the reverse concatenation of the first three triangular numbers.
- **691**: the only known prime which is a square 169 when turned upside down and another square 196 when reversed; moreover, 169 and 196 are the smallest consecutive squares using the same digits.
- **701**: the smallest prime whose square 491401 contains all of square digits only; it is equal to $5^4 + 4^3 + 3^2 + 2^1 + 1^0$.
- **786**: the largest known number n such that the binomial coefficient $\binom{2n}{n}$ is not divisible by the square of an odd prime.
- **900**: the smallest perfect square which is a sum of different primes by using all the ten digits: $900 = 503+241+89+67 = 509+283+61+47$.
- **1049**: the smallest square-digital prime whose square 1100401 contains only square digits; the smallest prime containing all of the square digits exactly once.

- **1493**: the largest known **Stern prime number**, i.e., a prime which is not a sum of a smaller prime and twice the square of a non-zero integer; all known Stern primes are 2, 3, 17, 137, 227, 977, 1187, 1493 (Sloane's A042978).
- **1759**: the smallest prime that is a sum of the first consecutive triangular numbers with prime indices: $1759 = 3 + 6 + 15 + \cdots + 496 = S_3(2) + S_3(3) + \cdots + S_3(31)$.
- **2861**: the largest known prime that can not be expressed as a sum of three pentagonal numbers.
- **3331**: the largest known prime p such that $p - t$ are all composite for $\frac{p}{2} < t < p$, where t is a triangular number.
- **4019**: a prime composed of square digits that remains prime if any of its $S_4(2)$ digits is deleted.
- **4939**: the smallest positive integer such that the concatenation of its prime factors forms a perfect square: $4939 = 11{\cdot}449$, $11, 449 \in P$, and $11449 = S_4(107)$; this concatenation consists from five square digits; also, a sum of digits of 4939 gives the square of a prime: $4 + 9 + 3 + 9 = S_4(5)$; moreover, the two leftmost digits of 4939 form another square of a prime.
- **5017**: the 58-th pentagonal number; both 5 and 17 are Fermat primes.
- **9041**: the largest prime whose digits are distinct square digits.
- **12758**: the largest integer, which is not a sum of distinct cubic numbers (cf. **128**).
- **13331**: a palindromic prime that is a sum of the squares of three consecutive odd triangular numbers: $13331 = S_4(45) + S_4(55) + S_4(91)$.
- **16361**: the only five-digital palindromic prime formed from all three triangular digits.
- **21679**: the largest known non-square integer which is not a sum of a square and a prime.
- **23397**: the largest prime whose square contains no duplicate digits: $S_4(21397) = 45783160$.
- **26951**: the largest known integer n, such that $n! + 1$ is a prime (cf. **94550**).

- **28541**: the smallest prime that results in a biquadratic number if added to a sum of its digits.
- **34847**: a sum of twin primes 34847 and 34849 is the only known palindromic square number ($S_4(264) = 69696$) which is a sum of a **twin prime pair**.
- **41041**: a triangular number, which is also a **Carmichael number**, i.e., an odd composite number n which is a **pseudoprime number** for any base, that is, satisfies *Fermat's Little Theorem* $a^n \equiv a(mod\ n)$ for any integer a; the number $S_3(41041)$, having the form $842202361 = 7 \cdot 11 \cdot 13 \cdot 41 \cdot 20521$, is also a Carmichael number; moreover, $842202361 = S_3(S_3(286))$ is a doubly triangular number.
- **61403**: the largest known prime that can not be expressed as a sum of three hexagonal numbers.
- **65537**: the smallest prime that is a sum of a non-zero perfect square and a non-zero cubic number in four different ways: $65537 = S_4(122) + C(37) = S_4(219) + C(26) = S_4(255) + C(8) = S_4(256) + C(1)$.
- **73037**: the smallest prime that can be represented as a sum of a perfect square and its reversal in two different ways.
- **81619**: the largest known prime whose (beastly) square 6661661161 is composed of only two different digits.
- **92239**: the smallest prime without digits 0 or 1 that forms a new prime by replacing each digit with its square, cube, 4-th, or 5-th power.
- **94550**: the largest known integer n, such that $n! - 1$ is a prime (cf. **26951**).
- **98689**: the first centered triangular number that is a palindromic prime.
- **101723**: the smallest prime whose square 10347568729 contains all of the digits from 0 to 9.
- **106721**: a prime which can be represented as $1^{15} + 3^{10} + 6^6 + 10^3 + 15^1$, where the first five triangular numbers in ascending order are bases, and in descending order are powers of the addendum.

- **123479**: the smallest prime formed from all of the non-zero digits that are expressible as sum of two triangular numbers in exactly one way.
- **149099**: the first occurrence of a twin prime pair (149099; 149101) that both primes consist of all square digits only.
- **248832**: the smallest five-dimensional hypercube number, which is a sum of six five-dimensional hypercube numbers: $C^5(12) = C^5(11) + C^5(9) + C^5(7) + C^5(6) + C^5(5) + C^5(4) + C^5(3)$.
- **835399**: a prime which can be represented as a sum of a triangular number and a perfect square in 27 distinct ways.
- **1092733**: the smallest prime p such that $p + 14, p + 15, p + 16, p + 17, p + 18, p + 19, p + 20$, and $p + 21$ are each divisible by the square of a prime.
- **1258723**: prime index of a 13-digital palindromic pentagonal number 2376574756732.
- **5134210**: the largest integer, which is not a sum of distinct biquadratic numbers (cf. **128**).
- **24710753**: this prime is a sum of all triangular numbers with prime indices less than 1000.
- **36101521**: the smallest prime formed from the concatenation of five consecutive triangular numbers.
- **136101521**: the prime formed by the concatenation of the first six consecutive triangular numbers (note that six itself is also triangular).
- **1882341361**: the smallest prime whose reversal is both square and triangular.
- **10123496857**: the smallest pandigital prime, whose square 102485188613688878449 is also **pandigital number**, i.e., a positive integer which contains all digits at least once.
- **3531577135439**: a sum of primes less than or equal to this number is a perfect square.
- **99999999944441**: the largest prime **emirp** that contains each non-zero square digit d exactly d times.
- **1000000000100011**: the smallest prime whose number of digits (16) is the square of the sum of its digits.

- **786655453628211510631**: a prime which is the reverse concatenation of the first twelve triangular numbers.
- **9561677372927686361**: 19-digital prime index of the 39-digital palindromic hexagonal number 182851348367914603505306419763843158281.
- $\approx \mathbf{4 \times 10^{22}}$: the **Sagan's number** is defined as the number of stars in the observable universe; it contains $3 \times 10^{22} - 5 \times 10^{22}$ stars (in at least 8×10^{10} galaxies), and at least 10^{80} atoms.
- $\mathbf{2^{157}}$: the smallest **apocalyptic number**, i.e., a number of the form 2^n that contains the beast number 666; the first few such powers are 157, 192, 218, 220, 222, ... (Sloane's A007356).
- **166...6699...991** (100 digits): a strobogrammatic prime of perfect square length.
- **99...99499999** (100 digits): the largest near-repdigit prime less than 10^{100} with only square digits.
- $\mathbf{10^{100}}$: **googol** (or **ten duotrigintillion**, **ten thousand sexdecillion**, **ten sexdecilliard**); a googol is of the same order of magnitude as the 70-th factorial number: $70! \approx 1.198 \times 10^{100}$.
- **100...00949** (101 digits): the smallest prime greater than 10^{100} and containing only square digits.
- $\mathbf{10^{123}}$: the **game tree size** (the total number of possible games that can be played) for Chess; it is estimated around 10^{123}, 10^{226}, and 10^{360} for Chess, Shogi, and Go, respectively.
- $\mathbf{10^{1000}}$: the 10-th power of the googol; the numbers with greater or equal to this number of decimal digits are called **titanic numbers**; the numbers with at least 10^{10000}, the 100-th power of googol, decimal digits are called **gigantic numbers**.
- **100...0019411** (1000 digits): the smallest titanic prime containing only square digits.
- $\mathbf{10^{10^{100}}}$: **googolplex**, i.e., 10^{googol}; its reciprocal, $10^{-\text{googol}}$, is called **googolminex**.
- **zillion**: a fictitious, indefinitely large number.

Chapter 7

Exercises

In this chapter we give a list of interesting facts and formulas pertaining to figurate numbers, as well as hints of their proofs (see [Dick05], [LeLi83], [Sier64], [CoGu96], [Sloa11], [Weis11], [Wiki11], [PrPe11], [Post11], etc.).

Exercises

(1) Prove the following identities, involving triangular numbers:

(a) $S_3(3n+1) = 9S_3(n) + 1$;
(b) $S_3(7n+3) = 49S_3(n) + 6$;
(c) $S_3(5n+1) = S_3(3n) + S_3(4n+1)$;
(d) $S_3(5n+3) = S_3(4n+2) + S_3(3n+2)$;
(e) $S_3(13n+10) = S_3(5n+4) + S_3(12n+9)$;
(f) $S_3(17n+10) = S_3(8n+4) + S_3(15n+9)$;
(g) $S_3(3n-1) = 2S_3(2n-1) + S_3(n)$;
(h) $S_3(3n) = 2S_3(2n) + S_3(n-1)$;
(i) $S_3(3n+1) = S_3(2n) + S_3(2n+1) + S_3(n)$;
(j) $3S_3(n) + 1 = S_3(n-1) + S_3(n) + S_3(n+1)$;
(k) $S_3(4n^2+5n+2) = S_3(4n^2+5n) + S_3(4n+2)$;
(l) $S_3(m+n) = S_3(m) + S_3(n) + mn$;
(m) $S_3(m+n+1) = S_3(m) + S_3(n) + (m+1)(n+1)$;
(n) $S_3(n-m) = S_3(m) + S_3(n) - m(n+1)$;

(o) $S_3(mn) = S_3(m)S_3(n) + S_3(n-1)S_3(m-1)$;

(p) $S_3(n^2) = (S_3(n))^2 + (S_3(n-1))^2$.

(2) Prove the following relations, involving triangular numbers:

(a) $S_3(k+n) = S_3(k) + S_3(2nm+n)$, where $k = 2nm^2 + (2n+1)m$;

(b) $S_3(n) = S_3(n-1) + S_3(m) + S_3(k)$, where $n = \frac{m^2+k^2+m+k}{2}$;

(c) $S_3(a+n) + S_3(b+n) + S_3(c+n) + S_3(d+n) = S_3(m-a) + S_3(m-b) + S_3(m-c) + S_3(m-d)$, where $m = n + \frac{a+b+c+d}{2}$;

(d) $S_3(a) + S_3(b) = \big(S_3(n^2+n-1) + S_3(n^2-n-1)\big) \times \big(S_3(m^2+m-1) + S_3(m^2-m-1)\big)$, where $a = n^2m^2 + nm - 1$, $b = n^2m^2 - nm - 1$;

(e) $S_3(7c+1) + S_3(c-1) = (S_3(7n+1) + S_3(n-1)) \times (S_3(7m+1) + S_3(m-1))$, where $c = 5nm + n + m$;

(f) $S_3(n) - S_3(m) = 3^{2\alpha}k$, where $n = 3^\alpha k + \frac{3^\alpha - 1}{2}$, $m = 3^\alpha k - \frac{3^\alpha+1}{2}$;

(g) $S_3(n) + S_3(m) = S_3(m - 3^\alpha k) + S_3(m + 3^\alpha k)$, where $n = 3^\alpha k^2 + \frac{3^\alpha - 1}{2}$, $m = 3^\alpha k^2 - \frac{3^\alpha + 1}{2}$.

(3) Prove the following identities, involving triangular and square numbers:

(a) $S_3(2n) = 2S_3(n) + S_4(n)$;

(b) $S_3(2n+1) = 2S_3(n) + S_4(n+1)$;

(c) $S_4(n+1) = 3S_3(n) - S_3(n-1) + 1$;

(d) $S_4(2n+1) = S_3(n-1) + 6S_3(n) + S_3(n+1)$;

(e) $S_4(2n+1) = S_3(3n+1) - S_3(n)$;

(f) $S_4(3n) = S_4(n) + 2S_4(2n)$;

(g) $S_4(3n+1) = S_4(n) + S_4(2n+1) + 2S_3(2n)$;

(h) $S_4(3n+2) = S_4(n+1) + S_4(2n+1) + 2S_3(2n+1)$;

(i) $S_4(5n+1) = S_3(n-1) + S_3(7n+1)$;

(j) $S_4(5n-1) = S_3(n) + S_3(7n-2)$;

(k) $S_4(6n+3) = S_3(9n+4) - S_3(3n+1)$;

(l) $S_4(n)+S_4(2n-1)+S_4(2n+1) = S_3(3n+1)+S_3(3n-2)$;
(m) $S_4(n+m) = S_4(n) + S_4(m) + 2nm$;
(n) $S_4(n-m) = S_4(n) + S_4(m) - 2nm$;
(o) $S_4(mn) = S_4(n)S_4(m)$;
(p) $S_4(n) + S_4(n+m) = S_3(2n+m) + S_3(m-1) - n$;
(q) $S_4(n+m+1) + S_4(n-m) = 4(S_3(n) + S_3(m)) + 1$;
(r) $S_4(n) = \frac{2S_3(n)S_3(2n)}{S_3(2n+1)}$;
(s) $S_4(2n+1) = \frac{S_3(2n(n+1))}{S_3(n)} - 1$;
(t) $S_4(n^2+1) = 1 + S_3(n^2+n) + S_3(n^2-n)$;
(u) $S_4(4S_3(n)+1) = S_4(2n+1) + S_4(4S_3(n))$;
(v) $S_4(4S_3(n) + 4S_3(m) + 1) = S_4(2n+1)S_4(2m+1) + S_4(4S_3(n) - 4S_3(m))$.

(4) Prove the following identities, involving polygonal numbers:
(a) $S_5(n) = \frac{1}{3}S_3(3n-1)$;
(b) $S_5(2n+1) = S_3(2n+1) + 2S_3(2n)$;
(c) $S_5(2n) = S_3(2n-1) + S_4(2n)$;
(d) $S_6(2n+1) = S_4(2n+1) + 2S_3(2n)$;
(e) $S_7(2n+1) = S_3(2n+1) + 4S_3(2n)$;
(f) $5S_7(n) + 1 = S_3(5n-2)$;
(g) $S_8(2n+1) = S_3(2n+1) + 5S_3(2n)$;
(h) $S_8(2n+1) = S_4(2n+1) + 4S_3(2n)$;
(i) $7S_9(n) + 3 = S_3(7n-3)$;
(j) $S_{10}(n) = S_4(n) + 3n(n-1)$.

(5) Prove the following identities, involving m-gonal numbers:
(a) $S_m(n+k) = S_m(n) + S_m(k) + nk(m-2)$;
(b) $S_m(a_1 + \cdots + a_n) = S_m(a_1) + \cdots + S_m(a_n) + (m-2) \times (a_1a_2 + \cdots + a_{n-1}a_n)$;
(c) $S_m(n) = 2S_s(n) + (m-2s+1)S_3(n-1) + S_3(n-2) - 1$;
(d) $S_m(n) = \frac{s-1}{2}S_3(n) + (m-s)S_3(n-1) + \frac{s-3}{2}S_3(n-2) - \frac{s-3}{2}$ for s odd;

(e) $S_m(n) = \frac{s-2}{2}S_3(n) + (m-s)S_3(n-1) + \frac{s-2}{2}S_3(n-2) + n - \frac{s-2}{2}$ for s even;
(f) $mS_m(n) - nS_n(m) = \frac{1}{2}mn(m-n)$.

(6) Prove the following identities, involving space figurate numbers:

(a) $S_3^3(1) + S_3^3(2) + S_3^3(3) + S_3^3(4) = S_3^3(5)$;
(b) $C(3) + C(4) + C(5) = C(6)$;
(c) $S_3^3(n) = \frac{(n+1)^3-(n+1)}{6}$;
(d) $S_5^3(n) = S_3^3(n) + 2S_3^3(n-1)$;
(e) $S_6^3(n) = S_5^3(n) + S_3^3(n-1)$;
(f) $S_6^3(n) = S_4^3(n) + 2S_3^3(n-1)$;
(g) $C(n+1) = (S_3(n+1))^2 - (S_3(n))^2$;
(h) $C(n) + 6S_3(n) + 1 = C(n+1)$.

(7) Prove the following identities, involving multidimensional figurate numbers:

(a) $C^5(n+1) - C^5(n) = S_3(n^2+n) + S_3(3n^2+3n+1)$;
(b) $C^5(n+1) - C^5(n) = 2S_3(n^2+n) + S_4(2n^2+2n+1)$.

(8) Prove the following identities, involving sums and triangular numbers:

(a) $S_3(1) + S_3(3) + S_3(6) + \cdots + S_3(2n-1) = \frac{n(n+1)(4n-1)}{6}$;
(b) $S_3(2) + S_3(4) + S_3(6) + \cdots + S_3(2n) = \frac{n(n+1)(4n+5)}{6}$;
(c) $S_3(3) + S_3(6) + S_3(9) + \cdots + S_3(3n) = \frac{3}{2}n(n+1)^2$;
(d) $S_3(1) - S_3(2) + S_3(4) - \cdots + S_3(2n-1) = S_4(n)$;
(e) $S_4(2n) - S_4(2n-1) + S_4(2n-2) - \cdots + S_4(2) - S_4(1) = S_3(2n)$;
(f) $S_4(2n+1) - S_4(2n) + S_4(2n-1) - \cdots - S_4(2) + S_4(1) = S_3(2n+1)$;
(g) $1 + 3 + \cdots + (2n-1) = S_3(n) + S_3(n-1)$;
(h) $S_5(1) + S_5(2) + \cdots + S_5(n) = nS_3(n)$;
(i) $1^3 + 2^3 + \cdots + n^3 = (S_3(n))^2$;

(j) $1^3 + 3^3 + 5^3 + \cdots + q^3 = S_3(n)$, where q is odd, and $n = \frac{1}{2}(q^2 + 2q - 1)$;

(k) $(S_3(k+1) - S_3(k))^3 + (S_3(k+2) - S_3(k+1))^3 + \cdots + (S_3(n) - S_3(n-1))^3 = (S_3(n))^2 - (S_3(k))^2$;

(l) $S_3(1) + \cdots + S_3(n-1) + S_3(n) + S_3(n-1) + \cdots + S_3(1) = S_4(1) + S_4(2) + \cdots + S_4(n)$;

(m) $S_3(1) + S_3(2) + \cdots + S_3(2n) = S_4(2) + S_4(4) + \cdots + S_4(2n)$;

(n) $S_3(1) + S_3(2) + \cdots + S_3(2n+1) = S_4(1) + S_4(3) + \cdots + S_4(2n+1)$;

(o) $S_3(2n+1) + S_4(2n+3) + S_4(2n+5) + \cdots + S_4(4n+1) = S_4(2n+2) + S_4(2n+4) + \cdots + S_4(4n) + S_3(4n+1)$;

(p) $S_3(1) + S_3(2) + \cdots + S_3(2n+1) = (2n+1)S_3(n+1) + S_4(1) + S_4(2) + \cdots + S_4(n)$;

(q) $S_3(k+1) + S_3(k+2) + \cdots + S_3(k+2n+1) = (2n+1) \times S_3(k+n+1) + S_4(1) + \cdots + S_4(n)$;

(r) $(S_3(2n))^2 + (S_3(2n)+1)^2 + \cdots + (S_3(2n)+n)^2 = (S_3(2n) + n + 1)^2 + (S_3(2n) + n + 2)^2 + \cdots + (S_3(2n) + 2n)^2$;

(s) $\frac{1}{S_1(1)} + \frac{1}{S_3(2)} + \frac{1}{S_3(3)} \cdots + \frac{1}{S_3(n)} + \cdots = 2.$

(9) Prove the following identities, involving sums of m-gonal numbers:

(a) $3(S_m(n) + S_m(2n) + \cdots + S_m(nk)) = S_m(n)S_3(k) + (k+1)S_m(nk)$;

(b) $S_m(n) + S_m(2n) + \cdots + S_m(kn) = S_m(n)S_3(k) + n^2(m-2) \times (S_3(1) + S_3(2) + \cdots + S_3(k-1))$;

(c) $S_4(a-b) + S_4(a-3b) + S_4(a-5b) + \cdots + S_4(a-(2k-1)b) = (2k)^3(r-2)S_r(n) + \frac{1}{3}(4k^3 - k)b^2$, where $a = 2nk(r-2)$, $b = r - 4$.

(10) Prove the following identities, involving sums and space figurate numbers:

(a) $S_3(1) + S_3(3) + \cdots + S_3(2n-1) = S_6^3(n)$;

(b) $S_3(2) + S_3(4) + \cdots + S_3(2n) = 3S_3^3(n) + S_3^3(n-1)$;

(c) $S_3(1) + S_3(3) + \cdots + S_3(2n+1) = 3S_3^3(n) + S_3^3(n+1)$;

(d) $S_4(3) + S_4(5) + \cdots + S_4(2n+1) = 8S_3^3(n) + n$;

(e) $S_3(2) + S_3(5) + \cdots + S_3(3n-1) = 3S_3^3(n) + 6S_3^3(n-1)$;
(f) $1\cdot n+2\cdot(n-1)+3\cdot(n-2)+\cdots+(n-1)\cdot 2+n\cdot 1 = S_3^3(n)$;
(g) $S_3^3(1)+S_3^3(2)+\cdots+S_3^3(n) = nS_3(1)+(n-1)S_3(2)+\cdots+2S_3(n-1)+S_3(n)$;
(h) $C(n)+C(2n)+\cdots+C(kn) = C(n)(S_3(k))^2 = n(n+2n+\cdots+kn)^2$;
(i) $\frac{1}{S_3^3(1)}+\frac{1}{S_3^3(2)}+\frac{1}{S_3^3(3)}+\cdots+\frac{1}{S_3^3(n)}+\cdots=\frac{3}{2}$;
(j) $\frac{C(1)}{2^0}+\frac{C(1)+C(3)}{2^1}+\frac{C(1)+C(3)+C(6)}{2^2}+\frac{C(1)+C(3)+C(6)+C(10)}{2^3}+\cdots+\frac{C(1)+C(3)+\cdots+C(S_3(n))}{2^{n-1}}+\cdots = 6416$.

(11) Prove the following identities, involving sums of multidimensional figurate numbers:

(a) $S_3^4(1)+S_3^4(2)+\cdots+S_3^4(n) = nS_3^3(1)+(n-1)S_3^3(2)+\cdots+S_3^3(n)$;
(b) $S_3^k(1)+S_3^k(2)+\cdots+S_3^k(n) = nS_3^{k-1}(1)+(n-1)S_3^{k-1}(2)+\cdots+S_3^{k-1}(n)$;
(c) $\frac{1}{S_3^4(1)}+\frac{1}{S_3^4(2)}+\frac{1}{S_3^4(3)}+\cdots+\frac{1}{S_3^4(n)}+\cdots=\frac{4}{3}$;
(d) $\frac{1}{S_3^5(1)}+\frac{1}{S_3^5(2)}+\frac{1}{S_3^5(3)}+\cdots+\frac{1}{S_3^5(n)}+\cdots=\frac{5}{4}$.

(12) Find the 8-th heptagonal number, using only triangular and square numbers.
(13) Find the 21-th square number in six ways.
(14) Find the triangular number, which is: the 6-th hexagonal number; the 21-th hexagonal number; the 39-th hexagonal number.
(15) Using the *Bachet de Méziriac* formula (i.e., as if only triangular numbers are known), find: the 15-th pentagonal number; the 9-th hexagonal number; the 12-th heptagonal number; the 10-th octagonal number; the 13-th decagonal number.
(16) Using the table below, check that *each polygonal number is equal to the sum of the polygon just above it in the table and the triangle in the preceding column.* Check, that each vertical column is an arithmetic progression whose common difference is the triangle in the preceding column.

triangle	1	3	6	10	15	21	28	36	45	55
squares	1	4	9	16	25	36	49	64	81	100
pentagons	1	5	12	22	35	51	70	92	117	145
hexagons	1	6	15	28	45	66	91	120	153	190
heptagons	1	7	18	34	55	81	112	148	189	235

(17) Prove the *upstairs-downstairs rule*: n-th square number $S_4(n)$ can be written as

$$S_4(n) = 1 + 2 + \cdots + (n-1) + n + (n-1) + \cdots + 2 + 1.$$

(18) Prove that $S_4(n) = S_3(n) + S_3(n-1)$, using the *upstairs-downstairs rule.*

(19) Prove that a positive integer n is a triangular number if and only if $8n + 1$ is a square number.

(20) Prove that n-th non-square number is given by $n + \lfloor \frac{1}{2} + \sqrt{n} \rfloor$, where $\lfloor x \rfloor$ is the floor function.

(21) Prove that the nested expression $9(9 \cdots (9(9(9(9(9S_3(n)+1)+1)+1)\cdots)+1)+1)+1$ gives only triangular numbers.

(22) Prove that the following sequences of positive integers consist only of triangular numbers:

(a) 21, 2211, 222111, . . . ;
(b) 55, 5050, 500500, . . . ;
(c) 5151, 501501, 50015001, . . . ;
(d) 78, 8778, 887778, 88877778, . . . ;
(e) 45, 4950, 499500, 49995000, . . . ;
(f) 45, 2415, 224115, 22241115, ...

(23) Prove that n-th hexagonal number can be rearranged into a rectangular number n long and $2n - 1$ tall, or vice versa.

(24) Check the *Diophantus' rule* for finding n-th m-gonal number: $S_m(n) = \frac{((m-2)(2n-1)+2)^2-(m-4)^2}{8(m-2)}$.

(25) Prove that the index n of a given n-th m-gonal number $S_m(n)$ can be found by the formula $n = \frac{\sqrt{(8m-16)S_m(n)+(m-4)^2}+m-4}{2m-4}$.

(26) Find first ten **myriagonal numbers** $n(4999n - 4998)$, corresponding to *myriagon*, i.e., 10000-gon.

(27) Consider (a, n)-**generalized figurate numbers**

$$a\frac{x(x+1)\cdots(x+n-1)}{n!} + (1-a)\frac{x(x+1)\cdots(x+n-2)}{(n-1)!}.$$

Show that, for $n = 2$ and $n = 3$, they give $(a + 2)$-gonal numbers, and $(a + 2)$-gonal pyramidal numbers, respectively.

(28) Find how many times a given number N is contained among all polygonal numbers, if $N \in \{5, 6, \ldots, 20\}$; $N = 35$; $N = 54$; $N = 88$; $N = 200$.

(29) Check the following rule of finding the number of ways a given number N can be polygonal: *express* $2N$ *as a product of two factors* >1 *in all possible ways; call the smaller factor* n; *subtract* 2 *from the larger factor and find whether or not the difference is divisible by* $n - 1$; *if it is, the quotient is* $m - 2$, *and* N *is an* m*-gonal number.*

(30) Find first ten **tricapped prism numbers** $TP(n) = \frac{n(3n^2-2n+1)}{2}$; show that they can be obtained by the recurrent equation $TP(n+1) = TP(n) + \frac{9n^2+5n+2}{2}$, $TP(1) = 1$; check that their generating function is $f(x) = \frac{x(1+5x+3x^2)}{(1-x)^4}$; find the number of points on the surface of tricapped prism; find first ten **centered tricapped prism numbers**; show that their generating function is $f(x) = \frac{x(1+x)(1+5x+x^2)}{(1-x)^4}$.

(31) Find first ten **centered hexagonal prism numbers**; show that they can be obtained by the recurrent equation $\overline{PCS}_6(n+1) = \overline{PCS}_6(n) + 12n^2 + 2$, $\overline{PCS}_6(1) = 1$; check that their generating function is $f(x) = \frac{x(1+x)(1+10x+x^2)}{(1-x)^4}$.

(32) Find first ten **doubly triangular numbers** $SS_3(n) = S_3(S_3(n)) = \frac{n(n+1)(n^2+n+2)}{8}$; show that they can be calculated by the recurrent equation $SS_3(n+1) = SS_3(n) + \frac{n^3+3n+2+4n+2}{2}$, $SS_3(1) = 1$; check that their generating function is $f(x) = \frac{x(1+x+x^2)}{(1-x)^5}$.

(33) Find first five **iterated triangular numbers** $S_3(1)$, $S_3(S_3(2))$, $S_3(S_3(S_3(3)))$, ...

(34) Show, that n-th square pyramidal number $S_4^3(n)$ can be obtained as the sum of the numbers in the following triangular array:

$$
\begin{array}{ccccccccccc}
 & & & & & 1 & & & & & \\
 & & & & 2 & & 2 & & & & \\
 & & & 3 & & 3 & & 3 & & & \\
 & & 4 & & 4 & & 4 & & 4 & & \\
 & & & & & \cdots & & & & & \\
 & n-1 & & n-1 & & n-1 & & n-1 & & n-1 & \\
n & & n & & n & & n & & n & & n
\end{array}
$$

(35) Show, that n-th octahedral number $O(n)$ can be obtained as the sum of the numbers in the following square array:

$$\begin{array}{ccccccccccc} n & n-1 & n-2 & n-3 & n-4 & \cdots & 5 & 4 & 3 & 2 & 1 \\ n-1 & n & n-1 & n-2 & n-3 & \cdots & 6 & 5 & 4 & 3 & 2 \\ n-2 & n-1 & n & n-1 & n-2 & \cdots & 7 & 6 & 5 & 4 & 3 \\ & & & & & \cdots & & & & & \\ 3 & 4 & 5 & 6 & 7 & \cdots & n-2 & n-1 & n & n-1 & n-2 \\ 2 & 3 & 4 & 5 & 6 & \cdots & n-3 & n-2 & n-1 & n & n-1 \\ 1 & 2 & 3 & 4 & 5 & \cdots & n-4 & n-3 & n-2 & n-1 & n \end{array}$$

(36) Prove the formula $S_m^3(n) = \frac{n+1}{6}(2S_m(n) + n)$ for n-th m-pyramidal number using the formula $S_m(n) = \frac{(m-2)}{2}n^2 - \frac{(m-4)}{2}n$ for n-th m-gonal number and the identities $1 + 2 + \cdots + n = \frac{n(n+1)}{2}$, and $1^2 + 2^2 \cdots + n^2 = \frac{n(n+1)(2n+1)}{6}$.

(37) Prove that the sequence $1, \frac{n+a}{1}, \frac{(n+1)(n+2a)}{2!}, \frac{(n+1)(n+2)(n+3a)}{3!}, \ldots$ gives, for $n = 1$, 2, and 3, the sequences of gnomonic numbers, $(a+2)$-gonal numbers, and $(a+2)$-gonal pyramidal numbers, respectively.

(38) Prove that the sum $T_k = (S_3(1))^k + (S_3(2))^k + \cdots + (S_3(n))^k$ of k-th powers of the first n triangular numbers is symbolically equal to $\frac{S^k(S+1)^k}{2^k}$, where, after expansion, S^t is replaced by $S_t = 1^t + 2^t + \cdots + n^t$, the sum of the t-th powers of $1, 2, \ldots, n$.

(39) Prove, that the generating function of the sequence $1, 1 + 2^k, 1 + 2^k + 3^k, 1 + 2^k + 3^k + 4^k, \ldots, 1 + 2^k + \cdots + n^k, \cdots$ for $k = 2, 3, 4, 5, 6, 7$ is $\frac{x(1+x)}{(1-x)^4}$, $\frac{x(1+4x+x^2)}{(1-x)^5}$, $\frac{x(1+11x+11x^2+x^3)}{(1-x)^6}$, $\frac{x(1+26z+66x^2+26x^3+x^4)}{(1-x)^7}$, $\frac{x(x+1)(1+56x+246x^2+56x^3+x^4)}{(1-x)^8}$, $\frac{x(1+120x+1191x^2+2416x^3+1191x^4+120x^5+x^6)}{(1-x)^9}$, respectively.

(40) Using the *Euler's formula* for square triangular numbers, find: 5-th square number, which is a triangular number; 7-th triangular number, which is a square number.

(41) Find several numbers which are simultaneously triangular, pentagonal and hexagonal.

(42) Prove that n-th hexagonal square number $S_{6,4}(n)$ can be calculated by the formula $S_{6,4}(n) = \frac{1}{32}(-2 + (17 - 12\sqrt{2}) \times (3 - 2\sqrt{2})^{4n} + (17 + 12\sqrt{2})(3 + 2\sqrt{2})^{4n})$.

(43) Check the following equality: $S_{4,3}(n) = 2^{2n-5}\prod_{k=1}^{2n}(3+\cos\frac{k\pi}{n})$.

(44) Prove the *Euler's rule* to find m-gomal numbers, which are simultaneously square numbers: *set* $2(m-2)p^2+1 = q^2$ *for some integers* p, q; *then* $4S_m(x)$ *is the square of* $0, (m-4)p, 2(m-4)pq, \ldots,$ *if* $x = 0, -\frac{(m-4)}{2(m-2)}(q-1)$, $-\frac{(m-4)}{(m-2)}(q^2-1), \cdots,$ *respectively.*

(45) Find a solution of the equation $ax^2 - a'x = by^2 - b'y$, where a, a', b, b' are given integers with no common factor.

(46) Prove that n-th triangular star number $ST(n)$ can be calculated by the formula $ST(n) = \frac{3((7+4\sqrt{3})^{2n-1}+(7-4\sqrt{3})^{2n-1})-10}{32}$, or by the recurrent equation $ST(n) = 194ST(n-1) - ST \times (n-2) + 60, ST(1) = 1$; check, that the generating function of this sequence is $f(x) = \frac{x(1+58x+x^2)}{(1-x)(x^2-194x+1)}$.

(47) Prove that n-th square star number $SS(n)$ can be calculated by the formula $SS(n) = \frac{((5+2\sqrt{6})^n(\sqrt{6}-2)-(5-2\sqrt{6})^n(\sqrt{6}+2))^2}{4}$, or by the recurrent equation $SS(n+1) = 98SS(n) - SS(n-1) + 24$, $SS(1) = 1$, $SS(2) = 121$; check that the generating function of this sequence is $f(x) = \frac{x(1+22x+x^2)}{(1-x)(x^2-98x+1)}$.

(48) Find first five numbers, which are both centered triangular and centered pentagonal.

(49) Prove, that there is no triangular number except 1, which is a biquadratic number.

(50) Prove that elements of any arithmetic progression $ax + 1$, $a \in \mathbb{N}$, $x = 1, 2, 3, \ldots,$ can be considered as gnomonic numbers.

(51) Find several examples of three polygonal numbers in an arithmetic progression.

(52) Prove that every three-term arithmetic progression of square numbers r^2, s^2 and t^2 can be associated with a *Pythagorean triple* (x, y, z) by $x = \frac{r+t}{2}, y = \frac{t-r}{2}, z = s$.

(53) Find several examples of three triangular numbers in a geometric progression.

(54) Find several examples of four-term arithmetic progression, such that the product of its entries is a square.

(55) Let, in the arithmetic progression $1, 1 + (m-2), 1 + 2(m-2), \ldots$, leading to the m-gonal numbers, the first term 1 gives the *first colonne*; the sum of the next two terms diminished by $m-4$ times the first triangular number 1 gives the *second colonne* $2m$; the sum of the fourth, fifth and sixth terms diminished by $m-4$ times the second triangular number 3 gives the *third colonne* $9m-9$. Find the *fourth colonne*. Prove that r-th colonne is the product of r-th m-gonal number by r.

(56) For a given pentagonal number $S_5(n) = \frac{3n^2-n}{2}$ of k digits, $k = 1, 2, 3$, find another number p also of k digits such that if p is prefixed to $S_5(k)$ there results a pentagonal number.

(57) Check that the number of three-digital numbers having $n = 0, 1, 2, \ldots, 27$ as a sum of their digits, is the triangular number $S_3(n+1)$.

(58) Find first ten octagonal numbers for which the sum of the digits is also an octagonal number.

(59) The *digital root* (or *repeated digital sum*) of a number is the number obtained by adding all the digits, then adding the digits of that number, and then continuing until an one-digital number is reached. Prove that in base 10 it holds:

(a) the digital root of a triangular number is always 1, 3, 6, or 9;
(b) the digital root of a square number is always 1, 4, 7, or 9;
(c) the digital root of a hexagonal number is always 1, 3, 6, or 9;
(d) the digital root of a heptagonal number is always 1, 4, 7, or 9;
(e) the digital root of a star number is always 1, or 4;
(f) the digital root of a cubic number is always 1, 8, or 9.

(60) Find first ten squares for which the digital sum is also a square; the digital root is also a square; the multiplicative digital root is also a square; the sum of the digits, the product of the digits, the digital root and the multiplicative digital root are all squares.

(61) Find first ten squares for which the sum of the digits are cubes; the sum of the digits and the product of the digits are cubes; both, the sum of the digits and the product of the digits, are triangular numbers.
(62) Find first ten cubes such that all their digits are cubes; the product of their digits is a cube; the sum of their digits is a square; the product of their digits is a square; both, the sum of their digits and the product of their digits, are squares.
(63) Find first ten squares which are a non-trivial concatenation of other squares; first ten cubes formed by concatenating other cubes.
(64) Prove that a cubic number can not be a concatenation of two cubes.
(65) Check the identities

$$1233 = 12^2 + 33^2, \quad 8833 = 88^2 + 33^2,$$

$$10100 = 10^2 + 100^2, \quad 5882353 = 588^2 + 2353^2,$$

giving the numbers that are equal to the sum of the squares of their two "halves".
(66) Find first ten squares which are divisible by the sum of their digits; by the product of their digits.
(67) Find first ten squares having exactly two distinct digits; exactly two distinct non-zero digits.
(68) Find first ten squares whose digits are squares.
(69) Find first few numbers which, when squared, give numbers composed of digits of only three types.
(70) Find smallest number whose digit sum is n^3, $n = 0, 1, 2, 3, 4, \ldots$.
(71) Find first few positive integers equal to the sum of the digits of their cubes; to the sum of the cubes of their digits.
(72) Find first few **Armstrong numbers** (or **plus perfect numbers**, **narcissistic numbers**): n-digital numbers equal to the sum of n-th powers of their digits.
(73) Check that

  (a) the smallest and the largest square numbers containing the digits 1 to 9 are $11826^2 = 139854276$, and $30384^2 = 923187456$;

(b) the smallest and the largest square numbers containing the digits 0 to 9 are $32043^2 = 139854276$, and $99066^2 = 9814072356$;

(c) the smallest and the largest square numbers containing the digits 1 to 9 twice each are $335180136^2 = 112345723568978496$, and $999390432^2 = 998781235573146624$;

(d) the smallest and the largest square numbers containing the digits 1 to 9 three times each are $10546200195312^2 = 111222338559598866946777344$, and $31621017808182^2 = 999888767225363175346145124$.

(74) Check the following relations:

(a) $gcd(S_m(2n), S_m(2n+1)) = gcd(2n+1, m-3)$;

(b) $gcd(S_m(2n-1), S_m(2n)) = gcd(n, m-3)$;

(c) $gcd(S_m(2n+1), S_{m+1}(2n+1)) = 2n+1$;

(d) $gcd(S_m(2n), S_{m+1}(2n)) = n$.

(75) Check that the following propositions hold:

(a) the parity of heptagonal numbers follows the pattern odd-odd-even-even;

(b) octagonal numbers have consistently alternate parity;

(c) the parity of nonagonal numbers follows the pattern odd-odd-even-even;

(d) decagonal numbers have consistently alternate parity;

(e) all centered square numbers are odd;

(f) the parity of centered pentagonal numbers follows the pattern even-even-odd-odd;

(g) all centered hexagonal numbers are odd;

(h) all centered octagonal numbers are odd;

(i) all star numbers are odd;

(j) the parity of tetrahedral numbers follows the pattern odd-even-even-even.

(76) Prove that the following propositions hold:

(a) each odd square number has a remainder of 1 when divided by 8;

(b) each centered triangular number has a remainder of 1 when divided by 3;
(c) each centered square number has a reminder of 1 or 2 when divided by 3;
(d) each centered hexagonal number has a reminder of 1 when divided by 3 (or 6);
(e) each star number has a reminder of 1 when divided by 3 (4, 6, or 12);
(f) each centered square number has a remainder of 1 when divided by 4;
(g) each centered octagonal number has a remainder of 1 when divided by 4 (or 8);
(h) each centered square number has a reminder of 1 or 5 when divided by 6 (8 or 12).

(77) Prove that in the sequence $S_m(1), S_m(2), \ldots, S_m(n), \ldots$ of m-gonal numbers, 3-rd term $S_m(3)$ is divisible by 3, 5-th term $S_m(5)$ is divisible by 5, 7-th term $S_m(7)$ is divisible by 7, in general, p-th term $S_m(p)$ is divisible by odd prime p.

(78) Check that in base 10 it holds:
(a) a triangular number can end only with digits 0, 1, 3, 5, 6, or 8;
(b) a square number can end only with digits 0, 1, 4, 5, 6, or 9;
(c) the one's digits of the centered square numbers follow the pattern 1-5-3-5-1;
(d) the one's digits of the centered pentagonal numbers follow the pattern 6-6-1-1;
(e) the one's digits of the centered hexagonal numbers follow the pattern 1-7-9-7-1;
(f) the one's digits of the centered octagonal numbers follow the pattern 1-9-5-9-1;
(g) the one's digits of the star numbers follow the pattern 1-3-7-3-1.

(79) Check that in base 10 it holds:
(a) the final two digits of a square number are one of 00, 01, 04, 09, 16, 21, 24, 25, 29, 36, 41, 44, 49, 56, 61, 64, 69, 76, 81, 84, 89, or 96;

(b) the final two digits of a star number are one of 01, 13, 21, 33, 37, 41, 53, 61, 73, 81, or 93;
(c) any pair of digits with the last digit odd can be the last two digits of an odd cubic number, except for cubes divisible by 5, where only 25 and 75 can be the last two digits;
(d) the only 00, $o2$, $e4$, $o6$, or $e8$ can be the last two digits of an even cubic number, where o stands for any odd digit, and e for any even digit.

(80) Construct the table of the possible residues modulo $n = 2, 3, \ldots, 10$ for square numbers; for cubic numbers.
(81) Find two consecutive integers, one of which is a square number, and the other triangular.
(82) Check up to 100, that the difference of the indices of two successive triangular numbers, each a square, is equal to the sum of two successive integers the sum of whose squares is a square.
(83) Find a triangular number which is

(a) a sum of the squares of two consecutive integers;
(b) a sum of the squares of two consecutive odd integers;
(c) a product of two consecutive integers;
(d) a sum of two consecutive primes;
(e) a sum of three consecutive triangular numbers.

(84) Find first five squares which are sums of three consecutive triangular numbers; sums of four consecutive triangular numbers.
(85) Check up to 100 the following propositions:

(a) there is at least one and at most two triangular numbers between any two consecutive non-zero square numbers;
(b) there is at most one square number between two consecutive triangular numbers;
(c) if there are exactly two triangular numbers between $S_4(a+1)$ and $S_4(a+2)$, where $a > 0$, there is exactly one triangular number between $S_4(a)$ and $S_4(a+1)$, and exactly one triangular number between $S_3(a+2)$ and $S_4(a+3)$.

(86) Find two triangular numbers whose sum and difference are triangular numbers; prove that there are infinitely many pairs

of triangular numbers, the sum and the difference of which are triangular.

(87) Find a solution of the following equation:

(a) $S_4(x) = S_3(y) + 1$;
(b) $3S_3(x) = S_3(y)$;
(c) $S_3(x) = p \cdot S_3(y)$, $p \neq a^2$;
(d) $\mu S_3(x) = \nu S_3(y)$;
(e) $S_3(x) + S_3(y) = S_3(z)$;
(f) $S_3(x-1) + S_3(x) = S_4(y)$;
(g) $S_3(x+y) + y = a$;
(h) $S_3(x) + S_3(y) = S_4(z)$;
(i) $S_3(x) = S_4(x) + S_4(y)$;
(j) $S_3(x) + S_3(y) = S_4(a) + 2S_3(b)$;
(k) $S_4(x) + S_3(x) = S_4(y)$;
(l) $(S_3(x))^2 + (S_3(z))^2 = 2(S_3(y))^2$;
(m) $(S_3(x))^2 = S_3(y)$;
(n) $S_3(x)S_3(y) = S_3(z)$;
(o) $S_3(a)S_3(x) = S_3(b)S_3(y)$;
(p) $S_3(x)S_3(y) = S_3(z^2 + z)$;
(q) $S_3(x)S_3(y) = (S_3(z))^2$;
(r) $S_3(x-1)S_3(x)S_3(x+1) = S_4(y)$;
(s) $C(y) \pm 1 = S_3(z)$;
(t) $C(x) - 13 = 4S_3(y)$;
(u) $S_3(y) - S_3(z) = C(x)$;
(v) $(S_3(x))^2 - (S_3(y))^2 = C(z)$;
(w) $BC(x) - BC(y) = S_3(z)$;
(x) $S_3(x) + S_3(y) = C^5(z)$.

(88) Find a solution of the equation $S_4(1) + S_4(2) + \cdots + S_4(n) = S_4(x)$.

(89) Find several solutions of the equation $(S_3(a) + S_3(b) + S_3(c))(S_3(\alpha) + S_3(\beta) + S_3(\gamma)) = S_3(x) + S_3(y) + S_3(z)$ for integer values of arguments.

(90) Prove the following identities, involving binomial coefficients:
(a) $S_m(n+1) = m + \binom{n-1}{1}(1+2(m-2)) + \binom{n-1}{2}(m-2)$;
(b) $S^3_m(n+1) = S^3_m(1) + \binom{n}{1}m + \binom{n}{2}(1+2(m-2)) + \binom{n}{3}(m-2)$.
Using these identities, obtain the general formula for n-th m-gonal pyramidal number.

(91) Show, that tritriangular numbers $TT(n) = \binom{\binom{n+2}{2}}{2}$ can be given by the formula $\frac{n(n+1)(n+2)(n+3)}{8}$, or by the recurrent equation $TT(n+1) = TT(n) + \frac{n(n+1)(n+2)}{2}$, $TT(1) = 3$; check, that their generating function is $f(x) = \frac{3x}{(1-x)^5}$.

(92) Find a Mersenne number M_k such that $M_k + 1$ is a difference of a square and a triangular number.

(93) Prove that every number of the form $M_{4m+1} - M_{2m} + 1$, $m \in \mathbb{N}$, is a product of a square and a triangular number.

(94) Find first five palindromic triangular numbers with prime indices.

(95) Check that the sum of indices of the first thirteen palindromic triangular numbers is itself palindromic.

(96) Check that any even palindromic triangular number greater than six is divisible by 11.

(97) Check that n-th pronic number $P(n)$ is palindromic for any palindromic index of the form $n = 255(4554)_k552$, where $k = 1, 2, \ldots, 6$; for any non-palindromic index of the form $n = 255(4554)_k45447$, where $k = 1, 2, \ldots, 5$.

(98) Find first five positive integers, which are palindromic cubes of prime numbers; first ten positive integers, which are palindromic prime powers of prime numbers.

(99) Find first ten cubic numbers, whose digit reversal is also a cubic number.

(100) Check that any square number $S_4(n)$, $n = 1, 2, \ldots, 13$, produces a palindrome after at most two iterations of the process of repeatedly reversing its digits and adding the resulting numbers.

(101) Check that the beast number 666 is a **Smith number**, i.e., a composite number, the sum of whose digits is equal to the sum of the digits of its prime factors.

(102) The **factorial numbers** $n! = 1 \cdot 2 \cdot \ldots \cdot n$, $n \in \mathbb{N}$, are multiplicative analogues of the triangular numbers $S_3(n) = 1 + 2 + \cdots + n$. Show, that second differences of the sequence 1, 4, 9, 16, 25... of square numbers give second factorial number $2! = 2$; third differences of the sequence 1, 8, 27, 64, 125, ... of cubic numbers give third factorial number $3! = 6$; fourth differences of the sequence 1, 16, 81, 256, 625, ... of biquadratic numbers give fourth factorial number $4! = 24$; in general, n-th differences of the sequence $C^n(1), C^n(2), C^n(3), C^n(4), C^n(5), \ldots$ of n-dimensional hypercube numbers give n-th factorial number $n!$.

(103) Find first five positive integers which are simultaneously: a sum of factorials of distinct integers and of the form a^2; a sum of distinct factorials and of the form a^b, $b \geq 3$; a sum of distinct factorials and of the form a^b, $b \geq 2$.

(104) Check that **Lah number** $L(n, k)$ counting the number of ways a set of n elements can be partitioned into k nonempty linearly ordered subsets, has the form $L(n, k) = S_3^{k-1}(n - k + 1)\frac{n!}{k!}$; prove that $L(n, 1) = n!$, $L(n, n) = 1$, and $L(n, n - 1) = P(n - 1)$.

(105) Check that $2S_3(n) - 1$ is a prime for $n \leq 9$.

(106) Find a positive integer of the form $4S_3(n)+5$, which is a square of a prime and not a sum of two squares.

(107) Arrange five consecutive primes 61, 67, 71, 73, and 79 in an $3{\times}3$ square whose columns and rows sum to consecutive primes 199, 211, 223, respectively.

(108) Find an 9-digital palindromic prime such that if it is placed in an 3×3 square, then each column and diagonal, sums of each column and diagonal, as well as the total sum of digits, are primes.

(109) Check that the number

999999999 9999933333 3333339933 3333333339 9337777777 3399337999
9973399337 9333973399 3379373973 3993379333 9733993379 9999733993
3777777733 9933333333 3339933333 3333339999 9999999999

is an *2-deletable square-congruent prime*, i.e., removing the outer two largest shells in its square-representation yields another square-congruent prime, and this can be repeated until one is down to the center nut.

(110) Find an 7-digital triangle-congruent prime number, such that if the corresponding triangle is rotated, two new primes are obtained.

(111) Find three prime numbers, which give an 7-digital repunit hex-congruent prime in bases 3, 5, and 6, respectively.

(112) A *prime magic square* is a square array of numbers consisting of the distinct primes arranged such that the sum of the numbers in any horizontal, vertical, or main diagonal line is always the same number, called *magic constant.* Construct an 5×5 prime magic square, containing all 25 primes less than 100; an 5×5 prime magic square with prime magic constant, including all 24 odd primes less than 100.

(113) Prove and apply the *Conway's rule* to remember the *Euler's pentagonal number theorem*, using relations between triangular and generalized pentagonal numbers: *divide triangular numbers 0, 3, 6, 10, 15,... by 3 (when you can exactly), and you get generalized pentagonal numbers in pairs, one of positive rank and the other negative; append signs according as the pair have the same (+) or opposite (−) parity.*

(114) Prove the following formal identities:

(a) $\left(\sum_{m=-\infty}^{+\infty}(-1)^m x^{S_5(-m)}\right)^3 = \sum_{n=0}^{\infty}(-1)^n(2n+1)x^{S_3(n)}$;

(b) $\prod_{r=1}^{\infty}(1 - x^{(m-2)r})(1 + x^{(m-2)r-m+3})(1 + x^{(m-2)r-1}) = \sum_{r=-\infty}^{+\infty} x^{S_m(r)},\ m \geq 3$;

(c) $\prod_{r=1}^{\infty}(1 - x^{(m-2)r})(1 - x^{(m-2)r-m+3})(1 - x^{(m-2)r-1}) = \sum_{r=-\infty}^{+\infty}(-1)^r x^{S_m(r)},\ m \geq 3$;

(d) $\prod_{r=1}^{\infty}(1 - x^r)^{(-1)^r} = \sum_{r=-\infty}^{+\infty} x^{S_5(r)}$;

(e) $\prod_{r=1}^{\infty}(1 - x^r) = \sum_{r=-\infty}^{+\infty}(-1)^r x^{S_5(r)}$;

(f) $\ln \sum_{r=-\infty}^{+\infty}(-1)^r x^{S_5(r)} = -\sum_{n=1}^{\infty} \frac{\sigma(k)x^k}{k}$, where $\sigma(k)$ is the sum of the divisors of k;

(g) $\ln \sum_{r=0}^{+\infty} x^{S_3(r)} = \sum_{n=1}^{\infty} \frac{\psi(k)x^k}{k}$, where $\psi(k)$ gives the excess of the sum of the odd divisors of k over the sum of the even divisors.

(115) Prove that the number of off-diagonal entries in a square matrix is always a pronic number.

(116) Prove that the value of the Möbius function $\mu(x)$ for any pronic number $P(n) = n(n+1)$ can be calculated as $\mu(P(n)) = \mu(n)\mu(n+1)$; each distinct prime factor of a pronic number $P(n)$ is present in only one of its factors n, $n+1$; a pronic number $P(n)$ is square-free if and only if n and $n+1$ are; the number of distinct prime factors of the pronic number $P(n)$ is the sum of the number of distinct prime factors of n and $n+1$.

(117) Prove the following identity: $\int_0^1 \int_0^1 |x-y|^n dxdy = \frac{2}{(n+1)(n+2)} = \frac{1}{S_3(n+1)}$.

(118) Find the minimum number of squares needed to represent the numbers $n = 1, 2, 3, \ldots, 10$.

(119) Find first ten positive integers that can be expressed as a sum of two squares.

(120) Check that 188 can be represented as a sum of seven distinct squares.

(121) Find the least numbers that are the sum of two squares in exactly w different ways for $w = 1, 2, \ldots, 10$.

(122) Find the number of distinct ways to represent the numbers $n = 1, 2, \ldots, 10$ as sums of squares.

(123) The number of representations of a non-negative integer n as a sum of k squares, distinguishing signs and order, is denoted $r_k(n)$, and is called the *sum of squares function*. Find the values $r_2(n)$ for $n = 0, 1, 2, \ldots, 9$.

(124) Find first ten numbers that are not a difference of two squares.

(125) Find first ten numbers that are neither square, nor a sum of a square and a prime.

(126) Find the number of positive cubes needed to represent the numbers $n = 1, 2, 3, \ldots, 16$.

(127) Find the number of distinct ways to represent the numbers $n = 1, 2, \ldots, 30$ as sums of positive cubes.

(128) Prove that any number of the form $6x$, $6x+3$, $18x+1$, $18x+7$, $18x+8$, $54x+2$, $54x+20$, $216x-16$, $216x+92$ can be represented as a sum of four signed cubes.

(129) Represent the numbers 30, 52, 75, 84, 110, 195, 290, 435, 44, 452, 462, 478 as sums of three signed cubes.

(130) Represent 1 as a sum of three signed cubes in five distinct ways.

(131) Check that the Hardy-Ramanujan number 1729 is a centered cube number, as well as dodecagonal, 24-gonal, and 84-gonal number.

(132) Check that in base 10 the Hardy-Ramanujan number 1729 is divisible by the sum of its digits, i.e., is a **Harshad number**.

(133) Show that the digits of the number 1729, added together, produce a sum which, when multiplied by its reversal, yields the original number.

(134) Find the minimum number of biquadratic numbers needed to represent the numbers $n = 1, 2, \ldots, 20$.

(135) Represent the number N, $N \in \{1, 2, \ldots, 20\}$, as a sum of m m-gonal numbers, $m \in \{3, 4, \ldots, 10\}$.

(136) Prove that there are numbers, which can not be represented as a sum of less then m non-zero m-gonal numbers.

(137) Prove that following positive integers can not be represented as sums of two triangular numbers: $9n + 5$, $9n + 8$; $49n + 5$, $49n + 19$, $49n + 26$, $49n + 33$, $49n + 40$, $49n + 47$; $81n + 47$, $81n + 74$.

(138) Find the minimum number of tetrahedral numbers needed to represent the numbers $n = 1, 2, \ldots, 10$.

(139) Prove that any positive integer N is an algebraic sum of four tetrahedral numbers: $N = S_3^3(N) - 2S_3^3(N-1) + S_3^3(N-2)$.

(140) Check that every positive integer up to 100 can be represented as:

(a) a sum of at most 10 odd squares;

(b) a sum of at most 11 triangular numbers 1, 10, 28, 55, ... of the form $S_3(3n + 1)$;

(c) a sum of at most 5 tetrahedral numbers;

(d) a sum of at most 7 octahedral numbers;

(e) a sum of at most 9 cubic numbers;

(f) a sum of at most 13 icosahedral numbers;

(g) a sum of at most 21 dodecahedral numbers;

(h) a sum of at most 11 squares of triangular numbers.

(141) Check that every positive integer up to 50 can be represented as a sum of at most $a + 2n - 2$ terms of the series $1, \frac{n+a}{1}, \frac{(n+1)(n+2a)}{2!}, \frac{(n+1)(n+2)(n+3a)}{3!}, \dots$. Is it true for every positive integer up to 100?

(142) Call a number N *decomposed into its maximum triangular numbers*, and m the index of N, if $N = A_1 + \cdots + A_m$ is a sum of m triangular numbers $A_1, \dots, A_m$, where the first number A_1 is the largest triangular number, $\leq N$, the second A_2 is the largest triangular number $\leq A - A_1$, the third A_3 is the largest triangular number $\leq N - A_1 - A_2$, etc. Find several such numbers. Find the least numbers of index m, $m = 1, 2, 3, 4, 5$.

(143) Check that there are no positive rational solutions x, y, z of $\frac{1}{2} = \frac{x^2+x}{2} + \frac{y^2+y}{2} + \frac{z^2+z}{2}$, i.e., the rational number $\frac{1}{2}$ is not a sum of three "rational" triangular numbers.

(144) Prove that every positive integer is a sum of four "rational" pentagonal numbers $\frac{3z^2-z}{2}$, $z \in \mathbb{Q}$.

(145) Check that every positive integer up to 100 can be written in any of the following forms:

(a) $2S_3(x) \pm S_4(y)$;

(b) $S_4(x) + S_3(y) + S_3(z)$;

(c) $S_4(x) + S_4(y) + S_3(z)$;

(d) $S_4(x) + S_4(y) + 2S_3(z)$;

(e) $S_4(x) + 2S_3(y) + S_3(z)$;

(f) $S_4(x) + 2S_4(y) + S_3(z)$;

(g) $S_3(x) + 2S_4(y) + 2S_3(z)$;

(h) $S_4(x) + 2S_4(y) + 2S_3(z)$;

(i) $S_3(x) + S_3(y) + S_3(z) + S_3(u)$;

(j) $S_3(x) + S_3(y) + S_3(z) + S_3(z + 1)$;

(k) $S_3(x) + S_3(y) + 2S_3(z)$.

(146) Prove that every positive integer can be represented in each of the following forms:

(a) $S_3(x) + S_3(y) + cS_3(z)$, $c = 1, 2, 4$;
(b) $S_3(x) + 2S_3(y) + dS_3(z)$, $d = 2, 4$.

(147) Check up to 100 that any positive integer between two consecutive triangular numbers is the sum of four triangular numbers, the sum of whose indices is constant.

(148) Check up to 100 the following propositions:

(a) if an odd number is represented as a sum of an even square and two triangular numbers, then it has a similar representation in which the square is odd;
(b) if a number is represented as a sum of four distinct odd squares, then it is represented as a sum of four distinct even squares.

(149) Check up to 100 the following propositions:

(a) every triangular number except 1 and 6 is a sum of three non-zero triangular numbers;
(b) every triangular number is a sum of a square and two equal triangular numbers;
(c) every triangular number is a sum of another triangular number and either a square or a doubled square;
(d) every triangular number is a sum of at most six pentagonal numbers;
(e) every square number is a sum of three squares or a sum of two squares and two triangular numbers;
(f) every square number is a sum of at most four tetrahedral numbers.

(150) Check that 2^3 is a sum of two non-zero triangular numbers; 3^4 is a sum of four non-zero triangular numbers; 5^5 is a sum of five non-zero triangular numbers.

(151) Prove the following propositions:

(a) if $4N + 1 = S_4(x) + S_4(y)$, then $N = S_3(u) + S_3(v)$, where $u = \frac{x+y-1}{2}$, and $v = \frac{x-y-1}{2}$.
(b) if $N = S_3(x) + S_3(y) + S_3(z)$, where $x = 2a$, $y = 2b$, $z = 2c$, then $2N + 1 = S_4(u) + S_4(v) + S_4(w) + S_4(s)$,

where $u = a + b + c + 1$, $v = a - b - c$, $w = a + b - c$, and $a - b + c$;

(c) if $N - ab = S_3(p) + S_3(q) + S_3(r)$, and $p - q = a - b$, then $N = S_3(p + b) + S_3(p - a) + S_3(r)$;

(d) if $N - p = S_3(a) + S_3(b) + S_3(c)$, and $p = (b - a)n + n^2$, then $N = S_3(a - n) + S_3(b + n) + S_3(c)$.

(152) Prove the following propositions:

(a) a positive integer N is a sum of five pentagonal numbers $\frac{3u_i^2 - u_i}{2}$ if and only if $24N + 5$ is a sum of five square numbers $(6u_i - 1)^2$, $i = 1, 2, \ldots, 5$;

(b) a positive integer N is a sum of six hexagonal numbers $2x_i^2 - x_i$ if and only if $8N+6$ is a sum of six square numbers $(4x_i - 1)^2$, $i = 1, 2, \ldots, 6$;

(c) in general, a positive integer N is a sum of $m+2$ $(m+2)$-gonal numbers $\frac{m-2}{2}(x_i^2 - x_i) + x_i$ if and only if $8mN + (m+2)(m-2)^2 = \sum_{i=1}^{m+2} = (2mx_i - m + 2)^2$, $i = 1, 2, \ldots, m+2$.

(153) Prove the following propositions:

(a) if $N \neq S_4(x) + S_3(y)$, then $8N + 1 \in S$;

(b) if $N \neq 2S_3(x) + S_3(y)$, then $8N + 3 \in S$;

(c) if $N \neq S_4(x) + 2S_3(y)$, then $4N + 1 \in S$.

(154) Prove that Fermat's m-gonal number theorem follows if we could prove that every integer occurs among the exponents in the expansion of $(\sum_{n=0}^{\infty} x^{S_m(n)})^m$.

(155) Check that every positive integer N is a sum of four triangular numbers in $\sigma(2N + 1)$ ways, where $\sigma(n)$ is the sum of the positive integer divisors of k.

Solutions

(1) For instance, $4S_3(m)S_3(n) + 4S_3(m - 1)S_3(n - 1) = m(m + 1) n(n + 1) + m(m - 1)n(n - 1) = mn(mn + m + n + 1 + mn - m - n + 1) = 2mn(mn + 1) = 4S_3(mn)$.

(2) For instance, $2(S_3(n) - S_3(n - 1)) = 2n = m^2 + k^2 + m + k = m(m + 1) + k(k + 1) = 2(S_3(m) + S_3(k))$.

(3) For instance, $\frac{2S_3(n)S_3(2n)}{S_3(2n+1)} = \frac{2\cdot\frac{n(n+1)}{2}\cdot\frac{2n(2n+1)}{2}}{\frac{(2n+1)(2n+2)}{2}} = n^2 = S_4(n)$.

(4) For instance, $\frac{1}{3}S_3(3n-1) = \frac{1}{3}(3S_3(n)+6S_3(n-1)) = S_3(n)+2S_3(n-1) = S_5(n)$.

(5) For instance, $S_m(n)+S_m(k)+nk(m-2) = \left(\frac{(m-2)}{2}(n^2-n)+n\right)+\left(\frac{(m-2)}{2}(k^2-k)+k\right)+nk(m-2) = \frac{(m-2)}{2}((n^2+k^2+2nk)-(n+k))+(n+k) = \frac{(m-2)}{2}((n+k)^2-(n+k))+(n+k) = S_m(n+k)$.

(6) For instance, $(S_3(n+1))^2-(S_3(n))^2 = \frac{1}{4}(n+1)^2(n+2)^2 - \frac{1}{4}n^2(n+1)^2 = \frac{1}{4}(n+1)^2((n+2)^2-n^2) = \frac{1}{4}(n+1)^2(4n+4) = (n+1)^3 = C(n+1)$.

(7) For instance, $(n+1)^5-n^5 = 1+5n+10n^2+10n^3+5n^4 = (n^4+2n^3+2n^2+n)+(4n^4+8n^3+8n^2+4n+1) = (n^2+n)(n^2+n+1)+(2n^2+2n+1)^2 = 2S_3(n^2+n)+S_4(2n^2+2n+1)$.

(8) For instance, it holds $1^3+\cdots+q^3 = \frac{q^2(q+1)^2}{4}$, and $1^3+\cdots+(\frac{q-1}{2})^3 = \frac{(\frac{q-1}{2})^2(\frac{q-1}{2}+1)^2}{4}$; then $2^3+4^3+\cdots+(q-1)^3 = 2(\frac{q-1}{2})^2(\frac{q-1}{2}+1)^2$, and $1^3+3^3+\cdots+q^3 = \frac{q^2(q+1)^2}{4} - 2(\frac{q-1}{2})^2(\frac{q-1}{2}+1)^2 = \frac{2q^2(q+1)^2-(q-1)^2(q+1)^2}{8} = \frac{(q+1)^2(q^2+2q-1)}{8} = \frac{(\frac{q^2+2q-1}{2})(\frac{q^2+2q-1}{2}+1)}{2} = S_3(n)$.

(9) For instance, $3(S_m(n)+S_m(2n)+\cdots+S_m(nk)) = 3\left(\frac{(m-2)}{2}(n^2-n)+n+\frac{(m-2)}{2}((2n)^2-2n)+2n+\cdots+\frac{(m-2)}{2}((kn)^2-kn)+kn\right) = 3\left(\frac{(m-2)}{2}(n^2(1+2^2+\cdots+k^2)-n(1+2+\cdots+k))+n(1+2+\cdots+k)\right) = \frac{3(m-2)}{2}(n^2\cdot\frac{k(k+1)(2k+1)}{6}-n\cdot\frac{k(k+1)}{2})+3n\cdot\frac{k(k+1)}{2} = \frac{(m-2)(k+1)}{2}(n^2\frac{k(2k+1)}{2}-3n\frac{k}{2})+3n\cdot\frac{k(k+1)}{2} = \frac{(m-2)(k+1)}{2}((nk)^2+n^2\frac{k}{2}-3n\frac{k}{2})+3n\cdot\frac{k(k+1)}{2} = \frac{(m-2)(k+1)}{2}((nk)^2+n^2\frac{k}{2}-nk-n\frac{k}{2})+nk(k+1)+n\frac{k(k+1)}{2} = (\frac{(m-2)}{2}(n^2-n)+n)\frac{k(k+1)}{2}+(k+1)(\frac{m-2}{2}((nk)^2-nk)+nk) = S_m(n)S_3(k)+(k+1)S_m(nk)$.

(10) For instance, it holds $S_3(2k+1) = 3S_3(k)+S_3(k+1)$. Then $\sum_{k=1}^{n} S_3(2k+1) = 3\sum_{k=1}^{n} S_3(k)+\sum_{k=1}^{n} S_3(k+1)$, i.e., $S_3(3)+S_3(5)+\cdots+S_3(2n+1) = 3S_3^3(n)+S_3^3(n+1)-1$.

(11) For instance, $S_3^4(1) + \cdots + S_3^4(n) = S_3^3(1) + (S_3^3(1) + S_3^3(2)) + \cdots + (S_3^3(1) + S_3^3(2) + \cdots + S_3^3(n)) = nS_3^3(1) + (n-1)S_3^3(2) + \cdots + S_3^3(n)$.

(12) In fact, $S_7(8) = S_6(8) + S_3(7) = \cdots = S_4(8) + 3S_3(7) = 64 + 84 = 148$.

(13) In fact, $S_4(21) = 21^2 = 441$; $S_4(21) = S_3(21) + S_3(20) = 210 + 231 = 441$; $S_4(21) = 8S_3(10) + 1 = 8 \cdot 55 + 1 = 441$; $S_4(21) = 3S_3(20) - S_3(19) + 1 = 3 \cdot 210 - 190 + 1 = 441$; $S_4(21) = \frac{2S_3(21)S_3(42)}{S_3(43)} = \frac{2 \cdot 231 \cdot 903}{946} = 441$; $S_4(21) = \frac{S_3(2 \cdot 10(10+1))}{S_3(10)} - 1 = \frac{S_3(220)}{S_3(10)} - 1 = \frac{24310}{55} - 1 = 441$.

(14) For instance, $S_6(6) = S_3(2 \cdot 6 - 1) = S_3(11) = 66$.

(15) For instance, $S_5(15) = S_3(15) + 2S_3(14) = \frac{15 \cdot 16}{2} + 2\frac{14 \cdot 15}{2} = 15 \cdot 8 + 7 \cdot 15 = 15 \cdot 15 = 225$.

(16) It follows from the Nicomachus formula $S_{m+1}(n) = S_m(n) + S_3(n-1)$.

(17) In fact, $S_4(n) = 1 + (2 \cdot 1 + 1) + (2 \cdot 2 + 1) + \cdots + (2 \cdot (n-1) + 1) = 2(1 + 2 + \cdots + (n-1)) + (1 + \cdots + 1) = 1 + 2 + \cdots + (n-1) + n + (n-1) + \cdots + 1$. It is easy to obtain a geometrical illustration of this property, placing the first n odd numbers $1, 3, \ldots, 2(n-1) + 1$, giving n^2, in the form

of the equilateral triangle, which can be seen as *upstairs-downstairs* construction; so, we can find n^2, consecutively adding the elements $1, 2, 3, \ldots, n-1, n, n-1, \ldots, 2, 1$ of the upstairs-downstairs construction of the hight n. This simple fact is useful for finding the square of a big number quickly. For instance, 52^2 is equal to $50^2 + 50 + 51 + 51 + 52 = 2500 + 204 = 2704$.

(18) In fact, $S_4(n) = (1 + 2 + \cdots + (n-1) + n) + ((n-1) + \cdots + 2 + 1) = S_3(n) + S_3(n-1)$. See also the following geometrical illustration

for $n = 4$:

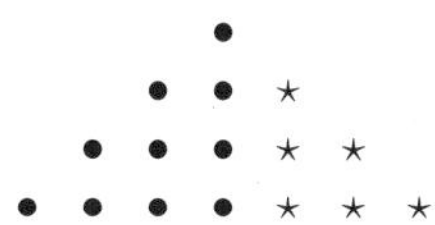

(19) If $n = S_3(k) = \frac{k(k+1)}{2}$, then $8n + 1 = 4k(k + 1) + 1 = 4k^2 + 4k + 1 = (2k + 1)^2$; on the other hand, if $8n + 1 = t^2$, then t is odd, and $n = \frac{t^2-1}{8} = \frac{\frac{t-1}{2}\frac{t+1}{2}}{2} = S_3(\frac{t-1}{2})$.

(20) The first few non-square numbers are 2, 3, 5, 6, 7, 8, 10, 11, 12, 13, ... (Sloane's A000037). For a given positive integer k there exists exactly $\lfloor\sqrt{k}\rfloor$ squares $1, 4, \ldots, \lfloor\sqrt{k}\rfloor^2$ before k, since $\lfloor\sqrt{k}\rfloor \leq \sqrt{k} < \lfloor\sqrt{k}\rfloor + 1$, and $\lfloor\sqrt{k}\rfloor^2 \leq (\sqrt{k})^2 = k < (\lfloor\sqrt{k}\rfloor + 1)^2$. So, in order to obtain n-th non-square number, one should add to n the value $\lfloor\sqrt{n+1}\rfloor$, or the value $\lfloor\frac{1}{2}+\sqrt{n}\rfloor$.

(21) In fact, $9S_3(n) + 1 = S_3(3n + 1)$, $9S_3(3n + 1) + 1 = S_3(3(3n + 1) + 1) = S_3(9n + 4)$, etc.

(22) For instance, $21 = \frac{6{\cdot}7}{2}, 2211 = \frac{66{\cdot}67}{2}, \ldots, 22\ldots211\ldots1 = 2 \cdot (10^{2n} + \cdots + 10^n) + (10^{n-1} + \cdots + 1) = 2 \cdot 10^n \frac{10^n-1}{9} + \frac{10^n-1}{9} = 2\frac{(10^n-1)^2}{9} + 3\frac{10^n-1}{9} = \frac{6{\cdot}\frac{10^n-1}{9}(6{\cdot}\frac{10^n-1}{9}+1)}{2} = \frac{66\ldots6{\cdot}66\ldots7}{2}$.

(23) It follows form the formula $S_6(n) = 2n^2 - n = n \cdot (2n - 1)$.

(24) It is Diophantus' generalization of the theorem that $8S_3(n)+1$ is a square. He proved it by a geometric method and spoke of this result as a new definition of polygonal numbers equivalent to that of Hypsicles.

(25) Using this formula, we can define *polygonal* (in fact, m-gonal) *root* of a given positive integer x as $\frac{\sqrt{(8m-16)x+(m-4)^2}+m-4}{2m-4}$. In particular, the *triangular root* of x has the form $\frac{\sqrt{8x+1}-1}{2}$, as well a the square root has the ordinary form $\sqrt{x}$.

(26) These numbers are 1, 10000, 29997, 59992, 99985, 149976, 209965, 279952, 359937, 449920, ... (Sloane's A167149).

(27) In fact, for $n = 2$ and $m = a + 2$, we get $a\frac{x(x+1)}{2} + (1 - a)\frac{x}{1} = \frac{x}{2}(ax+a+2-2a) = \frac{x}{2}(ax+2-a) = \frac{x}{2}((m-2)x-m+4) = S_m(x)$. For $n = 3$ and $m = a+2$, we get $a\frac{x(x+1)(x+2)}{6} + (1-a)\frac{x(x+1)}{2} = \frac{x(x+1)}{6}(ax + 2a + 3 - 3a) = \frac{x(x+1)}{6}(ax + 3 - a) = \frac{x}{2}((m - 2)x - m + 5) = S_m^3(x)$.

(28) For instance, let $N = 10$. All divisors of $2N = 20$ are $1, 2, 4, 5, 10, 20$. All divisors of $2N - 2 = 18$ are $1, 2, 3, 9, 18$. Choose from first set numbers $2 = 1 + 1$, $4 = 3 + 1$ and $10 = 9 + 1$. So, $n \in \{2, 4, 10\}$. For $n = 2$, one has $m = \frac{18}{1} - \frac{20}{2} + 2 = 10$, and $S_m(n) = S_{10}(2) = 10$; for $n = 4$, $m = \frac{18}{3} - \frac{20}{4} + 2 = 3$, and $S_m(n) = S_3(4) = 10$. For $n = 10$, $m = \frac{18}{9} - \frac{20}{10} + 2 = 2$, and $S_m(n) = S_2(10) = 10$.

(29) We use the formula $2S_m(n) = n((m-2)(n-1)+2)$. Since $m \geq 3$, we get $(m-2)(n-1)+2 \geq (n-1)+2 \geq n+1 > n$. For instance, if $N = 36$, then $2N = 2 \cdot 36 = 3 \cdot 24 = 4 \cdot 18 = 6 \cdot 12 = 8 \cdot 9$, and $n \in \{2, 3, 4, 6, 8\}$. For $n = 2$, one has $(m-2)(n-1) = 34$, $m-2 = 34$ and $m = 36$; for $n = 3$, one has $(m-2)(n-1) = 22$, $m-2 = 11$, and $m = 13$; for $n = 4$, one has $(m-2)(n-1) = 16$, i.e., $3(m-2) = 16$, a contradiction; for $n = 6$, one has $(m-2)(n-1) = 10$, $m-2 = 2$, and $m = 4$; for $n = 8$, one has $(m-2)(n-1) = 7$, $m-2 = 1$, and $m = 3$. So, we get $36 = S_{36}(2) = S_{13}(3) = S_4(6) = S_3(8)$.

(30) The first tricapped prism numbers are 1, 9, 33, 82, 165, 291, 469, 708, 1017, 1405, ... (Sloane's A005920). The number of points on surface of tricapped prism is $7(n-1)^2 + 2$, $n \geq 2$, giving 1, 9, 30, 65, 114, 177, 254, 345, 450, 569, ... (Sloane's A005919). The centered tricapped prism numbers $\frac{(2n-1)(7n^2-7n+6)}{6}$ are the partial sums of the above sequence, giving 1, 10, 40, 105, 219, 396, 650, 995, 1445, 2014, ... (Sloane's A063490). They coincide with centered octagonal pyramid numbers.

(31) The number of points on surface of hexagonal prism is $12(n-1)^2 + 2$, $n \geq 2$, giving 1, 14, 50, 110, 194, 302, 434, 590, 770, 974, ... (Sloane's A005914). The centered hexagonal prism numbers can be obtained as partial sums of the above sequence, giving 1, 15, 65, 175, 369, 671, 1105, 1695, 2465, ... (Sloane's A005917); they coincide with rhombic dodecahedral numbers and have the form $n^4 - (n-1)^4$.

(32) The first doubly triangular numbers are 1, 6, 21, 55, 120, 231, 406, 666, 1035, 1540, ... (Sloane's A002817). The recurrent equation can be proven by induction, while the generating

function can be obtained using the standard procedure, leading to the linear recurrent equation $SS_3(n+5) - 5SS_3(n+4) + 10SS_3(n+3) - 10SS_3(n+2) + 5SS_3(n+1) - SS_3(n) = 0$.

(33) The first iterated triangular numbers are 1, 6, 231, 1186570, 347357071281165, ... (Sloane's A099129).

(34) In fact, $S_4^3(n) = 1^2 + 2^2 + \cdots + (n-1)^2 + n^2 = 1 + 2 \cdot 2 + \cdots + (n-1) \cdot (n-1) + n \cdot n$.

(35) For instance, the sum of the elements

$$\begin{matrix} 4 & 3 & 2 & 1 \\ 3 & 4 & 3 & 2 \\ 2 & 3 & 4 & 3 \\ 1 & 2 & 3 & 4 \end{matrix}$$

gives $O(4) = 44$. The proof follows from the previous result, since the above square can be obtained as sum of two pyramids

$$\begin{matrix} & & & 1 & & & \\ & & 2 & & 2 & & \\ & 3 & & 3 & & 3 & \\ 4 & & 4 & & 4 & & 4 \end{matrix} \qquad \text{and} \qquad \begin{matrix} & & 1 & & \\ & 2 & & 2 & \\ 3 & & 3 & & 3 \end{matrix}$$

relating to $S_4^3(4)$ and $S_4^3(3)$. In general, $O(n) = S_4^3(n) + S_4^3(n-1)$, and the corresponding square array can be obtained as the sum of two pyramids, relating to $S_4^3(n)$ and $S_4^3(n-1)$.

(36) By definition, $S_m^3(n) = S_m(1) + \cdots + S_m(n) = \frac{(m-2)}{2}(1^2 + \cdots + n^2) - \frac{(m-4)}{2}(1 + \cdots + n) = \frac{(m-2)}{2} \cdot \frac{n(n+1)(2n+1)}{6} - \frac{(m-4)}{2} \cdot \frac{n(n+1)}{2} = \frac{(n+1)}{6}\left(\frac{(m-2)}{2} \cdot n(2n+1) - \frac{(m-4)}{2} \cdot 3n\right) = \frac{(n+1)}{6}\left(2\left(\frac{(m-2)}{2}n^2 - \frac{(m-4)}{2}n\right) + n\right) = \frac{(n+1)}{6}(2S_m(n) + n)$.

(37) For $n = 1$, we have the series $1, 1+a, 1+2a, 1+3a, \ldots$ of gnomonic numbers; for $n = 2$, we have the series $1, 2+a, 3+3a, \ldots, \frac{3 \cdot 4 \cdot \cdots \cdot k(2+(k-1)a)}{(k-1)!} = \frac{k}{2}(2+(k-1)a) = \frac{a}{2}(k^2 - k) + k = S_{a+2}(k), \ldots$ of $(a+2)$-gonal numbers, etc.

(38) For instance, in the case of $k = 2$, we have $\frac{S^2(S+1)^2}{4} = \frac{S^4+2S^3+S^2}{4}$, i.e., $(S_3(1))^2 + (S_3(2))^2 + \cdots + (S_3(n))^2 = \frac{1}{4}\left((1^4 + 2^4 + \cdots + n^4) + 2(1^3 + 2^3 + \cdots + n^3) + (1^2 + 2^2 + \cdots + n^3)\right)$.

(39) It can be obtained using generating functions for the sequences $1, 2^k, 3^k, \ldots, n^k, \ldots$, $k = 2, 3, 4, 5, 6, 7$ (see Chapter 3).

(40) As $v_5 = \frac{(3+2\sqrt{2})^5-(3-2\sqrt{2})^5}{4\sqrt{2}} = 1189$, then $S_{4,3}(5) = S_4(1189) = 1189^2 = 1413721$; As $u_7 = \frac{(3+2\sqrt{2})^7+(3-2\sqrt{2})^7-2}{4} = 57121$, then $S_{4,3}(7) = S_3(57121) = \frac{57121 \cdot 57122}{2} = 1631432881$.

(41) The first such numbers are 1, 40755, 1533776805, 5772215 6241751, 21723156264682834 65, ... (Sloane's A046180). In fact, let $\frac{m(m+1)}{2} = \frac{3n^2-n}{2} = 2p^2 - p$. Then $m = 2p - 1$, $n = \frac{1+R}{6}$, where $R^2 = 48p^2 - 24p + 1$. Thus, $1 + R = 6kp$, whence $p = \frac{2-k}{4-3k^2}$. Take $k = \frac{b}{a}$. Then p is an integer if $(2a)^2 - 3b^2 = 1$. By the continued fraction for $\sqrt{3}$, we get $(2a, b) = (2,1), (26,15), (362,209), (5042,2911), \ldots$, whence $p = 1, 143, 27693, \ldots$, so that answers are 1, 40755, 1533776801, ...

(42) The solutions of the equation $v^2 = u(2u-1)$ are $v_n = \frac{(3+2\sqrt{2})^{2n+1}-(3-2\sqrt{2})^{2n+1}}{4\sqrt{2}}$, $u_n = \frac{1}{4}(1 + \frac{1}{2}((3-2\sqrt{2})^{2n+1} + (3+2\sqrt{2})^{2n+1}))$, giving n-th hexagonal square number as $S_{6,4}(n) = v_n^2 = \frac{1}{32}(-2+(17-12\sqrt{2})(3-2\sqrt{2})^{4n}+(17+12\sqrt{2})(3+2\sqrt{2})^{4n})$.

(43) For instance, in the case of $n = 1$, we get $2^{2-5}(3 + cos\pi)(3 + \cos 2\pi) = \frac{1}{8} \cdot 2 \cdot 4 = 1 = S_{4,3}(1)$; in the case of $n = 2$, we get $2^{4-5}(3+\cos\frac{\pi}{2})(3+\cos\pi)(3+\cos\frac{3\pi}{2})(3+\cos 2\pi) = \frac{1}{2} \cdot 3 \cdot 2 \cdot 3 \cdot 4 = 36 = S_{4,3}(2)$.

(44) The problem is a special case of that to make a quadratic function a square. The x-th m-gonal number is $\frac{(m-2)x^2-(m-4)x}{2}$. To make it a square, set $2(m-2)p^2 + 1 = q^2$. Then $2(m-2)x^2 - 2(m-4)x = 0$ iff $x = 0$; $2(m-2)x^2 - (m-4)x = (m-4)^2p^2$ iff $2(m-2)x^2 - 2(m-4)x - (m-4)^2p^2 = 0$ iff $x = \frac{(m-4)\pm(m-4)\sqrt{1+2(m-2)p^2}}{2(m-2)} = \frac{(m-4)(1\pm q)}{2(m-2)}$, and we get the desired result for sign minus; $2(m-2)x^2 - 2(m-4)x = 4(m-4)^2p^2q^2$ iff $2(m-2)x^2 - 2(m-4)x - 4(m-4)^2p^2q^2 = 0$ iff $x = \frac{(m-4)\pm(m-4)\sqrt{1+8(m-2)p^2q^2}}{2(m-2)} = \frac{(m-4)(1\pm(2q^2-1))}{2(m-2)}$, and we get the desired result for sigh minus, etc. It remains to make the above expressions integers. For $m = 5$, q is to be chosen from 1, 5, 49, ... and hence, p from 0, 2, 20, ... The first fraction $-\frac{(m-4)}{2(m-2)}(q-1)$ is here $\frac{1-q}{6}$ and is an integer for $q = 49$,

whence $x = -8$. However, for $m > 4$, q can be taken negative. The value $q = -5$ gives $x = 1$, and the pentagonal number 1.

(45) It is a generalization of the method to find numbers both m- and n-gonal. Take $ax - a' = \frac{yp}{q}$, $x = \frac{(by-b')q}{p}$, so that $x = \frac{q(b'p+a'bq)}{N}$, $y = \frac{q(a'p+b'aq)}{N}$, where $-N = p^2 - abq^2$. Let p', q' give a particular solution of the last equation such that $A = \frac{a'p'+ab'q'}{N}$, $B = \frac{b'p'+ba'q'}{N}$ are integers. Take $p = p't + abq'u$, $q = q't + p'u$. Then $p^2 - abq^2 = -NF$, where $F = t^2 - abu^2$, and $x = \frac{q(Bt+Abu)}{F}$, $y = \frac{q(At+Bau)}{F}$. Now it is easy to find a set of solutions (t, u) of the equation $F = 1$, i.e., of the Pell's equation $t^2 - abu^2 = 1$, from the initial solution $t_0 = 1$, $u_0 = 0$. In fact, we get the solutions $t_i = 2t_1t_{i-1} - t_{i-2}$, $u_i = 2t_1u_{i-1} - u_{i-2}$. It remains only to find a solution p', q' of $p^2 - abq^2 = -N$. While one may to use the continued fraction for $\sqrt{ab}$, it suffices for our initial problem to note the solution $p' = a - a'$, $q' = 1$, for the case $a - a' = b - b'$; then $N = ab' + ba' - a'b'$, and $A = B = 1$. Now we can go back to the problem of finding numbers T_i both m- and n-gonal. First, if m and n are both odd, we may take $a = m-2$, $a' = m-4$, $b = n-2$, $b' = n-4$, which have no common factor. Then $a - a' = b - b' = 2$. In this case we can obtain for $P_i = \frac{1}{2}T_i$ the following recurrent equation: $P_0 = 1$, $P_i = 2t_4P_{i-1} - P_{i-2} + (2d-1)(t_4-1)$, $d = \frac{(m+n-4)(mn-2m-2n+8)}{16(m-2)(n-2)}$. But if m and n are both even, take $a = \frac{1}{2}m - 1$, $a' = \frac{1}{2}m - 2$, $b = \frac{1}{2}n - 1$, $b' = \frac{1}{2}n - 2$, whence $a - a' = b - b' = 1$. In this case the value $P_i = T_i$ satisfies the above recurrent equation. Also, $P_1 = \frac{1}{2}(t_4 - 1) + d(t_4 - 1) + \frac{1}{w}mnu_4$, where $w = 8$ in the first case, and $w = 16$ in the second case. For instance, 1, 210, 40755 are both, triangular and pentagonal.

(46) It can be obtained following the standard procedure: the Diophantine equation $u(u+1) = 12v(v-1) + 2$ leads to the Pell's equation $x^2 - 3y^2 = 1$, $x = 4v - 2$, $y = \frac{2u+1}{3}$.

(47) It can be obtained following the standard procedure: the Diophantine equation $u^2 = 6v(v-1) + 1$ leads to the Pell-like equation $2x^2 - 3y^2 = -1$, $x = u$, $y = 2v - 1$.

(48) Such numbers are $1, 31, 1891, 117181, 7263301, \ldots$ (Sloane's A131751). They can be obtained by the recurrent equation $a(n+1) = 62a(n) - a(n-1) - 30$, $a(1) = 1$, $a(2) = 31$, while their generating function is $f(x) = \frac{x(1-32x+x^2)}{(1-x)(1-62x+x^2)}$.

(49) Let $S_3(x) > 1$ be a fourth power. Since $gcd(x, x+1) = 1$, then according as x is even or odd, $\frac{x}{2}$ and $x+1$, or $\frac{x+1}{2}$ and x must be equal to a fourth power m^4. Thus, $2m^4 \pm 1 = n^4$, or $2n^4 \mp 2$ is a square. But $2n^4 \mp 2$ is a square only when $n = 1$, whence $m = 0$ or 1, and $x = 0$ or 1. Therefore, there are no triangular numbers greater than 1, which are equal to a biquadratic number. However, it is easy to check that $\frac{x^2+x}{2} = (\frac{6}{7})^4$ for $x = \frac{32}{49}$. In other words, there exist "rational" triangular numbers, which are equal to a "rational" biquadratic number. It is known that a triangular number greater than 1 is never a cube, biquadrate or fifth power. Hence, it can not be an k-dimensional hypercube number for any $k \in \{3, 4, 5, 6, 8, 9, 10, 12, 15\}$ (see [Dick05]).

(50) These numbers are gnomons of $(a+2)$-gonal numbers.

(51) Let $S_m(\lambda) + S_m(\nu) = 2S_m(\mu)$. Multiply each term by $8(m-2)$ and add $(m-4)^2$ to each product. By the *Diophantus' formula*, we get $P_\lambda^2 + P_\nu^2 = 2P_\mu^2$, where $P_\lambda = (m-2)(2\lambda - 1) + 2$, $P_\mu = (m-2)(2\mu - 1) + 2$, $P_\nu = (m-2)(2\nu - 1) + 2$. Take $P_\lambda = \pm(x^2 - 2xy - y^2)$, $P_\mu = x^2 + y^2$, $P_\nu = x^2 + 2xy - y^2$. Then λ, μ, ν are found in terms of x, y, m by use of the above equations defining P_λ, P_μ, and P_ν. So, for $m = 3$, $x = 3$, and $y = 4$, we get a progression $S_3(8) = 36$, $S_3(12) = 78$, and $S_3(15) = 120$.

(52) If numbers r^2, s^2 and t^2 form an arithmetic progression, then it holds $r^2 + t^2 = 2s^2$. Therefore, $2|(r^2 + t^2)$, i.e., the numbers r and t have the same parity. Then numbers $\frac{t+r}{2}$ and $\frac{t-r}{2}$ are positive integers, and we get the equation $(\frac{t+r}{2})^2 + (\frac{t-r}{2})^2 = s^2$. So, the numbers r, t, s produce a Pythagorean triple (x, y, z) with $x = \frac{t+r}{2}$, $y = \frac{t-r}{2}$, and $z = s$. Conversely, any pythagorean triple (x, y, z) produces a three-term arithmetic progression of square numbers r^2, s^2, t^2 by $r = x - y, s = z, t = x + y$.

However, it is known, that there exists no four-term arithmetic progression of square numbers (see [Dick05]).

(53) For instance, the numbers 1, 6 and 36 form a three-term geometric progression of triangular numbers. However, it was proven by Fang and Chen ([ChFa07]) that there are no four triangular numbers in a geometric progression.

(54) For instance, the product of all entries of the four-term arithmetic progression $(-2, 0, 2, 4)$ is a square. However, the product of four distinct non-zero integers in an arithmetic progression is square number only for $(-3, -1, 1, 3)$, giving 9 (see [LeLi83]).

(55) The fourth colonne is $8(m-4)$; r-th collone is $r^2 + \frac{r^2(r-1)(m-2)}{2} = r \cdot (\frac{m-2}{2}(r^2 - r) + r) = r \cdot S_m(r)$. In particular, for $m = 4$, r-th colonne is equal to r^3. So, this *Fermat's collone construction* is a generalization of the property $1 = 1^1, 3 + 5 = 2^3, 7 + 9 + 11 = 3^3, \ldots$ of cubic numbers.

(56) For $k = 1$ we have pairs $(1, 51)$, and $(5, 35)$; for $k = 2$ we have pairs $(12, 6112)$, $(22, 5922)$, $(35, 1335)$, $(51, 3151)$, $(70, 5370)$, $(92, 5192)$, etc.

(57) In fact, exactly 1 number (0) has 0 as a sum of their digits; exactly 3 numbers (1, 10, 100) have the sum 1; exactly 6 (2, 20, 200, 11, 110, 101) have the sum 2; exactly 10 (3, 30, 300, 21, 12, 102, 201, 111, 120, 210) have the sum 3; exactly 15 (4, 40, 400, 31, 13, 130, 310, 301, 103, 202, 22, 220, 211, 212, 122) have the sum 4, etc.

(58) These numbers are 0, 1, 8, 341, 1160, 4485, 6816, 9633, 12936, 16725, ... (Sloane's A117082).

(59) Since $a = c_n \cdot 10^n + \cdots + c_1 \cdot 10 + c_0 \equiv c_n + \cdots + c_1 + c_0 (mod\, 9)$, the digital root of a given positive integer a is equal to the smallest positive residue a modulo 9. Since n has residues 0, $1, \ldots, 8$ modulo 9, then $n+1$ has residues $1, 2, \ldots, 0$ modulo 9, and $n(n+1)$ has residues 0, 2, 6, 3, 2, 3, 6, 2, 0 modulo 9, respectively. As $(2, 3) = 1$, then $\frac{n(n+1)}{2}$ is divided by 3 in all cases, except $n = 1, 4, 7 (mod\, 9)$ or, that is the same, $n \equiv 1 (mod\, 3)$, in which case $n(n+1) \equiv 2 (mod\, 9)$, or $\frac{n(n+1)}{2} \equiv$

$1(mod\,9)$. So, every triangular number is either divisible by three or has a remainder of 1 when divided by nine. It means that its digital root is 1, 3, 6, or 9.

The table below gives similar results for square numbers n^2 and hexagonal numbers $2n^2 - n$.

n	0	1	2	3	4	5	6	7	8
n^2	0	1	4	0	7	7	0	4	1
$2n^2 - n$	0	1	6	6	1	0	3	1	3

Digital root equal to 1, 4, 7 or 9 is a necessary, but not sufficient condition for a number to be a square. The digital roots of the first few squares are 1, 4, 9, 7, 7, 9, 4, 1, 9, 1, 4, 9, 7,... (Sloane's A056992), while the list of numbers having digital roots 1, 4, 7, or 9 is 1, 4, 7, 9, 10, 13, 16, 18, 19, 22, 25,... (Sloane's A056991).

(60) These numbers are 0, 1, 4, 9, 36, 81, 100, 121, 144, 169,... (Sloane's A053057), 0, 1, 4, 9, 36, 49, 64, 81, 100, 121,... (Sloane's A117676), 0, 1, 4, 9, 25, 100, 169, 196, 225, 256,... (Sloane's A117678), and 0, 1, 4, 9, 100, 400, 900, 2304, 2601, 3600,... (Sloane's A117680), respectively.

(61) These numbers are 0, 1, 100, 3969, 7569, 8649, 10000, 12996, 13689, 15876,... (Sloane's A117685), 0, 1, 100, 8649, 10000, 59049, 88209, 91809, 104976, 106929,... (Sloane's A117687), and 0, 1, 100, 3025, 5041, 6400, 10000, 21025, 23104, 26569,... (Sloane's A118490), respectively.

(62) These numbers are 0, 1, 8, 1000, 8000, 10^6, $8 \cdot 10^6$, 10^9, $8 \cdot 10^9$, 10^{12},... (Sloane's A019545), 1, 8, 1000, 4096, 8000, 10648, 24389, 27000, 39304, 50653, ... (Sloane's A118545), 0, 1, 27, 216, 1000, 27000, 216000, 287496, 884736, 970299, ... (Sloane's A117688), 0, 1, 343, 1000, 1331, 4096, 8000, 10648, 19683, 27000, ... (Sloane's A117689), and 0, 1, 1000, 27000, 216000, 970299, 1000000, 1860867, 2146689, 4019679,... (Sloane's A117690), respectively.

(63) These numbers are 49, 100, 144, 169, 361, 400, 441, 900, 1225, 1369, 1444,... (Sloane's A019547), and 1000, 8000, 10648, 27000, 64000, 125000, 216000, 343000, 512000, 729000,... (Sloane's A019548), respectively.

(64) In fact, if c^3 is the concatenation of a^3 and b^3, then $c^3 = 10^k a^3 + b^3$, where k is the number of digits in b^3. After shifting any powers of 1000 in 10^k into a^3, the original problem is equivalent to finding a solution to one of the Diophantine equations $c^3 - b^3 = a^3$, $c^3 - b^3 = 10a^3$, $c^3 - b^3 = 100a^3$. None of these have solutions in integers (see [Dick05]).

(65) It can be checked by direct computation.

(66) These numbers are 1, 4, 9, 36, 81, 100, 144, 225, 324, 400, 441,... (Sloane's A118547), and 1, 4, 9, 36, 144, 1296, 2916, 11664, 41616, 82944,... (Sloane's A118548), respectively.

(67) The values n of such n^2 that contains exactly two different digits are given by 4, 5, 6, 7, 8, 9, 10, 11, 12, 15, 20, ... (Sloane's A016069), whose squares are 16, 25, 36, 49, 64, 81, 100, 121, 144, 225, ... (Sloane's A018885). It is conjectured that, other than 10^{2n}, $4 \cdot 10^{2n}$ and $9 \cdot 10^{2n}$, there are only a finite number of squares n^2 having exactly two distinct non-zero digits (see [Guy94a]). The first few such n are 4, 5, 6, 7, 8, 9, 11, 12, 15, 21, ... (Sloane's A016070), corresponding to squares 16, 25, 36, 49, 64, 81, 121, ... (Sloane's A018884).

(68) These numbers are 0, 1, 4, 9, 49, 100, 144, 400, 441, 900,... (Sloane's A019544). It is conjectured that there are only finitely many square numbers composed only of the square digits 1, 4, and 9, and the largest known is

$$648070211589107021^2 = 419994999149149944149149944191494441.$$

(69) These numbers are listed in the table below.

digits	n, n^2	Sloane
1, 2, 3	1, 11, 111, 36361, 363639, ...	*A*030175
	1, 121, 12321, 1322122321, 132233322321, ...	*A*030174
1, 4, 6	1, 2, 4, 8, 12, ...	*A*027677
	1, 4, 16, 64, 144, ...	*A*027677
1, 4, 9	1, 2, 3, 7, 12, ...	*A*027675
	1, 4, 9, 49, 144, ...	*A*006716
2, 4, 8	2, 22, 168, 478, 2878, ...	*A*027679
	4, 484, 28224, 228484, 8282884, ...	*A*027678
4, 5, 6	2, 8, 216, 238, 258, ...	*A*030177
	4, 64, 46656, 56644, 66564, ...	*A*030176

Note, that the only known square number composed only of the digits 7, 8, and 9 is 9.

(70) These numbers are 0, 1, 8, 999, 19999999,... (Sloane's A061105).

(71) There are six positive integers equal to the sum of the digits of their cubes: 1, 8, 17, 18, 26, and 27 (Sloane's A046459). There are five positive integers equal to the sum of the cubes of their digits: 1, 153, 370, 371, 407 (Sloane's A046197).

(72) These numbers are 1, 2, 3, 4, 5, 6, 7, 8, 9, 153, 370, 371, 407, 1634, 8208, 9474, 54748, 92727, 93084, 548834, ... (Sloane's A005188). It is a finite sequence, the last term being

$$115132219018763992565095597973971522401.$$

See also the sequence 0, 1, 2, 3, 4, 5, 6, 7, 8, 9, 153, 370, 371, 407, 1634, 4150, 4151, 8208, ... (Sloane's A023052) of numbers that are the sum of some fixed power of their digits.

(73) It can be checked by direct computation.

(74) So, $gcd(S_6(8), S_6(9)) = gcd(120, 153) = gcd(9, 3) = 3$; $gcd(S_5(9), S_5(10)) = gcd(117, 145) = gcd(5, 2) = 1$; $gcd(S_4(7), S_5(7)) = gcd(49, 112) = 7$; $gcd(S_4(8), S_5(8)) = gcd(64, 92) = 4$.

(75) (a) if $n \equiv 1, 2 (mod\, 4)$, then $5n - 3 \equiv 2, 3 (mod\, 4)$, and $n(5n - 3) \equiv 2 (mod\, 4)$, i.e., $S_7(n) \equiv 1 (mod\, 2)$; if $n \equiv 3, 0 (mod\, 4)$, then $5n - 3 \equiv 0, 1 (mod\, 4)$, and $n(5n - 3) \equiv 0 (mod\, 4)$, i.e., $S_7(n) \equiv 0 (mod\, 2)$;

(b) for any n, it holds $3n^2 - 2n \equiv n^2 \equiv n (mod\, 2)$;

(c) if $n \equiv 1, 2 (mod\, 4)$, then $7n - 5 \equiv -n - 1 \equiv 2, 1 (mod\, 4)$, and $n(7n - 5) \equiv 2 (mod\, 4)$, i.e., $S_9(n) \equiv 1 (mod\, 2)$; if $n \equiv 3, 0 (mod\, 4)$, then $7n - 5 \equiv 0, 3 (mod\, 4)$, and $n(7n - 5) \equiv 0 (mod\, 4)$, i.e., $S_9(n) \equiv 0 (mod\, 2)$.

(d) for any n, it holds $4n^2 - 3n \equiv n (mod\, 2)$;

(e) for any n, it holds $2n^2 - 2n + 1 \equiv 1 (mod\, 2)$;

(f) for $n \equiv 2, 3 (mod\, 4)$, $5n^2 - 5n + 2 \equiv n^2 - n + 2 \equiv 0 (mod\, 4)$, and $CS_5(n) \equiv 0 (mod\, 2)$; for $n \equiv 0, 1 (mod\, 4)$, $5n^2 - 5n + 2 \equiv n^2 - n + 2 \equiv 2 (mod\, 4)$, and $CS_5(n) \equiv 1 (mod\, 2)$;

(g) it is easy to see that $CS_6(n) \equiv 1 (mod\, 6)$;

(h) it is easy to see that $CS_8(n) \equiv 1 (mod\, 8)$;

(i) it is easy to see, that $CS_{12}(n) \equiv 1 (mod\, 12)$;

(j) more exactly, for $n \equiv 0, 2, 3 (mod\, 4)$ n-th tetrahedral number is even, and for $n \equiv 1 (mod\, 4)$ it is odd, i.e., all odd tetrahedral numbers are $S_3^3(4k + 1)$: 1, 35, 165, ...; one can check it, using the fact, that the tetrahedral numbers can be obtained by adding the entries on the rising diagonals of the multiplication table: $S_3^3(n) = 1 \cdot n + 2 \cdot (n-1) + \cdots + (n-1) \cdot 2 + n \cdot 1$; since the multiplication table is symmetric, and the squares down the main diagonal are alternatively odd and even, it makes every fourth tetrahedral number odd, counting from the first 1.

(76) (a) as $n(n+1)$ is divisible by 2, then $(2n+1)^2 = 4n^2+4n+1 = 4n(n+1)+1 \equiv 1(mod\, 8)$;

(b) as $CS_3(n) = 3S_3(n-1)+1$, then each centered triangular number has a remainder of 1 when divided by three, and the quotient (for $n \geq 2$) is the previous regular triangular number;

(c) in fact, $2n^2 - 2n + 1 \equiv -n^2 + n + 1(mod\, 3)$, that gives 1 modulo 3 for $n \equiv 0, 1(mod\, 3)$, and 2 modulo 3 for $n \equiv 2(mod\, 3)$;

(d) in fact, $CS_6(n) \equiv 1(mod\, 6)$;

(e) in fact, $6n^2 - 6n + 1 \equiv 6n(n-1) + 1 \equiv 1(mod\, 12)$;

(f) in fact, $CS_4(n) = 4S_3(n-1)+1$, i.e., $CS_4(n) \equiv 1(mod\, 4)$;

(g) similarly, $CS_8(n) = 8S_3(n-1)+1$, i.e., $CS_8(n) \equiv 1(mod\, 8)$;

(h) as $CS_4(n) \equiv 1(mod\, 4)$, and $CS_4(n) \equiv 1, 2(mod\, 3)$, then $CS_4(n) = 1, 5(mod\, 6,\ 8,\ 12)$.

(77) In fact, for a given odd prime p one gets $S_m(p) = \frac{m+2}{2}(p^2-p)+p = p(\frac{m+2}{2}(p-1)+1) = p((m+2)\frac{p-1}{2}+1)$, where $(m+2)\frac{p-1}{2}+1$ is an integer since $p-1$ is even.

(78) In fact, if $n \equiv 0, -1, 4, 5(mod\, 10)$, then $n+1 \equiv 1, 0, 5, 6(mod\, 10)$, and $n(n+1) \equiv 0(mod\, 10)$; therefore, $\frac{n(n+1)}{2} \equiv 0(mod\, 5)$, i.e., $S_3(n) \equiv 0(mod\, 10)$, or $S_3(n) \equiv 5(mod\, 10)$; if $n \equiv 1,\ -2,\ 3, -4(mod\, 10)$, then $n+1 \equiv 2, -1, 4, -3\ (mod\, 10)$, and $n(n+1) \equiv 2(mod\, 10)$; therefore, $\frac{n(n+1)}{2} \equiv 1(mod\, 5)$, i.e., $S_3(n) \equiv 1(mod\, 10)$, or $S_3(n) \equiv 6(mod\, 10)$; if $n \equiv 2, -3(mod\, 10)$, then $n+1 \equiv 3, -2(mod\, 10)$, and $n(n+1) \equiv 6(mod\, 10)$; therefore, $\frac{n(n+1)}{2} \equiv 3(mod\, 5)$, i.e., $S_3(n) \equiv 3(mod\, 10)$, or $S_3(n) \equiv 8(mod\, 10)$. The table below gives possible residues modulo 10 for square numbers n^2, and

centered square numbers $2n^2 - 2n + 1$.

n	0	1	2	3	4	5	6	7	8	9
n^2	0	1	4	9	6	5	6	9	4	1
$2n^2 - 2n + 1$	1	1	5	3	5	1	1	5	3	5

(79) The last two digits of a positive integer is a residue of the number modulo 100. If $n = 100a + 10b + c$, then $n^2 = 100(100a^2 + 20ab + 2ac + b^2) + (20bc + c^2)$, and the last two digits of n^2 must be the same as the last two digits of $20bc+c^2$. Furthermore, the last two digits can be obtained by considering only $b = 0, 1, 2, 3$, and 4, since $20(b+5)c+c^2 = 100c+(20bc+c^2)$ has the same last two digits as $20bc+c^2$, with the one additional possibility that $c = 0$ in which case the last two digits are 00. So, the following table (with the addition of 00) exhausts all possible last two digits.

	1	2	3	4	5	6	7	8	9
0	01	04	09	16	25	36	49	64	81
1	21	44	69	96	25	56	89	24	61
2	41	84	29	76	25	76	29	84	41
3	61	24	89	56	25	96	69	44	21
4	81	64	49	36	25	16	09	04	01

The only 22 possibilities are therefore 00, 01, 04, 09, 16, 21, 24, 25, 29, 36, 41, 44, 49, 56, 61, 64, 69, 76, 81, 84, 89, and 96, which can be summarized as 00, $e1$, $e4$, 25, $o6$, and $e7$, where e stands for an even number, and o for an odd number.

(80) The following tables gives the possible residues mod n for square and cubic numbers for $n = 2, 3, \ldots, 10$. The quantity $s(n)$ gives the number of distinct residues for a given n.

n	$s(n)$	$x^2 (mod\, n)$
2	2	0, 1
3	2	0, 1
4	2	0, 1
5	3	0, 1, 4
6	4	0, 1, 3, 4
7	4	0, 1, 2, 4
8	3	0, 1, 4
9	4	0, 1, 4, 7
10	6	0, 1, 4, 5, 6, 9

n	$s(n)$	$x^3 (mod\, n)$
2	2	0, 1
3	3	0, 1, 2
4	3	0, 1, 3
5	5	0, 1, 2, 3, 4
6	6	0, 1, 2, 3, 4, 5
7	3	0, 1, 6
8	5	0, 1, 3, 5, 7
9	3	0, 1, 8
10	10	0, 1, 2, 3, 4, 5, 6, 7, 8, 9

(81) These numbers are $(3, 4)$, $(9, 10)$, $(15, 16)$, etc. In general, the equation $x^2 - \frac{y(y+1)}{2} = \pm 1$ set $2x = z$, $2y + 1 = t$, whence $2z^2 - t^2 = 7$ or -9, which are reduced to the Pell's equations $u^2 - 2v^2 = 1$, or -1, and solved.

(82) For instance, $S_{4,3}(1) = S_3(1) = 1$, $S_{4,3}(2) = S_3(8) = 36$, $8 - 1 = 7$, and $7 = 3 + 4$, where $3^2 + 4^2 = 5^2$.

(83) It is known that unity is the only triangular number which is equal to a sum of the squares of two consecutive integers, and 10 is the only triangular number equal to a sum of the squares of two consecutive odd integers. If a triangular number is a product of two consecutive integers, it is twice other triangular number; the first few such numbers are 0, 6, 210, 7140, 242556, 8239770, 279909630, 9508687656, 323015470680, ... (Sloane's A029549). Moreover, if a triangular number $S_3(n)$ is a product of two consecutive integers, of which the least is a doubled triangular number, then $4S_3(n)+1$ and its square root is a sum of squares of two consecutive integers (see [Dick05]). Triangular numbers that are sums of two consecutive primes are 36, 78, 120, 210, 276, 300, 630, 946, 990, 1770, ... (Sloane's A111163), while triangular numbers, which are sums of three consecutive triangular numbers, are 10, 136, 1891, 26335, 366796, 5108806, 71156485, 991081981, 13803991246, ... (Sloane's A129803).

(84) These numbers are 4, 64, 361, 6241, 35344, ... (Sloane's A165516), and 100, 3364, 114244, 3880900, 131836324, ... (Sloane's A165518), respectively.

(85) For instance, there are exactly two triangular numbers 10 and 15 between two consecutive squares $S_4(3) = 9$ and $S_4(4) = 16$, while there is exactly one triangular number 6 between $S_4(2) = 4$ and $S_4(3) = 9$, as well as there is exactly one triangular number 21 between $S_4(4) = 16$ and $S_4(5) = 5$.

(86) They are $(15, 21)$: $S_3(5) + S_3(6) = 15 + 21 = 36 = S_3(8)$, $S_3(6) - S_3(5) = 21 - 15 = 6 = S_3(3)$. Similar pairs are $(105, 171) = (S_3(14), S_3(18))$, $(378, 703) = (S_3(27), S_3(37))$, $(780, 990) = (S_3(39), S_3(44)), \ldots$, $(1747515, 2185095) = (S_3(1869), S_3(2090))$. There are infinitely many pairs of triangular numbers, the sum and the difference of which are

triangular numbers. In fact, the system

$$\begin{cases} S_3(m) + S_3(2n) = \quad S_3(3n) \\ S_3(m) - S_3(2n) = S_3(n-1) \end{cases}$$

has infinitely many solutions (m, n) in positive integers. Each equation of the system is equivalent to the equation $m^2+m = 5n^2+n$. As the equation $u^2+u-5v^2-v = m^2+m-5n^2-n$ is true for $u = 161m+360n+116$, $v = 72m+61$, it follows, that if pair (m, n) is a solution of the equation above, then the pair (u, v) is also a solution of it. As the pair $(2, 1)$ is a solution, then one obtains infinitely many such pairs (see [Sier64]). However, it was proven, that there are no square numbers, the sum and the difference of which give square numbers.

(87) (a) For instance, it holds $S_4(2) = S_3(2) + 1$, and $S_4(4) = S_3(5) + 1$. A first series of solutions is given by

$$x = \frac{(2\sqrt{2}+1)(3+2\sqrt{2})^n + (2\sqrt{2}-1)(3-2\sqrt{2})^n}{4\sqrt{2}},$$

$$y = \frac{2\sqrt{2}+1)(3+2\sqrt{2})^n - (2\sqrt{2}-1)(3-2\sqrt{2})^n - 2}{4}.$$

For $n = 0$, we get $x = 1, y = 0$; for $n = 1$, it holds $x = 4, y = 5$. The next solutions can be obtained by the recurrent equation $x_n = 6x_{n-1} - x_{n-2}$, $y_n = 6y_{n-1} - y_{n-2} + 2$. A second series of solutions is obtained by using these formulas for negative n. In fact, $x_{-1} = 2, y_{-1} = -3$; $x_{-2} = 11, y_{-2} = -16$, etc.

(b) For instance, $3 \cdot S_3(1) = S_3(2)$, and $3S_3(5) = S_3(9)$. It is proved that the equation $3(x^2 + x) = y^2 + y$ has only the solutions $x = \frac{r+s}{4\sqrt{3}} - \frac{1}{2}$, $y = \frac{r-s}{4} - \frac{1}{2}$, where $r = \frac{(3\sqrt{3}+5)(2+\sqrt{3})^n}{2}$, $s = \frac{(3\sqrt{3}-5)(2-\sqrt{3})^n}{2}$, with $n = 0, \pm 1, \pm 2, \ldots$.

(c) It is known, that the equation $S_3(x) = p \cdot S_3(y)$ has infinitely many solutions if p is not a square. Let $2x = k-1$, $2y = z-1$. Then $k^2 - pz^2 = 1-p$. Let $k = \alpha + p\beta$, $z = \beta + \alpha$. Then the equation $\alpha^2 - p\beta^2 = 1$ has infinitely many solutions if p is not a square. If $p = r^2$, the problem has only a finite number of solutions if any. It is impossible if $r = 3, 4, 5, 7, 8, 9, 11, 12, 13, 15, 16, 17$. If $r = 4\lambda + 2$, $x = 4\lambda^2 2 + 4\lambda$, $y = \lambda$ is a solution.

(d) This equation is equivalent to the equation $\mu(2x+1)^2 - \nu(2y+1)^2 = \mu - \nu$, which can be rewrite as the Pell-like equation $u^2 - \mu\nu t^2 = \mu(\mu - \nu)$, $u = \mu(2x+1)$, $t = 2y+1$.

(e) For instance, $S_3(5) + S_3(6) = S_3(8)$, and $S_3(4) + S_3(9) = S_3(11)$. It can be solved by setting $y = z - \frac{xr}{s}$ and finding x rationally.

(f) For instance, $S_3(1) + S_3(2) = S_4(2)$, and $S_3(2) + S_3(3) = S_4(3)$.

(g) This equation has one and but one set of integral solutions. Solving for y we see that $8x + 8a + 9$ must be a square u^2; thus, x is integral only if $u^2 - 1 = 8\theta$, whence $\theta = \frac{t(t+1)}{2}$.

(h) For instance, $S_3(1) + S_3(5) = S_4(4)$, and $S_3(5) + S_3(6) = S_4(6)$. This equation is equivalent to the equation $(2x+1)^2 + (2y+1)^2 = (2z+1)^2 + (2z-1)^2$, which, by *Euler's formula for the product of two sums of two squares*, has the solution $2z+1 = ac+bd$, $2y+1 = bc-ad$, if $bc+ad = ac-bd+2$.

(i) For instance, $S_3(4) = S_4(1) + S_4(3)$, and $S_3(8) = S_4(3) + S_4(5)$.

(j) For instance, $S_3(1)+S_3(5) = S_4(2)+2\cdot S_3(3)$, and $S_3(2)+S_3(7) = S_4(5)+2\cdot S_3(2)$. The examples are also $x = 2s+1$, $y = 4s$, $b = 3s$, $a = s+1$; $x = 6s+2$, $y = 4s-1$, $b = 5s+1$, $a = s-1$.

(k) For instance, $S_4(8) + S_3(8) = S_4(10)$, and $S_4(800) + S_3(800) = S_4(980)$.

(l) For instance, $(S_2(3))^2 + (S_3(3))^2 = 2 \cdot (S_3(3))^2$, and $(S_3(6))^2 + (S_3(2))^2 = 2 \cdot (S_3(5))^2$. In general, consider $S_3(x) = \xi S_3(z)$, $S_3(y) = \nu S_3(z)$, where $(\xi, \nu) = (1, 1)$, $(7, 5)$, $(41, 29)$, $(239, 169)$.

(m) It is known that 0, 1, 6 are the only triangular numbers whose squares are triangular.

(n) For instance, $S_3(1) \cdot S_3(4) = S_3(4)$, and $S_3(2) \cdot S_3(5) = S_3(9)$. In general, it is satisfied if $px(y+1) = 2qz$, $qy(x+1) = p(z+1)$. The resulting values of z are equal if $((2q^2 - p^2)x + 2q^2)y = p^2x + 2pq$.

(o) For instance, $S_3(1) \cdot S_3(9) = S_3(2) \cdot S_3(5)$, and $S_3(8) \cdot S_3(4) = S_3(2) \cdot S_3(15)$. This equation is equivalent to $S_3(a)((2x+1)^2-1) = S_3(b)((2y^2+1)^2-1)$, which can be rewrite as the Pell-like equation $u^2 - S_3(a)S_3(b)t^2 = S_3(a)(S_3(a) - S_3(b)$, $u = S_3(a)(2x-1)$, $t = 2y+1$.

(p) For instance, $S_3(4) \cdot S_3(6) = S_3(4^2+4)$, and $S_3(9) \cdot S_3(13) = S_3(9^2+9)$. This equation can be reduced to $2S_3(x)+1 = S_3(y)$, giving the solutions $S_3(x) = 10, 45$, $S_3(y) = 21, 91$.

(q) For instance, $S_3(1) \cdot S_3(8) = (S_3(3))^2$, and $S_3(8) \cdot S_3(49) = (S_3(20))^2$. There exists a series for the above equation with the law of recurrence $z_{n+1} = 6z_n - z_{n-1} + 2$, $z_0 = 3$, $z_1 = 20$.

(r) For instance, $S_3(3) \cdot S_3(4) \cdot S_3(5) = S_4(30)$. In fact, this equation holds if $(x-1)(x+2) = 2z^2$, whence $u^2 - 8v^2 = 1$, where $2x+1 = 3u$, $z = 3v$. The solutions are known to be $u = 1, 3, 17, \dots$, $u_n = 6u_{n-1} - u_{n-2}$.

(s) It is easy to check that $y^3 \pm 1 = S_3(z)$ for $y = 1, 3, 16, 20$.

(t) It is known that 5 and 17 are the only integers whose cubes diminished by 13 are quadruples of triangular numbers.

(u) For instance, $S_3(7) - S_3(1) = C(3)$, and $S_3(8) - S_3(7) = C(3)$. This equation holds if $(y-z)(y+z+1) = 2x^3$, whence $2x^3$ is to be expressed as a product of two distinct factors, one even and one odd.

(v) For instance, $(S_3(69))^2 - (S_3(5))^2 = C(18)$. In general, it is true for $x, y = \frac{8m^4 \pm 12m^3 - 4m^2 - 1}{3}$.

(w) For instance, $2^4 - 1^4 = S_3(5)$, and $7^4 - 4^4 = S_3(65)$. In fact, this equation holds if $z = x^2 + y^2$, and $x^2 - 3y^2 = 1$, or if $z = \frac{x^2+y^2}{\lambda}$, $z + 1 = 2\lambda(x^2 - y^2)$, or vice versa, whence $(2\lambda^2 - 1)x^2 - (2\lambda^2 + 1)$ $y^2 = \pm\lambda$.

(x) For instance, $S_3(0) + S_3(1) = 1^5$, and $S_3(12) + S_3(43) = 4^5$. The equation $S_3(x) + S_3(y) = z^5$ is equivalent to $(2x+1)^2 + (2y+1)^2 = 2(4z^5+1)$, a necessary and sufficient condition for which is that every prime factor of $4z^5+1$ be of the form $4n+1$ (see [Dick05]).

(88) It is true for $n = 1$ and $n = 24$. In fact, the sum is $\frac{n(n+1)(2n+1)}{6}$. The case $n = 1$ is obvious. Let now $n = 6r^2 > 1$. Then the

condition gives that $(6r^2+1)(12r^2+1)$ is a square. Thus, $(9r^2+1)^2-(3r^2)^2$ is a square, so that $9r^2+1$ and $3r^2$ are the hypotenuse and one leg of a right triangle. Thus, the other leg is $9r^2-1$, whence $r=2$, and $n=24$. It is proven also that $1^2+\cdots+n^2=ky^2$ and $(S_3(x_1))^2+\cdots+(S_3(x_n))^2=ky^2$ are impossible if $k=2,3,6$ (see [Dick05]).

(89) It is easy to see that if $a+b+c=\alpha+\beta+\gamma=0$, then $P=(S_3(a)+S_3(b)+S_3(c))(S_3(\alpha)+S_3(\beta)+S_3(\gamma))$ is a sum of three triangular numbers. In fact, take $a=y-z$, $b=z-x$, $c=x-y$, $\alpha=\nu-\xi$, $\beta=\xi-\zeta$, $y=\zeta-\nu$, $X=x\zeta+z\nu+y\xi$, $Y=y\zeta+z\xi+z\xi$, $Z=y\zeta+x\nu+z\xi$, and get $P=S_3(Y-Z)+S_3(Z-X)+S_3(X-Y)$.

(90) Note that $S_m(2)=m$, and let $S_m(3)-S_m(2)=1+2(m-2)=r$. Then $S_m(n)-S_m(n-1)=1+(n-1)(m-2)=r+(n-3)(m-2)$. Now it is easy to show that $S_m(n+1)=m+\binom{n-1}{1}r+\binom{n-1}{2}(m-2)$. In fact, for $n=3$, one has $S_m(4)=S_m(3)+r+(m-2)=S_m(2)+(r+0\cdot(m-2))+(r+1\cdot(m-2))=m+2r+1\cdot(m-2)=m+\binom{2}{1}r+\binom{2}{2}(m-2)$. Then $S_m(n+1)=S_m(n)+r+(n-2)(m-2)=m+(\binom{n-2}{1}r+r)+(\binom{n-2}{2}+\binom{n-2}{1})(m-2)=m+\binom{n-1}{1}r+\binom{n-1}{2}(m-2)$. One can show now, that $S_m^3(n+1)=S_m^3(1)+\binom{n}{1}m+\binom{n}{2}r+\binom{n}{3}(m-2)$. In fact, for $n=4$, one has $S_m^3(5)=S_m(1)+\cdots+S_m(5)=S_m^3(1)+m+(m+r)+(m+2r+(m-2))+(m+3r+3(m-2))=S_m^3(1)+\binom{4}{1}m+\binom{4}{2}r+\binom{4}{3}(m-2)$. It implies $S_m^3(n+1)=S_m^3(n)+S_m(n+1)=S_m^3(1)+(\binom{n-1}{1}m+m)+(\binom{n-1}{2}r+\binom{n-1}{1}r)+(\binom{n-1}{3}(m-2)+\binom{n-1}{2}(m-2))=S_m^3(1)+\binom{n}{1}m+\binom{n}{2}r+\binom{n}{3}(m-2)$. Noting that $S_m^3(1)=1$, and $r=1+2(m-2)$, we obtain $S_m^3(n)=1+(n-1)m+\frac{(n-1)(n-2)}{2}(1+2(m-2))+\frac{(n-1)(n-2)(n-3)}{6}(m-2)=\frac{n(n+1)((m-2)n-m+5)}{6}$.

(91) This result can be obtained by the standard procedure, leading to the linear recurrent equation $TT(n+5)-5TT(n+4)+10TT(n+3)-10TT(n+2)+5TT(n+1)-TT(n)=0$.

(92) It is easy to see that $a^2-\frac{b(b+1)}{2}=2^{2n}$ for $a=2^{n+1}+2^n+1$, and $b=2^{n+2}+1$; for $a=2^{n+1}+2^n-1$, and $b=2^{n+2}-2$; for $a=\frac{9\cdot 2^n-2}{7}$, and $b=\frac{8(2^n-1)}{7}$.

(93) Since $\frac{2^{2m}-1}{3}$ is an integer N, then $\frac{2^{4m+1}-2^{2m}-1}{9} = \frac{(2N)(2N+1)}{2}$ is triangular.

(94) These numbers are 3, 6, 66, 5995, 15051, ... (Sloane's A050722), with corresponding indices 2, 3, 11, 109, 173, ... (Sloane's A050721).

(95) In fact, $1+2+3+10+11+18+34+36+77+109+132+173+363 = 969$. Moreover, the last index 363 is itself palindromic and can be expressed as the sum of consecutive powers of the base 3: $363 = 3^1 + 3^2 + 3^3 + 3^4 + 3^5$.

(96) The palindromic triangular numbers are 1, 3, 6, 55, 66, 171, 595, 666, 3003, 5995, 8778, 15051, 66066, 617716, 828828, ... (Sloane's A003098). All even such number is divisible by 11. This number 11 is itself a palindromic index. And 11 is the only existing palindromic prime with an even number of digits.

(97) It can be checked by direct computation. Thus, $P(2554554552) = 6525748961698475256$, and $P(25545544554552) = 652574846588626885648475256$.

(98) These numbers are 1 8, 343, 1331, 1030301 (Sloane's A135067), and 4, 8, 9, 121, 343, 1331, 10201, 94249, 1030301, 900075181570009, ... (Sloane's A076703), respectively.

(99) These numbers are 0, 1, 8, 343, 1000, 1331, 8000, 343000, 1000000, 1030301, ... (Sloane's A061458).

(100) In fact, the numbers 1, 4, and 9 are palindromes themselves. The number 16 (as well as 25, 36, 81, 100, 121, and 144) becomes palindromic after one iteration: $16 + 61 = 77$. The number 49 (as well as 64 and 169) becomes palindromic after two iterations: $49 + 94 = 133$; $133 + 331 = 464$. However, the next square number 196 does not yield a palindrome even after 700 000 000 iterations. It is a smallest candidate **Lychrel number**: a natural number which can not form a palindrome through the above iterative process, which is called 196-*algorithm*. The first few candidate Lychrel numbers are 196, 295, 394, 493, 592, 689, 691, 788, 790, 879, ... (Sloane's A023108).

(101) The beast number $666 = 2 \cdot 3^2 \cdot 37$ is an example of Smith number, since $6+6+6 = 2+3+3+(3+7) = 18$. The first few Smith numbers are 4, 22, 27, 58, 85, 94, 121, 166, 202, 265, ... (Sloane's A006753). The corresponding digits sums are 4, 4, 9, 13, 13, 13, 4, 13, 4, 13, ... (Sloane's A050218). It is proven that there are an infinite number of Smith numbers. A Smith number can be constructed from every factored repunit R_n. The largest known Smith number is $9 \cdot R_{1031}(10^{4594} + 3 \cdot 10^{2291} + 1) \cdot 10^{3913210}$ (see [Weis11]).

(102) The corresponding computations for squares and cubes are given in the tables below.

squares	1		4		9		16		25		36		49	...
first differences		3		5		7		9		11		13		...
second differences			2		2		2		2		2			...

cubes	1		8		27		64		125		216		343	...
first differences		7		19		37		61		91		127		...
second differences			12		18		24		30		36			...
tird differences				6		6		6		6				...

(103) These numbers are 1, 4, 9, 25, 121, 144, 729, 841, 5041, 5184, 45369, 46225, 363609, 403225, 3674889, 1401602635449 (Sloane's A025494, probably finite), 1, 8, 27, 32, 128, 729 (Sloane's A051760), and 0, 1, 4, 8, 9, 25, 27, 32, 121, 128, 144, 729, 841, 5041, 5184, 45369, 46225, 363609, 403225, 3674889, 1401602635449 (Sloane's A051761), respectively.

(104) It follows from the formula $L(n,k) = \binom{n-1}{k-1}\frac{n!}{k!}$.

(105) It can be checked by direct computation: $2S_3(2) - 1 = 5$, $2S_3(3) - 1 = 11$, etc.

(106) It is proven that 9 is the only such number (see [Dick05]).

(107) The corresponding 3×3 square is given below.

61	67	71
67	71	73
71	73	79

(108) The corresponding square-representation of the prime 733353337 is given below.

$$\begin{matrix} 7 & 3 & 3 \\ 3 & 5 & 3 \\ 3 & 3 & 7 \end{matrix}$$

(109) It can be checked by direct computation.

(110) The rotation of the triangle below gives primes 100153, 150301, and 305101.

$$\begin{array}{ccccc} & & 1 & & \\ & 0 & & 0 & \\ 1 & & 5 & & 3 \end{array}$$

(111) These numbers are $1111111_3 = 1093 \in P$, $1111111_5 = 19531 \in P$, and $1111111_6 = 55987 \in P$. Their hex-representation is given below.

$$\begin{array}{ccccc} & 1 & & 1 & \\ 1 & & 1 & & 1 \\ & 1 & & 1 & \end{array}$$

(112) These squares, with magic constants $213 \in S$, and $233 \in P$, are given below.

41	79	17	13	61
53	03	83	67	07
59	97	05	23	29
11	31	37	89	43
47	02	71	19	73

41	11	79	19	83
31	67	29	89	17
61	59	05	71	37
97	53	13	47	23
03	43	107	07	73

(113) An illustration of the Conway's rule is given below.

0	1	3	6	10	15	21	28	36	45	55	66	78	91	105	120	136	153	...
0	–	1	2	–	5	7	–	12	15	–	22	26	–	35	40	–	51	...
		−	−		+	+		−	−		+	+		−	−		+	

Then the *Euler's pentagonal number theorem* is easy to remember: $p(n-0) - p(n-1) - p(n-2) + p(n-5) + p(n-7) - p(n-12) - p(n-15) + \cdots = 0^n$, where $p(n)$ is the partition function, the left side terminates before the argument becomes negative, while $0^n = 1$ if $n = 0$, and $0^n = 0$ if $n > 0$. Thus, $p(0) = 1$, while $p(7) = p(7-1) + p(7-2) - p(7-5) - p(7-7) + 0^7 = 11 + 7 - 2 - 1 + 0 = 15$.

(114) The first identity was proven by Jacobi, 1829. The other relations were proven by Berger, 1898 (see [Dick05]).

(115) In fact, $n \cdot n - n = n(n-1) = P(n-1)$.

(116) It holds since $gcd(n, n+1) = 1$.

(117) It can be obtained by direct computation.

(118) These numbers for $n = 1, 2, 3, \ldots, 10, \ldots$ are 1, 2, 3, 1, 2, 3, 4, 2, 1, 2, ... (Sloane's A002828).

(119) These numbers are 1, 2, 4, 5, 8, 9, 10, 13, 16, 17, ... (Sloane's A001481).

(120) In fact, $188 = 1 + 4 + 9 + 25 + 36 + 49 + 64$.

(121) These numbers are 2, 50, 325, 1105, 8125, 5525, 105625, 27625, 71825, 138125, ... (Sloane's A016032).

(122) These numbers for $n = 1, 2, \ldots, 10, \ldots$ are 1, 1, 1, 2, 2, 2, 2, 3, 4, 4, ... (Sloane's A001156).

(123) These values are 1, 4, 4, 0, 4, 8, 0, 0, 4, 4, ... (Sloane's A004018). For instance, $r_2(5) = 8$, as $5 = 1^2 + 2^2 = 2^2 + 1^2 = (-1)^2 + 2^2 = 2^2 + (-1)^2 = 1^2 + (-2)^2 = (-2)^2 + 1^2 = (-1)^2 + (-2)^2 = (-2)^2 + (-1)^2$.

(124) These numbers are 2, 6, 10, 14, 18, 22, 26, 30, 34, 38, ... (Sloane's A016825). They have the form $4k + 2$. In fact, any odd number $n = ab$ can be represented as $\left(\frac{a+b}{2}\right)^2 - \left(\frac{a-b}{2}\right)^2$. Futhermore, if $n = x^2 - y^2$, then $4n = (2x)^2 - (2y)^2$. But a number $4k + 2$ can not be represented in the form $x^2 - y^2$, since this difference has the resudue 0, 1, or 3 modulo 4.

(125) These numbers are 10, 34, 58, 85, 91, 130, 214, 226, 370, 526, ... (Sloane's A020495).

(126) These numbers are 1, 2, 3, 4, 5, 6, 7, 1, 2, 3, 4, 5, 6, 7, 8, 2, ... (Sloane's A002376).

(127) These numbers are 1, 1, 1, 1, 1, 1, 1, 2, 2, 2, 2, 2, 2, 2, 2, 3, 3, 3, 3, 3, 3, 3, 3, 4, 4, 4, 5, 5, 5, 5, ... (Sloane's A003108).

(128) In fact, $6x = (x+1)^3 + (x-1)^3 - x^3 - x^3$, $6x+3 = x^3 + (-x+4)^3 + (2x-5)^3 + (-2x+4)^3$, $18x+1 = (2x+14)^3 + (-2x-23)^3 + (-3x-26)^3 + (3x+30)^3$, $18x+7 = (x+2)^3 + (6x-1)^3 + (8x-2)^3 + (-9x+2)^3$, $18x+8 = (x-5)^3 + (-x+14)^3 + (-3x+29)^3 + (3x-30)^3$, $54x+2 = (29484x^2 + 2211x + 43)^3 + (-29484x^2 - 2157x - 41)^3 + (9828x^2 + 485x + 4)^3 + (-9828x^2 - 971x - 22)^3$, $54x + 20 = (3x - 11)^3 + (-3x + 10)^3 + (x + 2)^3 + (-x + 7)^3$, $216x - 16 = (14742x^2 - 2157z + 82)^3 + (-14742x^2 + 2211x - 86)^3 + (4914x^2 - 971x + 44)^3 + (-4914x^2 + 485x - 8)^3$, $216x+92 = (3x - 164)^3 + (-3x + 160)^3 + (x - 35)^3 + (-x + 71)^3$. These identities show that all positive integers, excluding numbers of the forms $9x \pm 4$, and $108x \pm 38$, can be represented as a sum of four signed cubes ([Demj66]).

(129) For instance, $30 = (-283059965)^3 + (-2218888517)^3 + 2220422932^3$.

(130) In fact, $1 = n^3 + (-n)^3 + 1^3$ for any integer n. Therefore, 1 has infinitely many representations as a sum of three signed cubes.

(131) In fact, $1729 = \overline{C}(10) = S_{12}(19) = S_{24}(13) = S_{84}(7)$.

(132) In fact, $1 + 7 + 2 + 9 = 19$, and $19|1729$. It also has this property in octal ($1729 = 3301_8$, $3+3+0+1 = 7$, $7|1729$) and hexadecimal, but not in binary. 1729 has another interesting property: the 1729-th decimal place is the beginning of the first occurrence of all ten digits consecutively in the decimal representation of the transcendental number e.

(133) In fact, $1+7+2+9 = 19$, and $19 \cdot 91 = 1729$. It is the smallest such product that's one away from a third or higher power: $19 \cdot 91 = 12^3 + 1$. The number 1729 is one of four such positive integers; the others are 1, 81 and 1458 (Sloane's A110921).

(134) These numbers are 1, 2, 3, 4, 5, 6, 7, 8, 9, 10, 11, 12, 13, 14, 15, 1, 2, 3, 4, 5, ... (Sloane's A002377).

(135) For instance, in the case of hexagonal numbers, it holds $11 = 1+1+1+1+1+6$, and $26 = 1+1+6+6+6+6$. Note, that 11 and 26 are the only numbers, which need for the representation six hexagonal numbers.

(136) For instance, the number $2m - 1$ can be represented only in the form $2m - 1 = m + (m-1) = S_m(2) + (m-1)S_m(1)$, that requires exactly m non-zero m-gonal numbers.

(137) For instance, $49n + 19 = S_3(x) + S_3(y)$ would imply $(2x+1)^2 + (2y+1)^2 = 8(49n+19) + 2$, whereas the factor 7 of the right side is not a divisor of a sum of two squares.

(138) These numbers for $n = 1, 2, \ldots 10, \ldots$ are 1, 2, 3, 1, 2, 3, 4, 2, 3, 1, ... (Sloane's A104246).

(139) In fact, $S_3^3(N) - 2S_3^3(N-1) + S_3^3(N-2) = (S_3^3(N-1) + S_3(N)) - S_3^3(N-1) - (S_3^3(N-2) + S_3(N-1)) + S_3^3(N-2) = S_3(N) - S_3(N-1) = N$.

(140) For instance, $15 = S_4(3) + 6 \cdot S_4(1) = S_3(4) + 5 \cdot S_3(1) = S_3^3(1) + S_3^3(2) + S_3^3(3) = 2 \cdot O(2) + 3 \cdot O(1) = C(2) + 7 \cdot C(1) =$

$I(2)+3\cdot I(1) = 15\cdot D(1) = (S_3(2))^2+6\cdot(S_3(1))^2$. It was stated for any positive integer by Pollock (see [Dick05]).

(141) It holds for every positive integer up to 50. However, for $n = 4$ and $a = 1$ we get the series $1, 5, 15, 35, 70, \ldots$ of polytope numbers, and $a + 2n - 2 = 7$, whereas 8 terms are evidently required to produce the sum 64, since four terms must be unity (see [Dick05]).

(142) In fact, $1 = S_3(1)$ is the smallest number of index 1; $2 = S_3(1) + S_3(1)$ is the smallest number of index 2; $5 = S_3(2) + S_3(1) + S_3(1)$ is the smallest number of index 3; $20 = S_3(5) + S_3(2) + S_3(1) + S_3(1)$ is the smallest number of index 4; $230 = S_3(20) + S_3(5) + S_3(2) + S_3(1) + S_3(1)$ is the smallest number of index 5. It can be shown that, if N_m is the least number of index m, then $N_m = \frac{1}{2}N_{m-1}(N_{m-1} + 3)$. So, it holds $2^{m-1}N_m = (N_1 + 3)(N_2 + 3)\cdots(N_{m-1} + 3)$.

(143) It proves that an algebraic discussion of the theorem, that any number n is a sum of three triangular numbers, is of no help, since the theorem is not true if n is fractional, in particular, if $n = \frac{1}{2}, \frac{3}{2}, \frac{5}{2}, \frac{7}{2}$. Thus, for $\frac{1}{2}$ it is true since 7 is not a sum of three odd squares.

(144) It is easy to see that $N = \sum_{i=1}^{4} \frac{3x_i^2 - x_i}{2}$ iff $24N + 4 = \sum_{i=1}^{4}(6x_i - 1)^2$. So, it is sufficient to represent $24N + 4$ as a sum of four squares $\sum_{i=1}^{4} a_i^2$ and take $x_i = \frac{a_i+1}{6}$.

(145) For instance, $15 = 2\cdot S_3(2) + S_4(3) = S_4(2) + S_3(1) + S_3(4) = S_4(1) + S_4(2) + S_3(4) = S_4(2) + S_4(3) + 2S_3(1) = S_4(3) + 2\cdot S_3(2) + S_3(0) = S_4(1) + 2\cdot S_4(2) + S_3(3) = S_3(1) + 2\cdot S_4(2) + 2\cdot S_3(2) = S_4(1) + 2\cdot S_4(2) + 2\cdot S_3(2) = S_3(1) + S_3(1) + S_3(2) + S_3(4) = S_3(2) + S_3(2) + S_3(2) + S_3(3) = S_3(0) + S_3(2) + 2\cdot S_3(3)$. It is easy to show that these relations hold for any positive integer. For example, it is known that $4N + 1 = a^2 + b^2 + c^2$. Then a and b are even and c is odd, so $4N + 1 = 4x^2 + 4y^2 + 4z(z+1) + 1$, and $N = x^2 + y^2 + z(z+1)$. Therefore, any positive integer N can be represented in the form $S_4(x) + S_4(y) + 2S_3(z)$. Replacing N by $2M$, we get $2M = x^2 + y^2 + 2\frac{z(z+1)}{2}$; so, x and y are even, and $2M = 4u^2 + 4v^2 + 2\frac{z(z+1)}{2}$. Hence, it holds

$M = u^2 + v^2 + \frac{z(z+1)}{2}$, i.e., any positive integer M can be represented as $S_4(u) + S_4(v) + S_3(z)$.

(146) It is proven that the only forms $aS_3(x) + bS_3(y) + cS_3(z)$, $a, b, c \in \mathbb{N}$, which represent all numbers, are $S_3(x) + S_3(y) + cS_3(z)$, $c = 1, 2, 4, 5$, and $S_3(x) + 2S_3(y) + dS_3(z)$, $d = 2, 3, 4$. That conversely, each of these seven forms represents all numbers, can be proven by use of the Legendre's theorem that any $N \neq 4^{\alpha}(8m + 7)$ is a sum of three squares (see [Dick05]).

The case $c = 1$ is just Gauss' three-triangular-number theorem. Next, using Legendre's three-square theorem, we get a representation $2(2N + 1) = 4u^2 + (2t + 1)^2 + (2z + 1)^2$, and, hence, the representation $8N + 4 = (2u + 2t + 1)^2 + (2t - 2u + 1)^2 + 2(2z + 1)^2$. Therefore, it holds $N = S_3(u + t) + S_3(t - u) + 2S_3(z)$, that proves the case $c = 2$. Furthermore, using Legendre's three-square theorem, we get a representation $8N + 6 = (2x + 1)^2 + (2y + 1)^2 + 4(2z + 1)^2$. Hence, it holds $N = S_3(x) + S_3(y) + 4S_3(z)$, that proves the case $c = 4$. Next, using Legendre's three-square theorem, we get a representation $8N + 5 = (2x + 1)^2 + 4(2s + 1)^2 + 16t^2 = (2x + 1)^2 + 2(2s + 1 + 2t)^2 + 2(2s + 1 - 2t)^2$. Therefore, it holds $N = S_3(x) + 2S_3(s + t) + 2S_3(s - t)$, that proves the case $d = 2$. Finally, as was shown by Gauss, $8N + 7 = S_4(x) + S_4(y) + 2S_4(z) = (2x + 1)^2 + 4(2z + 1)^2 + 2(2y + 1)^2$, and it holds $N = S_3(x) + 2S_3(y) + 4S_3(z)$, that proves the case $d = 4$. The proofs for the remaining cases $c = 5$ and $d = 3$ are longer.

(147) For instance, $4 = 4 \cdot S_3(1)$, $5 = S_3(2) + 2 \cdot S_3(1) + S_3(0)$, and $1+1+1+1 = 2+1+1+0 = 4$; $7 = 2 \cdot S_3(2) + S_3(1) + S_3(0)$, $8 = S_3(3) + 2 \cdot S_3(1) + S_3(0)$, $9 = S_3(3) + S_3(2) + 2 \cdot S_3(0)$, and $2 + 2 + 1 + 0 = 3 + 1 + 1 + 0 = 3 + 2 + 0 + 0 = 5$; $11 = S_3(1)+S_3(1)+S_3(2)+S_3(3)$, $12 = S_3(3)+2S_3(2)+S_3(0)$, $13 = 2 \cdot S_3(3)+S_3(1)+S_3(0)$, $14 = S_3(4)+S_3(2)+S_3(1)+S_3(0)$, and $1+1+2+3 = 3+2+2+0 = 3+3+1+0 = 4+2+1+0 = 7$.

(148) (a) For instance, $S_4(4) + S_3(1) + S_3(7) = S_4(3) + S_3(8) + S_3(0)$. It holds since $S_4(x) + S_3(y) + S_3(z) = S_4(\frac{z-y}{2}) +$

$S_3(\frac{y+z}{2} - x) + S_3(\frac{y+z}{2} + x)$ for y, z both even or both odd, with a similar identity if one is even and the other odd.

(b) It holds since $(4a+1)^2 + (4b+1)^2 + (4c+1)^2 + (4d+1)^2 = 4(a+b+c+d+1)^2 + 4(a-b-c+d)^2 + 4(a-b+c-d)^2 + 4(a+b-c-d)^2$.

(149) For instance, $15 = 3S_3(2) = S_4(3) + 2 \cdot S_3(2) = S_3(6) + S_4(3) = 3 \cdot S_5(2) + 3 \cdot S_5(0)$; $25 = S_4(2) + S_4(3) + 2S_3(3) = 2S_3^3(3) + S_3^3(2) + S_3^3(1)$.

In general, every triangular number except 1 and 6 is a sum of three non-zero triangular numbers, since it holds $S_3(3n-1) = 2S_3(2n-1) + S_3(n)$, $S_3(3n) = 2S_3(2n) + S_3(n-1)$, $S_3(3n+1) = S_3(2n) + S_3(2n+1) + S_3(n)$. Furthermore, every triangular number is a sum of a square and two equal triangular numbers, since it holds $S_3(2n) = 2S_3(n) + S_4(n)$, $S_3(2n+1) = 2S_3(n) + S_4(n+1)$. Moreover, it holds $6(2n+1)^2 = (6x\mp1)^2 + (6y\mp1)^2 + 4(6z\mp1)^2$, whence $\frac{n(n+1)}{2} = \frac{3x^2\mp x}{2} + \frac{3y^2\mp y}{2} + 4(\frac{3z^2\mp z}{2})$. So, every triangular number is a sum of six (generalized) pentagonal numbers.

Since $S_4(n) = S_3(n) + S_3(n-1)$, one can use the above formulas for triangular numbers in order to obtain the similar relations for square numbers. In fact, it holds $S_4(3n) = S_4(n) + 2S_4(2n)$, $S_4(3n+1) = S_4(n) + S_4(2n+1) + 2S_3(2n)$, $S_4(3n+2) = S_4(n+1) + S_4(2n+1) + 2S_3(2n+1)$. Therefore, every square number is a sum of three squares, or a sum of two squares and two triangular numbers. Moreover, it is conjectured, that any perfect square is a sum of at most four tetrahedral numbers. This conjecture is checked for all squares up to 10^6.

(150) In fact, $2^3 = 1 + 1 + 6$; $3^4 = 55 + 15 + 10 + 1$; $5^5 = 2850 + 210 + 45 + 2 \cdot 10 = 3003 + 105 + 10 + 6 + 1$. It is stated that every n-th power is a sum of n non-zero triangular numbers (see [Dick05]).

(151) For instance, it is easy to see that if $4N + 1 = x^2 + y^2$, then $8N + 2 = 2x^2 + 2y^2 = (x-y)^2 + (x+y)^2$, where the numbers x and y have different parity. Therefore, $8N + 2$ is a sum of two odd squares $(2u+1)^2$ and $(2v+1)^2$, $2u+1 = x-y$, $2v+1 = x+y$.

It implies that $N = \frac{u(u+1)}{2} + \frac{v(v+1)}{2} = S_3(u) + S_3(v)$, $u = \frac{x-y-1}{2}$, $v = \frac{x+y+1}{2}$. Conversely, this expression for N implies $4N+1 = (u+v+1)^2 + (u-v)^2$.

(152) It can be checked by direct computation.

(153) It was proven, that any prime number of the form $8N+1$ can be represented as $x^2 + 2y^2$. It is easy to check that x is odd and y is even. So, it holds $8N+1 = (4u+1)^2 + 2(2v)^2 = 4u(u+1) + 8v^2 + 1$, and hence $N = \frac{u(u+1)}{2} + v^2$, i.e., $N = S_3(u) + S_4(v)$. Similarly, any prime number of the form $8N+3$ can be represented as $x^2 + 2y^2$. In this case both, x and y, are odd, and we get $8N+3 = (2u+1)^2 + 2(2v+1)^2 = 4u(u+1) + 8v(v+1) + 3$. Therefore, it holds $N = \frac{u(u+1)}{2} + 2\frac{v(v+1)}{2}$, i.e., $N = S_3(u) + 2S_3(v)$. Finally, it is known that any prime of the form $4N+1$ can be represented as $x^2 + y^2$. It is easy to see that numbers x and y have different parity, i.e., it holds $4N+1 = (2u+1)^2 + (2v)^2 = 4u(u+1) + 4v^2 + 1$. So, we get $N = 2\frac{u(u+1)}{2} + v^2$, i.e., $N = 2S_3(u) + S_4(v)$.

(154) Similarly, the Gauss three-triangular-number theorem would follow if we could prove that in the decomposition

$$\frac{1}{(1-z)(1-xz)(1-x^3z)(1-x^6z)\cdots} = 1 + Pz + Qz^2 + Rz^3 + \cdots$$

all integers occur as exponents of x in the series for R.

(155) For instance, in the case of $N = 2$ we can use only two triangular numbers, $S_3(0) = 0$, and $S_3(1) = 1$, both exactly twice. Then we get $2 = 0+0+1+1 = 0+1+0+1 = 0+1+1+0 = 1+0+0+1 = 1+0+1+0 = 1+1+0+0$, i.e., there are exactly $6 = \sigma(2 \cdot 2 + 1)$ ways to represent 2 as a sum of four triangular numbers. Legendre, 1828, concluded this fact from the formula $(1+q+q^3+q^6+q^{10}+\cdots)^4 = \frac{1}{1-q} + \frac{3q}{1-q^3} + \frac{5q^2}{1-q^5} + \cdots$. He gave also an identity which shows the number of ways N is a sum of eight triangular numbers. Plana, 1863, wrote

$(1+q+q^3+q^6+q^{10}+\cdots)^4 = 1+\sum_{n=1}^{\infty} q^n\sigma(2n+1)$ for the left member of the above formula, by expanding the second member as a power series in q and examining the earlier terms (see [Dick05]).

Bibliography

Andrews G. E. *The Theory of Partitions*, Cambridge, England: Cambridge University Press, 1998.

Andrews W. S. *Magic Squares and Cubes*, 2-nd rev. ed., New York: Dover, 1960.

Ankeny N. C. *Sums of Three Squeres*, Proc. Amer. Math. Soc., 8, 1957.

Avanesov E. T. *Solution of a Problem on Figurate Numbers*, Acta Arithmetica, 12, 1967.

Balasubramanian R., Deshouillers J.-M. and Dress F. *Problème de Waring pour les Bicarrés*, C. R. Acad. Sci., Paris Ser. I Math., 303, 1986.

Ball W. W. R. and Coxeter H. S. M. *Mathematical Recreations and Essays*, 13-th ed., New York: Dover, 1987.

Ballew D. and Weger R. *Repdigit Triangular Numbers*, J. Rec. Math., 8(2), 1976.

Beukers F. *On Oranges and Integral Points on Certain Plane Cubic Curves*, Nieuw Arch. Wisk., 6, 1988.

Brocard H. *Question 166*, Nouv. Corres. Math., 2, 1876.

Buchstab A. A. *Number Theory*, Moscow, 2009.

Cantor M. *Die Romischen Agrimensoren*, Leipzig, 1875.

Catalan E. *Note Extraite d'une Lettre Adressée à l'Éditeur*, J. Reine Angew. Math., 27, 1844.

Cauchy A. *Demonstration du Theoreme General de Fermat sur les Nombres Polygones*, Mém. Sci. Math. Phys. Inst. France (1), 14, 1813–15.

Chen J.-R. *Waring's Problem for* $g(5) = 37$, Sci. Sinica, 13, 1964.

Chen Y.-G. and Fang J.-H. *Triangular Numbers in Geometrical Progression*, Electronic Journal of Combinatorial Number Theory (♯ A19), 7, 2007.

Clark P. L. *A Theorem of Minkowski; the Four Square Theorem*, http://math.uga.edu/ pete/4400Minkowski.pdf.

Cohn J. H. E. *On Square Fibonacci Numbers*, J. London Math. Soc., 39, 1964.

Conway J. H. and Guy R. K. *The Book of Numbers*, New York: Springer-Verlag, 1996.

Coxeter H. S. M. *Regular Polytopes*, 3-rd ed., New York: Dover, 1973.

Davenport H. *On Waring's Problem for Fourth Powers*, Ann. Math., 40, 1939.

Demjanenko V. *On Sums of Four Cubes*, Izvesti Vischik Outchetsnik Zavedenii Math., 54, 1966.

Deza E. and Model D. *Elements of Discrete Mathematics*, Moscow: URSS, 2010.
Dickson L. E. *All Positive Integers are Sums of Values of a Quadratic Function of x*, Bull. Amer. Math. Soc., 33, 1927.
Dickson L. E. *Generalizations of the Theorem of Fermat and Cauchy on Polygonal Numbers*, Bull. Amer. Math. Soc., 34 (1), 1928.
Dickson L. E. *Extended Polygonal Numbers*, Bull. Amer. Math. Soc., 34 (2), 1928.
Dickson L. E. *History of the Theory of Numbers*, New York: Dover, 2005.
Diophantus *Arithmetics and the Book of Polygonal Numbers*, Moscow: Nauka, 1974.
Duke W. and Schultze-Pillot R. *Representations of Integers by Positive Ternary Quadratic Forms and Equidistribution of Lattice Points on Ellipsoids*, Invent. Math., 99, 1990.
Edwards H. M. *Fermat's Last Theorem: A Genetic Introduction to Algebraic Number Theory*, New York: Springer-Verlag, 1977.
Elkies N., Kuperberg G., Larsen M. and Propp J. *Alternating-Sign Matrices and Domino Tilings*, J. of Alg. Comb., 1, 1992.
Erdös P. and Obláth R. *Über diophantische Gleichungen der Form $n! = x^p \pm y^p$ und $n! \pm m! = x^p$*, Acta Szeged., 8, 1937.
Euler L. *Evolutio producti infiniti $(1-x)(1-x^2)(1-x^3)(1-x^4)$ etc. in Seriem Simplicem*, Acta Acad. Scient. Imp. Petrop. 1780, Opera Omnia, Series Prima, 3, 1783.
Franklin F. *Sur le développement du produit infini $(1-x)(1-x^2)(1-x^3)$...*, Comptes Rendus, 82, 1881.
Gardner M. *Second Book of Mathematical Puzzles and Diversions*, New York, 1961.
Gardner M. *Mathematical Carnival: A New Round-Up of Tantalizers and Puzzles from Scientific American*, New York: Vintage Books, 1977.
Gardner M. *Time Travel and Other Mathematical Bewilderments*, New York: Freeman, 1988.
Gauss C. F. *Disquisitiones Arithmeticae*, Leipzig, 1801.
De Geest P. *Palindromic Numbers and Other Recreational Topics*, http://www.worldofnumbers.com/
Graham R. L., Knuth D. E. and Patashnik O. *Concrete Mathematics: A Foundation for Computer Science*, 2-nd ed., Reading, MA: Addison-Wesley, 1994.
Guy R. K. *Every Number Is Expressible as the Sum of How Many Polygonal Numbers?*, Amer. Math. Monthly, 101, 1994.
Guy R. K. *Unsolved Problems in Number Theory*, 2-nd ed., New York: Springer-Verlag, 1994.
Hardy G. H. *Ramanujan: Twelve Lectures on Subjects Suggested by His Life and Work*, 3-rd ed., New York: Chelsea, 1999.
Hardy G. H. and Littlewood J. E. *Some Problems of Partitio Numerorum (VI): Further Researches in Waring's Problem*, Math. Z., 23, 1925.
Hardy G. H. and Wright E. M. *An Introduction to the Theory of Numbers*, 5-th ed., Oxford, England: Clarendon Press, 1979.

Heath T. L. *Diophantus of Alexandria: A Study in the History of Greek Algebra*, General Books LLC, 2010.

Hoggatt V. E. Jr. *The Fibonacci and Lucas Numbers*, Boston, MA: Houghton Mifflin, 1969.

Honsberger R. *Mathematical Gems III*, Washington, DC: Math. Assoc. Amer., 1985.

Honsberger R. *More Mathematical Morsels*, Washington, DC: Math. Assoc. Amer., 1991.

Jones Ch. and Lord N. *Characterising Non-trapezoidal Numbers*, Math. Gazette, 83 (497), 1999.

Karatsuba A. A. *Basic Analytic Number Theory*, Moscow: Nauka, 1983.

Kempner A. *Bemerkungen zum Waringschen Problem*, Math. Ann., 72, 1912.

Kim H. K. *On Regular Polytope Numbers*, Proc. of Amer. Math. Soc., 131 (1), 2003.

Knuth D. *Mathematics and Computer Science. Coping with Finiteness*, Science, 1976.

Kubina J. M. and Wunderlich M. C. *Extending Waring's Conjecture to 471600000*, Math. Comput., 55, 1990.

Lagrange J. L. *Démonstration d'un Téorème d'Arithmétique*, Nouveaux Mémoires de l'Acad. Royale des Sci. et Belles-L. de Berlin, 1770.

Landau E. *Handbuch der Lehre von der Verteilung der Primzahlen*, Leipzig, Berlin: Teubner, 1909.

Legendre A.-M. *Théorie des Nombres*, 3-th ed., Paris: Chez Firmin Didot Frères, Libraires, 1830.

Legendre A.-M. *Théorie des Nombres*, 4-th ed., Paris: A. Blanchard, 1979.

Le Lionnais F. *Les Nombres Remarquables*, Paris: Hermann, 1983.

Luca F. *Fermat Numbers in the Pascal Triangle*, Divulgac. Matem., 9 (2), 2001.

Lucas É. *Question 1180*, Nouv. Ann. Math., Ser. 2, 14, 1875.

Madachy J. S. *Madachy's Mathematical Recreations*, New York: Dover, 1979.

McDaniel W. L. *Pronic Fibonacci Numbers*, Fib. Quart., 36, 1998.

McDaniel W. L. *Pronic Lucas Numbers*, Fib. Quart., 36, 1998.

Meyl A.-J.-J. *Solution de Question 1194*, Nouv. Ann. Math., 17, 1878.

Ming L. *On Triangular Fibonacci Numbers*, Fib. Quart., 27(2), 1989.

Ming L. *On Triangular Lucas numbers*, in *Applications of Fibonacci Numbers*, Vol. 4, Kluwer Academic Pub., 1991.

Ming L. *Nearly Square Numbers in the Fibonacci and Lucas Sequences*, J. Chongqing Teachers College, 4, 1995.

Ming L. *Pentagonal Numbers in the Lucas Sequence*, Portug. Math., 533, 1996.

Minkowski H. *Geometrie der Zahlen*, Leipzig, Berlin: Teubner, 1911.

Minkowski H. *Diophantische Approximationen*, 2-nd ed., Leipzig, Berlin: Teubner, 1927.

Mozer L. *An Introduction to the Theory of Numbers*, The Trillia Group West Lafayette, IN, 2009.

Nagell T. *Introduction to Number Theory*, New York: Wiley, 1951.

Nagell T. *The Diophantine Equation* $x^2 + 7 = 2^n$, Ark. Mat., 30, 1961.

Natanson M. B. *A Short Proof of Cauchy's Polygonal Number Theorem*, Proc. Amer. Math. Soc., 99, 1987.

Nicomachus of Geraza *Introduction in Arithmetic*, translated by Martin Luther D'Ooge, Macmillan, 1926.

Olds C.D., Lax A. and Davidoff G. P. *The Geometry of Numbers*, Washington, DC: Math. Assoc. Amer., 2000.

Pascal B. *Traité du Triangle Arithmétique, Avec Quelques Autres Petits Traitez Sur la Mesme Matière*, Paris, 1654.

Pepin T. *Démonstration du Théorème de Fermat sur les Nomhres Polygones*, Atti Accad. Pont. Nuovi Lincei, 46, 1892–93.

Pietenpol J. L., Sylwester A. V., Just E. and Warten R. M. *Elementary Problems and Solutions: E 1473, Square Triangular Numbers*, Amer. Math. Month. (Math. Assoc. Amer.), 69 (2), 1962.

Pillai S. S. *On Waring's problem* $g(6) = 73$, Proc. Indian Acad. Sci., 12A, 1940.

Plouffe S. *1031 Generating Functions and Conjectures*, Université du Québec à Montréal, 1992.

Plutarch *Platonic questions*, in *The Morals*, Vol. 5, Boston: Little, Brown, and Co., 1878.

Post J. *π: Math Pages of Jonathan Vos Post*, http://www.magicdragon.com/math.html.

The Prime Pages (prime number research, records and resources), http://primes.utm.edu/

Rademacher H. and Toeplitz O. *The Enjoyment of Mathematics: Selections from Mathematics for the Amateur*, Princeton, NJ: Princeton University Press, 1957.

Ramanujan S. *Collected Papers of Srinivasa Ramanujan*, Ed. Hardy G.H., Aiyar P. V. S. and Wilson B.M., Providence, RI: Amer. Math. Soc., 2000.

Rao B. S. *Heptagonal Numbers in the Lucas Sequence and Diophantine Equations* $x^2(5x-3)^2 = 20y^2 \pm 16$, Fib. Quart., 40 (4), 2002.

Rao B. S. *Heptagonal Numbers in the Fibonacci Sequence and Diophantine Equations* $4x^2 = 5y^2(5y-3)^2 \pm 16$, Fib. Quart., 41 (5), 2003.

Ribenboim P. *Catalan's Conjecture*, Amer. Math. Monthly, 103, 1996.

Ribenboim P. *The New Book of Prime Number Records*, New York: Springer-Verlag, 1996.

Ribenboim P. *Fermat's Last Theorem for Amateurs*, New York: Springer-Verlag, 1999.

Robertson J. P. *Magic Squares of Squares*, Math. Mag., 69, 1996.

Sierpiński W. *Elementary Theory of Numbers*, Warszawa, 1964.

Sierpiński W. *Pythagorean Triangles*, New York: Dover, 2003.

Sloane N. J. A. and Plouffe S. *The Encyclopedia of Integer Sequences*, San Diego: Academic Press, 1995.

Sloane N. J. A. at all *On-line encyclopedia of integer sequences*, http://oeis.org/

Smith D. E. *A Source Book in Mathematics*, New York: Dover, 1984.

Smith J. *Trapezoidal Numbers*, Math. in School, 11, 1997.

Sylvester J. J. and Franklin F. *A Constructive Theory of Partitions, Arranged in Three Acts, an Interact and an Exodion*, Amer. J. of Math., 5 (1), 1882.

Theonia Smyrnaei *Platonici*, Latin transl. by Ismael Bullialdi, 1644.

Tijdeman R. *On the Equation of Catalan*, Acta Arith., 29, 1976.

Trigg C. W. *A Unique Magic Hexagon*, Recr. Math. Mag., 46, 1964.

Uspensky V. A. *Pascal's Triangle: Certain Applications of Mechanics to Mathematics*, Moscow: Mir, 1976.

Vorob'ev N. N. *Fibonacci Numbers*, New York: Blaisdell, 1961.

Wantzel M. L. *Recherches sur les Moyens de Reconnaître si un Problème de Géométrie Peut se Pésoudre avec la Règle et le Compas*, J. Math. pures appliq., 1, 1836.

Waring E. *Meditationes Algebraicae*, Cambridge, 1770. Reprinted as *Meditationes Algebraicae: An English Translation of the Work of Edward Waring*, Providence, RI: Amer. Math. Soc., 1991.

Watson G. N. *The Problem of the Square Pyramid*, Messenger. Math., 48, 1918.

Weisstein E. W. *MathWorld — A Wolfram Web Resource*, http://mathworld.wolfram.com/

Wells D. *The Penguin Dictionary of Curious and Interesting Numbers*, Middlesex: Penguin Books, 1986.

Wells D. *The Penguin Dictionary of Curious and Interesting Geometry*, London: Penguin, 1991.

Wieferich A. *Beweis des Satzes, Dass Sich Eine Jede Ganze Zahl als Summe von Höchstens Neun Positiven Kuben Darstellen Lässt*, Math. Ann., 66, 1909.

Wikipedia, the free Encyclopedia, http://en.wikipedia.org

Index